Lecture Notes in Computer Science 9084

Commenced Publication in 1973
Founding and Former Series Editors:
Gerhard Goos, Juris Hartmanis, and Jan van Leeuwen

Editorial Board

David Hutchison
Lancaster University, Lancaster, UK
Takeo Kanade
Carnegie Mellon University, Pittsburgh, PA, USA
Josef Kittler
University of Surrey, Guildford, UK
Jon M. Kleinberg
Cornell University, Ithaca, NY, USA
Friedemann Mattern
ETH Zürich, Zürich, Switzerland
John C. Mitchell
Stanford University, Stanford, CA, USA
Moni Naor
Weizmann Institute of Science, Rehovot, Israel
C. Pandu Rangan
Indian Institute of Technology, Madras, India
Bernhard Steffen
TU Dortmund University, Dortmund, Germany
Demetri Terzopoulos
University of California, Los Angeles, CA, USA
Doug Tygar
University of California, Berkeley, CA, USA
Gerhard Weikum
Max Planck Institute for Informatics, Saarbrücken, Germany

More information about this series at http://www.springer.com/series/7410

Said El Hajji · Abderrahmane Nitaj
Claude Carlet · El Mamoun Souidi (Eds.)

Codes, Cryptology, and Information Security

First International Conference, C2SI 2015
Rabat, Morocco, May 26–28, 2015, Proceedings
In Honor of Thierry Berger

 Springer

Editors
Said El Hajji
University of Mohammed V
Rabat
Morocco

Abderrahmane Nitaj
University of Caen
Caen
France

Claude Carlet
LAGA, Universities of Paris 8 and Paris 13,
    France
Saint-Denis Cedex 02
France

El Mamoun Souidi
University of Mohammed V
Rabat
Morocco

ISSN 0302-9743   ISSN 1611-3349   (electronic)
Lecture Notes in Computer Science
ISBN 978-3-319-18680-1   ISBN 978-3-319-18681-8   (eBook)
DOI 10.1007/978-3-319-18681-8

Library of Congress Control Number: 2015938320

LNCS Sublibrary: SL4 – Security and Cryptology

Springer Cham Heidelberg New York Dordrecht London

Printed on acid-free paper

Springer International Publishing AG Switzerland is part of Springer Science+Business Media
(www.springer.com)

Preface

This volume contains the papers accepted for presentation at C2SI-Berger2015, in honor of Prof. Thierry Berger, from XLIM Laboratory, University of Limoges, France. C2SI-Berger2015 is an international conference on the theory, and applications of cryptographic techniques, coding theory, and information security. The first aim of this conference is to pay homage to Prof. Thierry Berger for his valuable contribution in teaching and disseminating knowledge in coding theory and cryptography in Morocco since 2003. The second aim of the conference is to provide an international forum for researchers from academia and practitioners from industry, from all over the world for discussion of all forms of cryptology, coding theory, and information security.

The initiative of organizing C2SI-Berger2015 has been started by the Moroccan Laboratory of Mathematics, Computing sciences and Applications (LabMiA) at Faculty of Sciences of the University Mohammed V in Rabat and performed by an active team of researchers from Morocco and France. The conference was organized in co-operation with the International Association for Cryptologic Research (IACR), and the proceedings were published in Springer's Lecture Notes in Computer Science series.

The C2SI-Berger2015 Program Committee consisted of 39 members. There were 59 papers submitted to the conference. Each paper was assigned to at least two members of the Program Committee and was refereed anonymously. The review process was challenging and the Program Committee, aided by reports from 17 external reviewers, produced a total of 130 reviews in all. After this period, 22 papers were accepted on March 20, 2015. Authors then had the opportunity to update their papers until March 25, 2015. The present proceedings include all the revised papers. We are indebted to the members of the Program Committee and the external reviewers for their diligent work.

The conference was honored by the presence of the invited speakers François Arnault, Ezedin Barka, Johannes A. Buchmann, Anne Canteaut, Claude Carlet, Jean Louis Lanet, Ayoub Otmani, and Felix Ulmer. They gave talks on various topics in cryptography, coding theory, and information security and contributed to the success of the conference.

We had the privilege to chair the Program Committee. We would like to thank all committee members for their work on the submissions, as well as all external reviewers for their support. We thank the invited speakers, and the authors of all submissions. They all contributed to the success of the conference.

We would also like to thank Prof. Saaid Amzazi, President of University Mohammed V in Rabat and Prof. Wail Benjelloun, former Head of University Mohammed V, Agdal in Rabat for their unwavering support to research and teaching in the areas of cryptography, coding theory, and information security.

We are deeply grateful to Prof. Thierry Berger and his laboratory XLIM of the University of Limoges for great services in contributing to the establishment of a successful master's degree in coding theory, cryptography, and information security at

University Mohammed V in Rabat. We would like to take this opportunity to acknowledge their professional work.

Finally, we heartily thank all the Local Organizing Committee members, all sponsors, and everyone who contributed to the success of this conference. We are also thankful to the staff at Springer for their help in producing the proceedings.

May 2015

Said El Hajji
Abderrahmane Nitaj
El Mamoun Souidi

Organization

C2SI-Berger2015 is organized by University Mohammed V, Rabat, Morocco, in cooperation with the International Association for Cryptologic Research (IACR).

Honorary Chairs

Saaid Amzazi   President of University Mohammed V,
                                           Rabat, Morocco

Thierry Berger   XLIM, University of Limoges, France

General Chair

Said El Hajji   University Mohammed V, Rabat, Morocco

Program Chairs

Said El Hajji   University Mohammed V, Rabat, Morocco
Abderrahmane Nitaj   University of Caen, France
Claude Carlet   Universities of Paris 8 and Paris 13, France
El Mamoun Souidi   University Mohammed V, Rabat, Morocco

Organization Committee

Said El Hajji (Chair)   LabMIA, University Mohammed V,
                                               Rabat, Morocco
Ghizlane Orhanou (Co-chair)   LabMIA, University Mohammed V,
                                               Rabat, Morocco
El Mamoun Souidi (Co-chair)   LabMIA, University Mohammed V,
                                               Rabat, Morocco
Anas Aboulkalam   University Cadi Ayyad, Marrakesh, Morocco
François Arnault   XLIM, University of Limoges, France
Abdelmalek Azizi   University Mohammed I, Oujda, Morocco
Hafssa Benaboud   University Mohammed V, Rabat, Morocco
Redouane Benaini   University Mohammed V, Rabat, Morocco
Youssef Bentaleb   University Ibn Tofail, Kenitra, Morocco
Souad EL Bernoussi   University Mohammed V, Rabat, Morocco

Sidi Mohamed Douiri	University Mohammed V, Rabat, Morocco
Caroline Fontaine	Télécom Bretagne, Rennes, France
Abelkrim Haqiq	University of Settat, Morocco
Hicham Laanaya	University Mohammed V, Rabat, Morocco
Jalal Laassiri	University Mohammed V, Rabat, Morocco
Mounia Mikram	École des Sciences de l'Information, Rabat, Morocco
Ayoub Otmani	University of Rouen, France
Faissal Ouardi	University Mohammed V, Rabat, Morocco

Program Committee

Anas Aboulkalam	University Cadi Ayyad, Marrakesh, Morocco
François Arnault	XLIM, University of Limoges, France
Abdelmalek Azizi	University Mohammed I, Oujda, Morocco and Académie Hassan II, Morocco
Ezedin Barka	College of IT, United Arab Emirates University, Al Ain, UAE
Hafssa Benaboud	University Mohammed V, Rabat, Morocco
Youssef Bentaleb	ENSA, Kenitra, Morocco
Thierry Berger	XLIM, University of Limoges, France
Mohammed Bouhdadi	University Mohammed V, Rabat, Morocco
Mohamed Boulmalf	Université Internationale de Rabat, Morocco
Anne Canteaut	Inria-Rocquencourt, France
Sidi Mohamed Douiri	University Mohammed V, Rabat, Morocco
Pierre Dusart	University of Limoges, France
Mohamed Essaaidi	IEEE Morocco Section, ENSIAS, Rabat, Morocco
Caroline Fontaine	Télécom Bretagne, Rennes, France
Philippe Gaborit	XLIM, University of Limoges, France
Sanaa Ghouzali	College of Computer and Information Sciences, King Saud University, Saudi Arabia
Kenza Guenda	University of Science and Technology, Houari Boumedienne, Algiers, Algeria
Abelkrim Haqiq	University of Settat, Morocco
Maria Isabel Garcia	University of Barcelona, Spain
Zoubida Jadda	St Cyr, France
Salahddine Krit	Ibn Zohr University Polydisciplinary, Ouarzazate, Morocco
Jalal Laassiri	Ibn Tofail University, Kenitra, Morocco
Jean Louis Lanet	Inria Bretagne Atlantique, France
Mounia Mikram	École des Sciences de l'Information, Rabat, Morocco
Marine Minier	INSA, Lyon, France
Ghizlane Orhanou	University Mohammed V, Rabat, Morocco
Ayoub Otmani	University of Rouen, France

Ali Ouadfel	University Mohammed V, Rabat, Morocco
Faissal Ouardi	University Mohammed V, Rabat, Morocco
Patrice Parraud	St Cyr, France
Mohammed Rziza	University Mohammed V, Rabat, Morocco
Abderrahim Saaidi	University Sidi Mohamed Ben Abdellah, Taza, Morocco
Tayeb Sadiki	Université Internationale de Rabat, Morocco
Felix Ulmer	University of Rennes, France
Fouad Zinoun	University Mohammed V, Rabat, Morocco

Additional Reviewers

Hussain Ben-Azza	Johan Nielsen
Delphine Boucher	Tajjeeddine Rachidi
Ilaria Cardinali	Netanel Raviv
Pascale Charpin	Nicolas Sendrier
Willi Geiselmann	Zhang Shiwei
Norafida Ithnin	Anna-Lena Trautmann
Vadim Lyubashevsky	Antonia Wachter-Zeh
Sihem Mesnager	

Invited Speakers

François Arnault	XLIM, University of Limoges, France
Ezedin Barka	College of IT, United Arab Emirates, Al Ain, UAE
Johannes A. Buchmann	Technische Universität Darmstadt, Germany
Anne Canteaut	Inria-Rocquencourt, France
Claude Carlet	LAGA, Universities of Paris 8 and Paris 13, France
Jean Louis Lanet	Inria Bretagne Atlantique, France
Ayoub Otmani	University of Rouen, France
Felix Ulmer	University of Rennes, France

Sponsoring Institutions

Ministère de l'Enseignement Supérieur, de la Recherche Scientifique et de la
    Formation des Cadres
Faculty of Sciences, Rabat, Morocco
University Mohammed V, Rabat, Morocco
Académie Hassan II des Sciences et Techniques, Morocco
Centre National de Recherche Scientifiques et Techniques, Morocco
IEEE Morocco Section

Association Marocaine de Confiance Numérique (AMAN), Morocco
Centre Marocain de Recherches Polytechniques et d'Innovation, Morocco
Equipe Protection de l'Information, Codage et Cryptographie du Laboratoire
    XLIM de Limoges, France
Laboratoire de Mathématiques, Informatique et Applications (LabMiA), Rabat,
    Morocco

Origin of Submissions

Algeria	Pakistan
Brazil	Russian Federation
Cameroon	Saudi Arabia
Canada	Senegal
France	Spain
Germany	Syrian Arab Republic
Mauritius	Tunisia
Mexico	Turkey
Morocco	UAE
Norway	

Biography of Thierry Berger

Thierry P. Berger received the Ph.D. degree and the French Habilitation (Mathematics) from the University of Limoges, France.

From 1992, he has been with the University of Limoges. He is currently Professor in the Department of Mathematics and Informatics, Xlim Laboratory. He is the scientific head of the Protection of Information, Coding and Cryptography group of this department. His research interests include finite algebra, automorphism group of codes, links between coding and cryptography, stream cipher and pseudorandom generators, design and cryptanalysis of lightweight block ciphers.

Invited Papers

Multidimensional Bell Inequalities
and Quantum Cryptography

François Arnault

Université de Limoges, Laboratoire XLIM/DMI, France
arnault@unilim.fr

Abstract. The laws of quantum physics allow the design of cryptographic protocols for which the security is based on physical principles. The main cryptographic quantum protocols are key distribution schemes, in which two parties generate a shared random secret string. The privacy of the key can be checked using Bell inequalities. However, the Bell inequalities initial purpose was a fundamental one, as they showed how quantum rules are incompatible with our intuition of reality.

This paper begins with an introduction about quantum information theory, Bell inequalities, quantum cryptography. Then it presents the use of qudits for Bell inequalities and cryptography.

Securing the Web of Things
with Role-Based Access Control

Ezedine Barka, Sujith Samuel Mathew, and Yacine Atif

College of IT, UAE University, Al Ain, UAE
ebarka@uaeu.ac.ae

Abstract. Real-world things are increasingly becoming fully qualified members of the Web. From, pacemakers and medical records to children's toys and sneakers, things are connected over the Web and publish information that is available for the whole world to see. It is crucial that there is secure access to this Web of Things (WoT) and to the related information published by things on the Web. In this paper, we introduce an architecture that encompasses Web-enabled things in a secure and scalable manner. Our architecture utilizes the features of the well-known role-based access control (RBAC) to specify the access control policies to the WoT, and we use cryptographic keys to enforce such policies. This approach enables prescribers to WoT services to control who can access what things and how access can continue or should terminate, thereby enabling privacy and security of large amount of data that these things are poised to flood the future Web with.

On the Security of Long-Lived Archiving Systems Based on the Evidence Record Syntax

Matthias Geihs, Denise Demirel, and Johannes Buchmann

Technische Universität Darmstadt, University in Darmstadt, Germany
mgeihs@cdc.informatik.tu-darmstadt.de

Abstract. The amount of security critical data that is only available in digital form is increasing constantly. The Evidence Record Syntax Specification (ERS) achieves very efficiently important security goals: integrity, authenticity, datedness, and non-repudiation. This paper supports the trustworthiness of ERS by proving ERS secure. This is done in a model presented by Canetti et al. that these authors used to establish the long-term security of the Content Integrity Service (CIS). CIS achieves the same goals as ERS but is much less efficient. We also discuss the model of Canetti et al. and propose new directions of research.

Differential Attacks Against SPN:
A Thorough Analysis

Anne Canteaut and Joëlle Roué

Inria, project-team SECRET, Rocquencourt, France
{Anne.Canteaut,Joelle.Roue}@inria.fr

Abstract. This work aims at determining when the two-round maximum expected differential probability in an SPN with an MDS diffusion layer is achieved by a differential having the fewest possible active Sboxes. This question arises from the fact that minimum-weight differentials include the best differentials for the AES and several variants. However, we exhibit some SPN for which the two-round MEDP is achieved by some differentials involving a number of active Sboxes which exceeds the branch number of the linear layer. On the other hand, we also prove that, for some particular families of Sboxes, the two-round MEDP is always achieved for minimum-weight differentials.

On the Properties of Vectorial Functions with Plateaued Components and Their Consequences on APN Functions

Claude Carlet

LAGA, UMR 7539, CNRS, Universities of Paris 8 and Paris 13,
Department of Mathematics, University of Paris 8, 2 rue de laliberté,
93526 Saint-Denis cedex 02, France
claude.carlet@univ-paris8.fr

Abstract. [This is an extended abstract of paper [15], which has been submitted to a journal] Boolean plateaued functions and vectorial functions with plateaued components, that we simply call plateaued, play a significant role in cryptography, but little is known on them. We give here, without proofs, new characterizations of plateaued Boolean and vectorial functions, by means of the value distributions of derivatives and of power moments of the Walsh transform. This allows us to derive several characterizations of APN functions in this framework, showing that all the main results known for quadratic APN functions extend to plateaued functions. Moreover, we prove that the APN-ness of those plateaued vectorial functions whose component functions are unbalanced depends only on their value distribution. This proves that any plateaued (n, n)-function, n even, having same value distribution as APN power functions, is APN and has same extended Walsh spectrum as the APN Gold functions.

Beyond Cryptanalysis Is Software Security the Next Threat for Smart Cards

Jean-Louis Lanet

INRIA, LHS-PEC,
263 Avenue Général Leclerc, 35042 Rennes, France
jean-louis.lanet@inria.fr
http://secinfo.msi.unilim.fr/lanet/

Abstract. Smart cards have been considered for a long time as a secure container for storing secret data and executing programs that manipulate them without leaking any information. In the last decade, a new form of attack that uses the hardware has been intensively studied. We have proposed in the past to pay attention also to easier attacks that use only software. We demonstrated through several proof of concepts that such an approach should be a threat under some hypotheses. We have been able to execute self-modifying code, return address programming and so on. More recently we have been able to retrieve secret keys belonging to another application. Then all the already published attacks should have been a threat but the industry increased the counter measures to mitigate for each of the published attack. In such a sensitive domain, we always submit the attacks to the industrial partners but also national agencies before publishing any attack. Within such an approach, they have been able to patch their system before any vulnerabilities should be exploited.

Key-Recovery Techniques in Code-Based Cryptography

Ayoub Otmani

University of Rouen, LITIS, 76821 Mont-Saint-Aignan, France
ayoub.otmani@univ-rouen.fr

Abstract. An important step in the design of secure cryptographic primitives consists in identifying hard algorithmic problems. Despite the fact that several problems have been proposed as a foundation for public-key primitives, those effectively used are essentially classical problems coming from integer factorisation and discrete logarithm. On the other hand, coding theory appeared with the goal to solve the challenging problem of decoding a random linear code. It is widely admitted as a hard problem that has led McEliece in 1978 to propose the first code-based public-key encryption scheme. The key concept is to focus on codes that come up with an efficient decoding algorithm. McEliece recommended the use of binary Goppa codes which proved to be, up to now, a secure choice.

This talk will explore the important notion underlying code-based cryptography in order to understand its strengths and weaknesses. We then look at different extensions that lead to a wide range of variants of the McEliece scheme. This will give the opportunity to describe efficient and practical key-recovery cryptanalysis on these schemes, and to show the large diversity in the design of these attacks.

Extended Abstract:
Codes as Modules over Skew Polynomial Rings

Felix Ulmer

IRMAR, CNRS, UMR 6625, Université de Rennes 1,
Université Européenne de Bretagne, France
felix.ulmer@univ-rennes1.fr

Abstract. This talk is an overview of codes that are defined as modules over skew polynomial rings. These codes can be seen as a generalization of cyclic codes or more generally polynominal codes to a non commutative polynomial ring. Most properties of classical cyclic codes can be generalized to this new setting and self-dual codes can be easily identified. Those rings are no longer unique factorization rings, therefore there are many factors of $X^n - 1$, each generating a "skew cyclic code". In previous works many new codes and new self-dual codes with a better distance than existing codes have been found. Recently cyclic and skew-cyclic codes over rings have been extensively studied in order to obtain codes over subfields (or subrings) under mapping with good properties.

Contents

Invited Papers

Multidimensional Bell Inequalities and Quantum Cryptography

François Arnault[✉]

Université de Limoges, Laboratoire XLIM/DMI, France
arnault@unilim.fr

Abstract. The laws of quantum physics allow the design of crypto-graphic protocols for which the security is based on physical principles. The main cryptographic quantum protocols are key distribution schemes, in which two parties generate a shared random secret string. The privacy of the key can be checked using Bell inequalities. However, the Bell inequalities initial purpose was a fundamental one, as they showed how quantum rules are incompatible with our intuition of reality.

This paper begins with an introduction about quantum information theory, Bell inequalities, quantum cryptography. Then it presents the use of qudits for Bell inequalities and cryptography.

Keywords: Bell inequalities · Quantum cryptography · Key distribution schemes · Random numbers

I can give at least three reasons to be interested with quantum theory when working in the theory of information processing.
(a) Cryptography. Quantum physics may eventually broke most present public key protocols. But even more importantly, does provide very new cryptographic protocols for which security is based on physics postulates.
(b) Computing. Quantum computers, if can eventually be built, will oblige us to change our standard models of computers.
(c) Random generation. Random evolution is a fundamental feature in quantum physics. Hence true random generation is possible, while it is only approachable in a classical world.

In this paper we review the use of Bell inequalities and recent progresses in their use for cryptography. Section 1 exposes the notion Local Realism and how it is characterized by Bell inequalities. Section 2 introduces the use of multidimensional quantum states (in opposition with only qudits). Section 3 considers violations by quantum rules. Section 4 is devoted to key exchange, including a qutrit protocol we proposed in collaboration with Zoé Amblard.

1 Local Realism and CHSH Inequalities

Bell inequalities provide evidence of the incompatibility of the classical description of the world and its quantum description. The existence of *entangled*

© Springer International Publishing Switzerland 2015
S. El Hajji et al. (Eds.): C2SI 2015, LNCS 9084, pp. 3–13, 2015.
DOI: 10.1007/978-3-319-18681-8_1

systems, predicted by quantum physics and experimentally observed, is a manifestation of this incompatibility. When some particles are entangled, measurements on them have results that cannot be explained by classical rules. In particular, classical rules do conform with *Local Realism*

1.1 Local Realism

Let consider a physical system made of different (spatially separated) parts, denoted **A, B, C**... For each part, an experimenter (Alice, Bob, Charlie...) is invited to make an experiment of his choice. Classical physics have implicitly assumed during centuries that:

Objectivism: Measurable quantities are defined even when not measured.
Locality: Distant places can be causally separated.

These two assumptions are the two ingredients of *Local Realism* [12]. Bell inequalities are relations which are satisfied by systems which obey Local Realism rules, but are violated by some quantum systems.

1.2 CHSH Inequalities

Probably the nicest Bell inequalities are the so-called CHSH inequalities (after Clauser, Horne, Shimony, Holt) [8].

Alice and Bob make measurements on a system constituted of two distant parties **A** and **B**. Alice has the choice to measure X_A or Z_A on **A**, and Bob has the choice between X_B and Z_B over **B**. Measurements X_A, Z_A, X_B, Z_B are dichotomic ones: their issues belong to $\{\pm 1\}$. Assuming Local Realism, the value

$$T := X_A X_B + X_A Z_B + Z_A X_B - Z_A Z_B \qquad (1)$$

is well defined, and it is easy seen equal to ± 2. When repeating the experiment with many identically prepared systems, the expected value of T satisfies:

$$-2 \le E(T) \le 2. \qquad (2)$$

These are the CHSH inequalities.

1.3 Quantum World

In a quantum world, we can consider two half spin particles in a state usually denoted $|\psi\rangle = \frac{1}{\sqrt{2}}(|01\rangle - |10\rangle)$. This system is said to be *entangled* because of the observed correlations between the issues reported by Alice and Bob. For example, for this state $|\psi\rangle$, if Alice and Bob measurements are spin measurements with same direction, the issues ± 1 obtained by Alice and Bob are always opposite.

More generally, quantum formalism shows that when Alice and Bob carry spin measurements $S_{\vec{a}}$ and $S_{\vec{b}}$ with respective directions given by unitary vectors

$\overrightarrow{a}$ and $\overrightarrow{b}$, then the expected value of the product of their issues is given by $E(S_{\overrightarrow{a}} S_{\overrightarrow{b}}) = -\overrightarrow{a} \cdot \overrightarrow{b}$. Hence, for the configuration shown in Figure 1:

$$E(T) = -\overrightarrow{a}_1 \cdot \overrightarrow{b}_1 - \overrightarrow{a}_1 \cdot \overrightarrow{b}_2 - \overrightarrow{a}_2 \cdot \overrightarrow{b}_1 + \overrightarrow{a}_2 \cdot \overrightarrow{b}_2 \qquad (3)$$
$$= -\sqrt{2}/2 - \sqrt{2}/2 - \sqrt{2}/2 - \sqrt{2}/2 = -2\sqrt{2}. \qquad (4)$$

The value obtained for $E(t)$ does not belong to the interval $[-2, 2]$ as required by (2). This is **not compatible** with the assumptions of Local Realism. Experiments have been done to know which rules are obeyed by nature, and all have confirmed quantum rules instead of Local Realism.

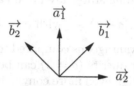

Fig. 1. Spin measurement directions chosen by Alice and Bob

When a Bell inequality is not satisfied for some measurements on a quantum state, it is said that this inequality is *violated*. The *violation factor* is given by the value obtained divided by the maximum compatible with Local Realism. For example, the expected value obtained in (4) corresponds to a violation factor $\sqrt{2}$.

1.4 Complete Set of Inequalities

If we swap parties and/or measurements in the two CHSH inequalities defined by (1) and (2), we obtain eight inequalities which, as shown in [15], form a *complete* set. This means that the four expected values $E(X_A)$, $E(Z_A)$, $E(X_B)$ and $E(Z_B)$ observed in some experiment made by Alice and Bob are compatible with Local Realism if and only if all eight inequalities are satisfied. These inequalities define in $\mathbf{C}^4$ a polytope Ω with eight facets, which is the set of values attainable by the four-vector of expected values assuming Local Realism.

1.5 Generalization to n Parties

Authors of [24] and [25] obtained a generalization of this complete set to the n parties case. Their inequalities involve the Walsh-Hadamard transform $\hat{f}$ of (multiplicative) Boolean functions $f : \{0, 1\}^n \to \{\pm 1\}$:

$$\sum_{s \in \{0,1\}^n} \hat{f}(s) E(M_s) \leq 2^n \qquad \text{where } M_s = \prod_{i=1}^n X_i Z_i^{1-s_i}.$$

There are 2^{2^n} such inequalities and they define a polytope Ω in $\mathbf{R}^{2^n}$. They form a complete set.

2 Multidimensional Inequalities

The above Bell inequalities have been obtained using dichotomic measurements. Non degenerate dichotomic measurements are described, in quantum physics, by operators on two-dimensional Hilbert spaces. These operators act on state vectors, which describe elementary quantum systems designed as *qubits*, and usually denoted

$$\alpha|0\rangle + \beta|1\rangle \quad \text{with } \alpha, \beta \in \mathbf{C} \text{ such that } |\alpha|^2 + |\beta|^2 = 1.$$

The multidimensional case corresponds to measurement with d possible issues where $d > 2$. These measurements are described using operators on d dimensional Hilbert spaces. The corresponding states are often called *qudits* and denoted

$$\alpha_0|0\rangle + \alpha_1|1\rangle + \cdots + \alpha_{d-1}|d-1\rangle \quad \text{with } \alpha_i \in \mathbf{C} \text{ and } \sum_{i=0}^{d-1} |\alpha_i|^2 = 1.$$

The use of qudits can be advantageous compared to the use of only qubits. For example, Bell inequalities over qutrits ($d = 3$) can be more noise resistant. This means that it is possible to design even more convincing experiments to check for (non) Local Realism. Also, qudits are useful to design better cryptographic protocols. Moreover, general entanglement remains incompletely understood, and multidimensional Bell inequalities may provide useful tools to give insight over it.

Some Bell inequalities had been obtained for the multidimensional $d > 2$ case. But the search for a complete set had not been successful before. We got such a complete set in [3]. The inequalities of this set are obtained using Discrete Fourier Transform.

2.1 Discrete Fourier Transform

Multidimensional discrete Fourier transform is a generalization of the Walsh-Hadamard transform. The "number of points" of the transform will be denoted $d \geq 2$ ($d = 2$ is the Walsh-Hadamard case). Boolean functions are replaced with functions from $\mathbf{Z}_d^n$ to the set $\mathbf{U}$ of complex d-roots of unity. There are d^{d^n} such functions. We denote $\mathcal{F}_{n,d}$ the set of functions from $\mathbf{Z}_d^n$ to $\mathbf{C}$, and ω a complex primitive d-root of 1 (say $\omega = \exp(2i\pi/d)$).

The discrete Fourier transform of $f \in \mathcal{F}_{n,d}$ is defined by

$$\hat{f}(r) = \sum_{s \in \mathbf{Z}_d^n} \omega^{r \cdot s} f(s).$$

Values of $\hat{f}$ can be obtained from the column vector of the values of f applying the matrix $H_d^{\otimes n} = \left(\omega^{r \cdot s}\right)_{r,s \in \mathbf{Z}_d^n}$. In other words, the Fourier transform is the isomorphism of the vector space $\mathcal{F}_{n,d}$, with matrix $H^{\otimes n}$. Let $H_d^{*\otimes n}$ the matrix $(\omega^{-r \cdot s})_{r,s \in \mathbf{Z}_d^n}$. Then $H_d^{\otimes n} H_d^{*\otimes n} = d^n I$. Hence it is possible to retrieve f from $\hat{f}$:

$$f(s) = \frac{1}{d^n} \sum_{r \in \mathbf{Z}_d^n} \omega^{-r \cdot s} \hat{f}(r).$$

2.2 Homogeneous Inequalities

Assume that a physical system is distributed to n parties. For each party i, two measurements X_i and Z_i are considered. These measurements are assumed to have $d \geq 2$ possible issues and, without loss in generality, this issues are assumed to be the powers of ω.

We denote Hull $\mathbf{U}$ the convex hull of the set $\mathbf{U}$ of the d-roots of unity. In [3] we obtained, assuming Local Realism, the following property:

$$\sum_{r \in \mathbf{Z}_d^n} \hat{f}(r) \, E\Big(\prod_{i=1}^n X_i^{r_i} Z_i^{d-1-r_i}\Big) \in d^n \cdot \text{Hull } \mathbf{U}, \tag{5}$$

where f is any function from $\mathbf{Z}_d^n$ to $\mathbf{U}$. This can be expressed by the inequalities:

$$\text{Re}\left(\frac{\exp(i\pi/d)}{\cos(\pi/d)} \sum_{r \in \mathbf{Z}_d^n} \hat{f}(r) \, E\Big(\prod_{i=1}^n X_i^{r_i} Z_i^{d-1-r_i}\Big)\right) \leq d^n. \tag{6}$$

We named homogeneous Bell inequalities these d^{d^n} inequalities (because they involve homogeneous polynomials). They define a polytope Ω in $\mathbf{C}^{d^n}$.

For $r \in \mathbf{Z}_d^n$, we abbreviate M_s the monomial $\prod_{i=1}^n X_i^{r_i} Z_i^{d-1-r_i}$. The expected values $E(M_s)$, for $s \in \mathbf{Z}_d^n$, form a vector in $\mathbf{C}^{d^n}$. The Local Realistic domain is the subset of $\mathbf{C}^{d^n}$ allowed to this vector under the Local Realistic assumptions. We shown in [3] that this domain is exactly the polytope Ω defined by the homogeneous Bell inequalities. These d^{d^n} inequalities form a complete set.

In the (very) special case $d = 2$, they are just the inequalities and the polytope found by Werner & Wolf and Žukowski & Brukner. But, because the convex hull of the square roots of 1 is contained in the real field, their polytope can indeed be considered in a real space, as they did.

3 Violation by Quantum Systems

The first purpose of Bell inequalities is to identify when measurement probabilities are compatible with Local Realism. This is exactly what the complete set of homogeneous inequalities does. However, it is also important to check and evaluate violations of Bell inequalities by quantum mechanics.

A difficulty appeared here for $d > 2$. Homogeneous Bell inequalities are formed with some monomials in which two measurements X_i and Z_i associated to the same party appear. But in quantum physics, such measurements are in general incompatible and cannot be separately carried.

We have addressed with concern using unitary measurements, instead of Hermitian ones as more frequently preferred in the literature. This is coherent with our approach, where the issues of measurements are assumed to be complex roots of unity, instead of real numbers. With unitary measurement operators, the product of two of them is also unitary, and can be considered as a separate measurement. The set of unitary operators we considered, the generalized Pauli group, is well known in quantum information theory. The Pauli operators are the products

$$Z, \ X, \ XZ, \ \ldots, \ XZ^{d-1}$$

where X and Z are given by

$$X = \begin{pmatrix} 0 & 0 & \cdots & 1 \\ 1 & \ddots & \ddots & \vdots \\ \vdots & \ddots & \ddots & 0 \\ 0 & \cdots & 1 & 0 \end{pmatrix} \quad \text{and} \quad Z = \begin{pmatrix} 1 & 0 & \cdots & 0 \\ 0 & \omega & \ddots & 0 \\ \vdots & \ddots & \ddots & \vdots \\ 0 & 0 & \cdots & \omega^{d-1} \end{pmatrix}.$$

With these operators, is is possible to compute violations by quantum physics. Moreover, the violations computed could be experimentally checked. In such an experiment, the issues of the measurement corresponding to an operator $X^i Z^i$ must not be considered as a product of the issues of two measurements but as the result of a single measurement. We make explicit in the following section how this can be done, when using tritters.

3.1 Measurements with Tritters

Measurement on qutrits (i.e. when $d = 3$) are often implemented with tritters [26]. Note that they can be easily generalized for any d.

A tritter is parametrized by a triplet $(\varphi_0, \varphi_1, \varphi_2)$ of phase shifts. For readability we put $\theta_j = \exp(i\varphi_j)$ (for $j = 0, 1, 2$) and $\Theta = (\theta_0, \theta_1, \theta_2)$. A tritter performs over a qutrit the unitary transformation $U_\Theta := H D_\Theta$ where the matrices H and D_Θ are $H = (\omega^{kl})_{0 \le k, l \le 2}$ and $D_\Theta = \mathrm{diag}(\theta_0, \theta_1, \theta_2)$. After the transformation performed by the tritter, a measurement is made using three detectors. This measurement is represented by the observable

$$Z = \sum_{k=0}^{2} \omega^k |k\rangle\langle k| = \begin{pmatrix} 1 & 1 & 1 \\ 1 & \omega & \omega^2 \\ 1 & \omega^2 & \omega \end{pmatrix}.$$

Thus, the measurement obtained by the combination of the tritter and the detectors corresponds to the following observable

$$Z_\Theta := D_{\Theta^*} H^\dagger Z H D_\Theta = \begin{pmatrix} 0 & 0 & \theta_2 \theta_0^* \\ \theta_0 \theta_1^* & 0 & 0 \\ 0 & \theta_1 \theta_2^* & 0 \end{pmatrix}. \tag{7}$$

Suppose now that we have two tritters, which implement the observables Z_Θ and Z_Λ described by Equation (7), with $\Theta = (\theta_0, \theta_1, \theta_2)$ and $\Lambda = (\lambda_0, \lambda_1, \lambda_2)$. Then we need to implement the product observable $Z_\Theta Z_\Lambda$. But

$$Z_\Theta Z_\Lambda = \begin{pmatrix} 0 & 0 & \theta_2 \theta_0^* \\ \theta_0 \theta_1^* & 0 & 0 \\ 0 & \theta_1 \theta_2^* & 0 \end{pmatrix} \begin{pmatrix} 0 & 0 & \lambda_2 \lambda_0^* \\ \lambda_0 \lambda_1^* & 0 & 0 \\ 0 & \lambda_1 \lambda_2^* & 0 \end{pmatrix} = \begin{pmatrix} 0 & \gamma_0^* \gamma_1 & 0 \\ 0 & 0 & \gamma_1^* \gamma_2 \\ \gamma_2^* \gamma_0 & 0 & 0 \end{pmatrix}$$

where
$$(\gamma_0, \gamma_1, \gamma_2) = (\theta_2^* \lambda_1^*, \theta_0^* \lambda_2^*, \theta_1^* \lambda_0^*).$$

Hence, $Z_\Theta Z_\Lambda = Z_\Gamma^\dagger$ where Γ has the components $(\gamma_0, \gamma_1, \gamma_2)$ just given. From $Z_\Gamma = D_\Gamma^* H^\dagger Z H D_\Gamma$, we obtain $Z_\Theta Z_\Lambda = Z_\Gamma^\dagger = D_\Gamma^* H^\dagger Z^\dagger H D_\Gamma$. The product observable $Z_\Theta Z_\Lambda$ can consequently also be implemented by a tritter and a detector, but with the detector performing a measurement corresponding to the observable $Z^\dagger$ instead of Z.

4 Quantum Keys Exchange

In a quantum key exchange protocol, two parties use a quantum channel to obtain a shared secret. The security of these protocols rely on physical postulates: any attack can be detected with some probability. These two most famous protocols are the one by Bennett & Brassard [5] and the one by Ekert [13]. We focused on the Ekert'91 protocol and variants.

4.1 Ekert'91 Protocol

The protocol relies on pairs of entangled qubits, say in state $(|01\rangle - |10\rangle)/\sqrt{2}$. This state is usually realized with pairs of polarized photons, routed in two optical fibers to their respective parties Alice and Bob. In the original protocol [13], Alice and Bob each have the choice between three measurement bases, but it is better to allow four different bases for each party. These measurements are denoted A_k (for Alice) and B_k (for Bob, with $0 \le k \le 3$). In practice these measurements are polarization measurements with directions given by angles $k\pi/4$ (much as shown in Figure 1 but each party can choose between the four measurements).

Independently, Alice and Bob choose their measurements (for each received pair of entangled states). Let A_a and B_b the chosen bases. When $a = b$, the issues obtained by Alice and Bob are opposite, hence they obtained shared keybits. The issues obtained when the parities of a and b differ can be used to detect the presence of an attacker. For this, the two following expected values are evaluated:

$$E(A_0 B_1) + E(A_0 B_3) + E(A_2 B_1) - E(A_2 B_3)$$
$$E(A_1 B_0) + E(A_1 B_2) + E(A_3 B_0) - E(A_3 B_2).$$

They correspond to two configurations of CHSH experiments which, by quantum rules, are predicted to reach $-\sqrt{2}$. A deviation from this value can be used to detect the presence of an attacker. Other pairs (a, b) are ignored. The following array summarizes the situation. The pairs which provide keybits are marked with k, and the pairs used to check CHSH violations are marked c_1 and c_2.

	B_0	B_1	B_2	B_3
A_0	k	c_1		c_1
A_1	c_2	k	c_2	
A_2		c_1	k	c_1
A_3	c_2		c_2	k

If no disturbance affects the measurements, the two checks for CHSH must give violations factors near $v = \sqrt{2}$. In practice, imperfections in apparatus will lower this value. If disturbance is approximated with a random noise, the resulting violation will be $(1 - F)v$ where F is the amount of noise. While this resultant violation remains greater than 1, the presence of an attacker can be detected. Hence, F has to remain lower than $1 - 1/v$ in order to keep the protocol secure. With Ekert protocol, $F = 1 - 1/\sqrt{2} \simeq 0.293$. If we can modify the violation factor v in order to make this threshold for F larger, the resulting protocol will allow to detect even more discreet attackers.

4.2   The Inequality CHSH-3

The use of qutrits allows a larger noise proportion F. This was explained in [10] where the 3DEB protocol was defined. This protocol uses an inequality similar to CHSH but involving 3-issues measurements.

This 3-issues variant of CHSH appeared in [20]. It has been rewritten in [6] in terms of correlation functions, in the form $S \leq 2$ with

$$S = \mathrm{Re}\left(E(A_1B_1) + E(A_1B_2) - E(A_2B_1) + E(A_2B_2)\right)$$
$$+ \tfrac{1}{\sqrt{3}}\,\mathrm{Im}\left(E(A_1B_1) - E(A_1B_2) - E(A_2B_1) + E(A_2B_2)\right).$$

But we can remark that $S = -\frac{2}{9}\,\mathrm{Re}\,T$ with

$$T = 3\big((\omega^2-1)E(A_1^2B_1^2)+(\omega-1)E(A_1^2B_2^2)+(1-\omega^2)E(A_2^2B_1^2)+(\omega^2-1)E(A_2^2B_2^2)\big).$$

Hence, the CHSH-3 inequality can finally be written $\mathrm{Re}(-T) \leq 9$.

The state $\frac{1}{\sqrt{3}}(|00\rangle + |11\rangle + |22\rangle)$ achieves violations of CHSH-3 with a $v = (6 + 4\sqrt{3})/9 \simeq 2.873/2$ factor. This corresponds to a noise level $F = 1 - 1/v = (11 - 6\sqrt{3})/2 \simeq 0.304$. It is even possible [17] to obtain a CHSH-3 violation factor $(1 + \sqrt{11/3})/2 \simeq 1.457$ with a non maximally entangled state. The noise threshold allowed is in this case $F = (7 - \sqrt{33})/4 \simeq 0.314$.

4.3   The 3DEB Protocol

This qutrits protocol appeared in [10] (but see also [21]). Alice uses measurement bases A_a (with $a = 0, 1, 2, 3$) which are obtained for example using tritters with parameters $\Theta_a = (1, \zeta^a, \zeta^{2a})$ where $\zeta = \exp(2i\pi/12)$. Bob uses measurement bases B_b (with $b = 0, 1, 2, 3$) obtained using tritters with parameters $\Theta_b = (1, \zeta^{-b}, \zeta^{-2b})$. When $a = b$, Alice and Bob obtain keybits because their respective issues are opposite. Pairs $(a, b) = (0, 1), (0, 3), (2, 1), (2, 3)$ can be used to check violations of CHSH-3, which without any disturbance must equal $v = (6 + 4\sqrt{3})/9 \simeq 2.873/2$. The same is true for pairs $(a, b) = (1, 0), (1, 2), (3, 0), (3, 2)$. Any presence of attacker will alter the observed violation. As remarked above, the threshold of admissible noise is $F \simeq 0.314$.

4.4   The Homogeneous Qutrits Protocol

Homogeneous Bell inequalities have allowed us to define in [2] an even better protocol, where the admissible noise threshold is larger. We have chosen to use the inequality $-\frac{2}{9}\operatorname{Re}(T_1) \leq 1$ found in [3], where

$$
\begin{aligned}
T_1 = \quad & -(2\omega + 4)E(A_1^2 B_1^2) + (\omega - 1)E(A_1^2 B_1 B_2) + (4\omega + 2)E(A_1^2 B_2^2) \\
& +(\omega - 1)E(A_1 A_2 B_1^2) - (2\omega + 1)E(A_1 A_2 B_1 B_2) + (4\omega - 1)E(A_1 A_2 B_2^2) \\
& +(\omega + 5)E(A_2^2 B_1^2) + (\omega + 2)E(A_2^2 B_1 B_2) + (\omega - 1)E(A_2^2 B_2^2).
\end{aligned}
$$

The state $\frac{1}{\sqrt{3}}\big(|00\rangle + |11\rangle + |22\rangle\big)$ achieves a much better violation factor $\simeq 1.693$ when using the same measurements bases as in 3DEB. Hence the threshold F is considerably improved because it reaches now $\simeq 0.409$.

The following array details which pairs of measurements provide key trits and which pairs are used to detect attacker (here $A_{00} := A_0^2$, $A_{02} := A_0 A_2, \dots$).

	B_{00}	B_{02}	B_{22}	B_{11}	B_{13}	B_{33}
A_{00}	k			c	c	c
A_{02}		k		c	c	c
A_{22}			k	c	c	c
A_{11}	c	c	c	k		
A_{13}	c	c	c		k	
A_{33}	c	c	c			k

Note that this protocol requires to be able to measure the product of two observables, such A_{02}. We have shown above how this can be done using tritters.

5   Conclusion

John Bell wrote the inequalities which now bear his name with a very theoretical aim in mind. This has been a great success and, together with experiments of Aspects and al, we have learned much of his work.

Near thirty years later, Bell inequalities have proven to have also a practical interest, in cryptography, with the work of Ekert. They have also applications in true random generation: they are useful to certify that a certain amount of entropy has been created in some quantum processes [23]. And many other uses are possible.

We have insisted on the use of multidimensional states (qudits). The description of Local Realism has been possible with the use of complex correlations functions, and of unitary observables. We have confirmed that the use of qudits can improve some protocols.

Bell inequalities, with the use of discrete mathematics to study quantum peculiarities, can be viewed as the start of emergence of quantum information theory. This exciting field will keep a double interest, as theoretical concerns and applications will remain quite interleaved. this makes even more interesting working in this field.

References

1. Acín, A., Durt, T., Gisin, N., Latorre, J.L.: Quantum non-locality in two three level systems. Physical Review A 65, 52325 (2002)
2. Amblard, Z., Arnault, F.: A qutrit quantum key distribution protocol with better noise resistance (Submitted)
3. Arnault, F.: A complete set of multidimensional Bell inequalities. Journal of Physics A, Mathematical and Theoretical 45, 255304 (2012)
4. Bell, J.S.: On the Einstein Podolsky Rosen paradox. Physics 1, 195–200 (1964)
5. Bennett, C.H., Brassard, G.: Quantum cryptography: public key distribution and coin tossing. In: Proceedings of the IEEE International Conference on Computers, Systems, and Signal Processing, Bangalore, pp. 175–179 (1984)
6. Chen, J.-L., Kaszlikowski, D., Kwek, L.C., Oh, C.H.: Wringing out new Bell inequalities for three-dimensional systems (qutrits). Modern Physics Letters A 17, 2231 (2002)
7. Chen, J.L., Kaszlikowski, D., Kwek, L.C., Oh, C.H., Żukowski, M.: Entangled three-state systems violate local realism more strongly than qubits: An analytical proof. Physical Review A 64, 052109 (2001)
8. Clauser, J.F., Horne, M.A., Shimony, A., Holt, R.A.: Proposed experiment to test local hidden variables theories. Physical Review Letters 23, 880 (1969)
9. Collins, D., Gisin, N., Linden, N., Massar, S., Popescu, S.: Bell inequalities for arbitrarily high-dimensional systems. Physical Review Letters 88, 040404 (2002)
10. Durt, T., Cerf, N.J., Gisin, N., Żukowski, M.: Security of quantum key distribution with entangled qutrits. Physical Review A 67, 012311 (2003)
11. Durt, T., Kaszlikowski, K., Żukowski, M.: Violations of local realism with quantum systems described by N-dimensional Hilbert spaces up to $N = 16$. Physical Review A 64, 024101 (2001)
12. Einstein, A., Podolsky, B., Rosen, N.: Can quantum-mechanical description of physical reality be considered complete? Physical Review 47, 777 (1935)
13. Ekert, A.K.: Quantum cryptography based on Bell's theorem. Physical Review Letters 67, 661 (1991)
14. Feynman, R.: Simulating physics with computers. International Journal of Theoretical Physics 21, 467–488 (1982)
15. Fine, A.: Hidden variables, joint probabilities, and the Bell inequalities. Physical Review Letters 48, 291 (1982)
16. Fu, L.-B.: General correlation functions of the Clauser-Horne-Shimony-Holt inequality for arbitrarily high-dimensional systems. Physical Review Letters 92, 130404 (2004)
17. Fu, L.-B., Chen, J.-L., Zhao, X.-G.: Maximal violation of the Clauser-Horne-Shimony-Holt inequality for two qutrits. Physical Review A 68, 022323 (2003)
18. Ji, S.-W., Lee, J., Lim, J., Nagata, K., Lee, H.-W.: Multisetting Bell inequality for qudits. Physical Review A 78, 052103 (2008)
19. Kaszlikowski, D., Gnaciński, P., Żukowski, M., Miklaszewski, W., Zeilinger, A.: Violations of local realism by two entangled N-dimensional systems are stronger than for two qubits. Physical Review Letters 85, 4418 (2000)
20. Kaszlikowski, D., Kwek, L.C., Chen, J.L., Żukowski, M., Oh, C.H.: Clauser-Horne inequality for three-state systems. Physical Review A 65, 032118 (2002)
21. Kaszlikowski, D., Oi, D.K.L., Christandl, M., Chang, K., Ekert, A., Kwek, L.C., Oh, C.H.: Quantum cryptography based on qutrit Bell inequalities. Physical Review A 67, 012310 (2003)

22. Masanes, L., Pironio, S., Acín, A.: Secure device-independent quantum key distribution with causally independent measurement devices. Nature Communications 2, 238, 1244 (2011)
23. Pironio, S., Acín, A., Massar, S., Boyer de la Giroday, A., Matsukevich, D.N., Maunz, P., Ohmschenk, S., Hayes, D., Luo, L., Manning, T.A., Monroe, C.: Random numbers certified by Bell's theorem. Nature 464, 1021 (2010)
24. Werner, R.F., Wolf, M.M.: All-multipartite Bell-correlation inequalities for two dichotomic observables per site. Physical Review A 64, 032112 (2001)
25. M. Żukowski, Brukner, Č.: Bell's theorem for general N-qubit states. Physical Review Letters 88, 210401 (2002)
26. Żukowski, M., Zeilinger, A., Horne, M.A.: Realizable higher-dimensional two-particle entanglements via multiport beam splitters. Physical Review A 55, 2564 (1997)

Securing the Web of Things with Role-Based Access Control

Ezedine Barka[✉], Sujith Samuel Mathew, and Yacine Atif

College of IT, UAE University, Al Ain, UAE
ebarka@uaeu.ac.ae

Abstract. Real-world things are increasingly becoming fully qualified members of the Web. From, pacemakers and medical records to children's toys and sneakers, things are connected over the Web and publish information that is available for the whole world to see. It is crucial that there is secure access to this Web of Things (WoT) and to the related information published by things on the Web. In this paper, we introduce an architecture that encompasses Web-enabled things in a secure and scalable manner. Our architecture utilizes the features of the well-known role-based access control (RBAC) to specify the access control policies to the WoT, and we use cryptographic keys to enforce such policies. This approach enables prescribers to WoT services to control who can access what things and how access can continue or should terminate, thereby enabling privacy and security of large amount of data that these things are poised to flood the future Web with.

Keywords: Web of Things · Privacy · Access Control · RBAC · UCON

1 Introduction

Today society is impacted by revolutionary innovations in information technology that are very pervasive and ubiquitous in nature. Along with these advances, particularly in communications technology, a series of new security threats and privacy issues arise. Among these technologies is the rapidly increasing Web of Things (WoT), where physical things are accessed and controlled via the Web. WoT has several methods that support a variety of applications such as subscribing to a service, notification of an event, status update, and location and presence services. WoT provides flexible, scalable, and real-time communications with the physical world in a ubiquitous way but additional security and privacy concerns result from its ubiquity and mobility.

Secure Web publishing approaches have been developed to allow authenticated users direct access to a dataset. In doing so, these appraoches provide users with a published, static "snapshot" of the dataset content. We follow this secure publishing paradigm [5] to enable a security framework for WoT.

Traditional access controls typically focus on the protection of data in closed environments, and the enforcement of control has been primarily based on identity and attributes of a known user. These types of access control lack a comprehensive, systematic approach to fulfill the security requirements of today's

© Springer International Publishing Switzerland 2015
S. El Hajji et al. (Eds.): C2SI 2015, LNCS 9084, pp. 14–26, 2015.
DOI: 10.1007/978-3-319-18681-8_2

pervasive and ubiquitous applications on the WoT. To address these issues, we introduce an architecture that implements role-based access control (RBAC) to check the access to datasets within WoT based environment. This enables publishers of things on the Web to control who can locate them, and subsequently access and use them. Furthermore, it enables the possibility of setting some attributes to determine whether certain accesses should proceed or be terminated.

The remainder of this paper is organized as follows. In Section 2, we provide some background on WoT and discusses our architecture and its role in the pervasive environment to address some security challenges. Section 3 provides an overview of the role-based access control (RBAC). Section 4 presents our architecture and explains the integration of WoT with RBAC. Section 5 describes how RBAC is used to specify the access policies to WoT datasets, and the cryptographic keys used to enforce these policies. Section 6 concludes the paper with some future work.

2 Overview of WoT

WoT is a platform where billions of physical things are interconnected over the World Wide Web. Researchers have successfully connected things over the Web and experimented with various applications in real-world scenarios [4]. The inevitable challenges lie in how to efficiently and effectively manage and secure the access to the informaiton hidden within these things, which is critical for a number of important applications. To address the management of heterogeneous and wide abundance of candidate things in WoT, the Ambient Space Manager (ASM) framework was suggested earlier by Mathew et. al [10]

2.1 Representation of Things on WoT

Mathew et. al. suggested a capability based classification, Fig 1 shows the Web Object Metadata (WOM) structure, which defines the ontological representation of a thing (Thing A) on the Web [6].

The *WOM-Profile* composes the semantic details from all ontologies of a thing that is revealed to external entities. The WOM-Profile is divided into two sections: the ¡preset¿ and ¡dynamic¿ sections. Preset describes static information about a thing like manufacturer, date of production, or country of production and the dynamic, describes information about a thing like cost, location, or owner, which changes. The WOM-Capability ontology classifies a thing based on its Identity, Processing, Communication, and Storage (IPCS) capabilities. The ontology classifies a thing to be *Web Smart* when these capabilities are Web related. Hence a Web Smart thing has a unique identity on the Web, processes Web requests, communicates via Web protocols, and has storage space on the Web. If any of the capabilities are missing, then the ontology recommends the augmentation of the missing capabilities.

Once things are Web Smart (i.e. they are participating members of the Web), they are grouped/clustered into an Ambient Spaces (AS) [10,9]. An AS is the

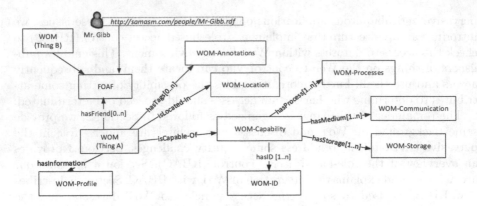

Fig. 1. Web Object Metadata (WOM) of a thing on the Web

virtual representation of a cluster of things i.e. the encapsulation of one or more real-world things that are Web Smart. An AS also represents the boundaries of a physical space. For example, Web Smart things in a classroom, or in a train compartment, or a hospital room, or a parking spot. These physical spaces are repeating patterns. Hence an AS provides a template to compose things and their containing physical spaces in a gradient to represent larger physical spaces like campuses, parking lots, airports, trains, and office buildings. Clustering things into an AS is done based on determining the similarities of things using similarity functions. The similarity functions are applied on all Web Smart things in an AS [8].

2.2   Ambient Space Stakeholders

In any fundamental computing setup, the main stakeholders are the providers and consumers of the services or infrastructure. The consumers use and update the system, while the providers deal with the manufacture, deployment and maintenance functions. The domain of WoT requires the addition of new stakeholders and redefinition of the traditional ones. The stakeholders within the WoT domain not only require providers and consumers but also needs to consider the role of owners and regulators who control the thing's inherent dynamic and proprietary state. Here, we briefly list the stakeholders, focusing on their contribution to the content of a thing's WOM-Profile.

Providers: The providers are essentially the manufacturers that create the WoT elements. The providers will also hold the responsibility of recycling or discarding a thing at the end of its lifespan [7]. The maintenance and upgrades to a thing are the responsibility of providers while a thing is used by other stakeholders. The providers hold the right to change the content of a thing while maintaining history of changes. The providers contribute to the preset content ¡wom:preset¿ of a thing's prole and are responsible for ensuring the presentation of thing's composition, use, and disposal. The preset content of a thing's

WOM-Profile is fixed and not changeable by other actors. Contact information of the providers needs to be provided, for the use of thing itself or any of the other stakeholders. The links to the user manual and the conditions of thing's usage are provided by the providers. The providers may also contribute to the dynamic content ¡wom:dynamic¿ of a thing's prole. Annotations for branding, price composition and marketing are initially added by the providers. The providers initiate the history of a thing's existence.

Consumers: The consumers of a Web Smart thing are its users. These users could be other *things* or people. Unlike other domains, consumers are not owners here and are bound to access restrictions that are controlled by the present owner of a thing. The contribution of consumers populates the dynamic content ¡wom:dynamic¿ of thing's prole. The consumers provide rich semantics to thing's use and add to the history of a thing. The content that the consumers provide to a thing essentially creates links with other *things* or people that are connected to the consumer. Thus the consumers play an important role in promoting *thing's* social connectivity.

Owners: The owners are consumers but have more rights to a thing's usage and content. The owners provide access restriction to a thing's operations and can loan or lease a thing. With proper authorization from regulators and providers, the owners can alter the dynamic content ¡wom:dynamic¿ of a thing and therefore change history. The options to re-brand or marketing a thing allows owners to change the value of a thing and promote its acceptance among other *things* or people.

Regulators: While the other stakeholders provide content to value a thing, the role of the regulators prevails over other stakeholders. For example, government authorities or regulatory authorities that ensures the safe, sustainable, and judicious use of Web Smart *things*. The regulators provide details on rights and obligations of other stakeholders. They provide contractual details wherein other stakeholders and authorities are informed if there is a breach of contract. Because of the wide spread implication of the virtual use of physical things, liabilities and exceptions are to be clearly defined by regulators. For the WOM-Profile, the regulators provide content that are both preset and dynamic related to issues like privacy, trust, cyber-attack and legal implications. The role of regulators needs to be actively researched, investigated, and formulated with government and international bodies so as to ensure the secure and sustainable use of *things* on the Web.

Manufacturers follow a structured product labeling standard to provide consumers with the information of a thing's content and usage. The process of monitoring and regulating these standards become easier when the information is digitally embeded or appended to products. The benefit of using the WOM-Profile as a digital standard for communicating product infomration is two-fold. Firstly the standard information can be included in the ¡wom:preset¿ part and secondly user experiences can be included in the ¡wom:dynamic¿ part of the WOM-Profile. While it is important to understand the semantic structure of

Web Smart thing's information and the major stakeholders, it is also important to realize how the information is stored and retrieved from real-world things.

2.3 WoT Framework

The AS enables real-world things to be imbibed into the WoT ensuring seamless communication between people and things. This opens up many social applications that is bound to enhance business and industry. Some applications were suggested based on the ASM framework [5,6]. Here, we take an example of how classrooms are virtually represented as Ambient Spaces, to describe the framework. Fig 2, depicts each classroom in a school campus as an Ambient Space (AS).

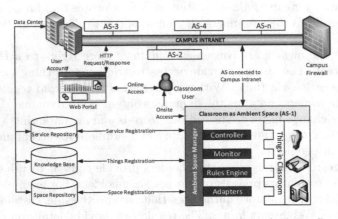

Fig. 2. Subsuming classrooms into the WoT using Ambient Spaces

Each AS is controled by an Ambient Space Manager (ASM) which includes the Controller, Monitor, Rules Engine, and Adapters. These modules provide essential management functionalities that provide the access and control of things in an AS. The Service repository, Knowledge Base, and Space repository contain the information that is relevant to all AS. The users has both onsite and online access to things in an AS.

The ASM framework creates a hierarchical structure for representing physical spaces and the things therein. Fig 3, provides a general depiction of the structure and also an example. Similar structure is suggestive to represent hospital rooms, train compartments, seats on an international flight, or in a movie theatre. Thus the ASM framework provides a scalable structure to represent physical things on the Web and populate the WoT.

2.4 WoT Security Challenges

Openness and sharing are always contradictory when it comes to security and privacy. A practical consideration for enabling widespread adoption of WoT is

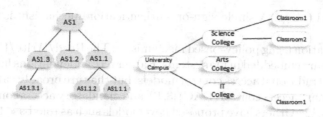

Fig. 3. Representing repeating patterns of physical spaces and things in them with Ambient Spaces

the security and privacy vulnerabilities of shared resources of things and related data. Moreover, how does the framework verify Web services and estimate their reliability against malicious intervention or inadvertent errors. Although security solutions and related technologies have been developed to protect systems against many vulnerabilities, most of these technologies do not have a cohesive structure to deal with the security issues specifically related to the WoT, and advocate ad-hoc approaches instead. This is because WoT introduces new dimensions of risk, due to its heterogeneous and ubiquitous nature. Some of the threats that are inherent to the use of WoT are listed as follows:

- Impersonating a server: A WoT user contacts a Proxy server to deliver requests. The server could be impersonated by an attacker. The mobility of things further complicates this scenario.
- Tampering with message bodies that contain requests.
- Tearing down sessions – insert a disconnect command.
- Denial of Service attacks - Denial of service attacks focus on rendering a thing on the Web unavailable, usually by directing an excessive amount of network traffic to its interfaces. The WoT face the public Internet in order to accept requests from worldwide IP endpoints, which creates a number of potential opportunities for distributed denial of service attacks that must be recognized and addressed by the implementers and operators of this ecosystem.

Therefore, the security challenges facing WoT is to ensure the following:

- Data Security and Privacy: How to protect the thing's data and private information and locations? In WoT, addressing the issue of data security is particularly challenging, due to the unique features of the network, such as mobility of the entities and the size of the network. It is essential that thing's critical information is protected from being inserted or modified by attackers. For privacy, the challenge is on how to ensure a conditional privacy in the sense that thing's private information like identity, speed, or location are protected from unauthorized access while access should always be granted when needed by authorities.
- Authentication: Most technologies use Web services today and have the HTTP style access mechanism which is not foolproof when dealing with

real-world things. A single sign-on authentication mechanism is at-least required.

– Authorization using policy-based mechanisms: The Read/Write/Execute controls that are embedded in file systems. Earlier recommendations have tried implement, traditional access control models, but they are broadly categorized as discretionary access control (DAC) [3,12] and mandatory access control (MAC) models [3,12]. Others have proposed new models such as role-based access control (RBAC) and task-based access control (TBAC) to address thee security requirements [13,16].

None of the above mentioned solutions are sufficient in isolation for providing security for a large-scale, distributed and sometimes resource constrained pervasive environment like in WoT context. Hence, our approach utilizes the well-known Role-Based Access Control (RBAC) to control access to things on the Web.

There are many benefits to adapting RBAC to WoT context. RBAC supports data abstractions which enables subscribers to WoT services to control who can identify the locations of the things, to approve or disapprove subsequent access, and to also set parameters to determine whether a certain accessn can continue or should terminate. RBAC also enforces other security concepts that are specific to some applications such as lease privileges or separation of privileges. In this case, RBAC may deny the access or connection when the requested authorization of the prescriber does not meet the access control policy requirement or the thing's attribute changes.

However, RBAC is susceptible to role proliferation. For example, thousands of users may be granted access to various parts of a thing's dataset. The access permission my differ depending upon each user's affiliation with the system. This scenario my demand that role-based policy assigns one role to each user, which can be too much to handle. Therefore, the concept of role parameterization, developed by [3], has shown to be an effective way to deal with the issue of role proliferation. The following section provides an overview of the RBAC model.

3   Overview of Role Based Access Control (RBAC) Model

In this section we briefly review the general ideas of RBAC and the core authorization models. The details of these models can be found in [2,14,1].

RBAC is proven to be a good alternative to traditional discretionary and mandatory access controls. It ensures that access to certain data or resources is given to authorized users only [14]. It also supports some important security principles such as least privilege, separation of duties, and data abstraction. Least privilege is supported, because RBAC is configurable such that only those permissions are assigned to the role required for the tasks conducted by members of the role. Separation of duties is achieved by ensuring that mutually exclusive roles must be invoked to complete a sensitive task, such as requiring an accounting clerk and account manager to participate in issuing a check. Data abstraction is supported by means of abstract permissions. Instead of the read, write, and execute permissions typically provided by the operating system. Other permissions such as join, leave, join as a sender, or join as a receiver, are also be expressable.

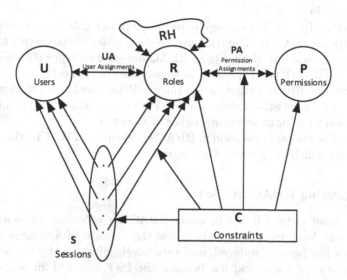

Fig. 4. Basic RBAC Model

A general RBAC model was defined by Sandhu [14] and is summarized in Fig 4. The model is based on three sets of entities called users (U), roles (R), and permissions (P). A user is a human being (an entity that seeks access). A role is a function with some associated semantics regarding the authority and responsibility conferred on a member of the role. Permission is an approval of a particular mode of access to one or more users in the system. The user assignment (UA) and permission assignment (PA) relations of Fig 4 are both many-to-many relationships (indicated by the double-headed arrows). A user can be a member of many roles, and a role can be assigned to many users. Similarly, a role can have many permissions, and the same permission can be assigned to different roles.

Role hierarchy (RH) in RBAC is a natural way of organizing roles to reflect the lines of authority and responsibility. The hierarchy is partially ordered, so it is reflexive, transitive, and anti-symmetric. Inheritance is reflexive because a role inherits its own permissions. Transitivity is a natural requirement in this context, and anti-symmetry rules out roles that inherit from one another and are therefore redundant.

4   Security Architecture for WoT

Integrating the RBAC technology into ubiquitous WoT-based environment requires a careful mapping between the entities of RBAC and those entities and components of the WoT. Following is a list of integrated components which require such mapping:

- User/Subjects: The concept of participants in WoT is represented as a user component in the RBAC.
- Permissions/Rights: The concept of permissions in RBAC is captured through the privileges that a WoT participant needs in order to complete a task.
- Objects: the concept of objects in RBAC are used to represent all resources *things* that a WoT participant seeks to access or to connects to.
- Authorization Rules: Authorization rules in RBAC are the set of requirements that should be satisfied before any WoT user be permitted to establish any connection with, or to access any other WoT entity.
- Session: The concept of session in RBAC is captured in WoT by the set of durations for which WoT entities are active.

4.1   Integrating RBAC in WoT

One of the most critical issues in using RBAC for enforcing the specified access policies in WoT environment is to use the concept of a reference monitor (RM), which has been introduced, and extensively discussed by the access control community for years, and has become the ISO standard for access control framework [15].

The RM concept has been considered as the core control mechanism for access and usage of digital information. In classical access control, subjects access digital objects only through the reference monitor, which is a process inside the trusted computer base that is always running and is a tamper proof.

The following section discusses our conceptual structure of RBAC/WoT access control domains, based on the reference monitor.

4.2   Policy Enforcement Facitilies

In our architecture, we use a customized version of the well-known ISO reference monitor standard [9].

According to this ISO standard, the reference monitor consists of two facilities: Access Control Enforcement Facility (AEF) and Access Decision Facility (ADF). The AEF and ADF interact with each other in such way that every request by a subject to access an object in the system get intercepted by AEF. The AEF in turn asks the ADF for a decision on whether to approve or disapprove the request, and subsequently the ADF returns either 'yes' or 'no' as appropriate. The enforcement of this decision takes place at the AEF.

In our architecture, the reference monitor is similar but differs in the details from that of ISO reference monitor. We incorporate the role-based access control to handle the "pre-decision" authorization rule. Fig 5 shows the conceptual structure of the RBAC/WoT reference monitor.

As the Fig 5 shows, any request to access any WoT resource "thing" is intercepted by the AEF. Before making any decisions, the AEF forwards the request

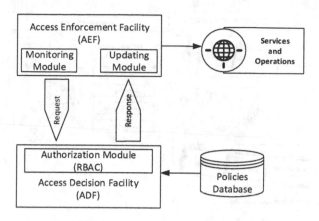

Fig. 5. Conceptual Structure for RBAC/WoT Reference Monitor

to the ADF, which in turn adheres to the RBAC policy decision of whether to grant or reject the authorization request. RBAC will allow authorization of an active (subject) entity to execute a certain right on a passive (resource) entity only if the subject belongs to a role that RBAC has previously assign that right to.

The rest of the decision process by AEF would continue only if RBAC grants authorization, otherwise the process is stopped and response by ADF is negative (no authorization). Furthermore, RBAC allows authorization after it tests other decision factors, mainly, hierarchal relationships and constraints. For example, if the condition for granting authorization is met (i.e., the request is within the range of the allowed operating time), and also the requester agrees to accept to perform a certain obligation, then the ADF returns a positive response "Authorize" to the AEF, otherwise request is denied.

4.3   Areas of Control Architecture

To control the access to the WoT environment, our architecture considers one area of control, based on the location of the reference monitor, which is located at the space manager. We refer to this set up as the server side control domain (SCD), because this is the area where the reference monitor is located and where the access policy to the system resources (things) is enforced. Fig 6 below depicts this architecture.

Fig 6 shows that the control of subject's access to objects is done centrally. In this setup, the subject can either be located within the network or outside, and the objects may or may not be stored in the client's storage, depending upon the criticality and sensitivity of the content of the object. If it is not that sensitive, then it can be allowed to reside outside of the server-side storage. However, if the content is very critical or very sensitive, the object must stay within the server-side storage.

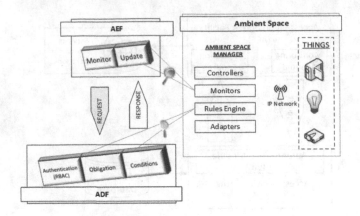

Fig. 6. Integrating RM into SM

5   WOT Resources Protection

In this section, we reveal details of the RBACprocess for protecting WOM-Profiles of things on the Web. To do this, we adopt the method described by Muldner, Mizilek and Leighton [11]. In this paper, RBAC specifies rules that consist of pairs of the form (role, resources), where a resource is a document fragment specified using an XPath expression (XPath, 2008). RBAC's data abstraction feature allows us to consider any permission needed to control access to the different fragments of an XML document.

5.1   Documents and Views

In this paper, access rights are defined using Access Control Policies (ACPs). In other words, ACPs are defined for fragments of XML documents, which we refer to as views. Each WoT activity is published as a single XML document.

Views are specified using a subset of XPath expression referred to as document paths as follows:

Definition 1. *A local document path is a document path with no free variables. A free variables are those variables that represent systems variable and their names start with $. A global document path is a document path which is not local, and considered instantiated when each occurrence of free variables is replaced by some value. For a document D, $P_{D,loc}$ denotes the set of local paths in D. Each local document path defines a fragment of the document D. Similarly, $P_{D, glob}$ denotes the set of global paths in D. Hence, the set of all document paths is denoted by $PD = P_{D, loc} \cup P_{D, glob}$. RBAC is susceptible to role proliferation. Parameterization has been used in the literature to address this problem (REF), and is out of the scope of this paper.*

Definition 2. *Let $\Delta\tau$ denotes the language for all roles then, for a WOM, D, and a finite set of simples roles $\psi \subset \Delta\tau$ the document-level ACP is a mapping $\Pi D : \psi \to P_D$ such that $\Pi D(\psi)$ covers the set D; i.e. each element of D belongs to at least one document path that occurs in the policy. Often, the ΠD mapping is tabulated and shown as tuple [(R1, P1), (R2, P2), ... , (Rm, Pn)].*

For a simple role $R \in \psi$, if $\Pi_D(R)$ is local document then it defines a view of D. If $\Pi_D(\psi)$ is global document path that contains free variables, then once path is instantiated, it defines a view of D. The designer of the RBAC policy for, WOM D may elect to leave some parts of D unencrypted or make them inaccessible to all users.

For a WOM D, a finite set of roles $\psi \subset \Delta\tau$, and the document–level ACP $\Pi_D : \psi \to P_D$ a user in role R can access precisely the set $\Pi_D(R)$ and those nodes in D which are not covered by any path.

5.2   Key Generation and Encryption

Let κ be a finite set of keys, where each key is a tuple made of ¡key name, symmetric key¿, and κ_D, $\Pi_D$denotes a document-level key ring for the WOM D and D's policy Π_D, then the key generation for a document-level policy ACP $\Pi_D : \psi \to P_D$ takes place as following: If the all paths are local, then each path can uniquely identify a fragment of D. However, if the paths are global, the issues of parameterization will complicate the case because condition of the path cannot evaluated before the values of the variables are known. For simplicity, we will consider only the local paths.

In this case, a key ring κ_D, Π_D is defined and for each $R \in \psi$, this key ring defines a set κ_D, $\Pi_D(R)$ of R-Accessible keys. A user in role R will be provided with R-Accessible keys allowing the decryption of the view $\Pi_D(R)$.

To decrypt the document, a user U will travers the document and use the names of the keys from κ_D, $\Pi_D(R)$ to extract the appropriate key to decrypt the accessible document.

To obtain a key ring that can be used to decrypt a fragment of an encrypted document, a user can request that key ring form the list of roles that user is a member of. Verification of membership can be achieved through presenting the certificate that user obtained membership to that role.

6   Conclusion and Future Work

In this paper we introduced a new architecture that encompasses WoT in a secure and scalable manner. Our architecture integrated the features of the well-known role-based access control (RBAC) to specify the access control policies to the WoT. More specifically, we showed how RBAC can be integrated to the WoT architecture to specify access control to the things, which are represented on the Web. We also showed how cryptographic keys are generated and used to enforce such access control policies for these documents. This enable prescribers of WoT services to control who can access their things and how, thereby enables privacy and the security of large amount of data that these things flood the Web with. Our future work will focus on implementing this architecture.

References

1. Ferraiolo, D., Cugini, J., Kuhn, D.R.: Role-based access control (RBAC): Features and motivations. In: Proceedings of 11th Annual Computer Security Application Conference, pp. 241–248 (1995)
2. Ferraiolo, D., Kuhn, D.R., Chandramouli, R.: Role-based access control. Artech House (2003)
3. Ferraiolo, D., Kuhn, D.R.: Role-based access controls, arXiv preprint arXiv:0903.2171 (2009)
4. Guinard, D., Trifa, V.: Towards the web of things: Web mashups for embedded devices, Workshop on Mashups, Enterprise Mashups and Lightweight Composition on the Web (MEM 2009). In: Proceedings of WWW (International World Wide Web Conferences), Madrid, Spain (2009)
5. Mathew, S.S., Atif, Y., Sheng, Q.Z., Maamar, Z.: Towards an Efficient Sales Pitch with the Web of Things. In: ICEBE, 2013, pp. 377–384 (2013)
6. Mathew, S.S., Atif, Y., Sheng, Q.Z., Maamar, Z.: Building sustainable parking lots with the Web of Things. In: Personal and Ubiquitous Computing, 2013, pp. 1–13. Springer, Heidelberg (2013)
7. Mathew, S.S., Atif, Y., Sheng, Q.Z., Maamar, Z.: Ambient things on the Web. Journal of Ubiquitous Systems and Pervasive Networks (JUSPN) 1(1), 1–8 (2010, 2013)
8. Mathew, S.S.: Classifying and Clustering the Web of Things, University of Adelaide, School of Computer Science (2013), http://hdl.handle.net/2440/83366
9. Mathew, S.S., Atif, Y., Sheng, Q.Z., Maamar, Z.: The Web of Things - Challenges and Enabling Technologies. In: Bessis, N., Xhafa, F., Varvarigou, D., Hill, R., Li, M. (eds.) Internet of Things & Inter-cooperative Comput. Technol. SCI, vol. 460, pp. 1–24. Springer, Heidelberg (2013)
10. Mathew, S.S., Atif, Y., Sheng, Q.Z., Maamar, Z.: Web of Things: Description, Discovery and Integration. In: International Conference on Internet of Things and Cyber, Physical and Social Computing (iThings/CPSCom), pp. 9–15. IEEE (2013)
11. Müldner, T., Miziolek, J.K., Leighton, G.: Succinct Access Control Policies for Published XML Datasets. In: ICEIS, vol. (1), pp. 380–385 (2008)
12. Osborn, S., Sandhu, R., Munawer, Q.: Configuring role-based access control to enforce mandatory and discretionary access control policies. ACM Transactions on Information and System Security (TISSEC) 3, 85–106 (2000)
13. Oh, S., Park, S.: Task–role-based access control model, Information Systems, vol. 28, pp. 533–562. Elsevier (2003)
14. Sandhu, R.S., Coyne, E.J., Feinstein, H.L., Youman, C.E.: Role-based access control models. IEEE Computer Society 29(2), 38–47 (1996)
15. Security frameworks for open systems: Access control framework, Technical Report ISO/IEC 10181-3, ISO (1996),
http://www.iso.org/iso/catalogue_detail.htm?csnumber=18199
16. Thomas, R.K., Sandhu, R.S.: Task-based authorization controls (TBAC): A family of models for active and enterprise-oriented authorization management. In: DBSec, 1997, vol. 113, pp. 166–181 (1997)

On the Security of Long-Lived Archiving Systems Based on the Evidence Record Syntax

Matthias Geihs[✉], Denise Demirel, and Johannes Buchmann

Technische Universität Darmstadt, University in Darmstadt, Germany
mgeihs@cdc.informatik.tu-darmstadt.de

Abstract. The amount of security critical data that is only available in digital form is increasing constantly. The Evidence Record Syntax Specification (ERS) achieves very efficiently important security goals: integrity, authenticity, datedness, and non-repudiation. This paper supports the trustworthiness of ERS by proving ERS secure. This is done in a model presented by Canetti et al. that these authors used to establish the long-term security of the Content Integrity Service (CIS). CIS achieves the same goals as ERS but is much less efficient. We also discuss the model of Canetti et al. and propose new directions of research.

1 Introduction

The amount of data that is only available in digital form is increasing constantly. Examples include scientific data, medical records, and land registries. Therefore, digital archives are needed that efficiently and securely preserve this information for a long period of time.

Important protection goals for archived data objects are *authenticity*, *integrity*, *non-repudiation*, and *datedness*. Integrity means that the data object has not been altered. Authenticity refers to the origin being identifiable. Non-repudiation prevents an originator from repudiating that he is the origin of a document. Datedness allows to identify a time reference when a document existed.

The Evidence Record Syntax Specification (ERS) [5, 2] achieves these protection goals efficiently and in the long-term. In fact, ERS focuses on datedness. This is sufficient as integrity follows from datedness. Also, if the data objects are digitally signed, then datedness also provides authenticity and non-repudiation.

To make ERS trustworthy it is desirable to have a security model and a corresponding security proof that establishes the security properties of ERS from a theoretical point of view. This is what we do in this paper. As a security model, we use the framework of Canetti et al. for analyzing computational security in long-lived systems [4]. Using their framework, they analyze the security of the Content Integrity Service (CIS) proposed by Haber et al. [6] that also ensures datedness in archives. ERS is a refined, more efficient variant of CIS. The main

---

This work has been co-funded by the DFG as part of project Long-Term Secure Archiving within the CRC 1119 CROSSING.

© Springer International Publishing Switzerland 2015
S. El Hajji et al. (Eds.): C2SI 2015, LNCS 9084, pp. 27–44, 2015.
DOI: 10.1007/978-3-319-18681-8_3

difference is the intelligent use of hash functions that allow for better performance. In this work, we extend their analysis of CIS to ERS. The main idea is to introduce hash services extending the signature services used by Canetti et al. They allow to model the ERS evidence records that are used to establish datedness at any point in time.

The structure of the paper is as follows. In Section 2, we describe the setup of long-term archiving systems and provide a summary of the ERS specification. In Section 3, we present the security framework of Canetti et al. and briefly explain their analysis of CIS. Using their framework, in Sections 4 and 5 we analyze the security of ERS. In Section 6, we draw conclusions and present future work.

2   ERS Archiving System

In this section, we describe the setup of secure archiving systems and provide a summary of the ERS specification.

2.1   Setup

A secure archiving system is used to store data objects for a long period of time while ensuring datedness of stored data. To achieve this, for each data object d stored at time t, the system maintains an evidence record e_d which allows to prove that data object d was archived at time t.

For maintaining evidence records, archiving systems typically rely on *timestamp services*. Timestamp services are trusted third parties which can be queried to issue a *timestamp* on a given bit string. When a timestamp service A is queried to timestamp bit string x at time t, it responds with timestamp θ. Afterwards, timestamp θ can be used to verify that timestamp service A indeed timestamped bit string x for time t.

In this work, we consider signature-based timestamp services. A timestamp for bit string x and time t issued by a signature-based timestamp service is a signature on $\langle x, t \rangle$.

2.2   ERS Specification

We give an overview of the ERS specification [5]. For a set of stored data objects $\{d_1, \ldots, d_n\}$, the ERS specification supports to maintain an *evidence record e*. For each data object $d \in \{d_1, \ldots, d_n\}$, evidence record e can be used to verify datedness of d.

When the ERS archiving system is initially asked to store a set of data objects $\{d_1, \ldots, d_n\}$, it generates a new evidence record for $\{d_1, \ldots, d_n\}$ and stores it together with the data objects. The generation of an evidence record uses cryptographic primitives. In particular, collision-resistant hash functions and signature schemes are used. The lifetime of those primitives is limited due to brute-force attacks, advances in cryptanalysis, or key compromise. Consequently, in order to remain valid, an evidence record needs to be refreshed periodically.

The ERS specification provides two methods of evidence record refresh, namely timestamp-refresh and hash-refresh. Timestamp-refresh protects against the expiration of a signature-based timestamp. Hash-refresh protects against the expiration of a hash value.

We describe the data structure of an evidence record and how it is generated, timestamp-refreshed, hash-refreshed and verified.

Structure. An evidence record consists of a list of timestamps and the verification information required for timestamp verification. We refrain from explicitly describing maintenance of verification information since it is not fundamental for our analysis of ERS. An initially generated evidence record contains a single timestamp. Upon evidence record refresh, new timestamps are added to the list.

Generation. Generation of an evidence record e for a set of data objects $\{d_1, \ldots, d_n\}$ is done as follows. First, a Merkle hash tree [7] is generated having the data objects as the leaves. Let r be the hash value corresponding to the root of that hash tree. A timestamp θ on r is requested from a timestamp service. The freshly generated evidence record e contains timestamp θ.

Timestamp-Refresh. An evidence record e is timestamp-refreshed as follows. Let $\theta_1, \ldots, \theta_n$ be the timestamps contained in e, where θ_n is the most recent timestamp. A new timestamp θ' on θ_n is requested. The timestamp-refreshed evidence record e' contains timestamps $\theta_1, \ldots, \theta_n, \theta'$.

Hash-Refresh. An evidence record e is hash-refreshed as follows. Let $\{d_1, \ldots, d_n\}$ be the data objects covered by e and let $\theta_1, \ldots, \theta_n$ be the timestamps contained in e. A new Merkle hash tree is built with $d_1, \ldots, d_n, \theta_1, \ldots, \theta_n$ as the leaves. Let r' be the root of that hash tree. A new timestamp θ' on r' is requested. The hash-refreshed evidence record e' contains timestamps $\theta_1, \ldots, \theta_n, \theta'$.

Verification. Datedness verification of data object d for time t_1 using evidence record e is done as follows. Let $\theta_1, \ldots, \theta_n$ be the timestamps of e and for $i = 1, \ldots, n$, let t_i be the time when θ_i was issued. Check the following.
- For $i = 2, \ldots, n$, verify if timestamp θ_i covers timestamp θ_{i-1} for time t_i. If θ_i results from hash-refresh, additionally verify if it covers data object d and timestamps $\theta_1, \ldots, \theta_{i-2}$ for time t_i.
- Verify if θ_1 covers data object d for time t_1.

3   Security Framework

In this section, we provide a high level description of the security framework of Canetti et al. for modeling computational security in long-lived systems [4]. We refer to the framework as the long-lived computational security framework, or short, LCS framework.

In this paper, our goal is to analyze the security of the ERS archiving system. In cryptography, the security of a system is typically defined in the presence of a resource bounded adversary, often modeled as a polynomial-time machine.

We must allow the adversary to be active during the whole lifetime of the system. However, long-lived systems, like the ERS system, are potentially running for super-polynomial time. Modeling the adversary as a polynomial-time machine is too restrictive for analyzing the security of systems with super-polynomial lifetime.

In the context of long-lived systems, we want to allow entities to be active for unbounded lifetime, while bounding their computational power at any point in time. To model this behavior, a special kind of automaton model is used, namely the task-PIOA model [3], augmented with a notion of real time. Combining the task-PIOA model with a notion of real time allows to put in relationship the number of automaton steps and the duration of real time required to complete a task. Computational restrictions on a task-PIOA are imposed in terms of computation rates, i.e. number of computation steps per unit of real time.

By its nature, a polynomial-time machine uses only a polynomial-bounded amount of space. There is no such implicit space bound for a machine with unbounded lifetime, such as a task-PIOA. In addition to specifying a bound on the computation rate of bounded task-PIOAs, we impose a bound on the space consumed by a bounded task-PIOA. We allow a bounded task-PIOA to only use a bounded amount of space at any point in time.

Note that, with respect to the security parameter k, computational bounds are fixed over the lifetime of the whole protocol. In particular, the LCS framework does not allow to model systems whose computational power increases over time.

Using the LCS framework, a security proof of a cryptographic system is done in style of the *real-ideal paradigm*. In this style, an ideal version of the system and a real version of the system are defined. Here, the *ideal system* represents the *functionality* of the system, which is secure by definition and usually relies on a trusted party. The *real system* represents the *implementation* of the system, which uses cryptography to mimic the ideal system's behavior. To prove the implementation secure, it is shown that a computationally bounded environment interacting with the two systems cannot distinguish them. Since the ideal system implicitly defines the functionality of the secure system, this suffices to show the security of the real system.

The LCS framework provides a mechanism for long-lived systems to recover from past security failures. Therefore, an ideal system is allowed to take designated failure steps. For any polynomial-bounded time interval, the real system will only have to approximate the ideal system if no failure tasks occur in that interval.

In Section 3.1 we introduce the task-PIOA model. In Section 3.2 we introduce the long-term implementation relation which allows to compare an ideal system to a real system in the presence of a long-lived environment. In Section 3.3 we briefly describe the CIS archiving system model from [4].

3.1   Task-PIOAs

If we say, a system is described within the LCS framework, we mean that it is modeled as a *task-PIOA* [3], which is a version of a *probabilistic input/output automaton (PIOA)*.

A PIOA $\mathcal{A}$ is defined by a tuple $\langle V, S, s^{init}, I, O, H, \Delta \rangle$. Here, V is a set of state variables, S is a set of states, $s^{init} \in S$ is the initial state, I is a set of input actions, O is a set of output actions, H is a set of hidden actions, and Δ is a transition relation. The transition relation describes how the automaton, for a given action, transitions from one state into another. An action transition can be viewed as an atomic computation step of a PIOA.

Multiple PIOA actions can be grouped into a *task*. Formally, a task-PIOA is a pair $\langle \mathcal{A}, \mathcal{R} \rangle$, where $\mathcal{A}$ is a PIOA and $\mathcal{R}$ is a partition of locally-controlled actions (i.e., output and hidden actions) of $\mathcal{A}$. The equivalence classes in $\mathcal{R}$ are called tasks. For notational simplicity, we often omit $\mathcal{R}$ and refer to the task-PIOA $\mathcal{A}$.

Computational bounds on a task-PIOA are three-fold. Firstly, a step bound on a task-PIOA limits the turing complexity of every single task-PIOA step. Secondly, in the LCS framework, task-PIOAs are augmented with a real-time scheduling mechanism. This allows to impose real-time scheduling constraints on task schedules. More precisely, real-time scheduling constraints allow to limit the number of steps performed by a task-PIOA per fraction of real time. Thirdly, step bound and real-time scheduling constraints are combined to obtain an overall bound.

Operations. Task-PIOAs are subject to the composition and hiding operation.

Composition. Let $\mathcal{A}_1$ and $\mathcal{A}_2$ be two task-PIOAs. We say $\mathcal{A}_1$ and $\mathcal{A}_2$ are *compatible*, if they do not share any state variables or output actions, and hidden actions of the one automaton do not collide with any actions of the other automaton (and vice versa). If two task-PIOAs $\mathcal{A}_1$ and $\mathcal{A}_2$ are compatible, they can be composed into a new task-PIOA. We denote the composition of $\mathcal{A}_1$ and $\mathcal{A}_2$ by $\mathcal{A}_1 \| \mathcal{A}_2$. The composition $\mathcal{A}_1 \| \mathcal{A}_2$ is itself a task-PIOA which synchronizes on shared actions of $\mathcal{A}_1$ and $\mathcal{A}_2$.

Hiding Operator. We define a hiding operator for task-PIOAs. Let $\mathcal{A} := \langle V, S, s^{init}, I, O, H, \Delta \rangle$ be a task-PIOA and $X \subseteq O$ be a set of output actions. Then, $\text{hide}(\mathcal{A}, X)$ is the task-PIOA given by $\langle V, S, s^{init}, I, O \setminus X, H \cup X, \Delta \rangle$. This prevents other task-PIOAs from synchronizing with $\mathcal{A}$ via actions in X: any task-PIOA with an action in X is no longer compatible with $\mathcal{A}$.

Step Bound. The notion of a *step bound* is defined to limit the amount of computation a task-PIOA can perform, and the amount of space it can use, in executing a single step. For $p \in \mathbb{N}$, we say a task-PIOA $\mathcal{A}$ has step bound p, if for every single step of $\mathcal{A}$, p limits the complexity of a turing machine simulating the step.

Real-Time Scheduling Constraints. In the LCS framework, task-PIOAs are augmented with *real-time scheduling constraints*. This allows to model entities with unbounded lifetime but bounded processing rates. Therefore, a task schedule can be associated with a bound map $\langle rate, burst, lb, ub \rangle$. Here, $rate$ bounds the number of task executions per real time, $burst$ allows for a fixed violation of this bound, and lb and ub are lower and upper real time bounds for the first and last execution of a task, respectively. We say a real time task schedule is constrained by p, if it is valid under a p-bounded bound map.

Note that real time is only used to express constraints on task schedules. Computationally bounded system components are not allowed to maintain real time information in their states, nor to communicate real-time information to each other. System components that require knowledge of time will maintain discrete approximations of time in their states, based on inputs from a global task-PIOA *Clock*.

Overall Bound. Step bound and real time scheduling constraints are combined to obtain an overall bound on a task-PIOA $\mathcal{A}$. We say that a task-PIOA $\mathcal{A}$ is p-*bounded*, if $\mathcal{A}$ has step bound p and real time task scheduling is constrained by p. We say a task-PIOA $\mathcal{A}$ is *quasi-p-bounded* if $\mathcal{A}$ is of the form $\mathcal{A}' \| Clock$, where $\mathcal{A}'$ is p-bounded.

Task-PIOA Families. Task-PIOAs can be gathered into task-PIOA families, indexed by a security parameter k. A task-PIOA family $\bar{A}$ is an indexed set $\{\mathcal{A}_k\}_{k \in \mathbb{N}}$ of task-PIOAs. Given a function $p : \mathbb{N} \to \mathbb{N}$, we say that $\bar{A}$ is p-bounded if for all k, $\mathcal{A}_k$ is $p(k)$-bounded. If p is a polynomial, then we say $\bar{A}$ is polynomially bounded.

3.2   Longterm Implementation Relation

The LCS framework allows modeling computational security in long-lived systems. Traditionally, a system is considered secure if a polynomial-time environment cannot distinguish the ideal system model (i.e., the functionality) from the real system model (i.e., the implementation). Restricting environments to be polynomial-time bounded is not satisfactory in the context of long-lived systems which potentially run for super-polynomial time.

The LCS framework provides a notion of indistinguishability in the context of long-lived systems. The idea is to not limit the overall amount of computation performed by a long-lived environment, but to polynomially bound the amount of computation performed per fraction of time. Furthermore, long-lived systems are allowed to recover from past security failures. Therefore, an ideal system is allowed to take designated failure steps. For a polynomial-bounded time interval, the real system will only have to approximate the ideal system, if no failure tasks occur in that interval.

A long-term implementation relation defines indistinguishability of systems in the context of a long-lived environment. We sketch the definition of the long-term implementation relations $\leq_{p,q,\epsilon}$ and $\leq_{neg,pt}$ given in [4], Section 5. Task-PIOAs can only be put in relationship by a long-term implementation relation if they are *comparable*. We say task-PIOAs $\mathcal{A}^1$ and $\mathcal{A}^2$ are comparable, if they have the same external interface, that is, they have the same input and output actions. We say task-PIOA families $\bar{\mathcal{A}}^1$ and $\bar{\mathcal{A}}^2$ are comparable if for every k, $(\bar{\mathcal{A}}^1)_k$ is comparable to $(\bar{\mathcal{A}}^2)_k$.

Let $\mathcal{A}^1$ and $\mathcal{A}^2$ be comparable task-PIOAs. Let F^1 and F^2 be sets of designated failure tasks associated with $\mathcal{A}^1$ and $\mathcal{A}^2$, respectively. Let $p, q \in \mathbb{N}$ and $\epsilon \in \mathbb{R}_{\geq 0}$. If for every q-bounded time window in which no failure tasks F^1 and F^2 occur, any quasi-p-bounded environment cannot distinguish $\mathcal{A}^1$ and $\mathcal{A}^2$ with probability at most ϵ, we write $(\mathcal{A}^1, F^1) \leq_{p,q,\epsilon} (\mathcal{A}^2, F^2)$.

The $\leq_{p,q,\epsilon}$ definition is extended to task-PIOA families. Let $\bar{\mathcal{A}}^1$ and $\bar{\mathcal{A}}^2$ be comparable task-PIOA families. Let $\bar{F}^1$ and $\bar{F}^2$ be sets of designated failure tasks associated with $\bar{\mathcal{A}}^1$ and $\bar{\mathcal{A}}^2$, respectively. Let p, q be polynomials and $\epsilon : \mathbb{N} \to \mathbb{R}_{\geq 0}$ be a function. We say $(\bar{\mathcal{A}}^1, \bar{F}^1) \leq_{p,q,\epsilon} (\bar{\mathcal{A}}^2, \bar{F}^2)$, if $\forall k : ((\bar{\mathcal{A}}^1)_k, (\bar{F}^1)_k) \leq_{p(k),q(k),\epsilon(k)} ((\bar{\mathcal{A}}^2)_k, (\bar{F}^2)_k)$.

We write $(\bar{\mathcal{A}}^1, \bar{F}^1) \leq_{neg,pt} (\bar{\mathcal{A}}^2, \bar{F}^2)$, if $\forall p, q \exists \epsilon : (\bar{\mathcal{A}}^1, \bar{F}^1) \leq_{p,q,\epsilon} (\bar{\mathcal{A}}^2, \bar{F}^2)$, where p, q are polynomials and ϵ is a negligible function. In this case we say $\bar{\mathcal{A}}^1$ implements $\bar{\mathcal{A}}^2$ in the sense of the long-term implementation relation. Here, $\bar{\mathcal{A}}^1$ is usually referred to as the real system (i.e., the implementation), and $\bar{\mathcal{A}}^2$ is usually referred to as the ideal system (i.e., the functionality).

Composition Theorems. We quote the following statement regarding composition theorems from [4], Section 7.

> In practice, cryptographic services are seldom used in isolation. Usually, different types of services operate in conjunction, interacting with each other and with multiple protocol participants. For example, a participant may submit a bit string to an encryption service to obtain a ciphertext, which is later submitted to a timestamping service. In such situations, it is important that the services are provably secure even in the context of composition.

Indeed, as described in Section 3.1, single task-PIOAs (e.g., encryption or timestamp services) can be composed to obtain more complex task-PIOAs (e.g., a system composed of communicating services). The following composition theorems allow to preserve the longterm implementation relation $\leq_{neg,pt}$. For a formal definition of the composition theorems see [4], Section 7.

Parallel Composition Theorem. The *Parallel Composition Theorem* allows for the parallel composition of polynomially many components.

Sequential Composition Theorem. The *Sequential Composition Theorem* allows for the sequential composition of exponentially many components. We say task-PIOAs are sequential if for every real time t at most one of the task-PIOAs is not dormant at time t.

d-**Bounded Composition Theorem.** The *d-Bounded Composition Theorem* allows for the d-bounded concurrent composition of exponentially many components, where d is a positive integer. We say task-PIOAs are d-bounded concurrent if for every real time t at most d of the task-PIOAs are not dormant at time t.

We describe application of a composition theorem to sequences of task-PIOAs associated with a sequence of designated failure task families. Let $\bar{\mathcal{A}}_1^1, \bar{\mathcal{A}}_2^1, \ldots$ and $\bar{\mathcal{A}}_1^2, \bar{\mathcal{A}}_2^2, \ldots$ be comparable sequences of compatible task-PIOA families associated with sequences of failure task set families $\bar{F}_1^1, \bar{F}_2^1, \ldots$ and $\bar{F}_1^2, \bar{F}_2^2, \ldots$, respectively. Let $C := \{1, 2, \ldots, n\}$ be a set of indices. Define the compositions of task-PIOA families $\hat{\mathcal{A}}^1 := \|_{i \in C} \bar{\mathcal{A}}_i^1$ and $\hat{\mathcal{A}}^2 := \|_{i \in C} \bar{\mathcal{A}}_i^2$, and the unions of failure task set families $\hat{F}^1 := \{\bigcup_{i \in C} (\bar{F}_i^1)_k\}_{k \in \mathbb{N}}$ and $\hat{F}^2 := \{\bigcup_{i \in C} (\bar{F}_i^2)_k\}_{k \in \mathbb{N}}$. Note that index set C is subject to the composition theorem to be applied. Then, $(\hat{\mathcal{A}}^1, \hat{F}^1) \leq_{neg,pt} (\hat{\mathcal{A}}^2, \hat{F}^2)$, if $\forall p, q \exists \epsilon \forall i : (\bar{\mathcal{A}}_i^1, \bar{F}_i^1) \leq_{neg,pt} (\bar{\mathcal{A}}_i^2, \bar{F}_i^2)$, where p, q are polynomials and ϵ is a negligible function.

3.3 CIS System Model

In [4], Canetti et al. propose a model for another long-lived archiving system, namely the content integrity service (CIS) [6]. We explain briefly how the CIS system is modeled as the composition of task-PIOAs.

The CIS system model is composed of a dispatcher component and a sequence of timestamp services. The dispatcher component accepts various timestamp requests and forwards them to the appropriate timestamp service. In [4], Section 8, it is shown that the composition of the dispatcher and real timestamp services is indistinguishable from an ideal system, composed of the same dispatcher and corresponding ideal timestamp services. Specifically, this guarantees that the probability of a new forgery is small at any given point in time, regardless of any forgeries that may have happened in the past.

We sketch some of the technicalities of the CIS analysis from [4]. The dispatcher component, the real timestamp services and the ideal timestamp services are modeled as task-PIOAs. It is shown that a real timestamp service implements its ideal timestamp service counterpart in the sense of $\leq_{neg,pt}$. Using the d-bounded composition theorem, it is shown that the d-bounded composition of real timestamp services implements the d-bounded composition of ideal timestamp services. Using the parallel composition theorem, it is shown that the parallel composition of the dispatcher and the real timestamp services (i.e., the real system) implements the parallel composition of the dispatcher and the ideal timestamp services (i.e., the ideal system).

4 ERS System Model

In this section, we propose a task-PIOA model of the ERS archiving system by extending the CIS system model (cf. Section 3.3).

The ERS system extends the CIS system as follows. The CIS system supports one method for evidence refresh, where data object and evidence are timestamped together. In particular, in the CIS system model, no hash functionality is described. The ERS system supports two methods for evidence refresh, namely timestamp-refresh and hash-refresh (cf. Section 2). The hash-refresh method is similar to CIS evidence refresh (i.e., data object and evidence are timestamped together). The timestamp-refresh method is special to ERS as it allows to refresh the evidence while only part of the current evidence needs to be hashed and timestamped. This makes ERS more efficient compared to CIS.

4.1 Construction Overview

We give an overview of the ERS model construction. The ERS system is modeled as the composition of a dispatcher component, a sequence of timestamp services, and, in particular, a sequence of hash services. The dispatcher component accepts various evidence record requests and uses appropriate hash and timestamp services to answer them.

A timestamp service can be queried to produce a timestamp for a bit string. Here, we consider signature-based timestamp services. When a signature-based timestamp service is queried for a timestamp on bit string x, it responds with a signature on $\langle x, t \rangle$, where t is the time at timestamp request. Each timestamp service wakes up at a certain time and is active for a specified amount of time before becoming dormant again. This can be viewed as a regular update of the service, which may entail a simple refresh of the timestamp key, or the adoption of a new timestamp algorithm.

A hash service can be queried to produce a hash of a bit string. When a hash service is queried for a hash of bit string x, it responds with a fixed-length hash $H(x)$, where H is a collision-resistant hash function. Because the hash service offers a collision-resistant hash functionality, it is hard to find a bit string x', such that $x \neq x'$ and $H(x) = H(x')$. Each hash service starts being available at a certain time and is available for a specified amount of time before becoming unavailable again. This can be viewed as a regular update of the hash algorithm.

The real ERS model consists of the dispatcher component, a collection of hash services, and a collection of real timestamp services. Similarly, the ideal ERS model consists of the same dispatcher component, a collection of hash services, and a collection of ideal timestamp services. Note that we do not distinguish between real and ideal hash services. This is due to the fact that we model the functionality of a collision-resistant hash algorithm using the random oracle methodology (cf. Section 4.4).

4.2 Signature Service

We describe the signature service model from [4]. A signature service is identified by its service identifier. We denote the domain of signature service identifiers by $\mathsf{SID_{sign}}$. A signature service is constructed using a signature scheme.

Definition 1 (Signature Scheme). *A signature scheme consists of three algorithms* KeyGen, Sign, *and* Verify. KeyGen *is a probabilistic algorithm that outputs a signing-verification key pair* $\langle sk, vk \rangle$. Sign *is a probabilistic algorithm that produces a signature* σ *from a message* m *and the key* sk. *Finally,* Verify *is a deterministic algorithm that maps* $\langle m, \sigma, vk \rangle$ *to a boolean. The signature* σ *is said to be valid for* m *and* vk *if* $\text{Verify}(m, \sigma, vk) = 1$.

In the following, we describe the real signature service model, the ideal signature service model, and sketch the proof of Theorem 1 from [4]. According to this theorem, the real signature service, if instantiated with a complete and existentially unforgeable signature scheme, implements the corresponding ideal signature service in the sense of the $\leq_{neg,pt}$ definition (cf. Section 4.2).

For every $j \in \text{SID}_{\text{sign}}$, suppose that $\langle \text{KeyGen}_j, \text{Sign}_j, \text{Verify}_j \rangle$ is a signature scheme. We assume a function alive : $\mathbb{T} \to 2^{\text{SID}_{\text{sign}}}$ such that, for every t, alive(t) is the set of services alive at discrete time t. The lifetime of each service j is then given by aliveTimes$(j) := \{t \in \mathbb{T} | j \in \text{alive}(t)\}$.

Real Signature Service. For $k \in \mathbb{N}$ and $j \in \text{SID}_{\text{sign}}$, we define three task-PIOAs, KeyGen(k, j), Signer(k, j), and Verifier(k, j), representing the key generator, signer, and verifier, respectively.

KeyGen(k, j) chooses a signing key $mySK$ and a corresponding verification key $myVK$ by running the KeyGen$_j$ algorithm. It does this exactly once during its lifetime. It outputs the two keys separately, via actions signKey$(sk)_j$ and verKey$(vk)_j$. The signing key goes to Signer(k, j), while the verification key goes to Verifier(k, j). Signer(k, j) responds to signing requests by running the Sign$_j$ algorithm on message m and the signing key sk. Verifier(k, j) accepts verification requests and simply runs the Verify$_j$ algorithm.

For $k \in \mathbb{N}$ and $j \in \text{SID}_{\text{sign}}$, we define the real signature service as

$$\text{RealSig}(j)_k := \text{hide}(\text{KeyGen}(k, j) \| \text{Signer}(k, j) \| \text{Verifier}(k, j), \text{signKey}_j) .$$

Note that the hiding operator prevents the environment from learning the signing key (cf. Section 3.1).

Ideal Signature Service. We specify an ideal signature functionality SigFunc. As with KeyGen, Signer, and Verifier, each instance of SigFunc is parametrized with a security parameter k and an identifier j. The task-PIOA SigFunc(k, j) is very similar to the composition of Signer(k, j) and Verifier(k, j). The important difference is that SigFunc(k, j) maintains an additional internal variable history, which records the set of signed messages. In addition, SigFunc(k, j) has an interal action fail$_j$, which sets a boolean flag failed. If failed = false, then SigFunc(k, j) uses history to answer verification requests: a signature is rejected if the submitted message is not in history, even if Verify$_j$ returns 1. If failed = true, then SigFunc(k, j) bypasses the check on history, so that its answers are identical to those from the real signature service.

For $k \in \mathbb{N}$ and $j \in \mathsf{SID}_{\mathsf{sign}}$, we define the ideal signature service as

$$\mathsf{IdealSig}(j)_k := \mathsf{hide}(\mathsf{KeyGen}(k, j) \| \mathsf{SigFunc}(k, j), \mathsf{signKey}_j) \ .$$

Implementation Proof. We define standard properties of signature schemes, namely *completeness* and *existential unforgeability*. Afterwards, we show that if a real signature service is instantiated with a complete and existential unforgeable signature scheme, it implements the corresponding ideal signature service.

Definition 2 (Completeness). *A signature scheme* $\langle \mathsf{KeyGen}, \mathsf{Sign}, \mathsf{Verify} \rangle$ *is complete if* $\mathsf{Verify}(m, \sigma, vk) = 1$ *whenever* $\langle sk, vk \rangle \leftarrow \mathsf{KeyGen}(1^k)$ *and* $\sigma \leftarrow \mathsf{Sign}(sk, m)$.

Definition 3 (EUF-CMA). *We say a signature scheme* $\langle \mathsf{KeyGen}, \mathsf{Sign}, \mathsf{Verify} \rangle$ *is existentially unforgeable under adaptive chosen message attack if no polynomial-time forger has non-negligible success probability in the following game.*

Setup *The challenger runs* KeyGen *to obtain* $\langle vk, sk \rangle$ *and gives the forger* vk.

Query *The forger submits message* m. *The challenger responds with signature* $\sigma \leftarrow \mathsf{Sign}(m, sk)$. *This may be repeated adaptively.*

Output *The forger outputs a pair* $\langle m^*, \sigma^* \rangle$ *and he wins if* m^* *is not among the messages submitted during the query phase and* $\mathsf{Verify}(m^*, \sigma^*, vk) = 1$.

For $j \in \mathsf{SID}_{\mathsf{sign}}$, define the ideal signature service family

$$\overline{\mathsf{IdealSig}}(j) := \{\mathsf{IdealSig}(j)_k\}_{k \in \mathbb{N}}$$

and the real signature service family

$$\overline{\mathsf{RealSig}}(j) := \{\mathsf{RealSig}(j)_k\}_{k \in \mathbb{N}} \ .$$

Theorem 1 from [4] says that if a real signature service is instantiated with a complete and existentially unforgeable signature scheme, it implements the corresponding ideal signature service. We quote Theorem 1 from [4].

Theorem 1. *Let* $j \in \mathsf{SID}_{\mathsf{sign}}$ *be given. Suppose that* $\langle \mathsf{KeyGen}_j, \mathsf{Sign}_j, \mathsf{Verify}_j \rangle$ *is a complete and EUF-CMA secure signature scheme. Then* $(\overline{\mathsf{RealSig}}(j), \emptyset) \leq_{neg,pt} (\overline{\mathsf{IdealSig}}(j), \{\mathsf{fail}_j\})$.

To prove Theorem 1, one needs to show the following for every time t and polynomials p, q. If task fail_j is not scheduled in interval $[t, t + q(k)]$, then no p-bounded environment can distinguish $\mathsf{RealSig}(j)_k$ from $\mathsf{IdealSig}(j)_k$ with high probability between time t and time $t + q(k)$. The full proof of Theorem 1 can be found in [4].

4.3   Timestamp Service

A timestamp service can be queried to create a timestamp on a bit string. The timestamp can later be used to verify that the bit string was available at a certain point in time. More precisely, for bit string x, timestamp service j can be queried to create a timestamp θ on x. The timestamp θ issued by timestamp service j is associated with a certain point in time t. Timestamp θ can later be used to verify that x was in fact timestamped for time t by service j.

We augment signature services to support timestamping. For every security parameter k and signature service $j \in \mathsf{SID}_{\mathsf{sign}}$, we define task-PIOA $\mathsf{Stamper}(k,j)$. When $\mathsf{Stamper}(k,j)$ receives a timestamp request for bit string x via action $\mathsf{reqStamp}(rid, x)$, where rid is the request identifier, it computes a signature σ on $\langle x, t \rangle$, where t is the clock reading at $\mathsf{reqStamp}$. Then, $\mathsf{Stamper}(k,j)$ responds with timestamp $\theta := \langle \sigma, t \rangle$ via $\mathsf{respStamp}(rid, \theta)$.

When $\mathsf{Stamper}(k,j)$ receives a verification request for timestamp $\theta := \langle \sigma, t \rangle$ and bit string x via $\mathsf{reqVerTs}(rid, x, \theta)$, it verifies if signature σ is a valid signature for $\langle x, t \rangle$. If verification is successful, it answers with $\mathsf{respVerTs}(rid, \mathsf{true})$. Otherwise, it answers with $\mathsf{respVerTs}(rid, \mathsf{false})$.

We use $\mathsf{Stamper}(k,j)$ and the signature service task-PIOAs defined in Section 4.2 (i.e., $\mathsf{KeyGen}(k,j)$, $\mathsf{Signer}(k,j)$, $\mathsf{Verifier}(k,j)$, and $\mathsf{SigFunc}(k,j)$) to build the real and ideal timestamp service. For $k \in \mathbb{N}$ and $j \in \mathsf{SID}_{\mathsf{sign}}$, we define the real timestamp service $\mathsf{RealStamp}(j)_k$ as

$$\mathsf{RealStamp}(j)_k := \mathsf{hide}(\mathsf{KeyGen}(k,j)\|\mathsf{Signer}(k,j)\|\mathsf{Verifier}(k,j)\|$$
$$\mathsf{Stamper}(k,j), \mathsf{signKey}_j)$$

and the ideal timestamp service $\mathsf{IdealStamp}(j)_k$ as

$$\mathsf{IdealStamp}(j)_k := \mathsf{hide}(\mathsf{KeyGen}(k,j)\|\mathsf{SigFunc}(k,j)\|\mathsf{Stamper}(k,j), \mathsf{signKey}_j) \ .$$

We gather the real and ideal timestamp services into families. For $j \in \mathsf{SID}_{\mathsf{sign}}$, we define the real timestamp service family

$$\overline{\mathsf{RealStamp}}(j) := \{\mathsf{RealStamp}(j)_k\}_{k \in \mathbb{N}} \ ,$$

and the ideal timestamp service family

$$\overline{\mathsf{IdealStamp}}(j) := \{\mathsf{IdealStamp}(j)_k\}_{k \in \mathbb{N}} \ .$$

Theorem 2. *Let $j \in \mathsf{SID}_{\mathsf{sign}}$ be given. Suppose that $\langle \mathsf{KeyGen}_j, \mathsf{Sign}_j, \mathsf{Verify}_j \rangle$ is a complete and EUF-CMA secure signature scheme. Then $(\overline{\mathsf{RealStamp}}(j), \emptyset) \leq_{neg,pt} (\overline{\mathsf{IdealStamp}}(j), \{\mathsf{fail}_j\})$.*

Proof. By Theorem 1 we have $(\overline{\mathsf{RealSig}}(j), \emptyset) \leq_{neg,pt} (\overline{\mathsf{IdealSig}}(j), \{\mathsf{fail}_j\})$. Observe that $\overline{\mathsf{RealSig}}(j)$ and $\overline{\mathsf{IdealSig}}(j)$ are modified in the same way (i.e., point-wise composition with $\mathsf{Stamper}(k,j)$) to obtain $\overline{\mathsf{RealStamp}}(j)$ and $\overline{\mathsf{IdealStamp}}(j)$. It follows that $(\overline{\mathsf{RealStamp}}(j), \emptyset) \leq_{neg,pt} (\overline{\mathsf{IdealStamp}}(j), \{\mathsf{fail}_j\})$.

4.4 Hash Service

Generation of evidence records in the ERS system involves using hash algorithms. A hash algorithm $H : \mathcal{M} \to \mathcal{H}$ is an efficient deterministic algorithm mapping a message $m \in \mathcal{M}$ to a fixed-length hash $H(m) \in \mathcal{H}$. We call $\mathcal{M}$ the message space and $\mathcal{H}$ the hash space. We say a hash algorithm H is collision resistant if it is hard to find two messages m and m' such that $m \neq m'$ and $H(m) = H(m')$.

In order to model the functionality of a collision resistant hash algorithm we make use of the random oracle methodology [1]. A random oracle can be thought of as a public, randomly-chosen function $H : \mathcal{M} \to \mathcal{H}$ that can be evaluated only by querying an oracle that returns $H(x)$ when given input x. It can easily be seen that a random oracle serves as a collision-resistant hash algorithm. In the following, we use a random oracle in place of a collision-resistant hash functionality.

We identify a hash service by its hash service identifier. We denote the domain of hash algorithm identifiers by $\mathsf{SID}_{\mathsf{hash}}$. For security parameter $k \in \mathbb{N}$ and hash identifier $j \in \mathsf{SID}_{\mathsf{hash}}$, we define task-PIOA $\mathsf{Hasher}(k, j)$. $\mathsf{Hasher}(k, j)$ has access to a random oracle $H_{k,j} : \mathcal{M}_{k,j} \to \mathcal{H}_{k,j}$, where $|\mathcal{H}_{k,j}| \geq 2^k$. When $\mathsf{Hasher}(k, j)$ receives a hash request on message $m \in \mathcal{M}_{k,j}$ via input action $\mathsf{reqHash}(rid, m)$ it queries oracle $H_{k,j}$ with m and returns the hash $H_{k,j}(m) \in \mathcal{H}_{k,j}$ via output action $\mathsf{respHash}(rid, H_{k,j}(m))$.

In addition, $\mathsf{Hasher}(k, j)$ has an internal action fail_j, which sets a boolean flag failed. If failed = false, then $\mathsf{Hasher}(k, j)$ uses the random oracle to answer hash requests as specified above. If failed = true, then $\mathsf{Hasher}(k, j)$ denies to answer hash requests: in that case, to every request $\mathsf{reqHash}(rid, m)$, it responds with $\mathsf{respHash}(rid, \perp)$.

For $j \in \mathsf{SID}_{\mathsf{hash}}$ and security parameter k, define the hash service

$$\mathsf{Hash}(j)_k := \mathsf{Hasher}(j, k) \ .$$

For $j \in \mathsf{SID}_{\mathsf{hash}}$, define the hash service family

$$\overline{\mathsf{Hash}}(j) := \{\mathsf{Hasher}(j, k)\}_{k \in \mathbb{N}} \ .$$

4.5 Service Times

Hash services and timestamp services have limited lifetime. During protocol execution a service can be in various service states, namely being *alive*, being the *preferred* service, or being a *usable* service. Let $\mathbb{T} := \mathbb{N}$ be the domain of discrete time and define the union of all service identifiers as $\mathsf{SID} := \mathsf{SID}_{\mathsf{hash}} \cup \mathsf{SID}_{\mathsf{sign}}$. We assume the following.

- $\mathsf{alive} : \mathbb{T} \to 2^{\mathsf{SID}}$. For every t, $\mathsf{alive}(t)$ is the set of services alive at discrete time t.
- $\mathsf{aliveTimes} : \mathsf{SID} \to \mathbb{T}$. For every service j, $\mathsf{aliveTimes}(j)$ denotes the lifetime of service j, $\mathsf{aliveTimes}(j) := \{t \in \mathbb{T} : j \in \mathsf{alive}(t)\}$.

- $\mathsf{pref_{hash}} : \mathbb{T} \to \mathsf{SID_{hash}}$. For every $t \in \mathbb{T}$, the hash service $\mathsf{pref_{hash}}(t)$ is the designated hasher for time t, i.e., any hash request sent by the dispatcher at time t goes to hash service $\mathsf{pref_{hash}}(t)$.
- $\mathsf{pref_{sign}} : \mathbb{T} \to \mathsf{SID_{sign}}$. For every $t \in \mathbb{T}$, the signature service $\mathsf{pref_{sign}}(t)$ is the designated signer for time t, i.e., any signature request sent by the dispatcher at time t goes to signature service $\mathsf{pref_{sign}}(t)$.
- $\mathsf{usable} : \mathbb{T} \to 2^{\mathsf{SID}}$. For every $t \in \mathbb{T}$, $\mathsf{usable}(t)$ specifies the set of services that are accepting new requests.

4.6   Dispatcher

We describe the task-PIOA $\mathsf{Dispatcher}_k$ for each security parameter k. In particular, we describe evidence record generation, timestamp-refresh, hash-refresh, and verification. In our model, an evidence record is a tuple $\langle i, \chi, \theta, j \rangle$, where i is the currently used hash service, χ is the previously timestamped data, θ is the most recent timestamp, and j is the corresponding timestamp service.

Generation. If the environment requests evidence record generation for bit string x via action $\mathsf{reqEviGen}(rid, x)$, $\mathsf{Dispatcher}_k$ requests a hash of x from hash service $i = \mathsf{pref_{hash}}(t)$, where t is the clock reading at the time of the request. After hash service i returned hash h, $\mathsf{Dispatcher}_k$ requests a timestamp on $\langle i, h \rangle$ from service $j = \mathsf{pref_{sign}}(t)$. After timestamp service j returned timestamp θ, $\mathsf{Dispatcher}_k$ issues a new evidence record $\langle i, x, \theta, j \rangle$ via action $\mathsf{respEvi}(rid, \langle i, x, \theta, j \rangle)$.

Timestamp-Refresh. If the environment requests timestamp-refresh of evidence record $\langle i, \chi, \theta, j \rangle$ via action $\mathsf{reqEviTs}(rid, \langle i, \chi, \theta, j \rangle)$, $\mathsf{Dispatcher}_k$ first checks to see if hash service i and timestamp service j are still usable. If not, it responds with an error message. Otherwise, it requests a hash of χ from hash service i. After hash service i returned hash h, $\mathsf{Dispatcher}_k$ checks if θ is a valid timestamp for $\langle i, h \rangle$. If not, it responds with an error message. Otherwise, it requests a hash of $\langle i, \theta \rangle$ from hash service i. After hash service i returned hash h', $\mathsf{Dispatcher}_k$ requests a timestamp on $\langle i, h' \rangle$ from service $j' = \mathsf{pref_{sign}}(t)$, where t is the clock reading at the time of the request. After timestamp service j' returned timestamp θ', $\mathsf{Dispatcher}_k$ issues the refreshed evidence record $\langle i, \theta, \theta', j' \rangle$ via action $\mathsf{respEvi}(rid, \langle i, \theta, \theta', j' \rangle)$.

Hash-Refresh. If the environment requests hash-refresh of evidence record $\langle i, \chi, \theta, j \rangle$ via action $\mathsf{reqEviHash}(rid, \langle i, \chi, \theta, j \rangle)$, $\mathsf{Dispatcher}_k$ first checks to see if hash service i and timestamp service j are still usable. If not, it responds with an error message. Otherwise, it requests a hash of χ from hash service i. After hash service i returned hash h, $\mathsf{Dispatcher}_k$ checks if θ is a valid timestamp for $\langle i, h \rangle$. If not, it responds with an error message. Otherwise, it requests a hash of $\langle i, \langle x, \theta \rangle \rangle$ from hash service $i' = \mathsf{pref_{hash}}(t)$, where t is the clock reading at the time of the request. After hash service i' returned hash h', $\mathsf{Dispatcher}_k$ requests a timestamp on $\langle i', h' \rangle$ from service $j' = \mathsf{pref}_{\mathsf{sign}}(t)$. After timestamp service j' returned timestamp θ', $\mathsf{Dispatcher}_k$ issues the refreshed evidence record $\langle i', \langle x, \theta \rangle, \theta', j' \rangle$ via action $\mathsf{respEvi}(rid, \langle i', \langle x, \theta \rangle, \theta', j' \rangle)$.

Verification. If the environment requests evidence verification of evidence record $\langle i, \chi, \theta, j \rangle$ via action $\mathsf{reqCheck}(rid, \langle i, \chi, \theta, j \rangle)$, $\mathsf{Dispatcher}_k$ first checks to see if hash service i and timestamp service j are still usable. If not, it responds with $\mathsf{respCheck}(rid, \mathsf{false})$. Otherwise, it requests a hash of χ from hash service i. After hash service i returned hash h, $\mathsf{Dispatcher}_k$ checks if θ is a valid timestamp for $\langle i, h \rangle$. If the verification request fails, $\mathsf{Dispatcher}_k$ responds with $\mathsf{respCheck}(rid, \mathsf{false})$. Otherwise, $\mathsf{Dispatcher}_k$ responds via action $\mathsf{respCheck}(rid, \mathsf{true})$.

4.7 ERS Service

We describe how the ideal ERS service and the real ERS service are composed of the previously described components.

Let $\mathsf{SID}_{\mathsf{hash}}$, the domain of hash service names, be $\{\mathsf{hash}\} \times \mathbb{N}$. Likewise, let $\mathsf{SID}_{\mathsf{sign}}$, the domain of timestamp service names, be $\{\mathsf{sign}\} \times \mathbb{N}$. We limit the number of service components by some exponential in security parameter k. For every k and polynomial p, let $\mathbb{N}_{<2^{p(k)}} \subseteq \mathbb{N}$ denote the set of $p(k)$-bit numbers. For every k, define service identifier subsets $(\mathsf{SID}_{\mathsf{hash}})_k \subseteq \mathsf{SID}_{\mathsf{hash}}$ and $(\mathsf{SID}_{\mathsf{sign}})_k \subseteq \mathsf{SID}_{\mathsf{sign}}$ as $(\mathsf{SID}_{\mathsf{hash}})_k := \{\mathsf{hash}\} \times \mathbb{N}_{<2^{p(k)}}$ and $(\mathsf{SID}_{\mathsf{sign}})_k := \{\mathsf{sign}\} \times \mathbb{N}_{<2^{q(k)}}$, respectively, for some polynomials p and q.

For security parameter k, define the composition of hash services

$$\mathsf{Hash}_k := \|_{j \in (\mathsf{SID}_{\mathsf{hash}})_k} \mathsf{Hasher}(k, j) \ .$$

Ideal ERS Service. The ideal ERS service is composed of a dispatcher component, a sequence of hash services, and a sequence of ideal timestamp services. For security parameter k, define the composition of ideal timestamp services $\mathsf{IdealStamp}_k := \|_{j \in (\mathsf{SID}_{\mathsf{sign}})_k} \mathsf{IdealStamp}(j)_k$. The ideal ERS service $\mathsf{IdealSys}_k$ is defined as

$$\mathsf{IdealSys}_k := \mathsf{Dispatcher}_k \| \mathsf{Hash}_k \| \mathsf{IdealStamp}_k \ .$$

Real ERS Service. The real ERS service is composed of a dispatcher component, a sequence of hash services, and a sequence of real timestamp services. For security parameter k, define the composition of real timestamp services $\mathsf{RealStamp}_k := \|_{j \in (\mathsf{SID}_{\mathsf{sign}})_k} \mathsf{RealStamp}(j)_k$. The real ERS service $\mathsf{RealSys}_k$ is defined as

$$\mathsf{RealSys}_k := \mathsf{Dispatcher}_k \| \mathsf{Hash}_k \| \mathsf{RealStamp}_k \ .$$

5   ERS Security Proof

In Section 4.7, we specified the real ERS system and the ideal ERS system. In this section, we first define a concrete time scheme according to which hash and timestamp services are active. Then, we show that the real ERS system implements the ideal ERS system in the sense of the longterm-implementation relation $\leq_{neg,pt}$.

We assume a concrete time scheme for timestamp and hash services. Let $d \in \mathbb{N}_{>0}$. Each signature service $\langle\mathsf{sign}, j\rangle \in \mathsf{SID}_{\mathsf{sign}}$ is in $\mathsf{alive}(t)$ for $t = (j - 1)d, \ldots, (j + 2)d - 1$, is preferred signer for times $(j - 1)d, \ldots, jd - 1$, and is usable for times $(j - 1)d, \ldots, (j + 1)d - 1$. Each hash service $\langle\mathsf{hash}, j\rangle \in \mathsf{SID}_{\mathsf{hash}}$ is in $\mathsf{alive}(t)$ for $t = (j - 1)de, \ldots, (j + 2)de - 1$, is preferred hasher for times $(j - 1)de, \ldots, jde - 1$, and is usable for times $(j - 1)de, \ldots, (j + 1)de - 1$. Note that, at any real time t, at most three signature services and three hash services are concurrently alive.

Define the ideal ERS service family $\overline{\mathsf{IdealSys}} := \{\mathsf{IdealSys}_k\}_{k\in\mathbb{N}}$, and the real ERS service family $\overline{\mathsf{RealSys}} := \{\mathsf{RealSys}_k\}_{k\in\mathbb{N}}$. Let $\mathsf{SID}_k := (\mathsf{SID}_{\mathsf{hash}})_k \cup (\mathsf{SID}_{\mathsf{sign}})_k$. Define the family of empty failure sets as $\bar{\emptyset} := \{\emptyset\}_{k\in\mathbb{N}}$ and the family of signature failure sets as $\bar{F} := \{F_k\}_{k\in\mathbb{N}}$, where $F_k := \bigcup_{j\in\mathsf{SID}_k}\{\mathsf{fail}_j\}$.

Theorem 3 states that the real ERS system, $\overline{\mathsf{RealSys}}$, implements the ideal ERS system, $\overline{\mathsf{IdealSys}}$, in the sense of the long-term implementation relation $\leq_{neg,pt}$.

Theorem 3. *Assume the concrete time scheme described above and assume that every signature scheme used in the timestamping protocol is complete and existentially unforgeable. Then $(\overline{\mathsf{RealSys}}, \bar{\emptyset}) \leq_{neg,pt} (\overline{\mathsf{IdealSys}}, \bar{F})$.*

Proof. Observe that $\overline{\mathsf{RealSys}}$ and $\overline{\mathsf{IdealSys}}$ are 7-bounded concurrent and polynomially bounded. We apply the *d-Bounded Composition Theorem* to

$$\overline{\mathsf{Dispatcher}}, \overline{\mathsf{Hash}}(1), \overline{\mathsf{Hash}}(2), \ldots, \overline{\mathsf{RealStamp}}(1), \overline{\mathsf{RealStamp}}(2), \ldots$$

and

$$\overline{\mathsf{Dispatcher}}, \overline{\mathsf{Hash}}(1), \overline{\mathsf{Hash}}(2), \ldots, \overline{\mathsf{IdealStamp}}(1), \overline{\mathsf{IdealStamp}}(2), \ldots$$

to obtain $(\overline{\mathsf{RealSys}}, \bar{\emptyset}) \leq_{neg,pt} (\overline{\mathsf{IdealSys}}, \bar{F})$.

6   Conclusions

The Evidence Record Syntax specification allows to ensure datedness for data objects stored in a long-lived archiving system. We have described the Evidence Record Syntax specification and given a high level description of the LCS security framework, which is a framework for analyzing security properties of long-lived systems. Extending the CIS analysis by Canetti et al., we have analyzed the security of ERS using the LCS framework and obtained a security argument for

ERS analogous to the security argument for CIS given in [4]. This was possible because ERS is a refined, more efficient variant of CIS. In particular, we have extended the CIS analysis by introducing hash services and allowing cryptographic primitives with different lifetimes.

We now discuss in how far the security analysis of CIS and ERS establishes the expected security properties of these schemes. CIS and ERS allow for datedness verification of stored data objects. Verifiers just verify digital signatures on time stamps. They are required to trust the time stamping authorities to properly issue time stamps. They also need to trust the PKI to allow for correct signature verification.

However, the model of Canetti at al. [4] requires more trust by the retriever, namely in the archiving system to act as a trustworthy notary. This notary verifies previous time stamps and attests their validity by its signature while in the original versions of CIS and ERS all these time stamps are verified by the retrievers. Therefore, the security proof only refers to these modified versions of CIS and ERS. This is a big step forward as no security models for long-lived archiving systems were known previously. But it also raises the question of whether there is a model that allows a security proof for the original CIS and ERS. This is challenging, as the task-PIOA model only allows to process a polynomial amount of data at each point in time but over time, a super polynomial chain of time stamps may be generated.

We also discuss a few other research directions. As suggested in [4], it would be desirable to specify an abstract archiving system suiting the specification of various archiving systems such as the ERS system and the CIS system. This would allow to analyze security properties of archiving systems in a more generic way.

In this work we have been concerned with signature-based timestamping. However, other methods for timestamping exist, such as hash-linking-based timestamping. It would be worthwhile to analyze the security of such solutions.

As it has been stated in Section 8 of [4], the analysis of Canetti et al. and our results do not imply that any data object is reliably certified for super-polynomial time. This is closely related to the fact that the security parameter is fixed over the lifetime of the protocol. We would like to know if it is possible to reliably certify a document for super-polynomial time while keeping the security parameter fixed.

As it has been observed in [4] and we have stated in Section 3, the LCS framework does not allow to model components whose computational power increases over time. Since in reality, according to Moore's law and as observed over the last 40 years, computational power doubles roughly every 18 months, this seems to be a shortcoming of the framework. It might be useful to modify the framework such that it tolerates an increase of computational power over time.

Acknowledgments. This work has been co-funded by the DFG as part of project Long-Term Secure Archiving within the CRC 1119 CROSSING.

In addition, we thank Robert Künnemann for the interesting discussions.

References

[1] Bellare, M., Rogaway, P.: Random oracles are practical: A paradigm for designing efficient protocols. In: Denning, D.E., Pyle, R., Ganesan, R., Sandhu, R.S., Ashby, V. (eds.) Proceedings of the 1st ACM Conference on Computer and Communications Security, CCS 1993, Fairfax, Virginia, USA, November 3-5, pp. 62–73. ACM (1993), http://doi.acm.org/10.1145/168588.168596

[2] Blazic, A.J., Saljic, S., Gondrom, T.: Extensible Markup Language Evidence Record Syntax (XMLERS). RFC 6283 (Proposed Standard) (July 2011), http://www.ietf.org/rfc/rfc6283.txt

[3] Canetti, R., Cheung, L., Kaynar, D.K., Liskov, M., Lynch, N.A., Pereira, O., Segala, R.: Time-bounded task-pioas: A framework for analyzing security protocols. In: Dolev, S. (ed.) DISC 2006. LNCS, vol. 4167, pp. 238–253. Springer, Heidelberg (2006), http://dx.doi.org/10.1007/11864219_17

[4] Canetti, R., Cheung, L., Kaynar, D.K., Lynch, N.A., Pereira, O.: Modeling computational security in long-lived systems, version 2. IACR Cryptology ePrint Archive 2008, 492 (2008), http://eprint.iacr.org/2008/492

[5] Gondrom, T., Brandner, R., Pordesch, U.: Evidence Record Syntax (ERS) (2007), http://www.ietf.org/rfc/rfc4998.txt

[6] Haber, S.: Content Integrity Service for Long-Term Digital Archives. In: Archiving 2006, pp. 159–164. IS&T, Ottawa (2006)

[7] Merkle, R.C.: Protocols for public key cryptosystems. In: IEEE Symposium on Security and Privacy, pp. 122–134 (1980)

Differential Attacks Against SPN:
A Thorough Analysis

Anne Canteaut[✉]and Joëlle Roué

Inria, project-team SECRET, Rocquencourt, France
{Anne.Canteaut,Joelle.Roue}@inria.fr

Abstract. This work aims at determining when the two-round maximum expected differential probability in an SPN with an MDS diffusion layer is achieved by a differential having the fewest possible active Sboxes. This question arises from the fact that minimum-weight differentials include the best differentials for the AES and several variants. However, we exhibit some SPN for which the two-round MEDP is achieved by some differentials involving a number of active Sboxes which exceeds the branch number of the linear layer. On the other hand, we also prove that, for some particular families of Sboxes, the two-round MEDP is always achieved for minimum-weight differentials.

Keywords: Differential cryptanalysis · Linear layer · MDS codes · AES

1 Introduction

Since the design of the AES and the seminal related work [12], it is known that the mixing layer which aims at providing diffusion within a block cipher must have a high differential branch number [10]. This quantity corresponds to the smallest number of active Sboxes within a two-round differential characteristic. Indeed, for a given choice of the Sbox, the maximal probability for an r-round differential characteristic decreases when the number of active Sboxes within r rounds increases. For this reason, many security analyses focus on the minimal number of active Sboxes within r consecutive rounds when r varies, not only for AES-like designs but for some other types of ciphers, including Present [5] or Feistel ciphers [23]. This approach is rather natural since, in differential attacks, cryptanalysts usually start by searching for a differential characteristic with the fewest possible active Sboxes. Therefore, the construction of MDS diffusion layers with an efficient implementation has been investigated by several authors, e.g., [22,3,1].

However, the complexity of a differential attack depends on the probability of a *differential*, i.e., on the sum of the probabilities of all characteristics starting by a given input difference and ending by a given output difference. And, within two consecutive rounds of an SPN (Substitution-Permutation Networks), the

---

Partially supported by the French Agence Nationale de la Recherche through the BLOC project under Contract ANR-11-INS-011.

S. El Hajji et al. (Eds.): C2SI 2015, LNCS 9084, pp. 45–62, 2015.
DOI: 10.1007/978-3-319-18681-8_4

number of constituent characteristics increases with the Hamming weight of
the differential. Then, the maximum expected probability (MEDP) for a two-
round differential may result from a differential which contains a huge number
of characteristics each with a low but nonzero probability, rather than from a
differential which contains a few characteristics having a high probability. In
other words, for two rounds of an SPN, there is *a priori* no reason to believe
that the best differential corresponds to a differential with the lowest number
of active Sboxes. However, it appears to be the case for most known examples,
including the AES [17]. This aim of this paper is then to determine whether
this phenomenon is more general and whether there are some general situations
where it can be proved that the two-round MEDP is achieved by a differential
with the smallest number of active Sboxes.

Our Contributions. After recalling the main definitions in Section 2, we show
in Section 3 that the choice of the MDS diffusion layer may affect the two-
round MEDP even if the Sbox is fixed. In particular, we show that the form
of the minimum-weight codewords plays an important role. Also, we provide
some upper bound on the number of characteristics with nonzero probability
within a given differential for an MDS linear layer. Section 4 focuses on the case
where the Sbox is APN: in this case, it appears that the two-round MEDP is
usually achieved by minimum-weight differentials. We prove this result for any
APN Sbox over $\mathbf{F}_8$ and any $\mathbf{F}_8$-linear MDS diffusion layer. Finally, Section 5
exploits the previous analysis and exhibits some MDS mixing layers for which
the maximum EDP over two rounds is achieved by a differential in which the
number of active Sboxes exceeds the branch number.

2   Differential Attacks Against Substitution-Permutation Networks

2.1   Substitution-Permutation Networks

One of the most widely-used constructions for iterated block ciphers is the so-
called key-alternating construction [10,11], which consists of an alternation of
key-independent (usually similar) permutations and of round-key additions. The
round permutation is usually composed of a nonlinear substitution function Sub
which provides confusion, and of a linear permutation which provides diffusion.
In order to reduce the implementation cost of the substitution layer, which is
usually the most expensive part of the cipher in terms of circuit complexity, a
usual choice for Sub consists in concatenating several copies of a permutation
S which operates on a much smaller alphabet. In the whole paper, we will con-
centrate on such block ciphers, and use the following notation to describe the
corresponding round permutation.

Definition 1. *Let m and t be two positive integers. Let S be a permutation
of $\mathbf{F}_2^m$ and M be a linear permutation of $\mathbf{F}_2^{mt}$. Then, $\mathrm{SPN}(m, t, S, M)$ denotes*

any substitution-permutation network defined over $\mathbf{F}_2^{mt}$ whose substitution function consists of the concatenation of t copies of S and whose diffusion function corresponds to M.

For instance, up to a linear transformation, two rounds of the AES can be seen as the concatenation of four similar *superboxes* [13]. The superbox, depicted on Fig. 1, is linearly equivalent to a two-round permutation of the form $\mathrm{SPN}(8, 4, S, M)$ where the AES Sbox S corresponds to the composition of the inversion in $\mathbf{F}_{2^8}$ with an affine permutation A. More precisely, $S(x) = A \circ \varphi^{-1}\left(\varphi(x)^{254}\right)$ where φ is the isomorphism from $\mathbf{F}_2^8$ into $\mathbf{F}_{2^8}$ defined by the basis $\{1, \alpha, \alpha^2, \ldots, \alpha^7\}$ with α a root of $X^8 + X^4 + X^3 + X + 1$.

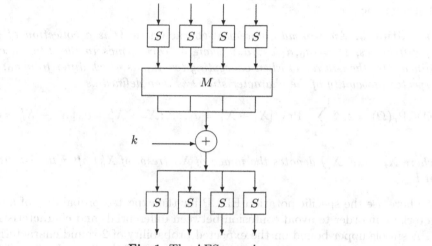

Fig. 1. The AES superbox

2.2 Differential Cryptanalysis

Differential [4] cryptanalysis is one of the most prominent statistical attacks. The complexity of differential attacks depends critically on the distribution over the keys k of the probability of the differentials (a, b), *i.e.*,

$$\mathrm{DP}(a, b) = \Pr_X[E_k(X) + E_k(X + a) = b]$$

where E_k corresponds to the (possibly round-reduced) encryption function under key k. Since computing the whole distribution of the probability of a differential is a very difficult task, cryptanalysts usually focus on its expectation.

Definition 2. *Let $(E_k)_{k \in \mathbf{F}_2^\kappa}$ be an r-round iterated cipher with key-size κ. Then, the* expected probability *of an r-round differential (a, b) is*

$$\mathrm{EDP}_r^E(a, b) = 2^{-\kappa} \sum_{k \in \mathbf{F}_2^\kappa} \Pr_X[E_k(X) + E_k(X + a) = b].$$

The maximum expected differential probability for r rounds *is*

$$\text{MEDP}_r^E = \max_{a \neq 0, b} \text{EDP}_r^E(a, b) .$$

The index in MEDP_r^E will be omitted when the number of rounds is not specified.

2.3   Expected Probability of a Differential Characteristic

Since computing the MEDP for most ciphers, even for a small number of rounds, is very difficult, most works focus on the expected probability of a differential characteristic.

Definition 3. *An r-round differential characteristic Ω is a collection of $(r + 1)$ differences, $\Omega = (a_0, a_1, \ldots, a_r)$ where a_i corresponds to the difference obtained after the i-th round when encrypting two inputs which differ from a_0. The expected probability of the characteristic Ω is then defined as*

$$\text{EDCP}_r(\Omega) = 2^{-\kappa} \sum_{k \in \mathbf{F}_2^\kappa} \text{Pr}_{X_0}[X_1 + X_1' = a_1; \ldots; X_r + X_r' = a_r \mid X_0 + X_0' = a_0] ,$$

where X_i (resp. X_i') denotes the image of X_0 (resp. of X_0') after the i-th round of E_k.

We here use the specific notation EDCP for the expected probability of a characteristic in order to avoid confusion between differentials and characteristics.

A simple upper-bound on the expected probability of 2-round characteristics can be derived from the *differential branch number* of the linear layer and from the *differential uniformity* of the Sbox, in the sense of the following definition.

Definition 4 (Differential Uniformity). *Let S be a function from $\mathbf{F}_2^m$ into $\mathbf{F}_2^m$. For any a and b in $\mathbf{F}_2^m$, we define*

$$\delta^S(a, b) = \#\{x \in \mathbf{F}_2^m, S(x + a) + S(x) = b\} .$$

The multi-set $\{\delta^S(a, b), a, b \in \mathbf{F}_2^m\}$ is the differential spectrum *of S and its maximum $\Delta(S) = \max_{a \neq 0, b} \delta^S(a, b)$ is the* differential uniformity *of S.*

Then, for any two-round characteristic $\Omega = (a, M(b), M(c))$, the Markov assumption [19] implies that

$$\text{EDCP}_2(\Omega) = \text{DP}_1^E(a, M(b)) \times \text{DP}_1^E(M(b), M(c))$$

$$= \left(\prod_{i=1}^t \text{DP}_1^S(a_i, b_i) \right) \left(\prod_{i=1}^t \text{DP}_1^S(M(b)_i, c_i) \right) . \tag{1}$$

Let $\text{Supp}(x)$ and $wt(x)$ denote the support and the weight of a vector $x \in \mathbf{F}_2^{mt}$ seen as an element in $(\mathbf{F}_2^m)^t$. Then, the previous equation shows that

$\mathrm{EDCP}_2(\Omega) = 0$ unless $\mathsf{Supp}(a) = \mathsf{Supp}(b)$ and $\mathsf{Supp}(M(b)) = \mathsf{Supp}(c)$. Using this relation, we deduce that

$$\mathrm{EDCP}_2(\Omega) \leq \left(2^{-m}\Delta(S)\right)^{wt(b)+wt(M(b))} .$$

It then appears that the lowest possible value for the weight of a nonzero word of the form $(b, M(b))$ plays a major role in the resistance against differential attacks. This criterion on the diffusion layer of the cipher corresponds to the notion of *differential branch number*.

Definition 5 (Differential Branch Number [10]). *Let M be a permutation of $(\mathbf{F}_2^m)^t$. We associate to M the code $\mathcal{C}_M$ of length $2t$ and size 2^t over $\mathbf{F}_2^m$ defined by*

$$\mathcal{C}_M = \{(c, M(c)), \ c \in (\mathbf{F}_2^m)^t\} .$$

The differential branch number *of M is the minimum distance of the code $\mathcal{C}_M$.*

The following upper bound on the expected probability of any 2-round differential characteristic then follows:

$$\max_{\Omega} \mathrm{EDCP}_2(\Omega) \leq \left(2^{-m}\Delta(S)\right)^{d} , \tag{2}$$

where d is the differential branch number of the linear layer.

It is worth noticing that a similar notion is considered in the case of linear cryptanalysis. The *linear branch number* is then the minimum distance of the dual code $\mathcal{C}_M^{\perp}$ but this quantity is out of the scope of this paper. For this reason, in the following, *branch number* always refers to the differential branch number.

From Singleton's bound, the highest possible value for the branch number of a permutation of $(\mathbf{F}_2^m)^t$ is $(t + 1)$ and it corresponds to the case where the associated code $\mathcal{C}_M$ is an MDS (maximum distance separable) code.

3   From Characteristics to Differentials

The problem with the previous result is that differential cryptanalysis exploits *differentials* and not *characteristics* since the differences obtained after each intermediate round do not matter in the attack. The probability of a differential $(a, M(b))$ then corresponds to the sum of the probabilities of all characteristics with input difference a and output difference $M(b)$. Then, the relevant quantity for two rounds is the maximum of

$$\mathrm{EDP}_2(a, M(b)) = \sum_{x \in \mathbf{F}_2^{mt}} \mathrm{EDCP}_2^E(a, x, M(b)) .$$

Determining the expected probability of a differential, rather than focusing on a single characteristic, is difficult in general.

3.1 Expected Probability of a 2-round Differential

From Equation (1), any element x of $(\mathbf{F}_2^m)^t$ verifies

$$\mathrm{EDCP}_2(a, M(x), M(b)) = \left(\prod_{i=1}^{t} \mathrm{DP}_1^S(a_i, x_i)\right) \left(\prod_{i=1}^{t} \mathrm{DP}_1^S(M(x)_i, b_i)\right) .$$

If this probability is different from zero, we have that $\mathsf{Supp}(a) = \mathsf{Supp}(x)$ and $\mathsf{Supp}(M(x)) = \mathsf{Supp}(b)$, implying that $(x, M(x)) \in (\mathbf{F}_2^m)^{2t}$ is a word of $\mathcal{C}_M$ having the same support as (a, b). Moreover, by definition of the differential spectrum, $\mathrm{DP}_1^S(\alpha, \beta) = 2^{-m}\delta^S(\alpha, \beta)$. Thus, the two-round probability of a differential is

$$\mathrm{EDP}_2(a, M(b)) = 2^{-mwt(a,b)} \sum_{\substack{c \in \mathcal{C}_M: \\ \mathsf{Supp}(c)=\mathsf{Supp}(a,b)}} \left(\prod_{i \in \mathsf{Supp}(a)} \delta^S(a_i, c_i)\right) \left(\prod_{j \in \mathsf{Supp}(b)} \delta^S(c_{t+j}, b_j)\right)$$

$$(3)$$

A simple upper bound for the two-round MEDP can then be derived from the branch number of M and from the differential uniformity of the Sbox (see [15] and [12, Section B.2]):

$$\mathrm{MEDP}_2 \leq \left(2^{-m}\Delta(S)\right)^t .$$

This result has then been refined in [9,21,8]. The bounds in [15,9,21] are invariant under affine equivalence, *i.e.*, their values are the same for two Sboxes S and S' when there exist two affine permutations A_1 and A_2 such that $S' = A_1 \circ S \circ A_2$. However, the exact values of MEDP_2 may differ for Sboxes in the same equivalent class, and there can be a gap between these bounds and the exact value of MEDP_2. In [8], a new upper bound is introduced, that enhances the previously known bounds in the sense that it may vary when the Sbox is composed by an affine permutation. This new bound only applies when the diffusion layer M is linear over the field $\mathbf{F}_{2^m}$, where m is the size of the Sbox, exactly as in the AES. In this case, the linear layer and the Sbox can be represented as functions over the field $\mathbf{F}_{2^m}$ and the representation does not change the MEDP. In particular, the choice of the isomorphism that identifies the vector space $\mathbf{F}_2^m$ with the finite field $\mathbf{F}_{2^m}$ has no influence on the differential properties of the cipher. For this reason, we use the following alternative notation to define an SPN with this representation.

Definition 6. *Let m and t be two positive integers. Let S be a permutation of $\mathbf{F}_{2^m}$ and $\mathcal{M}$ be a permutation of $(\mathbf{F}_{2^m})^t$ which is linear over $\mathbf{F}_{2^m}$. Then, we denote by $\mathrm{SPN}_F(m, t, S, \mathcal{M})$ a substitution-permutation network defined over $(\mathbf{F}_{2^m})^t$ whose substitution function consists of the concatenation of t copies of S and whose diffusion function corresponds to $\mathcal{M}$.*

When the Sbox is defined over $\mathbf{F}_{2^m}$, we equivalently define the differential spectrum as follows. Let $(\alpha_0, \ldots, \alpha_{m-1})$ be a basis of $\mathbf{F}_{2^m}$, and φ the corresponding isomorphism from $\mathbf{F}_2^m$ into $\mathbf{F}_{2^m}$. Let S be a mapping over $\mathbf{F}_2^m$, and $\mathcal{S} = \varphi \circ S \circ \varphi^{-1}$. Then, for any $(\alpha, \beta) \in \mathbf{F}_{2^m}$,

$$\delta_F^{\mathcal{S}}(\alpha, \beta) = \#\{x \in \mathbf{F}_{2^m}, \mathcal{S}(x + \alpha) + \mathcal{S}(x) = \beta\} = \delta^S(\varphi^{-1}(\alpha), \varphi^{-1}(\beta)) .$$

As the differential properties of any SPN(m, t, S, M) can be equivalently studied by considering the alternative representation SPN$_F(m, t, \mathcal{S}, \mathcal{M})$ [14], this paper focuses on the representation of an SPN$_F$ in the field $\mathbf{F}_{2^m}$. For the sake of clarity, all quantities related to the representation in the field $\mathbf{F}_{2^m}$ will be indexed by F, and all functions defined over $\mathbf{F}_{2^m}$ will be denoted by calligraphic letters.

The new bounds on MEDP$_2$ presented in [8] are derived from the particular structure of the set formed by all codewords in $\mathcal{C}_\mathcal{M}$ having a given support, when $\mathcal{M}$ is linear over $\mathbf{F}_{2^m}$. These bounds are expressed in terms of the following quantities. For any Sbox $\mathcal{S}$ over $\mathbf{F}_{2^m}$ with differential spectrum $(\delta_F(a, b))_{a,b \in \mathbf{F}_{2^m}}$ and any branch number d, we define for any $\mu \in \mathbf{F}_{2^m}$ and any integer $u > 0$,

$$\mathcal{B}_u(\mu) = \max_{\alpha, \beta, \lambda \in \mathbf{F}_{2^m}^*} \sum_{\gamma \in \mathbf{F}_{2^m}^*} \delta_F(\alpha, \gamma)^u \delta_F(\gamma\lambda + \mu, \beta)^{(d-u)}$$

$$\text{and} \quad \mathcal{B}(\mu) = \max_{1 \leq u < d} \mathcal{B}_u(\mu) .$$

In the rest of the paper, we will restrict ourselves to the case where *the diffusion layer is linear over $\mathbf{F}_{2^m}$ and MDS (i.e., with branch number $(t + 1)$)*. We also assume that the well-known MDS conjecture [20] is valid, *i.e.*, in our context, that $t \leq 2^{m-1}$ for $m > 3$ and $t \leq 3$ for $m = 2$. For such MDS diffusion layers, Theorem 2 and Proposition 3 in [8] can be expressed as follows.

Theorem 1. *Let $\mathcal{S}$ be a permutation of $\mathbf{F}_{2^m}$ and t be any integer such that $t \leq 2^{m-1}$.*

– *For any $\mathbf{F}_{2^m}$-linear diffusion layer $\mathcal{M}$ over $\mathbf{F}_{2^m}^t$ with maximal branch number, the block cipher E of the form SPN$_F(m, t, \mathcal{S}, \mathcal{M})$ satisfies*

$$\text{MEDP}_2^E \leq 2^{-m(t+1)} \max_{\mu \in \mathbf{F}_{2^m}} \mathcal{B}(\mu) .$$

– *There exists an $\mathbf{F}_{2^m}$-linear diffusion layer $\mathcal{M}$ over $\mathbf{F}_{2^m}^t$ with maximal branch number such that*

$$\text{MEDP}_2^E \geq 2^{-m(t+1)} \mathcal{B}(0) .$$

In most cases, the values of the two-round MEDP for two ciphers of the form SPN$_F(m, t, \mathcal{S}, \mathcal{M}_1)$ and SPN$_F(m, t, \mathcal{S}, \mathcal{M}_2)$ where $\mathcal{M}_1$ and $\mathcal{M}_2$ are different MDS linear layers differ. The minimum-weight codewords of $\mathcal{C}_\mathcal{M}$ have a large influence on this value, as shown in the following example.

Example 1. Let us study the two-round MEDP of the SPN with the same building blocks as the Prøst permutation, which is the core function of several AEAD-schemes submitted to the CAESAR competition [16]. It is worth noticing that the following results do not provide any direct information on the security of the Prøst permutation: indeed, we study the differential probabilities averaged over all keys while the key is fixed in the Prøst permutation. Two consecutive rounds of the Prøst permutation over $\mathbf{F}_2^{16d}$, $d \geq 1$, can be seen as the parallel application of d copies of a superbox defined over $\mathbf{F}_{16}$. This superbox is of the form $\mathrm{SPN}(4, d, S, M)$ where S is a 4-bit involution named SubRows and M corresponds to the so-called MixSlices transformation. It has been shown in [8] that MixSlices is linear over $\mathbf{F}_{16}$ for some particular isomorphism between $\mathbf{F}_2^4$ and $\mathbf{F}_{16}$. Then, Theorem 1 applies and we get that, for any $\mathbf{F}_{16}$-linear layer $\mathcal{M}$, the block cipher E of the form $\mathrm{SPN}_F(4, d, \mathcal{S}, \mathcal{M})$ where $\mathcal{S}$ corresponds to the Prøst Sbox satisfies

$$\mathrm{MEDP}_2^E \leq 2^{-8} .$$

But, when the diffusion layer corresponds to MixSlices, we have computed the exact value of the MEDP_2 and obtained that $\mathrm{MEDP}_2 = 3 \times 2^{-11}$, which is smaller than the general upper bound.

However, since both lower and upper bounds in Theorem 1 are equal, we deduce that there exists another diffusion layer $\mathcal{M}$ such that

$$\mathrm{MEDP}_2^E = 2^{-8} .$$

An example of such a diffusion layer is

$$\begin{pmatrix} \alpha^2 + \alpha + 1 & \alpha^3 + \alpha & \alpha^3 + \alpha + 1 & 1 \\ \alpha + 1 & \alpha^3 + \alpha^2 + \alpha & \alpha^2 + \alpha + 1 & 1 \\ \alpha^2 + 1 & \alpha^3 + \alpha^2 + 1 & \alpha^3 & 1 \\ \alpha^2 & \alpha^3 + \alpha^2 & \alpha^3 + 1 & 1 \end{pmatrix}$$

where α is a root of $X^4 + X^3 + 1$. Indeed, the set of codewords of the form $\{\lambda(0, 0, 0, 1, 1, 1, 1, 1), \lambda \in \mathbf{F}_{16}^*\}$ belongs to the code associated with this diffusion layer. Then, the differences $a = (0, 0, 0, 1)$ and $b = (1, 1, 1, 1)$ satisfy $\mathrm{EDP}_2(a, \mathcal{M}(b)) = 2^{-8}$.

3.2   Influence of the Weight of the Differential

The previous example shows that, in some cases, the form of the minimum-weight codewords in $\mathcal{C}_\mathcal{M}$ plays an important role when determining the two-round MEDP. We observe from Equation (3) that these codewords are involved in the computation of $\mathrm{EDP}_2(a, \mathcal{M}(b))$ when the weight of the corresponding pair (a, b) is equal to the branch number of $\mathcal{M}$. We then call such a differential a minimum-weight differential. The role played by minimum-weight differentials appears in a more direct way when the Sbox $\mathcal{S}$ has the following additional property [8, Definition 7]. A mapping S of $\mathbf{F}_{2^m}$ is said to have multiplicative-invariant derivatives if, for any $x \in \mathbf{F}_{2^m}^*$ there exists a permutation π_x of $\mathbf{F}_{2^m}^*$ such that

$$\delta_F(\alpha, xy) = \delta_F(\pi_x(\alpha), y), \quad \forall y \in \mathbf{F}_{2^m}^* .$$

Power permutations, and more generally any function resulting from the composition on the right of a power permutation with an $\mathbf{F}_2$-linear permutation, has multiplicative-invariant derivatives. Another example of functions with multiplicative-invariant derivatives are the crooked permutations, which include all APN permutations of degree 2. When an Sbox has this property, the expression of $\mathcal{B}(\mu)$ (including $\mathcal{B}(0)$) simplifies but, more interestingly, we get some universal lower bound on MEDP_2, i.e., which holds for any diffusion layer with maximal branch number. For instance, for all Sboxes S such that both S and S^{-1} have multiplicative-invariant derivatives, we obtain that, for any $\mathbf{F}_{2^m}$-linear diffusion layer $\mathcal{M}$ with maximal branch number, the corresponding block cipher satisfies

$$2^{-m(t+1)}\mathcal{B}(0) \leq \mathrm{MEDP}_2^E \leq 2^{-m(t+1)} \max_{\mu \in \mathbf{F}_{2^m}} \mathcal{B}(\mu) . \tag{4}$$

Moreover, $\mathrm{MEDP}_2^E = 2^{-m(t+1)}\mathcal{B}(0)$ if and only if the maximum expected differential probability is achieved by a minimum-weight differential.

Since the probability of a characteristic decreases when the weight of the underlying differential increases, a natural question is to determine in which situations the two-round MEDP is achieved by a minimum-weight differential. This is an important information: computing the two-round MEDP for a given cipher becomes obviously much easier once it is known that only the minimum-weight differentials need to be examined. Surprisingly enough, for all AES-like ciphers which have been investigated, the two-round MEDP is achieved by a minimum-weight differential. For instance, the bounds in [8] applied to the AES Sbox show that for any $\mathbf{F}_{2^8}$-linear layer $\mathcal{M}$, we have

$$53 \times 2^{-34} \leq \mathrm{MEDP}_2 \leq 55.5 \times 2^{-34} , \tag{5}$$

where the lower bound corresponds to some minimum-weight differentials. For the particular diffusion layer defined by MixColumns in the AES, the exact value of the two-round MEDP_2 computed by a pruning search algorithm [17], is $\mathrm{MEDP}_2 = 53 \times 2^{-34}$. It then corresponds to the lower bound of (5).

There also exist some SPN_F for which the exact value of the two-round MEDP can be directly deduced from the bounds in [8], for instance, when the Sbox S is an involution with multiplicative-invariant derivatives. In this case, the lower and upper bounds in (4) are equal and do not depend on the MDS diffusion layer. In other words, for any involution with multiplicative-invariant derivatives, the two-round MEDP is always achieved by a minimum-weight differential, for any choice of the MDS linear layer. This holds in particular for the so-called AES naive Sbox, i.e. the inversion in $\mathbf{F}_{2^m}$, which satisfies these conditions.

A natural question then arises from these examples: does there exist any cipher of the form SPN_F for which the two-round MEDP is not achieved by a minimum-weight differential? We now investigate this problem, and first exhibit some general families of ciphers for which this situation cannot occur.

3.3 Number of Characteristics Within a Given 2-round Differential

For the sake of simplicity, for any differential $(a, \mathcal{M}(b))$, we denote by $(a, c, \mathcal{M}(b))$ the corresponding characteristic where c is the codeword in $\mathcal{C}_\mathcal{M}$ defined by the concatenation of the input and output differences of the first diffusion layer. With this notation, we have

$$\text{EDP}_2(a, \mathcal{M}(b)) = \sum_{\substack{c \in \mathcal{C}_\mathcal{M}: \\ \text{Supp}(c) = \text{Supp}(a,b)}} \text{EDCP}_2(a, c, \mathcal{M}(b)) \, .$$

In this differential, each characteristic having a nonzero probability is defined by a codeword in $\mathcal{C}_\mathcal{M}$ whose support is equal $\text{Supp}(a, b)$. Therefore, we define the *weight of the differential* as the weight $w = wt(a) + wt(b)$. Then, the number of characteristics within a given differential $(a, \mathcal{M}(b))$ of weight w is defined by

$$\mathcal{A}_w(a, b) = \#\{c \in \mathcal{C}_\mathcal{M} : \text{Supp}(c) = \text{Supp}(a, b) \text{ and } \text{EDCP}_2(a, c, \mathcal{M}(b)) \neq 0\}$$
$$= \#\{c \in \mathcal{C}_\mathcal{M} : \text{Supp}(c) = \text{Supp}(a, b), \delta_F^\mathcal{S}(a_i, c_i) \neq 0, \forall i \in \text{Supp}(a)$$
$$\text{and } \delta_F^\mathcal{S}(c_{t+j}, b_j) \neq 0, \forall j \in \text{Supp}(b)\} \, .$$

A first criterion to determine whether the two-round expected differential probability is maximized by a minimum-weight differential or not consists in estimating the number of characteristics involved in a differential having a given weight w. Since we only consider diffusion layers which are linear over $\mathbf{F}_{2^m}$, the codewords in $\mathcal{C}_\mathcal{M}$ having a given support can be gathered in *bundles* as pointed out in [13]: if c belongs to $\mathcal{C}_\mathcal{M}$, then the whole bundle $\mathcal{P}(c) = \{\gamma c, \gamma \in \mathbf{F}_{2^m}^*\}$ is also included in $\mathcal{C}_\mathcal{M}$. It follows that the number of codewords in $\mathcal{C}_\mathcal{M}$ having a given support is always divisible by $(2^m - 1)$. Moreover, for any pair $(\alpha, \beta) \in (\mathbf{F}_{2^m}^*)^2$, the values $\delta_F^\mathcal{S}(\alpha, \gamma\beta)$, when γ varies in $\mathbf{F}_{2^m}^*$, correspond to a row of the difference table of $\mathcal{S}$. Since these coefficients are all even and sum to 2^m, we deduce that, for any permutation $\mathcal{S}$, at least $2^{m-1} - 1$ coefficients among all $(\delta_F^\mathcal{S}(\alpha, \gamma\beta), \gamma \in \mathbf{F}_{2^m}^*)$ vanish, with equality if and only if $\mathcal{S}$ is APN. It then follows that, for any $c \in \mathcal{C}_\mathcal{M}$,

$$\#\{c' \in \mathcal{P}(c) : \text{EDCP}_2(a, c', \mathcal{M}(b)) \neq 0\} \leq 2^{m-1} \, .$$

Differentials of Weight $w = t + 1$. Recall that we focus on the case where the diffusion layer has maximal branch number, *i.e.*, where $\mathcal{C}_\mathcal{M}$ is MDS. It is well-known (e.g. [20, Page 319]) that if $\mathcal{C}_\mathcal{M}$ is an MDS code of length $2t$ and dimension t over $\mathbf{F}_{2^m}$, then for each support of size $(t + 1)$, there exist exactly $(2^m - 1)$ codewords (i.e., one bundle) having this support. From the previous discussion, we deduce that, for any minimum-weight differential (a, b)

$$\mathcal{A}_{t+1}(a, b) \leq 2^{m-1} \, .$$

Differentials of Weight $w = t + 2$. We now provide a similar upper bound on the number of characteristics within a differential of weight $(t + 2)$.

Proposition 1. *Let $\mathcal{M}$ be an $\mathbf{F}_{2^m}$-linear MDS permutation of $\mathbf{F}_{2^m}^t$. Then, for any differential (a, b) of weight $(t + 2)$, we have*

$$\mathcal{A}_{t+2}(a, b) \leq 2^{m-1}(2^m - (t + 1)) \, .$$

Proof. From the previous discussion, we only have to prove that, for any support I of size $(t + 2)$ there exist exactly $(2^m - (t + 1))$ distinct bundles having I for support. Let $J = \{i_1, \ldots, i_{t-2}\}$ be the set formed by the $2t - (t + 2) = t - 2$ coordinates which do not belong to I. The codewords whose support is included in I then correspond to the codewords which vanish on J. Using that any t coordinates of $\mathcal{C}_\mathcal{M}$ is an information set [20, Page 321], we deduce that there are exactly $(2^{2m} - 1)$ nonzero codewords whose support is included in I. Since we count the number of codewords whose support is equal to I, we need to remove the codewords of weight $(t + 1)$ from the previous set. As previously mentioned, for any support of size $(t + 1)$, there exists one bundle having this support. Since I contains $(t + 2)$ subsets of size $(t + 1)$, we need to remove $(t + 2)(2^m - 1)$ codewords from the previous set. It follows that the number of codewords having I for support is

$$2^{2m} - 1 - (t + 2)(2^m - 1) = (2^m - 1)(2^m - (t + 1)) .$$

Therefore, $\mathcal{C}_\mathcal{M}$ contains exactly $(2^m - (t + 1))$ bundles having I for support, implying that

$$\mathcal{A}_{t+2}(a, b) \leq 2^{m-1}(2^m - (t + 1)) .$$

□

Most notably, we deduce from this formula that, when $t = 2^{m-1}$, $\mathcal{A}_{t+2}$ may be limited by the maximal value of $\mathcal{A}_{t+1}$. Some application of this result will be detailed in the next section.

4   SPN with an APN Sbox

In this section, we focus on the block ciphers SPN_F which use an APN Sbox. These ciphers are of particular interest in our context since the whole differential spectrum of the Sbox is known. It follows that, for any characteristic within a differential of weight w has probability either 0 or $2^{-w(m-1)}$. Then, we deduce that the expected probability of a differential of weight w only depends on the value of $\mathcal{A}_w(a, b)$:

$$\text{EDP}_2(a, \mathcal{M}(b)) = 2^{-w(m-1)} \mathcal{A}_w(a, b) .$$

It follows that there exists a differential (a, b) of weight $(t + 2)$ whose probability is higher than the probability of any minimum-weight differential if and only if, for any (α, β) of weight $(t + 1)$,

$$2^{-(t+2)(m-1)} \mathcal{A}_{t+2}(a, b) \geq 2^{-(t+1)(m-1)} \mathcal{A}_{t+1}(\alpha, \beta)$$

or equivalently

$$\mathcal{A}_{t+2}(a, b) \geq 2^{m-1} \mathcal{A}_{t+1}(\alpha, \beta) .$$

From Proposition 1, we know that $\mathcal{A}_{t+2}(a, b) \leq 2^{m-1}(2^m - (t + 1))$, implying that this situation can only occur if all minimum-weight differentials (α, β) satisfy

$$\mathcal{A}_{t+1}(\alpha, \beta) \leq (2^m - (t + 1)) . \tag{6}$$

For given parameters m and t, we can then directly deduce that, if the number of characteristics in a minimum-weight differential exceeds some bound, then the two-round MEDP cannot be achieved by a differential of weight $(t + 2)$.

4.1 APN Sboxes over $\mathbf{F_8}$

We now show that, if $\mathcal{S}$ is an APN permutation over $\mathbf{F}_{2^3}$ (*i.e.*, $m = 3$), then the maximum EDP is always achieved by a minimum-weight differential. This result is mainly due to the particular properties of 3-bit APN permutations.

Properties of APN Sboxes over $\mathbf{F_8}$. Since a permutation of $\mathbf{F}_{2^m}$ has degree at most $(m-1)$, all APN Sboxes over $\mathbf{F_8}$ are quadratic, and their inverses are also quadratic. Therefore, they are *crooked* [2,18], *i.e.*, for any nonzero $a \in \mathbf{F}_{2^3}$, the set $\{b \in \mathbf{F}_{2^3} : \delta_F^{\mathcal{S}}(a, b) = 2\}$ is an affine hyperplane of $\mathbf{F}_{2^3}$. Furthermore, it is known that all these affine hyperplanes are distinct [7, Lemma 5]. Since the inverse $\mathcal{S}^{-1}$ is also a crooked permutation, the same property holds for the columns of the difference table of $\mathcal{S}$: for any nonzero b, the set $\{a \in \mathbf{F}_{2^3} : \delta_F^{\mathcal{S}}(a, b) = 2\}$ is an affine hyperplane and all these hyperplanes are distinct.

Minimum-Weight Differentials. From this algebraic structure, we deduce the maximal value of the expected differential probability of the minimum-weight differentials.

Proposition 2. *Let $\mathcal{S}$ be an APN permutation of $\mathbf{F}_{2^3}$. For any integer t and any $\mathbf{F}_{2^3}$-linear MDS diffusion layer $\mathcal{M}$ over $(\mathbf{F}_{2^3})^t$, the block cipher of the form* $\mathrm{SPN}_F(3, t, \mathcal{S}, \mathcal{M})$ *satisfies*

$$\max_{\substack{a \neq 0, b \\ wt(a,b)=t+1}} \mathrm{EDP}_2(a, \mathcal{M}(b)) = 2^{-2t}.$$

Proof. Let $I = \{i_1, \ldots, i_{t+1}\}$ be a subset of $\{1, \ldots, 2t\}$ of size $(t + 1)$. Our aim is to exhibit a pair (a, b) whose support equals I and such that $\mathcal{A}_{t+1}(a, b) = 4$. Such a differential leads to the result since $\mathcal{A}_{t+1}(a, b) = 4$ is the highest value we can have for a minimum-weight differential. Let c be a codeword in $\mathcal{C}_{\mathcal{M}}$ with $\mathsf{Supp}(c) = I$ since such a codeword always exists. Let us choose some nonzero element $a_{i_1} \in \mathbf{F}_{2^3}$. Then, we consider the set $H = \{\beta : \delta_F^{\mathcal{S}}(a_{i_1}, \beta) = 2\}$. Then H is an affine hyperplane. We now define

$$\Gamma = \{c_{i_1}^{-1}\lambda, \lambda \in H\}.$$

Obviously, Γ is also an affine hyperplane. Then, the four codewords in the bundle of c, $c' = \gamma c$ with $\gamma \in \Gamma$, satisfy

$$\delta_F^{\mathcal{S}}(a_{i_1}, c'_{i_1}) = \delta_F^{\mathcal{S}}(a_{i_1}, \lambda c_{i_1}^{-1} c_{i_1}) = 2.$$

Moreover, for any position i_j in I with $i_j \leq t$, the coordinates of these four codewords at position i_j vary in the set $c_{i_j}\Gamma$ which is an affine hyperplane.

Therefore, there exists some a_{i_j} such that this set corresponds to $\{\beta : \delta_F^S(a_{i_j}, \beta) = 2\}$. Similarly, for any position $i_j \in I$ with $i_j > t$, there exists some b_{i_j} such that the affine hyperplane $c_{i_j}\Gamma$ corresponds to $\{\alpha : \delta(\alpha, b_{i_j}) = 2\}$. For this choice of (a, b), we get that, by construction,

$$\mathcal{A}_{t+1}(a, b) = 4 \,,$$

implying that

$$\mathrm{EDP}_2(a, \mathcal{M}(b)) = 4 \times 2^{-2(t+1)} = 2^{-2t} \,.$$

□

It is worth noticing that we have proved a more general result: for any bundle, we can find a pair (a, b) such that the corresponding differential includes four characteristics from this bundle having a nonzero probability. However, this does not enable us to determine the maximum EDP for higher-weight differentials since the involved codewords correspond to several bundles, and we cannot control the different bundles together.

Higher-Weight Differentials. Since $\mathcal{C}_\mathcal{M}$ is an MDS code over $\mathbf{F}_8$, we have that t is at most 4. Moreover, we deduce from (6) that the maximum two-round EDP cannot be achieved by a differential of weight $(t + 2)$ when $t = 4$ since it would imply that all minimum-weight differentials would satisfy $\mathcal{A}_{t+1}(a, b) \leq 3$ while we have proved that $\mathcal{A}_{t+1}(a, b)$ can be equal to 4.

Then, we need to examine all linear MDS codes of length $2t$ and dimension t over $\mathbf{F}_8$ for $t \in \{2, 3\}$. For each of these codes, we have computed the highest value of $\mathcal{A}_{t+2}(a, b)$ we can get for all (a, b) of weight $(t + 2)$. Since the difference tables of all crooked Sboxes over $\mathbf{F}_8$ have the same structure, the maximal value of $\mathcal{A}_{t+2}(a, b)$ over all (a, b) having a given support I corresponds to the largest set Γ of codewords c with support I such that, for each $i \in I$, c_i for all $c \in \Gamma$ belong to the same affine hyperplane.

For $t = 2$, the previous quantity has been computed for all $[4, 2, 3]$-codes over $\mathbf{F}_8$. For all of them, we get that the maximal value for $\mathcal{A}_4(a, b)$ is equal to 8. We then deduce that

$$\max_{\substack{a \neq 0, b \\ wt(a,b)=3}} \mathrm{EDP}_2(a, \mathcal{M}(b)) = 2^{-4} \text{ and } \max_{\substack{x \neq 0, y \\ wt(x,y)=4}} \mathrm{EDP}_2(x, \mathcal{M}(y)) = 2^{-8} \times 8 = 2^{-5}.$$

Then, the two-round MEDP is achieved by a minimum-weight differential only.

For $t = 3$, we have computed the highest possible value of $\mathcal{A}_5(a, b)$ for all $[6, 3, 4]$-codes over $\mathbf{F}_8$, and we have obtained that for all these codes, the maximal $\mathcal{A}_5(a, b)$ is 4, implying that

$$\max_{\substack{a \neq 0, b \\ wt(a,b)=4}} \mathrm{EDP}_2(a, \mathcal{M}(b)) = 2^{-6} \text{ and } \max_{\substack{x \neq 0, y \\ wt(x,y)=5}} \mathrm{EDP}_2(x, \mathcal{M}(y)) = 2^{-10} \times 4 = 2^{-8}.$$

Moreover, it can be checked that, for all these codes, the maximal $\mathcal{A}_6(a, b)$ is 32, implying that

$$\max_{\substack{x \neq 0, y \\ wt(x,y)=6}} \mathrm{EDP}_2(x, \mathcal{M}(y)) = 2^{-12} \times 32 = 2^{-7}.$$

We then deduce the following result.

Proposition 3. *Let S be an APN permutation of $\mathbf{F}_{2^3}$. For any integer t and any $\mathbf{F}_{2^3}$-linear MDS diffusion layer $\mathcal{M}$ over $(\mathbf{F}_{2^3})^t$, the block cipher of the form $\mathrm{SPN}_F(3, t, S, \mathcal{M})$ satisfies*

$$\mathrm{MEDP}_2 = 2^{-2t},$$

and this value is achieved by some minimum-weight differentials only.

4.2   APN Sboxes over $\mathbf{F}_{32}$

APN permutations over $\mathbf{F}_{32}$ have been classified in [6] up to equivalence. But since APN permutations over $\mathbf{F}_{32}$ do not have the same algebraic structure as APN permutations over $\mathbf{F}_8$, each function from this classification has to be studied. Moreover, the number of MDS codes with these parameters is also much higher than in the previous case.

We have then computed the maximal value for $\mathcal{A}_{t+1}$ for several APN permutations and MDS permutations with $t = 2, 3$. For $t = 2$, we have always observed that the maximal $\mathcal{A}_{t+1}$ is at least 10. We should then find some differential of weight 4 with $\mathcal{A}_4 \geq 10 \times 2^{5-1} = 160$ to reach the same EDP than the best minimum-weight differential. However, the highest values we have observed for $\mathcal{A}_4$ are between 83 and 92. In other words, the maximum EDP for a differential of weight 4 is slightly higher than half of the maximum EDP for a minimum-weight differential.

For $t = 3$, we have observed that the maximal $\mathcal{A}_{t+1}$ is at least 9. We should then find some differential of weight 5 with $\mathcal{A}_5 \geq 9 \times 2^{5-1} = 144$, while the highest values we have observed for $\mathcal{A}_5$ lie between 54 and 60.

5   MEDP$_2$ can be Tight for a Differential of Non-minimal Weight

It seems that the number $\mathcal{A}_w$ of characteristics having a nonzero probability in a differential of weight $w > t + 1$ cannot be large enough to achieve a two-round EDP higher than the one which can be obtained with minimal-weight differentials. However, in the previously studied cases, the highest probability of a minimal-weight characteristic is always equal to the maximal value $(\Delta(S)/2^m)^{t+1}$. If the probability EDP$_2$ is minimized for any minimal-weight differential, that is, if the number $\mathcal{A}_{t+1}$ is small and the probabilities of the constituent characteristics are different from $(\Delta(S)/2^m)^{t+1}$, it should be possible to have a differential of weight $w > t + 1$ which has a higher probability than all minimal-weight differentials.

5.1   Examples where MEDP$_2$ is Tight for a Differential of Weight $(t + 2)$

Sboxes such that only a few entries in the difference table are equal to $\Delta(S)$ are a good choice to avoid the existence of characteristics with probability $(\Delta(S)/2^m)^{t+1}$ within any given minimum-weight differential. But for differentials of weight $t + 2$, the probability of a characteristic also needs to be high. An Sbox with 4 to 6 entries in the difference table equal to $\Delta(\mathcal{S})$ seems to be a good tradeoff, as shown in the following examples. Note that the Sboxes are defined over the vectorial space $\mathbf{F}_2^m$ while the diffusion layer is defined over the field $\mathbf{F}_{2^m}$, as it is done in many concrete specifications (using the binary representation may be relevant to choose the Sbox, for instance in order to minimize the number of gates).

Let S be a permutation of $\mathbf{F}_2^3$ defined by

x	0 1 2 3 4 5 6 7
$S(x)$	0 1 2 3 4 6 7 5

Its differential uniformity is $\Delta(\mathcal{S}) = 4$ and there are 6 coefficients equal to 4 in its difference table. Then there exist some $\mathbf{F}_8$-linear permutations with maximal branch number such that there are differentials of weight $(t+2)$ having a higher probability than all minimum-weight differentials. An example of such a diffusion layer with $t = 2$ is

$$M = \begin{pmatrix} \alpha & \alpha + 1 \\ \alpha^2 & \alpha^2 + 1 \end{pmatrix}$$

where α is a root of $X^3 + X + 1$. We compute the exact value of EDP_2 for all minimum-weight differentials first and then for differentials with weight $d+1 = 4$. We obtain:

$$\max_{\substack{a \neq 0, b \\ wt(a,b)=3}} \mathrm{EDP}_2(a, M(b)) = 2^{-4}$$

as there is only one characteristic of probability 2^{-4} in the differentials having the highest probability, and

$$\max_{\substack{x \neq 0, y \\ wt(x,y)=4}} \mathrm{EDP}_2(x, M(y)) = 2^{-3}$$

as there are some differentials of weight 4 composed of two characteristics of probability 2^{-4}.

Let S be a permutation of $\mathbf{F}_2^4$ defined by

x	0 1 2 3 4 5 6 7 8 9 10 11 12 13 14 15
$S(x)$	0 2 1 5 4 9 15 8 12 11 6 7 3 14 10 13

Its differential uniformity is $\Delta(\mathcal{S}) = 6$ and there are 4 coefficients equal to 6 in its difference table. Then there exist some $\mathbf{F}_{16}$-linear permutations with maximal branch number such that there exist some differentials of weight $(t + 2)$ having a higher probability than all minimum-weight differentials. An example of such a diffusion layer with $t = 4$ is

$$M = \begin{pmatrix} 1 & 1 & \alpha^3 & \alpha^3 \\ \alpha^2 + \alpha + 1 & 1 & 1 & \alpha^2 + \alpha \\ \alpha^2 & \alpha^3 + 1 & 1 & \alpha^3 + \alpha^2 + 1 \\ \alpha^2 + 1 & \alpha^3 + \alpha^2 + \alpha & \alpha^3 + \alpha & 1 \end{pmatrix}$$

where α is a root of $X^4 + X + 1$.

We compute the exact value of EDP_2 for all differentials of a given weight. We obtain

$$\max_{\substack{a \neq 0,\, b \\ wt(a,b)=5}} EDP_2(a, \mathcal{M}(b)) = 1,2656 \times 2^{-8},$$

$$\max_{\substack{a \neq 0,\, b \\ wt(a,b)=6}} EDP_2(a, \mathcal{M}(b)) = 1,4238 \times 2^{-8},$$

$$\max_{\substack{a \neq 0,\, b \\ wt(a,b)=7}} EDP_2(a, \mathcal{M}(b)) = 1,0942 \times 2^{-10} \text{ and}$$

$$\max_{\substack{a \neq 0,\, b \\ wt(a,b)=8}} EDP_2(a, \mathcal{M}(b)) = 1,292 \times 2^{-12}.$$

5.2 Example where MEDP$_2$ is Tight for a Differential of Weight $(t + 3)$

Similarly, we can exhibit an SPN whose two-round MEDP is achieved by some differentials of weight $(t + 3)$ only.

Let S be a permutation of $\mathbf{F}_2^4$ defined by

x	0	1	2	3	4	5	6	7	8	9	10	11	12	13	14	15
$S(x)$	0	4	3	7	9	14	11	12	10	13	15	8	6	5	2	1

It has differential uniformity $\Delta(S) = 8$ and has 4 coefficients equal to 8 in its difference table. An example of an MDS diffusion layer with $t = 3$ such that there are differentials of weight $t + 3 = 6$ having a higher probability than all differentials of weight $(t + 1)$ or $(t + 2)$ is

$$\mathcal{M} = \begin{pmatrix} 1 & \alpha & \alpha^3 + \alpha^2 + \alpha \\ \alpha^2 & \alpha + 1 & \alpha^3 + \alpha^2 + \alpha + 1 \\ \alpha^2 + 1 & \alpha^2 + 1 & \alpha^2 + 1 \end{pmatrix}$$

where α is a root of $X^4 + X + 1$.

By computing the exact value of EDP_2 for differentials with the same weight, we obtain:

$$\max_{\substack{a \neq 0,\, b \\ wt(a,b)=4}} EDP_2(a, \mathcal{M}(b)) = \max_{\substack{a \neq 0,\, b \\ wt(a,b)=5}} EDP_2(a, \mathcal{M}(b)) = 2^{-6}$$

and

$$\max_{\substack{a \neq 0,\, b \\ wt(a,b)=6}} EDP_2(a, \mathcal{M}(b)) = 524288 \times 2^{-24} = 2^{-5}.$$

In these two examples, the Sboxes are such that there are only a few entries in their difference table which reach the maximum value $\Delta(S)$. Conversely, in the previous section, we have proved that the two-round MEDP is achieved by minimum-weight differentials when the Sbox is an APN permutation, that is, when all the nonzero coefficients of the difference table achieve the maximal value. Then we can wonder whether, when the number of entries in the difference table of the Sbox which are equal to the differential uniformity exceeds some bound, we can deduce that the two-round MEDP is tight for some minimum-weight differential only.

6   Conclusions

In this work, we have shown that the form of the minimum-weight codewords associated to the diffusion layer in an SPN_F affects the two-round MEDP. Moreover, we have exhibited for the first time some SPN such that the two-round MEDP is achieved by some differentials of weight higher than the branch number. On the other hand, we have also proved that this situation cannot occur in some cases, for instance when the Sbox is an APN permutation of $\mathbf{F}_8$. But, we give some concrete examples of round functions for which the highest differential probability is not achieved when the number of active Sboxes is minimized. This observation means that, while the branch number provides an upper bound on the two-round MEDP in any AES-like cipher [15,12], an attacker searching for the best two-round differential has to consider all possible number of active Sboxes.

Acknowledgments. The authors want to thank Thierry Berger for many stimulating discussions, including the discussions around the relevance of the rank minimum distance of the diffusion layer, which have initiated our work.

References

1. Augot, D., Finiasz, M.: Direct Construction of Recursive MDS Diffusion Layers using Shortened BCH Codes. In: Cid, C., Rechberger, C. (eds.) FSE 2014. LNCS, vol. 8540, pp. 3–17. Springer, Heidelberg (2015)
2. Bending, T.D., Fon-Der-Flaass, D.: Crooked Functions, Bent Functions, and Distance Regular Graphs. Electr. J. Comb. 5 (1998)
3. Berger, T.P.: Construction of recursive MDS diffusion layers from Gabidulin codes. In: Paul, G., Vaudenay, S. (eds.) INDOCRYPT 2013. LNCS, vol. 8250, pp. 274–285. Springer, Heidelberg (2013)
4. Biham, E., Shamir, A.: Differential cryptanalysis of DES-like cryptosystems. Journal of Cryptology, 3–72 (1991)
5. Bogdanov, A.A., Knudsen, L.R., Leander, G., Paar, C., Poschmann, A., Robshaw, M., Seurin, Y., Vikkelsoe, C.: PRESENT: An ultra-lightweight block cipher. In: Paillier, P., Verbauwhede, I. (eds.) CHES 2007. LNCS, vol. 4727, pp. 450–466. Springer, Heidelberg (2007)
6. Brinkmann, M., Leander, G.: On the classification of APN functions up to dimension five. Designs, Codes and Cryptography 49(1-3), 273–288 (2008)
7. Canteaut, A., Charpin, P.: Decomposing bent functions. IEEE Transactions on Information Theory 49(8), 2004–2019 (2003)

8. Canteaut, A., Roué, J.: On the behaviors of affine equivalent sboxes regarding differential and linear attacks. In: Oswald, E., Fischlin, M. (eds.) EUROCRYPT 2015. LNCS, vol. 9056, pp. 45–74. Springer, Heidelberg (2015)
9. Chun, K., Kim, S., Lee, S., Sung, S.H., Yoon, S.: Differential and linear cryptanalysis for 2-round SPNs. Inf. Process. Lett. 87(5), 277–282 (2003)
10. Daemen, J.: Cipher and hash function design strategies based on linear and differential cryptanalysis. Ph.D. thesis, K.U. Leuven (1995)
11. Daemen, J., Rijmen, V.: The Wide Trail Design Strategy. In: Honary, B. (ed.) Cryptography and Coding 2001. LNCS, vol. 2260, pp. 222–238. Springer, Heidelberg (2001)
12. Daemen, J., Rijmen, V.: The Design of Rijndael: AES - The Advanced Encryption Standard. Information Security and Cryptography. Springer (2002)
13. Daemen, J., Rijmen, V.: Understanding Two-Round Differentials in AES. In: De Prisco, R., Yung, M. (eds.) SCN 2006. LNCS, vol. 4116, pp. 78–94. Springer, Heidelberg (2006)
14. Daemen, J., Rijmen, V.: Correlation Analysis in $GF(2^n)$. In: Advanced Linear Cryptanalysis of Block and Stream Ciphers. Cryptology and information security, pp. 115–131. IOS Press (2011)
15. Hong, S.H., Lee, S.-J., Lim, J.-I., Sung, J., Cheon, D.H., Cho, I.: Provable Security against Differential and Linear Cryptanalysis for the SPN Structure. In: Schneier, B. (ed.) FSE 2000. LNCS, vol. 1978, pp. 273–283. Springer, Heidelberg (2001)
16. Kavun, E.B., Lauridsen, M.M., Leander, G., Rechberger, C., Schwabe, P., Yalçın, T.: Prøst v1.1. Submission to the CAESAR competition (2014), http://proest.compute.dtu.dk/proestv11.pdf
17. Keliher, L., Sui, J.: Exact maximum expected differential and linear probability for two-round Advanced Encryption Standard. IET Information Security 1(2), 53–57 (2007)
18. Kyureghyan, G.M.: Crooked maps in $\mathbf{F}_{2^n}$. Finite Fields and Their Applications 13(3), 713–726 (2007)
19. Lai, X., Massey, J.L., Murphy, S.: Markov ciphers and differential cryptanalysis. In: Davies, D.W. (ed.) EUROCRYPT 1991. LNCS, vol. 547, pp. 17–38. Springer, Heidelberg (1991)
20. MacWilliams, F., Sloane, N.: The Theory of Error-Correcting Codes, vol. 16. North-Holland (1977)
21. Park, S., Sung, S.H., Lee, S.-J., Lim, J.-I.: Improving the Upper Bound on the Maximum Differential and the Maximum Linear Hull Probability for SPN Structures and AES. In: Johansson, T. (ed.) FSE 2003. LNCS, vol. 2887, pp. 247–260. Springer, Heidelberg (2003)
22. Sajadieh, M., Dakhilalian, M., Mala, H., Sepehrdad, P.: Efficient recursive diffusion layers for block ciphers and hash functions. J. Cryptology 28(2), 240–256 (2015)
23. Shibutani, K., Bogdanov, A.: Towards the optimality of Feistel ciphers with substitution-permutation functions. Des. Codes Cryptography 73(2), 667–682 (2014), http://dx.doi.org/10.1007/s10623-014-9970-4

On the Properties of Vectorial Functions with Plateaued Components and Their Consequences on APN Functions

Claude Carlet[✉]

LAGA, UMR 7539, CNRS, Universities of Paris 8 and Paris 13,
Department of Mathematics, University of Paris 8, 2 rue de laliberté,
93526 Saint-Denis cedex 02, France
claude.carlet@univ-paris8.fr

Abstract. [This is an extended abstract of paper [15], which has been submitted to a journal] Boolean plateaued functions and vectorial functions with plateaued components, that we simply call plateaued, play a significant role in cryptography, but little is known on them. We give here, without proofs, new characterizations of plateaued Boolean and vectorial functions, by means of the value distributions of derivatives and of power moments of the Walsh transform. This allows us to derive several characterizations of APN functions in this framework, showing that all the main results known for quadratic APN functions extend to plateaued functions. Moreover, we prove that the APN-ness of those plateaued vectorial functions whose component functions are unbalanced depends only on their value distribution. This proves that any plateaued (n, n)-function, n even, having same value distribution as APN power functions, is APN and has same extended Walsh spectrum as the APN Gold functions.

1   Introduction

The notion of plateaued Boolean function, introduced in [29], is the widest known generalization of quadratic Boolean functions (i.e. of functions from $\mathbb{F}_2^n$ to $\mathbb{F}_2$ of algebraic degree 2, see e.g. [13]). It plays an important role in the cryptographic framework, still more when the notion is extended component wise to vectorial functions (from $\mathbb{F}_2^n$ to $\mathbb{F}_2^m$) used as substitution boxes in block ciphers. The set of vectorial functions whose components are plateaued, that we shall call plateaued, includes bent (n, m)-functions (n even, $m \leq n/2$), almost bent (AB) vectorial (n, n)-functions (n odd) and, for n even, some APN (n, n)-functions such as the Kasami APN functions. An illustration of the importance of these functions is that any plateaued APN (n, n)-function in odd number n of variables is AB.

However, little is known on (non-quadratic) plateaued Boolean and vectorial functions, except (1) a few characterizations given in [29] for Boolean functions, which are direct consequences of the definition, (2) a characterization valid for Boolean functions, obtained in [17], which will be a starting point for the present work, and (3) interesting but hardly usable in practice characterizations by the constance of the ratio of two consecutive Walsh power moments of even orders

© Springer International Publishing Switzerland 2015
S. El Hajji et al. (Eds.): C2SI 2015, LNCS 9084, pp. 63–73, 2015.
DOI: 10.1007/978-3-319-18681-8_5

[24]. There is a huge gap between the interest of the notion and the knowledge we have on it.

After recalling the necessary background in Section 2, we give in Section 3 new characterizations of plateaued Boolean functions and of plateaued vectorial functions, by means of the second-order and first-order derivatives in Subsection 3.1 (with a particular case when all component functions are unbalanced) and by means of power moments of the Walsh transform in Subsection 3.2. Then Section 4 applies these characterizations to the study of APN functions, gives new tools for constructing APN functions from known ones and generalizes the main properties of APN quadratic functions to plateaued functions. Subsection 4.3 studies the important sub-case where all component functions are unbalanced.

2   Preliminaries

A Boolean function $f : \mathbb{F}_2^n \mapsto \mathbb{F}_2$ is called *plateaued* if its Walsh transform $W_f(a) = \sum_{x \in \mathbb{F}_2^n} (-1)^{f(x) + a \cdot x}$, where "$\cdot$" is any inner product in $\mathbb{F}_2^n$ (for instance $a \cdot x = tr_n(ax)$ where tr_n is the trace function from $\mathbb{F}_{2^n}$ to $\mathbb{F}_2$) takes at most three values: 0 and $\pm \mu$ (where μ is some positive integer, called the *amplitude* of the plateaued function). Changing the inner product permutes the values of the Walsh transform but does not modify their distribution nor the notion. According to Parseval's Relation $\sum_{a \in \mathbb{F}_2^n} W_f^2(a) = 2^{2n}$, denoting by N_{W_f} the cardinality of the support $\{a \in \mathbb{F}_2^n / W_f(a) \neq 0\}$ of the Walsh transform, we have $N_{W_f} \times \max_{a \in \mathbb{F}_2^n} W_f^2(a) \geq 2^{2n}$ and therefore, the minimal Hamming distance to affine functions (called the nonlinearity of f, and equal to $2^{n-1} - \frac{1}{2} \max_{a \in \mathbb{F}_2^n} |W_f(a)|$) satisfies $nl(f) \leq 2^{n-1} \left(1 - \frac{1}{\sqrt{N_{W_f}}}\right)$. Equality is achieved if and only if f is plateaued.

Because of Parseval's relation, the amplitude μ of any plateaued function must be of the form 2^r where $r \geq n/2$ (since $N_{W_f} \leq 2^n$). Hence, the values of the Walsh transform of a plateaued function are divisible by $2^{n/2}$ if n is even and by $2^{(n+1)/2}$ if n is odd.

Vectorial functions $F : \mathbb{F}_2^n \mapsto \mathbb{F}_2^m$ are called (n, m)-functions and used as substitution boxes (S-boxes) in block ciphers. We shall use the notation $W_F(a, u)$ for $W_{u \cdot F}(a)$ and call *Walsh spectrum* (resp. *extended Walsh spectrum*) the value distribution of the Walsh transform (resp. of its absolute value).

Definition 1. *An (n, m)-function is called* plateaued *if all its* component functions $u \cdot F$; $u \in \mathbb{F}_2^m$, $u \neq 0$, are plateaued.

Definition 2. *An (n, m)-function is called* plateaued with single amplitude *if all its component functions are plateaued with the same amplitude.*

If the graphs $\{(x, F(x)); x \in \mathbb{F}_2^n\}$ and $\{(x, G(x)); x \in \mathbb{F}_2^n\}$ of two (n, m)-functions F, G correspond to each other by an affine permutation of $\mathbb{F}_2^n \times \mathbb{F}_2^m$ (we say then that F and G are CCZ-equivalent; the notion is from [16] and the term comes from [10]), then one is plateaued with single amplitude if and only if the other is. The set of plateaued vectorial functions with single amplitude is then CCZ-invariant. The larger set of plateaued vectorial functions is only EA-invariant, that is, if two (n, m)-functions are equivalent under composition on the left and on the right by affine permutations and under addition of an affine function, then one is plateaued if and only if the other is.

Plateaued Boolean functions and plateaued vectorial functions seem rare and they do not seem to have a simple structure. Another class whose structure is difficult to grasp is that of those APN (n, n)-functions, which oppose an optimal resistance to the differential attack, when used as S-boxes in block ciphers [2,25,26,16]. An (n, n)-function F is called Almost Perfect Nonlinear (APN) if, for every nonzero $a \in \mathbb{F}_2^n$ and every $v \in \mathbb{F}_2^n$, the equation $D_a F(x) := F(x) + F(x + a) = v$ has at most 2 solutions, or equivalently, if for every linearly independent elements a and b of $\mathbb{F}_2^n$, the second-order derivative $D_a D_b F$ does not vanish. More generally, given a positive integer δ, F is called differentially δ-uniform if the equation $F(x) + F(x + a) = v$ has at most δ solutions, for every $v \in \mathbb{F}_2^n$ and nonzero $a \in \mathbb{F}_2^n$. Any (n, n)-function is APN (that is, differentially 2-uniform) if and only if the set $\{(x, a, b) \in (\mathbb{F}_2^n)^3 \mid F(x) + F(x + a) + F(x + b) + F(x + a + b) = 0\}$ has the size $3 \cdot 2^{2n} - 2^{n+1}$ (i.e. contains only triples (x, a, b) such that a, b are linearly dependent). Equivalently the Walsh transform $W_F(a, u) = \sum_{x \in \mathbb{F}_2^n} (-1)^{u \cdot F(x) + a \cdot x}$, has fourth power moment $\sum_{a \in \mathbb{F}_2^n, u \in \mathbb{F}_2^n, u \neq 0} W_F^4(a, u)$ equal to $2^{3n+1}(2^n - 1)$, which is the smallest possible value. Few APN functions are known and it is important for cryptography to find more and to better understand their structure. A sub-class (see [19]) of APN functions is that of those plateaued functions with single amplitude called Almost Bent (AB) functions, whose Walsh transform $W_F(a, u)$ takes values 0 and $\pm 2^{\frac{n+1}{2}}$ only (n odd), when a and u range over $\mathbb{F}_2^n$ and u is nonzero. They oppose an optimal resistance to the linear attack [23], thanks to the fact that their nonlinearity (the minimum distance between their component functions and affine functions) is optimal (it achieves with equality the Sidelnikov-Chabaud-Vaudenay bound [19]). Their structure is a little better known than for APN functions, but much has still to be found on them as well. Surveys on APN and AB functions can be found in [3,14].

3   Characterizations of Plateaued Boolean and Vectorial Functions

3.1   Characterization by Means of the Derivatives

It is proved in [17] that any Boolean function f is plateaued on $\mathbb{F}_2^n$ if and only if the expression $\sum_{a,b \in \mathbb{F}_2^n} (-1)^{D_a D_b f(x)}$ does not depend on $x \in \mathbb{F}_2^n$, and that this constant expression then equals the square of the amplitude. We deduce:

Theorem 1. *Let F be an (n, m)-function. Then:*

- *F is plateaued if and only if, for every $v \in \mathbb{F}_2^m$, the size of the set*

$$\{(a, b) \in (\mathbb{F}_2^n)^2 \, ; \, D_a D_b F(x) = v\} \tag{1}$$

does not depend on $x \in \mathbb{F}_2^n$;

- *F is plateaued with single amplitude if and only if the size of the set (1) does not depend on x nor of v if v ≠ 0;*
- *Moreover, for every (n, m)-function F, the value distribution of $D_a D_b F(x)$ when $(a, b) \in (\mathbb{F}_2^n)^2$ equals the value distribution of $D_a F(b) + D_a F(x)$, and two plateaued functions having same such distribution have the same extended Walsh spectrum.*

Plateaued vectorial functions appear then as a natural generalization of quadratic vectorial functions (i.e. functions of algebraic degree at most 2), which are characterized by the fact that their second-order derivatives are constant. Note that the algebraic degree $d = 2$ is the only one for which all Boolean functions of degrees at most d are plateaued; cubic functions can have very diverse Hamming weights (see [12]).

Example 1. 1. Let F be AB, then, for every x, the number of solutions (a, b) of the equation $D_a D_b F(x) = 0$ equals (as for any APN function) the number $3 \cdot 2^n - 2$ of ordered pairs (a, b) of linearly dependent elements. We know (see [14, Proposition 9.12]) that, for every $v \neq 0$ and every x, the number of solutions (a, b) of $D_a D_b F(x) = v$ equals (uniformly) $2^n - 2$, and that conversely any APN function having this property is AB.

2. Let n be even and $F(x) = x^{2^i + 1}$ be a Gold APN function, $(i, n) = 1$. We have $D_a D_b F(x) = a^{2^i} b + a b^{2^i}$. The number of solutions (a, b) of $D_a D_b F(x) = 0$ equals again $3 \cdot 2^n - 2$, and for $v \neq 0$, the equation $D_a D_b F(x) = v$ has two solutions a for every $b \neq 0$ such that $\frac{v}{b^{2^i + 1}}$ has null trace. The number of such nonzero b equals $2^{n-1} \pm 2^{\frac{n}{2}} - 1$ when v is a cube and $2^{n-1} \pm 2^{\frac{n}{2}-1} - 1$ when v is not a cube. Hence the number of solutions (a, b) of $D_a D_b F(x) = v$ equals:

$$\begin{cases} 3 \cdot 2^n - 2 & \text{for } v = 0, \\ 2^n \pm 2^{\frac{n}{2}+1} - 2 & \text{for } v \text{ a nonzero cube } (\frac{2^n - 1}{3} \text{ cases}) \\ 2^n \pm 2^{\frac{n}{2}} - 2 & \text{for } v \text{ a non-cube } (2 \cdot \frac{2^n - 1}{3} \text{ cases}). \end{cases}$$

Since the number of all (a, b) equals 2^{2n}, we deduce that, among the two "$\pm$" above, one is a "$+$" and one is a "$-$". We shall see below that the Kasami APN functions (see definition below) have the same distribution.

It is deduced in [15] that:

Corollary 1. *Let n be any even integer, $n \geq 4$. Let F be an (n, n)-function CCZ-equivalent to a Gold APN function $G(x) = x^{2^i + 1}$ or to a Kasami APN function $G(x) = x^{4^i - 2^i + 1}$, $(i, n) = 1$. Then F is plateaued with single amplitude if and only if it is EA-equivalent to $G(x)$.*

The Case of Power Functions. It is often simpler to consider power functions than general functions. This has been illustrated for instance in the study of APN functions. The case of plateaued functions makes no exception.

Corollary 2. *Let $F(x) = x^d$ be any power function. Then, for every $v \in \mathbb{F}_{2^n}$, every $x \in \mathbb{F}_{2^n}$, and every $\lambda \in \mathbb{F}_{2^n}^*$, we have*

$$|\{(a, b) \in \mathbb{F}_{2^n}^2 \, ; \, D_a F(b) + D_a F(x) = v\}|$$
$$= |\{(a, b) \in \mathbb{F}_{2^n}^2 \, ; \, D_a F(b) + D_a F(x/\lambda) = v/\lambda^d\}|.$$

In particular, $|\{(a, b) \in \mathbb{F}_{2^n}^2 \, ; \, D_a F(b) + D_a F(0) = v\}|$ is invariant when v is multiplied by any d-th power in $\mathbb{F}_{2^n}^*$.
    Then:

– *F is plateaued if and only if, for every $v \in \mathbb{F}_{2^n}$:*

$$|\{(a, b) \in \mathbb{F}_{2^n}^2 \, ; \, D_a F(b) + D_a F(1) = v\}|$$
$$= |\{(a, b) \in \mathbb{F}_{2^n}^2 \, ; \, D_a F(b) + D_a F(0) = v\}|;$$

– *F is plateaued with single amplitude if and only if, for every v, $|\{(a, b) \in \mathbb{F}_{2^n}^2 \, ; \, D_a F(b) + D_a F(1) = v\}| = |\{(a, b) \in \mathbb{F}_{2^n}^2 \, ; \, D_a F(b) + D_a F(0) = v\}|$, and for every nonzero v, this size does not depend on v.*

If d is co-prime with n, then F is plateaued if and only if it is plateaued with single amplitude.

The Case of Unbalanced Components. If all component functions of F are unbalanced, then $W_F(0, u) \neq 0$ for every $u \neq 0$ (and therefore, for every u), and we know then that the amplitude of the component function $u \cdot F$ equals $|W_F(0, u)|$. Hence, F is plateaued if and only if, for every u, x, the sum $\sum_{a, b \in \mathbb{F}_2^n} (-1)^{u \cdot D_a D_b F(x)}$ equals $W_F^2(0, u) = \sum_{a, b \in \mathbb{F}_2^n} (-1)^{u \cdot (F(a) + F(b))}$. Conversely, if this equality is satisfied, then $W_F^2(0, u)$ equals the square of the amplitude of $u \cdot F$ and is then nonzero.

Theorem 2. *Let F be any (n, m)-function. Then F is plateaued with component functions all unbalanced if and only if, for every $v, x \in \mathbb{F}_2^n$, we have:*

$$|\{(a, b) \in (\mathbb{F}_2^n)^2 \, ; \, D_a D_b F(x) = v\}| = |\{(a, b) \in (\mathbb{F}_2^n)^2 \, ; \, F(a) + F(b) = v\}|.$$

Moreover, F is plateaued with single amplitude if and only if, additionally, this common value does not depend on v for $v \neq 0$.

3.2   Characterization by Means of Power Moments of the Walsh Transform

Theorem 3. *Any n-variable Boolean function f is plateaued if and only if, for every nonzero $\alpha \in \mathbb{F}_2^n$, we have*

$$\sum_{a \in \mathbb{F}_2^n} W_f(a + \alpha) \, W_f^3(a) = 0.$$

Any (n, m)-function F is plateaued if and only if:

$$\forall u \in \mathbb{F}_2^m, \forall \alpha \in \mathbb{F}_2^n, \alpha \neq 0, \sum_{a \in \mathbb{F}_2^n} W_F(a + \alpha, u) W_F^3(a, u) = 0.$$

And F is plateaued with single amplitude if and only if, additionally, $\sum_{a \in \mathbb{F}_2^n} W_F^4(a, u)$ does not depend on u for $u \neq 0$.

Corollary 3. *Any n-variable Boolean function f is plateaued if and only if, for every $b \in \mathbb{F}_2^n$:*

$$\sum_{a \in \mathbb{F}_2^n} W_f^4(a) = 2^n (-1)^{f(b)} \sum_{a \in \mathbb{F}_2^n} (-1)^{a \cdot b} W_f^3(a).$$

Any (n, m)-function F is plateaued if and only if, for every $b \in \mathbb{F}_2^n$ and every $u \in \mathbb{F}_2^m$:

$$\sum_{a \in \mathbb{F}_2^n} W_F^4(a, u) = 2^n (-1)^{u \cdot F(b)} \sum_{a \in \mathbb{F}_2^n} (-1)^{a \cdot b} W_F^3(a, u).$$

And F is plateaued with single amplitude if and only if, additionally, these sums do not depend on u, for $u \neq 0$.

Proposition 1. *For every n-variable Boolean function f, we have:*

$$\left(\sum_{a \in \mathbb{F}_2^n} W_f^4(a) \right)^2 \leq 2^{2n} \left(\sum_{a \in \mathbb{F}_2^n} W_f^6(a) \right), \tag{2}$$

with equality if and only f is plateaued.
For every (n, m)-function F, we have:

$$\sum_{u \in \mathbb{F}_2^m} \left(\sum_{a \in \mathbb{F}_2^n} W_F^4(a, u) \right)^2 \leq 2^{2n} \sum_{u \in \mathbb{F}_2^m} \left(\sum_{a \in \mathbb{F}_2^n} W_F^6(a, u) \right), \tag{3}$$

with equality if and only if F is plateaued, which is equivalent to the fact that the size of the set

$$\{(x_1, x_2, x_3, x_4, y_1, y_2, y_3, y_4) \in (\mathbb{F}_2^n)^8 \mid \sum_{i=1}^{4} F(x_i) + \sum_{i=1}^{4} F(y_i)) = \sum_{i=1}^{4} x_i = \sum_{i=1}^{4} y_i = 0\}$$

equals 2^n times the size of the set

$$\{(x_1, x_2, x_3, x_4, x_5, x_6) \in (\mathbb{F}_2^n)^6 \mid \sum_{i=1}^{6} F(x_i) = \sum_{i=1}^{6} x_i = 0\}.$$

For every (n, m)-function F, we have also:

$$\sum_{u \in \mathbb{F}_2^m} \sum_{a \in \mathbb{F}_2^n} W_F^4(a, u) \leq 2^n \sum_{u \in \mathbb{F}_2^m} \sqrt{\sum_{a \in \mathbb{F}_2^n} W_F^6(a, u)} \tag{4}$$

with equality if and only if F is plateaued.

The characterization by the equality in (2) is a particular case of the result of [24] mentioned in introduction. The work in [24] and the present work have been done independently.

4 Characterizations of the APN-ness of Componentwise Plateaued Vectorial Functions

If a function F is quadratic, then given $a \neq 0$, the property that all equations $F(x) + F(x + a) = v$ have at most 2 solutions is equivalent to the fact that the single homogeneous linear equation $F(x) + F(x + a) = F(0) + F(a)$ has 2 solutions. This is the main reason why many recent results of constructions of APN functions [4,6,7,8,9,18,27,28] produce quadratic functions. Unfortunately, quadratic functions are hardly usable as S-boxes [22,21]. Thanks to the results of the previous section, we shall show that the nice property of quadratic functions recalled above, and other properties as well, can be extended to all plateaued functions.

4.1 Characterization by the Derivatives

An (n, n)-function F is APN if and only if, for every $\mathbb{F}_2$-linearly independent a, b, the equation $D_a D_b F(x) = F(x) + F(x + a) + F(x + b) + F(x + a + b) = 0$ has no solution x. If F is plateaued, then according to Theorem 1, it is APN if and only if, for every $\mathbb{F}_2$-linearly independent a, b, we have $F(0) + F(a) + F(b) + F(a + b) \neq 0$. Then:

Theorem 4. *Any plateaued (n, n)-function F is APN if and only if, for every $a \neq 0$ in $\mathbb{F}_2^n$, the equation $F(x) + F(x + a) = F(0) + F(a)$ has the 2 solutions 0 and a only.*

4.2 Characterization by the Walsh Transform

In the next proposition, we assume that $F(0) = 0$, with no loss of generality.

Proposition 2. *Let F be any plateaued (n, n)-function. Assume that $F(0) = 0$. Then F is APN if and only if the set $\{(x, b) \in \mathbb{F}_{2^n}^2 \mid F(x) + F(x + b) + F(b) = 0\}$ has size $3 \cdot 2^n - 2$. Equivalently:*

$$\sum_{a \in \mathbb{F}_{2^n}, u \in \mathbb{F}_2^n, u \neq 0} W_F^3(a, u) = 2^{2n+1}(2^n - 1).$$

This necessary and sufficient condition was known until now only for quadratic functions, see [14] (of course, it was also known as a necessary condition for functions of unrestricted degree).

Proposition 3. *Let F be any (n, n)-function. Then F is APN and plateaued if and only if the Walsh transform of F satisfies:*

$$3 \cdot 2^{3n} - 2^{2n+1} = \sum_{u \in \mathbb{F}_2^n} \sqrt{\sum_{a \in \mathbb{F}_2^n} W_F^6(a, u)},$$

or equivalently

$$22^{2n+1}(2^n - 1) = \sum_{\substack{u \in \mathbb{F}_2^n \\ u \neq 0}} \sqrt{\sum_{a \in \mathbb{F}_2^n} W_F^6(a, u)}, \tag{5}$$

We give now a result which is new, even for quadratic functions, as far as we know. It depends on the amplitude of each component function, but has the interest of leading to a characterization involving a sum of squares of the Walsh values instead of sums of larger degrees as above.

Proposition 4. *Let F be a plateaued (n, n)-function. For every u, let 2^{λ_u} be the amplitude of $u \cdot F$. Then F is APN if and only if:*

$$\sum_{u \in \mathbb{F}_2^n, u \neq 0} 2^{2\lambda_u} \leq 2^{n+1}(2^n - 1), \tag{6}$$

or equivalently if, for every function $\psi : \mathbb{F}_2^n \mapsto \mathbb{F}_2^n$, we have:

$$\sum_{u \in \mathbb{F}_2^n, u \neq 0} W_F^2(\psi(u), u) \leq 2^{n+1}(2^n - 1). \tag{7}$$

Inequality (6), and Inequality (7) for some ψ, are then equalities.

Remark 1. As already recalled in introduction, it is known that for n odd, if F is APN and is plateaued, then F is AB (see e.g. [14]). Proposition 4 gives a new way of proving this result: we know that for n odd we have $2\lambda_u \geq n + 1$ and (6) implies that $2\lambda_u = n + 1$ for every $u \neq 0$. This proves that F is AB.   □

4.3   The Case of Unbalanced Component Functions

In the case that all component functions of a plateaued (n, n)-function F are unbalanced, we have simpler and more efficient characterizations of its APN-ness. From Theorem 2 and from the observation that if a and b are $\mathbb{F}_2$-linearly dependent, then we have $D_a D_b F(x) = 0$, we directly deduce the following theorem:

Theorem 5. *Let F be any plateaued (n, n)-function having all its component functions unbalanced, then*

$$\left| \{(a, b) \in (\mathbb{F}_2^n)^2, a \neq b ; F(a) = F(b)\} \right| \geq 2 \cdot (2^n - 1),$$

with equality if and only if F is APN.

Remark 2. Theorem 2 and Theorem 5 show that any APN function having all its component functions unbalanced is plateaued with single amplitude if and only if it is AB, and then n must be odd (note that this generalizes up to EA-equivalence). Indeed, the condition is clearly necessary and, according to these theorems, it is also sufficient because the size of $\{(a, b) \in (\mathbb{F}_2^n)^2 ; F(a) + F(b) = v\}$ equals then $\frac{2^{2n} - 2^n - 2 \cdot (2^n - 1)}{2^n - 1} = 2^n - 2$ for every $v \neq 0$, that is,

$|\{(a,b) \in (\mathbb{F}_2^n)^2 ; D_a D_b F(x) = v\}|$ equals $2^n - 2$, and this is equivalent to F AB. But in fact this result is true without the hypothesis that all component functions are unbalanced. In odd dimension (i.e. for n odd), we have already recalled that this is well-known. In even dimension, it is proved in [14] that for a plateaued APN function, at least two third of the component functions are bent; therefore, if F is plateaued with single amplitude, it is necessarily bent, a contradiction with Nyberg's result that bent (n,n)-functions cannot exist.

Remark 3. In the framework of Theorem 5, the number

$$Nb_F = \left|\{(a,b) \in (\mathbb{F}_2^n)^2 , a \neq b ; F(a) = F(b)\}\right|$$

is minimal for an APN function. In such case, since $Nb_F = \sum_{a \in \mathbb{F}_2^n ; a \neq 0} |(D_a F)^{-1}(0)|$ and each set $(D_a F)^{-1}(0)$ has size at most 2, each such set has size exactly 2. Such function F with the property that there exist exactly 2 solutions of the equation $F(x) + F(x + a) = 0$, for every $a \neq 0$, is called *zero-difference 2-balanced*, see [20]. It is proved in [18] that every quadratic zero-difference 2-balanced function is APN. With Theorem 5, we extend this result from the class of quadratic functions to the larger class of plateaued functions (and we also have its converse).   □

Theorem 5 may also lead to a way of constructing new APN functions from known ones, thanks to the following:

Corollary 4. *Let F be any plateaued APN (n,n)-function having all its component functions unbalanced. Let π be a permutation of $\mathbb{F}_2^n$ and G a function injective on the image set $F(\mathbb{F}_2^n) = \{F(x), x \in \mathbb{F}_2^n\}$ of F. Then if $G \circ F \circ \pi$ is plateaued, $G \circ F \circ \pi$ is APN. Moreover, if G is identity, F and $F \circ \pi$ have same extended Walsh spectrum.*

A case of application in which G is identity and for which we can characterize the fact that $F \circ \pi$ is plateaued is by taking for F a Gold function (see [15]):

Corollary 5. *Let n be an even positive integer. Let $d = 2^i + 1$, $(i, n) = 1$. Let π be the compositional inverse of a quadratic permutation Q of $\mathbb{F}_{2^n}$. For every $b \in \mathbb{F}_{2^n}$, let us denote by L_b the linear (n,n)-function such that $tr_n[(bQ(x + y) + bQ(x) + bQ(y) + bQ(0)] = tr_n[L_b(x)y]$ and by $E_{u,b}$ the vector subspace $\{x \in \mathbb{F}_{2^n} \mid ux^{2^i} + (ux)^{2^{n-i}} + L_b(x) = 0\}$ of $\mathbb{F}_{2^n}$. Then:*

1. *Function $F(x) = (\pi(x))^d$ is plateaued if and only if, for every $u \in \mathbb{F}_{2^n}^*$, the dimension of $E_{u,b}$ is the same for all b's such that function $tr_n(ux^{2^i+1} + bQ(x))$ is constant on $E_{u,b}$;*
2. *If this condition is satisfied, then F is APN.*

Dobbertin proved that if F is a power APN function in even dimension then it is 3-to-1 over $\mathbb{F}_{2^n}^*$ (see his proof reported in [14]) and $gcd(d, 2^n - 1) = 3$. Theorem 5 allows proving the converse for any plateaued power function:

Corollary 6. *Let n be even and $F(x) = x^d$ be any plateaued power function. Then F is APN if and only if $\gcd(d, 2^n - 1) = 3$.*

This applies to the Kasami functions for n even.

Question: Does there exist, for any APN plateaued function F, a linear function L such that $F + L$ has unbalanced components?

Acknowledgement. We wish to thank Alexander Pott for useful information and Lilya Budaghyan for very helpful discussions and information.

References

1. Berger, T., Canteaut, A., Charpin, P., Laigle-Chapuy, Y.: On almost perfect nonlinear functions. IEEE Trans. Inform. Theory 52(9), 4160–4170 (2006)
2. Biham, E., Shamir, A.: Differential Cryptanalysis of DES-like Cryptosystems. Journal of Cryptology 4(1), 3–72 (1991)
3. Budaghyan, L.: Construction and Analysis of Cryptographic Functions. Springer, (200 pages) (to appear)
4. Budaghyan, L., Carlet, C.: Classes of Quadratic APN Trinomials and Hexanomials and Related Structures. IEEE Trans. Inform. Theory 54(5), 2354–2357 (2008)
5. Budaghyan, L., Carlet, C.: CCZ-equivalence of Bent Vectorial Functions and Related Constructions. Designs, Codes and Cryptography 59(1-3), 69–87 (2011)
6. Budaghyan, L., Carlet, C., Felke, P., Leander, G.: An infinite class of quadratic APN functions which are not equivalent to power functions. In: Proceedings of IEEE International Symposium on Information Theory (ISIT) (2006)
7. Budaghyan, L., Carlet, C., Leander, G.: Two classes of quadratic APN binomials inequivalent to power functions. IEEE Trans. Inform. Theory 54(9), 4218–4229 (2008), This paper is a completed and merged version of [6] and [8]
8. Budaghyan, L., Carlet, C., Leander, G.: Another class of quadratic APN binomials over $\mathbb{F}_{2^n}$: the case n divisible by 4. In: Proceedings of the Workshop on Coding and Cryptography, WCC 2007, pp. 49–58 (2007)
9. Budaghyan, L., Carlet, C., Leander, G.: Constructing new APN functions from known ones. Finite Fields and Applications 15(2), 150–159 (2009)
10. Budaghyan, L., Carlet, C., Pott, A.: New Classes of Almost Bent and Almost Perfect Nonlinear Polynomials. In: Proceedings of the Workshop on Coding and Cryptography 2005, Bergen, pp. 306–315 (2005)
11. Budaghyan, L., Carlet, C., Pott, A.: New Classes of Almost Bent and Almost Perfect Nonlinear Functions. IEEE Trans. Inform. Theory 52(3), 1141–1152 (2006), This is a completed version of [10]
12. Carlet, C.: A transformation on Boolean functions, its consequences on some problems related to Reed-Muller codes. In: Charpin, P., Cohen, G. (eds.) EUROCODE 1990. LNCS, vol. 514, pp. 42–50. Springer, Heidelberg (1991)
13. Carlet, C.: Boolean Functions for Cryptography and Error Correcting Codes. Chapter of the monography. In: Crama, Y., Hammer, P. (eds.) Boolean Models and Methods in Mathematics, Computer Science, and Engineering, pp. 257–397. Cambridge University Press (2010), Preliminary version available at http://www.math.univ-paris13.fr/~carlet/pubs.html

14. Carlet, C.: Vectorial Boolean Functions for Cryptography. Chapter of the monography. In: Crama, Y., Hammer, P. (eds.) Boolean Methods and Models. Cambridge University Press, Preliminary version available at http://www.math.univ-paris13.fr/~carlet/pubs.html
15. Carlet, C.: Boolean and vectorial plateaued functions and APN functions. Preprint
16. Carlet, C., Charpin, P., Zinoviev, V.: Codes, bent functions and permutations suitable for DES-like cryptosystems. Designs, Codes and Cryptography 15(2), 125–156 (1998)
17. Carlet, C., Prouff, E.: On plateaued functions and their constructions. In: Johansson, T. (ed.) FSE 2003. LNCS, vol. 2887, pp. 54–73. Springer, Heidelberg (2003)
18. Carlet, C., Gong, G., Tan, Y.: Quadratic Zero-Difference Balanced Functions, APN functions and Strongly Regular Graphs. To appear in Designs, Codes and Cryptography
19. Chabaud, F., Vaudenay, S.: Links between Differential and Linear Cryptanalysis. In: De Santis, A. (ed.) EUROCRYPT 1994. LNCS, vol. 950, pp. 356–365. Springer, Heidelberg (1995)
20. Ding, C., Tan, Y.: Zero-Difference Balanced Functions With Applications. Journal of Statistical Theory and Practice 6(1), 3–19 (2012)
21. Knudsen, L.: Truncated and higher order differentials. In: Preneel, B. (ed.) FSE 1994. LNCS, vol. 1008, pp. 196–211. Springer, Heidelberg (1995)
22. Lai, X.: Higher order derivatives and differential cryptanalysis. In: Proceedings of the Symposium on Communication, Coding and Cryptography (1994), in Honor of J. L. massey on the occasion of his 60'th birthday
23. Matsui, M.: Linear cryptanalysis method for DES cipher. In: Helleseth, T. (ed.) EUROCRYPT 1993. LNCS, vol. 765, pp. 386–397. Springer, Heidelberg (1994)
24. Mesnager, S.: Characterizations of plateaued and bent functions in characteristic p. In: Schmidt, K.-U., Winterhof, A. (eds.) SETA 2014. LNCS, vol. 8865, pp. 71–81. Springer, Heidelberg (2014)
25. Nyberg, K.: Perfect nonlinear S-boxes. In: Davies, D.W. (ed.) EUROCRYPT 1991. LNCS, vol. 547, pp. 378–386. Springer, Heidelberg (1991)
26. Nyberg, K.: On the construction of highly nonlinear permutations. In: Rueppel, R.A. (ed.) EUROCRYPT 1992. LNCS, vol. 658, pp. 92–98. Springer, Heidelberg (1993)
27. Yu, Y., Wang, M., Li, Y.: A matrix approach for constructing quadratic APN functions. In: Proceedings of International Workshop on Coding and Cryptography, pp. 39–47 (2013)
28. Weng, G., Tan, Y., Gong, G.: On almost perfect nonlinear functions and their related algebraic objects. In: Proceedings of International Workshop on Coding and Cryptography, pp. 48–57 (2013)
29. Zheng, Y., Zhang, X.-M.: Plateaued functions. In: Varadharajan, V., Mu, Y. (eds.) ICICS 1999. LNCS, vol. 1726, pp. 284–300. Springer, Heidelberg (1999)

Beyond Cryptanalysis Is Software Security the Next Threat for Smart Cards

Jean-Louis Lanet(✉)

INRIA, LHS-PEC,
263 Avenue Général Leclerc, 35042 Rennes, France
jean-louis.lanet@inria.fr
http://secinfo.msi.unilim.fr/lanet/

Abstract. Smart cards have been considered for a long time as a secure container for storing secret data and executing programs that manipulate them without leaking any information. In the last decade, a new form of attack that uses the hardware has been intensively studied. We have proposed in the past to pay attention also to easier attacks that use only software. We demonstrated through several proof of concepts that such an approach should be a threat under some hypotheses. We have been able to execute self-modifying code, return address programming and so on. More recently we have been able to retrieve secret keys belonging to another application. Then all the already published attacks should have been a threat but the industry increased the counter measures to mitigate for each of the published attack. In such a sensitive domain, we always submit the attacks to the industrial partners but also national agencies before publishing any attack. Within such an approach, they have been able to patch their system before any vulnerabilities should be exploited.

Keywords: Smart Card · Attacks · Ethical Process

1 Introduction

Java Card is a kind of smart card that implements one of the two editions, "*Classic Edition*" or "*Connected Edition*", of the standard Java Card 3.0 [12]. Such a smart card embeds a virtual machine which interprets codes already romized with the operating system or downloaded after issuance. Due to security reasons, the ability to download code into the card is controlled by a protocol defined by Global Platform [7]. This protocol ensures that the owner of the code has the necessary authorization to perform the action. Java Card is an open platform for smart cards, *i.e.* able of loading and executing new applications after issuance. Thus, different applications from different providers run in the same smart card. Thanks to type verification, byte codes delivered by the Java compiler and the converter (in charge of giving a compact representation of class files) are safe, *i.e.* the loaded application is not hostile to other applications in the Java Card. Furthermore, the Java Card firewall checks permissions between applications in the card, enforcing isolation between them.

© Springer International Publishing Switzerland 2015
S. El Hajji et al. (Eds.): C2SI 2015, LNCS 9084, pp. 74–82, 2015.
DOI: 10.1007/978-3-319-18681-8_6

Java Cards have shown an improved robustness compared to native applications regarding many attacks. They are designed to resist to numerous attacks using both physical and logical techniques. Currently, the most powerful attacks are hardware based attacks and particularly fault attacks. A fault attack modifies parts of memory content or signal on internal bus and lead to deviant behavior exploitable by an attacker. A comprehensive consequence of such attacks can be found in [11]. Although fault attacks have been mainly used in the literature from a cryptanalytic point of view (see [1,9,13]), they can be applied to every code layers embedded in a device. For instance, while choosing the exact byte of a program the attacker can bypass counter-measures or logical tests.

The design of a Java Card virtual machine cannot rely on the environmental hypotheses of Java. In fact, physical attacks have never been taken into account during the design of the Java platform. To fill this gap, card designers developed an interpreter which relies on the principle that once the application has been linked to the card, it will not be modifiable again. The trade-off is between a highly defensive virtual machine which will be too slow to operate and an offensive interpreter that will expose too much vulnerabilities. The know-how of a smart card design is in the choice of a set of minimal counter-measures with high fault coverage.

Nevertheless some attacks have been successful in retrieving secret data from the card. Thus we will present here a survey of different approaches to get access to data, which should bypass Java security components. The aim of an attacker is to generate malicious applications which can bypass firewall restrictions and modify other applications, even if they do not belong to the same security package. Several papers were published and they differ essentially on the hypotheses of the platform vulnerabilities. After a brief presentation of the Java Card platform and its security functions, we will present attacks based on a faulty implementation of the transaction, due to ambiguities in the specification. Then we will describe the flaws that can be exploited with an ill-typed applet and we will finish with hostile applet that gain privilege to access the physical processor leading to the dump of the operating system and the crypto API.

2   Smart Card Security

Smart cards security depends on the underlying hardware and the embedded software. Embedded sensors (light sensors, heat sensors, voltage sensors, *etc.*) protect the card from physical attacks. While the card detects such an attack, it has the possibility to erase quickly the content of the EEPROM preserving the confidentiality of secret data or blocking definitely the card (Card is mute). In addition to the hardware protection, softwares are designed to securely ensure that application are syntactically and semantically correct before installation and also sometimes during execution. They also manage sensitive information and ensure that the current operation is authorized before executing it. The Byte Code Verifier guarantees type correctness of code, which in turn guarantees the Java properties regarding memory access. For example, it is impossible in Java to perform arithmetic on reference. Thus, it must be proved that the two elements on top of the stack are of primitive types before performing any arithmetic operation. On the Java platform, byte code verification is invoked at load time by the loader. Due to the fact that Java Card does not support dynamic class loading, byte code verification is performed at installation time *i.e.* before loading the Card APplet (CAP) onto the card. However, most of the Java Card smart cards do not have an on-card BCV as it is quite expensive in terms of memory consumption. Thus, a trusted third party performs an off-card byte code verification and sign it, and on card its digital signature is checked.

Moreover, the Firewall performs checks at runtime to prevent applets from accessing (reading or writing) data of other applets. When an applet is created, the system uses a unique applet identifier (AID) from which it is possible to retrieve the name of the package in which it is defined. If two applets are instances of classes coming from the same Java Card package, they are considered belonging to the same context. The firewall isolates the contexts in such a way that a method executing in one context cannot access any attribute or method of objects belonging to another context unless it explicitly exposes functionality *via* a Shareable Interface Object.

Smart card security is a complex problem with different points of view but products based on Java Card Virtual Machine (JCVM) have passed successfully real-world security evaluations for major industries around the world. It is also the platform that has passed high level security evaluations for issuance by banking associations and by leading government authorities, they have also achieved compliance with FIPS 140-1 certification scheme. Nevertheless implementations have suffered severals attacks either hardware or software based. Some of them succeeded into getting access to the EEPROM (code of the downloaded applets) but as far as we know nobody succeeded into reversing the code *i.e.* having access to the code of the virtual machine, the operating system and the cryptographic algorithm implementations. These latter are protected by the interpretation layer which denies access to other memories than the EEPROM.

3   Some Software Attacks Again Java Card

3.1   Ambiguity in the Specification: The Type Confusion

Erik Poll made a presentation at CARDIS'08 about attacks on smart cards. In his paper [10], he did a quick overview of the classical attacks available on smart cards and gave some counter-measures. He explained the different kinds of attacks and the associated counter-measures. He described four methods (1) CAP file manipulation, (2) Fault injection, (3) Shareable interfaces mechanisms abuse and (4) Transaction Mechanisms abuse.

He proposed a new way to abuse the Transaction mechanism (4). The purpose of transaction is to make a group of operations becomes atomic. Of course, it is a widely used concept, like in databases, but still hard to implement. By definition, the rollback mechanism should also deallocate any objects allocated during an aborted transaction, and reset references to such objects to null.

However, Erik Poll find some strange cases where the card keep the references of objects allocated during transaction even after a roll back.

If he can get the same behavior, it should be easy to get and exploit type confusion. A first example is to get two arrays of different types, for example a byte and a short array. One of them is a field (permanent storage) the second is a local variable. While aborting the transaction, the permanent reference must be nullified. But the specification do not explain what to do with local variables if they reference also a permanent object. Poll discovered that some cards cleared all the references while other let dangling pointers. In such a case reallocating the memory will let the dangling pointer referencing another object of potentially another type. If he declares a byte array of 10 bytes, and he has another reference as a short array, he will be able to read 10 shorts, so 20 bytes. With this method he can read the 10 bytes saved after the array. If he increases the size of the array, he can read as much memory as he wants. The main problem is more how to read memory before the array. The other confusion he used is an array of bytes and an object. If he puts a byte as first object attribute, it is bound to the array length. It is then really easy to change the length of the array using the reference to the object.

3.2 Weakness in the Linker Process

The Java Card Specification defines the linking step done during the loading of CAP file. When the software is downloading in the card, the Java Card Virtual Machine provides a way to link, the CAP file to install, with the installed Java Card API. This step is done thanks to a tokens link resolution references in the Constant Pool component. To friendly find where each token is used, the Reference Location component keeps a list of offsets, in the Method Component. So, in this loading step, the JCVM translates, with the help of the Constant Pool component and the Reference Location component, each reference to methods or fields use in the CAP file. To abuse the linking mechanism [14], [8] we modify the token following any natural instructions, as invokestatic, which are following by a token. If the card have not any BCV component, a modification may push the linked reference on the stack and returned at the end of the current function.

Using this approach we are able to use the on board linker to generate the correct information, to store it on top of the stack and to send it back to the reader. Thanks to this information leakage we are able to obtain all the linked address of the Java Card API for a given card. For retrieving one address we need to build one CAP file. Retrieving the complete API, need to generate 98 test cases for the methods of the classes and 60 test cases for the interfaces. All the test cases are valid whatever the card is tested. It means that the effort to design the test cases for retrieving the addresses will be reusable on all the cards. This attack is completely generic and independent of the platform.

3.3 Dumping the EEPROM

As said previously, the verifier must check several points. In particular: there are no violations of memory management and any stack underflow or overflow. This means that these checks are potentially not verified during run time and then can lead to vulnerabilities. The Java frame is a non persistent data structure but can be implemented in different manners and the specification gives no design direction for it. Getting access to the RAM provides information of other objects like the APDU buffer, return address of a method and so on. So, changing the return of a local address modifies the control flow of the call graph and returns it to a specific address.

The EMAN2 attack [3] allows to modify the value of the return address of a method by storing a short into a local. By choosing the right value for the local number we overwrite the return address. In a given card the return address register is stored at MAX_LOCAL + 2. The value stored in this register will be the address where Java PC will be updated while returning from the current method. We just need to define a static array which is stored close to the method area. Then after returning from the method, the JCVM will execute the content of the array. Due to the fact that `getstatic` and `putstatic` are not checked by the firewall, we can read the content of the memory. The shell code is presented in Listing 1.1.

Listing 1.1. Executing the basic shell code

```
7C 01 00     getStatic   0x0100
78           sreturn
```

This code puts on top of the stack, the content of the memory at the address 0x0100 and returns this value. The caller has just to store it into the APDU buffer and the value is send to the terminal. Then, the third byte of the static array must be incremented and the next call will return the value of the address 0x0101. We just need to manage the carry from the low byte to the high byte representing the address. Another way to update the return address is the `sinc` instruction. The `sinc` instruction aims to increase a local short variable by a constant value given in its parameter.

Recently, Faugeron [6] presented a way to fool the Java Card runtime based on the `dup_x` instruction. This instruction duplicates the top of operands stack words and inserts them below. This instruction takes two parameters encoded on 1-byte where the high nibble describes the number of words to duplicate and the low nibble defines where the duplicated words are placed. Since the Java Card operands stack does not contain enough elements, the runtime uses the system data as words for the `dup_x` instruction. Thus, an attacker can shift the value of the frame header by a custom words pushed on the stack.

3.4 Dumping the ROM

In [4] we demonstrated the ability to dump the content of the ROM and thus to get access to the implementation of the cryptographic functions. We used several weaknesses. During the analysis of EEPROM dump corresponding to a linked applets into the smart card memory, a method with an abnormal call has been noticed at the address 0xDBE6. This address corresponds to another EEP-ROM address and not a ROM address. At that address we found a table which corresponded to non standard method headers. The JCVM Specification [12] defines a method as a method_header_info, described in the listing 1.2, and its associated byte code.

Listing 1.2. Java Card Method Header Info

```
method_header_info {
  u1 bitfield {
    bit [4] flags // a mask of modifiers defined for the method
    bit [4] max_stack // max cells required during execution of
                      // the method
  }
  u1 bitfield {
    bit [4] nargs // number of parameters passed to the method
    bit [4] max_locals // number of local variables declared
                       // by the method
  }
}
```

For the flag value, three defined possibilities are expected:

- 0x0: it is a normal method;
- 0x8 (ACC_EXTENDED): the method represents an extended method;
- 0x4 (ACC_ABSTRACT): the method represents an abstract method;
- All other flag values are reserved.

Each methods of the table contains a non standardized flag value (*i.e.* : 2). Moreover, the associated byte code (1-byte) cannot be an instruction. On the other side, we also also have a set of interesting values in the EEPROM. We assumed that all these values are addresses that refer to the ROM, except one which refers to the EEPROM. To prove our hypothesis we checked the data contained at the address corresponds to a 8051 assembler language which corresponds to the native code for the targeted card. We reversed the code in order to verify the calling convention of this native Java Interface.

To exploit this weakness, we added to the method table a **fake method** (a method with a flag value equals to 2) contains an offset to an address in the indirection table. Each element in the indirection table refers to a native function. At this offset we put the address of our shellcode. Without integrity check, the Java Card Runtime execute the malicious code. Finally, to execute the native shell code the parameter of an **invokestatic** instruction, or another kind of call instruction should be changed by the address of our **fake method**. Thus, the faulty instruction provides a way to execute any shell code with native privilege. With this shell code, we have been able to do a memory dump of the ROM code. Examining carefully the code we discovered the cryptographic code corresponding to the embedded algorithms within this specific card.

3.5   A Complete Methodology to Attack Smart Card

In his PhD, Bouffard [2] applied the Attack Tree Analysis (ATA) to have a global view on the vulnerability of the smart card. Attack trees have been introduced by Schneier in[15], they represent a convenient approach to analyze the different ways in which a system can be attacked. It is an analytical technique (top-down) where an undesirable event is defined and the system is then analyzed to find the combinations of basic events that could lead to the undesirable event. Such an analysis is closed to the risk analysis community with the cause-effect diagrams. An attack tree is a tree in which the nodes represent attacks. The root node of the tree is the property that an attacker wants to break. Children of a node are refinements of this goal, and leafs therefore represent initial causes. An attack tree is not a model of all possible combination but a restricted set. It is related to the property evaluated. In this case, code integrity is the most sensible property because if not guaranteed, it enables the attacker to execute any arbitrary code.

The property we want to protect is the integrity of the code which can be violated by a Control Flow Transfer (CFT) attack. So one of the events which can transgress this property is the CFT attack which becomes the root of the subtree of the code integrity ATA. Until now, the control flow attack instance was only the EMAN2 attack. To mitigate such an attack, it was only required to either check at runtime the locals, pass the BCV or enable a frame integrity check. Such leaf requires to check the underflow of the stack on some instructions. Some of the cards now implement a frame integrity that disallows to arbitrary write into the frame. One can remark that the Frame Integrity detection mechanism covers both EMAN2 and Faugeron's attack, while the Check of Local Variables covers only the EMAN2.

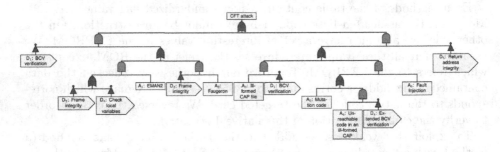

Fig. 1. Attack Tree

To succeed, detection event and mitigation event must be inhibited with a not gate. In this figure a **nand** gate plays this role. The CFT attack represented in Figure 1 will succeed if the adequate ill formed CAP is loaded and no integrity check or no local variable check are present on the card and the BCV is bypassed. When the event is detected, then the card is muted and the attack is stopped. We use this methodology to provide a clear overview on how different events can be combined to set up attacks that can break the integrity of the code. We do not pay attention here on the valuation of the effort of the attacker but on the efficiency of a counter measure. The minimal cut of an ATA defines the minimal sets of basic events determining an attack scenario. Closer to the root is the detection event or the mitigation event better is the coverage.

4   Conclusion and Future Works

We have presented here a set of attacks concerning the smart card world an in particular the Java Card domain. The abality to download application from an untrusted environment open the possibility to characterize the content of the smart card. In particular it allows the attacker to recover code from application (EEPROM) or from the system (ROM) but also to recover some of the data that do not belong him. integrity and confidentiality can be broken just using the techniques used in main stream IT programming. We proposed a methodology based on attack trees to model the knowledge of the attacker. By defining a minimal cut in such a tree, we define the scenario that could lead to the attack. Such a tree can also be used as a defensive means by defining close to the root the adequate counter measure. This optimize the coverage and thus the efficiency of the defense.

References

1. Aumller, C., Bier, P., Fischer, W., Hofreiter, P., Seifert, J.P.: Fault Attacks on RSA with CRT: Concrete Results and Practical Countermeasures. In: Kaliski Jr., B.S., Koç, Ç.K., Paar, C. (eds.) CHES 2002. LNCS, vol. 2523, pp. 260–275. Springer, Heidelberg (2003)
2. Bouffard, G.: A Generic Approach for Protecting Java Card Smart Card Against Software Attacks. Ph.D. thesis, University of Limoges, 123 Avenue Albert Thomas, 87060 LIMOGES CEDEX (October 2014)
3. Bouffard, G., Iguchi-Cartigny, J., Lanet, J.-L.: Combined Software and Hardware Attacks on the Java Card Control Flow. In: Prouff, E. (ed.) CARDIS 2011. LNCS, vol. 7079, pp. 283–296. Springer, Heidelberg (2011)
4. Bouffard, G., Lanet, J.L.: Reversing the operating system of a java based smart card. Journal of Computer Virology and Hacking Techniques 10(4), 239–253 (2014), http://dx.doi.org/10.1007/s11416-014-0218-7
5. Card, J.: 2.1. 1 virtual machine specification. SUN Microsystems Inc. (2000)
6. Faugeron, E.: Manipulating the Frame Information With an Underflow Attack. In: Francillon, A., Rohatgi, P. (eds.) CARDIS 2013. LNCS, vol. 8419, pp. 140–151. Springer, Heidelberg (2014)
7. GlobalPlatform: Card Specification. GlobalPlatform Inc., 2.2.1 edn. (January 2011)
8. Hamadouche, S., Bouffard, G., Lanet, J.L., Dorsemaine, B., Nouhant, B., Magloire, A., Reygnaud, A.: Subverting byte code linker service to characterize java card api. In: Seventh Conference on Network and Information Systems Security (SAR-SSI), May 22-25, pp 75–81 (2012)
9. Hemme, L.: A differential fault attack against early rounds of (triple-) DES. In: Joye, M., Quisquater, J.-J. (eds.) CHES 2004. LNCS, vol. 3156, pp. 254–267. Springer, Heidelberg (2004)

10. Hubbers, E., Poll, E.: Transactions and non-atomic API calls in Java Card: specification ambiguity and strange implementation behaviours. Tech. rep., University of Nijmegen (2004)
11. Iguchi-Cartigny, J., Lanet, J.L.: Developing a Trojan applets in a Smart Card. Journal in Computer Virology 6, 343–351 (2010)
12. Oracle: Java Card 3 Platform, Virtual Machine Specification, Classic Edition. No. Version 3.0.4, Oracle, Oracle America, Inc., 500 Oracle Parkway, Redwood City, CA 94065 (2011)
13. Piret, G., Quisquater, J.-J.: A differential fault attack technique against SPN structures, with application to the AES and KHAZAD. In: Walter, C.D., Koç, Ç.K., Paar, C. (eds.) CHES 2003. LNCS, vol. 2779, pp. 77–88. Springer, Heidelberg (2003)
14. Razafindralambo, T., Bouffard, G., Lanet, J.-L.: A friendly framework for hidding fault enabled virus for Java based smartcard. In: Cuppens-Boulahia, N., Cuppens, F., Garcia-Alfaro, J. (eds.) DBSec 2012. LNCS, vol. 7371, pp. 122–128. Springer, Heidelberg (2012)
15. Schneier, B.: Attack trees: Modeling security threat. Dr. Dobbs Journal (1999)

Extended Abstract: Codes as Modules over Skew Polynomial Rings

Felix Ulmer[✉]

IRMAR, CNRS, UMR 6625, Université de Rennes 1,
Université Européenne de Bretagne
felix.ulmer@univ-rennes1.fr

Abstract. This talk is an overview of codes that are defined as modules over skew polynomial rings. These codes can be seen as a generalization of cyclic codes or more generally polynominal codes to a non commutative polynomial ring. Most properties of classical cyclic codes can be generalized to this new setting and self-dual codes can be easily identified. Those rings are no longer unique factorization rings, therefore there are many factors of $X^n - 1$, each generating a "skew cyclic code". In previous works many new codes and new self-dual codes with a better distance than existing codes have been found. Recently cyclic and skew-cyclic codes over rings have been extensively studied in order to obtain codes over subfields (or subrings) under mapping with good properties.

In order to generalize cyclic codes (or more generally polynomial codes) we use a well known construction of a non commutative polynomial ring. Starting from the finite Ring A and an automorphism θ of A, we define a ring structure on the set

$$A[X; \theta] = \{a_n X^n + \ldots + a_1 X + a_0 \mid a_i \in A \text{ and } n \in \mathbf{N}\}.$$

The addition in $A[X; \theta]$ is defined to be the usual addition of polynomials and the multiplication is defined by the basic rule $X \cdot a = \theta(a)\, X$ ($a \in A$) and extended to all elements of $A[X; \theta]$ by associativity and distributivity. With this two operations $A[X; \theta]$ is a ring known as **skew polynomial ring** or Ore ring. If the leading coefficient of $g \in A[X; \theta]$ is invertible, then for any $f \in A[X; \theta]$ there exists a unique decomposition $f = qg + r$.

Definition 1. *[2–4] Let A be a ring, θ an automorphism of A and $f \in A[X; \theta]$ be of degree n. A* principal module θ-code $\mathcal{C}$ *is a left $A[X; \theta]$-submodule*

$$A[X; \theta]g/A[X; \theta]f \subset A[X; \theta]/A[X; \theta]f$$

in the basis $1, X, \ldots, X^{n-1}$ where g is a monic right divisor of f in $A[X; \theta]$. The length of the code is $n = \deg(f)$ and its dimension is $k = \deg(f) - \deg(g)$, we say that the code $\mathcal{C}$ is of type $[n, k]$. If the minimal Hamming distance of the code is d, then we say that the code $\mathcal{C}$ is of type $[n, k, d]_A$. We denote this code $\mathcal{C} = (g)_{n, \theta}$.

If there exists an $a \in A^*$ such that g divides $X^n - a$ on the right then the code $(g)_{n, \theta}$ is θ-constacyclic. We will denote it $(g)_{n, \theta}^a$. If $a = 1$, the code is θ-cyclic and if $a = -1$, it is θ-negacyclic.

© Springer International Publishing Switzerland 2015
S. El Hajji et al. (Eds.): C2SI 2015, LNCS 9084, pp. 83–86, 2015.
DOI: 10.1007/978-3-319-18681-8_7

Note that a submodule $A[X;\theta]g/A[X;\theta]f \subset A[X;\theta]/A[X;\theta]f$ where g is not monic will in general not be a free $A[X;\theta]$-module.

For a principal module θ-*constacyclic* of length n over a ring A generated by a right divisor $(g)_{n,\theta^a}$ of $X^n - a \in A[X;\theta]$, we have

$$(c_0,\ldots,c_{n-1}) \in (g)_n^{\theta,a} \Rightarrow (a \cdot \theta(c_{n-1}),\theta(c_0),\ldots,\theta(c_{n-2})) \in (g)_n^{\theta,a}.$$

When θ is the identity and $a = 1$ we obtain the classical cyclic codes, showing that principal module θ-cyclic codes are a natural generalization of cyclic codes. However, if θ is not the identity or if A is not a domain, then the ring $A[X;\theta]$ is not a unique factorization domain, leading to many right divisors of $X^n \pm 1$.

Example 1. The field $A = \mathbf{F}_{5^2}$ of order 25 has two automorphisms: the identity and the frobenius automorphisms $\sigma : y \mapsto y^5$. This leads to two skew polynomial rings, the standard commutative polynomial ring $\mathbf{F}_{5^2}[X;\mathrm{id}] = \mathbf{F}_{5^2}[X]$ and the non commutative skew polynomial ring $\mathbf{F}_{5^2}[X;\theta]$.

Table 1. Number of factors of $X^n + 1 \in \mathbf{F}_{5^2}[X;\theta]$ of degree $n/2$

	$n = 2$	$n = 4$	$n = 6$	$n = 8$
$\theta = \mathrm{id}$	2	6	20	6
$\theta = \sigma$	6	38	156	678

Example 2. The automorphism group of the ring $A = \mathbf{F}_5[x]/(x^2)$ of order 25 is isomorphic to the cyclic group C_4 of order 4 generated by γ. This leads to 4 skew polynomial rings, one of which is the standard commutative polynomial ring corresponding to the identity.

Table 2. Number of factors of $X^n + 1 \in \mathbf{F}_5[Y]/(Y^2)[X;\theta]$ of degree $n/2$

	$n = 2$	$n = 4$	$n = 6$	$n = 8$
$\theta = \mathrm{id}$	2	2	4	2
$\theta = \gamma^2$	10	2	500	2
$\theta = \gamma$	2	50	4	2
$\theta = \gamma^3$	2	50	4	2

Definition 2. *(cf. [5]) Let A be a commutative ring. The* **skew reciprocal polynomial** *of $h = \sum_{i=0}^m h_i X^i \in A[X;\theta]$ of degree m is*

$$h^* = \sum_{i=0}^m X^{m-i} \cdot h_i = \sum_{i=0}^m \theta^i(h_{m-i}) X^i.$$

The **left monic skew reciprocal polynomial** *of h is $h^\natural := (1/\theta^m(h_0)) \cdot h^*$.*

When θ is the identity we obtain again the classical reciprocal polynomial. Since θ is an automorphism, the map $*: A[X;\theta] \rightarrow A[X;\theta]$ given by $h \mapsto h^*$ is a bijection. In particular for any $g \in A[X;\theta]$ there exists a unique $h \in A[X;\theta]$ such that $g = h^*$ and, if g is monic, such that $g = h^\natural$.

Corollary 1. *(cf. [5]) Let A be a commutative ring. A module θ-code $(g)_{2k}^\theta$ with $g \in A[X;\theta]$ of degree k is self-dual if and only if there exists $h \in A[X;\theta]$ such that $g = h^\natural$ and $h^\natural h = X^{2k} - \varepsilon$ with $\varepsilon \in \{-1,1\}$.*

Table 3. Number of generators of self dual codes $g \in \mathbf{F}_{5^2}[X;\theta]$ of degree $\frac{n}{2}$

	$n = 2$	$n = 4$	$n = 6$	$n = 8$
$\theta = \mathrm{id}$	2	4	8	4
$\theta = \sigma$	2	8	12	28

Example 3. For $\mathbf{F}_{25} = \mathbf{F}_5(\alpha)$ where $\alpha^2 + 4\alpha + 2 = 0$ the polynomial $X^4 + \alpha^9 X^3 + \alpha^2 X^2 + \alpha X + \alpha^1 6 \in \mathbf{F}_{25}[X,\theta]$ is a right factor of $X^8 + 1 \in \mathbf{F}_{25}[X,\theta]$ and generates a self-dual code C over $\mathbf{F}_{25}$. For the $\mathbf{F}_5$-basis (α^5, α^7) of $\mathbf{F}_{5^2}$, the mapping $\Phi: (\mathbf{F}_{25})^n \rightarrow (\mathbf{F}_5)^{2n}$ given by

$$(a_0\alpha^5 + b_0\alpha^7, \ldots, a_{n-1}\alpha^5 + b_{n-1}\alpha^7) \mapsto (a_0, b_0, \ldots, a_{n-1}, b_{n-1})$$

has the property that a self dual code over $\mathbf{F}_{25}$ is mapped to a self-dual over $\mathbf{F}_5$ (cf. [7]). Under this map the code C is mapped to an optimal self-dual code $\Phi(C)$ over $\mathbf{F}_5$ with minimal distance 7 and whose generating matrix is

$$\begin{pmatrix} 4 & 3 & 2 & 1 & 0 & 1 & 2 & 3 & 1 & 0 & 0 & 0 & 0 & 0 & 0 & 0 \\ 3 & 0 & 1 & 4 & 1 & 2 & 3 & 3 & 0 & 1 & 0 & 0 & 0 & 0 & 0 & 0 \\ 0 & 0 & 0 & 2 & 4 & 4 & 2 & 4 & 3 & 2 & 1 & 0 & 0 & 0 & 0 & 0 \\ 0 & 0 & 2 & 4 & 4 & 2 & 4 & 0 & 2 & 2 & 0 & 1 & 0 & 0 & 0 & 0 \\ 0 & 0 & 0 & 0 & 4 & 3 & 2 & 1 & 0 & 1 & 2 & 3 & 1 & 0 & 0 & 0 \\ 0 & 0 & 0 & 0 & 3 & 0 & 1 & 4 & 1 & 2 & 3 & 3 & 0 & 1 & 0 & 0 \\ 0 & 0 & 0 & 0 & 0 & 0 & 0 & 2 & 4 & 4 & 2 & 4 & 3 & 2 & 1 & 0 \\ 0 & 0 & 0 & 0 & 0 & 0 & 2 & 4 & 4 & 2 & 4 & 0 & 2 & 2 & 0 & 1 \end{pmatrix}$$

The best cyclic code (over the classical commutative polynomial ring) is of minimal distance 4

Example 4. The polynomial $X^2 + 2xX + 3 \in (\mathbf{F}_5[x]/(x^2))[X,\gamma]$ is a right factor of $X^4 + 1 \in (\mathbf{F}_5[x]/(x^2))[X,\gamma]$ and generates a self-dual code C over $\mathbf{F}_5[x]/(x^2)$. For the $\mathbf{F}_5$-basis $(r+2, 1)$ of $\mathbf{F}_5[r]/(r^2)$, the mapping $\Phi: (\mathbf{F}_5[r]/(r^2))^n \rightarrow (\mathbf{F}_5)^{2n}$ given by

$$(a_0(x + 2) + b_0, \ldots, a_{n-1}(x + 2) + b_{n-1}) \mapsto (a_0, b_0, \ldots, a_{n-1}, b_{n-1})$$

Table 4. Number of generators of self dual codes $g \in \mathbf{F}_5[Y]/(Y^2)[X;\theta]$ of degree $\frac{n}{2}$

	$n=2$	$n=4$	$n=6$	$n=8$
$\theta = \mathrm{id}$	2	2	4	2
$\theta = \gamma^2$	2	2	20	2
$\theta = \gamma$	2	6	4	2
$\theta = \gamma^3$	2	6	4	2

has the property that a self dual code over $\mathbf{F}_{25}$ is mapped to a self-dual over $\mathbf{F}_5$ (cf. [7]). Under this map the code C is mapped to an optimal self-dual code $\Phi(C)$ over $\mathbf{F}_5$ with minimal distance 4 and whose generating matrix is

$$\begin{pmatrix} 3\,0\,4\,2\,1\,0\,0\,0 \\ 0\,3\,2\,1\,0\,1\,0\,0 \\ 0\,0\,3\,0\,3\,4\,1\,0 \\ 0\,0\,0\,3\,4\,2\,0\,1 \end{pmatrix}$$

The best cyclic code (over the classical commutative polynomial ring) is of minimal distance 2

Recently mapping of skew cyclic codes have received some attention [2, 1, 6, 8].

References

1. Abualrub, T., Aydin, N., Seneviratne, P.: On R-cyclic codes over $\mathbf{F}_2 + v\mathbf{F}_2$. Australas. J. Combin. 54, 115–126 (2012)
2. Boucher, D., Solé, P., Ulmer, F.: Skew constacyclic codes over galois rings. Advances in Mathematics of Communications 2, 273–292 (2008)
3. Boucher, D., Ulmer, F.: Codes as modules over skew polynomial rings. In: Parker, M.G. (ed.) Cryptography and Coding 2009. LNCS, vol. 5921, pp. 38–55. Springer, Heidelberg (2009)
4. Boucher, D., Ulmer, F.: A note on the dual codes of module skew codes. In: Chen, L. (ed.) IMACC 2011. LNCS, vol. 7089, pp. 230–243. Springer, Heidelberg (2011)
5. Boucher, D., Ulmer, F.: Self-dual skew codes and factorization of skew polynomials. Journal of Symbolic Computation 60, 47–61 (2014)
6. Bhaintwal, M.: Skew quasi-cyclic codes over Galois rings, Des. Codes Cryptogr. 62, 85–101 (2012)
7. Szabo, S., Ulmer, F.: Dualilty Preserving Gray Maps (Pseudo) Self-dual Bases and Symmetric Base (preprint) (2015)
8. Yildiz, B., Karadeniz, S.: Linear codes over $\mathbf{F}_2 + u\mathbf{F}_2 + v\mathbf{F}_2 + uv\mathbf{F}_2$. Des. Codes Cryptogr. 54(1), 61–81 (2010)

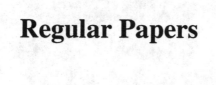

Regular Papers

CUBE Cipher: A Family of Quasi-Involutive Block Ciphers Easy to Mask

Thierry P. Berger[1], Julien Francq[2], and Marine Minier[3($\boxtimes$)]

[1] XLIM (UMR CNRS 7252), Université de Limoges, 123 avenue A. Thomas,
87060 Limoges Cedex, France
thierry.berger@xlim.fr
[2] Airbus Defence and Space - CyberSecurity, 1 Bd Jean Moulin, CS 40001,
MetaPole, 78996 Elancourt Cedex, France
julien.francq@cassidian.com
[3] Université de Lyon, INRIA - INSA-Lyon, CITI, F-69621, Villeurbanne, France
marine.minier@insa-lyon.fr

Abstract. This paper proposes a new quasi-involutive lightweight design called CUBE cipher family. The design has been carefully chosen to be easily masked. The basic building block is a cube of size $n \times n \times n$ on which are applied SPN transformations followed by a cube mapping.

We analyze the proposals from a security point of view and provide a full hardware implementation analysis.

Keywords: Involutive lightweight block cipher · Boolean masking · Design

Introduction

During the last decade, a part of the symmetric cryptographic community has focused its efforts on designing new lightweight primitives to fit with the hardware requirements of RFID tags. Among those primitives, we could cite some lightweight block ciphers: PRESENT [6], LED [15] or PRINCE [7] that are SPNs and TWINE [23], LBlock [24], SIMON [1] or *Piccolo* [22] that are Feistel-based constructions. More recently, some researchers try to add to these requirements one more constraint leading to build lightweight block ciphers that are by design easy to mask. In this last category, we could cite PICARO [20], Zorro [12] or Fantomas and Robin [13].

The aim of this paper is to bring grist to the mill in this research direction. Thus, we present a new family of lightweight block cipher called CUBE that is easy to mask. Moreover, the proposed family is built on a cube representation and is quasi-involutive to limit the hardware footprint required for encryption and decryption processes. The family is an SPN based framework where each component is involutive. Moreover, the presented family is generic in the sense that several possible plaintext sizes are proposed and that two particular cube mappings are investigated.

This work was partially supported by the French National Agency of Research: ANR-11-INS-011.

© Springer International Publishing Switzerland 2015
S. El Hajji et al. (Eds.): C2SI 2015, LNCS 9084, pp. 89–105, 2015.
DOI: 10.1007/978-3-319-18681-8_8

This paper is organized as follows: Sect. 1 describes the specifications of CUBE cipher families. Sect. 2 explains our design choices for those ciphers. In Sect. 3, we discuss their security while Sect. 4 deals with the hardware implementation results. Sect. 5 concludes this paper.

1   Specifications

The CUBE block cipher family is mainly a 3D view of a bit string of length n^3 dedicated to hardware applications. We will instantiate two different variants of the CUBE cipher family (called CUBEAES and CUBE) for the different n value $n = 4$, $n = 5$ and $n = 6$ focusing more particularly on the case $n = 4$ whereas the other instantiations are given in App. A.

The plaintext is seen as a CUBE of size $n \times n \times n$ as shown on Fig. 1. The CUBE is fulfilled beginning with its least significant bit at position $(0, 0, 0)$ according the reference (X, Y, Z), then the bits are written plane by plane (the first one is $(0, Y, Z)$) until the most significant bit fills the position $(n - 1, n - 1, n - 1)$.

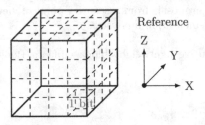

Fig. 1. Block representation

Then the CUBE block cipher family iterates the following round function on r rounds (the r value of course depends on n). The i-th round function is composed of the following quasi-involutive operations:

- **KeyAdd**: A subkey addition with the subkey K_i.
- **SbLayer**: A layer of involutive S-boxes that applies $n \times n$ a single involutive S-box on input/output of size n bits in the direction indicated in the left part of Fig. 2. We choose particular involutive S-boxes that are easy to mask as explained in Section 2.
- **MDSLayer**: On each plane $(0, Y, Z)$, $(1, Y, Z)$, $(2, Y, Z)$ and $(3, Y, Z)$, apply a quasi-involutive Feistel-MDS transformation on n words of size n bits as shown on the right part of Fig. 2. A quasi-involutive Feistel-MDS transformation, as introduced in [21], is a linear transformation that after n iterations gives an MDS code.
- **Permutation**: we define two different permutations to apply to the cube for the two families of block ciphers.

- For the CUBEAES family, `PermAES` rotates by 90° the reference (X, Y, Z) as shown on Fig. 3.
- For the CUBE family, `Perm` rotates the axes (X, Y, Z) as (Z, X, Y) as shown on Fig. 4.

A last `KeyAdd` operation with the subkey K_r is added at the end of the r rounds.

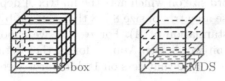

Fig. 2. The `SbLayer` on the left and the `MDSLayer` on the right

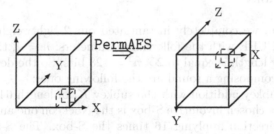

Fig. 3. The `PermAES` transformation

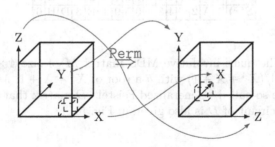

Fig. 4. The `Perm` transformation

1.1 Key Schedule

We define two possible key sizes for the master key K: n^3 bits or $2 \times n^3$ bits.

For a key of length n^3, the subkeys are computed as $K_0 = K$ and $K_{i+1} = K_i A \oplus (i + 1)$ for $i = 0, \cdots, (r - 1)$ where A is an invertible matrix of linear diffusion using a Feistel structure that will be detailed latter. The counter $(i+1)$ is added to the least significant bits.

For a key of length $2 \times n^3$, the subkeys are computed as $K = K_1 \| K_0$ and $K_1 \leftarrow K_1 \oplus 1$ where $\|$ denotes the concatenation and $K_{i+2} \leftarrow K_{i+1}A \oplus K_i \oplus (i+2)$ for $i = 0, \cdots, (r-2)$ where A is always an invertible matrix of linear diffusion. The counter $(i+2)$ is added to the least significant bits.

The size and the word size on which acts the matrix A depend on the value of n. For $n = 4$, we choose a matrix of size 8×8 that acts on bytes (see Subsection 1.2 for a complete instantiation of A). For $n = 5$, we choose a matrix of size 5×5 that acts on 25-bit words (see App. A for further details). For $n = 6$, we choose a matrix of size 12×12 that acts on 18-bit words (see App. A for further details).

1.2   Instantiations

In this subsection, we completely instantiated our 2 lightweight block ciphers CUBEAES and CUBE with the following parameters: $n = 4$, the cube is a 64-bit block, the key length is equal to $2 \times n^3 = 128$ bits. So the details of the four transformations composing a round are the following ones:

• KeyAdd: A subkey addition with the subkey K_i of length 64 bits.

• SbLayer: The chosen involutive S-box is the Noekeon one and acts at nibble level in the X direction applying 16 times the S-box. The S-box is given in Table 1.

Table 1. S-box in hexadecimal notation

x	0	1	2	3	4	5	6	7	8	9	A	B	C	D	E	F
$S(x)$	7	A	2	C	4	8	F	0	5	9	1	E	3	D	B	6

• MDSLayer: The quasi-involutive MDS matrix M of size 4×4 acts on the Field $\mathbb{F}_{16} = \mathbb{F}_2[X]/(X^4 + X + 1)$ with a a root of $X^4 + X + 1$. M is obtained as 4 iterations of the so-called "Generalized Feistel" D matrix that acts on nibbles (see Fig. 5). The circuit of D is also given in Fig. 5.

$$D = \begin{pmatrix} 0 & a^{13} & 1 & 0 \\ 1 & a & 0 & 1 \\ 1 & 0 & 0 & 0 \\ 0 & 1 & 0 & 0 \end{pmatrix} \text{ and } M = D^4$$

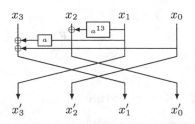

Fig. 5. On the left, the D matrix. On the right, the scheme of the D matrix. $x_3, \cdots, x_0$ and $x'_3, \cdots, x'_0$ are nibbles.

In this case, the multiplications by a and a^{13} are given by the following binary matrices: $M_a = \begin{pmatrix} 0 & 1 & 0 & 0 \\ 0 & 0 & 1 & 0 \\ 0 & 0 & 0 & 1 \\ 1 & 1 & 0 & 0 \end{pmatrix}$ and $M_{a^{13}} = \begin{pmatrix} 1 & 0 & 1 & 1 \\ 1 & 0 & 0 & 1 \\ 1 & 0 & 0 & 0 \\ 0 & 1 & 0 & 0 \end{pmatrix}$. If $y = (y_0, y_1, y_2, y_3)^T$ and $x = (x_0, x_1, x_2, x_3)^T$ are the binary representations of two nibbles then the binary matrix multiplications are $y = M_a x$ and $y = M_{a^{13}} x$.

• **Permutation:** CUBEAES uses the permutation **PermAES**. CUBE uses the **Perm** permutation.

The number of rounds is equal for the two instances to 15. Those 15 rounds are followed by a final key addition with the 64-bit subkey K_{15}.

The key schedule algorithm derives 16 subkeys $K_0, \cdots, K_{15}$ of 64 bits from the master key K of length 128 bits. The key schedule works as described in Subsection 1.1 for a key of length $2 \times n^3 = 128$ bits here. The 8×8 matrix A acts at byte level and is built using the matrix B given with its scheme in Fig. 6 and using the relation $A = B^3$.

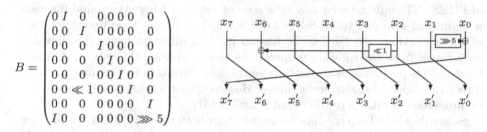

$$B = \begin{pmatrix} 0 & I & 0 & 0 & 0 & 0 & 0 & 0 \\ 0 & 0 & I & 0 & 0 & 0 & 0 & 0 \\ 0 & 0 & 0 & I & 0 & 0 & 0 & 0 \\ 0 & 0 & 0 & 0 & I & 0 & 0 & 0 \\ 0 & 0 & 0 & 0 & 0 & I & 0 & 0 \\ 0 & 0 & \lll 1 & 0 & 0 & 0 & I & 0 \\ 0 & 0 & 0 & 0 & 0 & 0 & 0 & I \\ I & 0 & 0 & 0 & 0 & 0 & 0 & \ggg 5 \end{pmatrix}$$

Fig. 6. On the left, the B matrix. On the right, the scheme of the B matrix. $x_7, \cdots, x_0$ and $x'_7, \cdots, x'_0$ are bytes.

The parameters for the other possible instantiations with $n = 5$ and $n = 6$ are given in App. A.

2 Design Rationale

Cube Structure. In the design of an SPN block cipher, non-linear and linear layers are successively applied to the current state. In a lightweight block cipher, the size of the state is generally 64 bits, while in a classical block cipher it is at least 128 bits.

For efficiency reasons, it is not possible to apply a single non-linear transformation simultaneously on the whole current state. So this state is usually divided into subblocks, for example in bytes for the AES or in nibbles for many lightweight block ciphers. So an S-box is applied to each byte or each nibble. The linear layer must mix the subblocks together in order to diffuse the non-linearity between the subblocks. One of the most efficient way to optimize the linear layer is to use an MDS matrix on subblocks. The MDS property ensures the optimality of linear and differential branch numbers, i.e. maximize the number of active S-boxes per round.

However, even if it is easy to construct some MDS matrices with some given parameters, its use is costly in terms of implementation. Indeed, it is linear on the size of subblocks, but quadratic on the number of subblocks. A classical tradeoff consists in choosing a number of subblocks equal to the size of subblocks or half of this size.

In the context of lightweight block ciphers, this constraint becomes more accurate if we use nibbles as inputs of S-boxes. In this case, the size of the MDS matrix is limited to 8 subblocks, so the size of the whole state is limited to 16 nibbles, and due to the bound given in [16], it is not possible to find an MDS code so long. Indeed, the MDS conjecture is the fact that, except for some trivial cases, there is no MDS codes of length greater than $q + 1$ defined over an alphabet of size q. In practice, this implies that there is no MDS diffusion matrix defined over nibbles of size strictly greater than 8.

The idea to use CUBE structure is not new and has been used in KECCAK [2] and also in the lightweight block cipher PRESENT [6]. Indeed, the structure of PRESENT could be seen as a cube seen at bit level where the round function is composed of a call to an S-box layer applied at nibble level and a rotation of the axes. However, PRESENT has no its own diffusion layer, the diffusion only comes from the rotation property. As shown in [8,9], this lake of diffusion layer creates particular statistical properties and "linear hulls" coming from the direct iterations of linear probabilities. However, this weakness is compensated by numerous iterations of the round function (31).

So, we decide to keep the cube structure at bit level used in PRESENT because if the size of the cube is n^3, then, we apply S-boxes on n-bit words and we could find MDS matrices of size $n \times n$ that work on n-bit words leading to apply the classical elementary operations on smaller words and to improve the latency of our proposals. The benefits of such an approach is preserved as long as the rotation/permutation layer is sufficiently well chosen to mix together the n-bit words coming from different planes and to break the internal n-bit word structure using rotation/permutation at bit level. The idea for CUBEAES is to be able to choose a rotation/permutation that will preserve the MDS property.

MDS Diffusion in Cube Structure. If the final rotation/permuta-tion is carefully chosen, the MDS diffusion property between the n-bit words of a plane becomes a diffusion from the n-bit words of a plane to the set of the planes of the next round. Indeed, suppose there is one active n-bit word in a given plane, the MDS property ensures the activation of n n-bit words in this plane. If the rotation sends each n-bit word in a distinct plane, all the planes will be activated after the application of the next S-box layer followed by the MDS diffusion.

This is the design choice we have made for the CUBEAES cipher. As it will be explained in Section 3, the transformation `PermAES` allows to maximize the number of active S-boxes per round as done for the AES case, leading to maximize the differential and the linear branch numbers.

The other proposal CUBE keeps the original cube permutation used in PRESENT. CUBE could be then seen as a proof of concept of a PRESENT like cipher that does not have the "bad PRESENT properties" in terms of "linear hulls" and "statistical saturations" due to the presence of a full diffusion layer induced by the MDS multiplication.

Recursive and Quasi-Involutive MDS Linear Parts Using Feistel Schemes. From an implementation point of view, there are two important linear operations: the MDS transformation applied just after the S-box layer, and the matrix multiplication used in the key schedule to derive the round subkeys from the master key.

We want to keep in mind when designing, two main requirements: a quasi-involutive structure to minimize the cost of the deciphering process and the use of elementary operations such as shifts, word rotations,... to minimize the hardware footprint of the used operations.

As explained in [21,14], MDS diffusion could be performed using an iterative approach and a kind of generalized Feistel scheme with elementary linear internal functions. We decide to use this approach that guarantees, due to the recursive implementation a minimal footprint and also the quasi-involutivity of our scheme due to the use of Feistel networks that are quasi-involutive scheme, i.e. the only part which is not involutive is the final permutation of subblocks of each round, which has a negligible hardware cost when implementing the deciphering process.

The MDS diffusion used in the CUBE cipher family has the same recursive structure that the one of PHOTON [14]. The only difference is that the D matrix is not a companion matrix. In practice, this difference reduces by a little the fan-in of our implementation. It is also up to our knowledge, the first example of recursive implementation of an MDS matrix which is not derived from a companion matrix.

Involutive S-box Suitable for Masking. We decide to use involutive S-boxes to build a complete quasi-involutive cipher. Such a choice leads to make our cipher quasi-involutive, i.e. the implementation cost for the deciphering process is really low.

We use the Noekeon S-box for the case $n = 4$ because this S-box is involutive, has optimal differential and linear probabilities (respectively equal to 2^{-2} and to 2^{-1}), an algebraic degree equal to 3 and a simple implementation circuit. Indeed, it is composed of 7 XORs, 2 ANDs and 2 NORs leading to a very compact hardware implementation. Moreover, this S-box is easy to mask because the masking cost mainly depends on the number of non-linear operations and we have 4 non-linear operations. Indeed, a boolean masking is quadratic in the number of shares for the 4 non linear operations and linear in the number of shares for the 7 linear operations using the method described in [17].

Key Schedule. Contrary to numerous lightweight block ciphers, we want to provide a key schedule which guarantees a good mixing between the key bits to maximize the uncertainty coming from the key at low hardware implementation cost. Our choice focuses on an algorithm which is, up to an XOR of a round counter that prevents slide attacks, linear and involutive. We also want to guarantee that each subkey contains the maximal possible master key entropy. In order to ensure a good diffusion of the master key randomness in any round, we require that this master key can be recovered from any non necessary consecutive pair of round subkeys in the case of a key of length $2 \times n^3$ and from each round subkey in the case of a master key length equal to n^3.

Thus the matrix A defined in Subsection 1.1 follows the previous rules and is an invertible binary matrix. For efficiency reasons, A is calculated by applying several iterations of a matrix B which has a Feistel structure. Moreover, in the different choices of A, we try to privilege matrices that do not act on the same word length than the round function to try to prevent attacks notably in the related or in the chosen key settings that exploit this kind of properties.

3   Security Analysis

We focus our security analysis on the two instantiations given above with $n = 4$, thus a block size of 64 bits and a key length equal to 128 bits.

Differential / Linear Cryptanalysis. Differential and linear cryptanalysis (respectively described in [5] and in [19]) are the most famous attacks on block ciphers. Since their discovery, many works have focused on the ways to prevent them from happening for a given cipher [11]. Usually, designers count the minimal number of active S-boxes crossed all along the ciphering process by differential and linear characteristics denoted here respectively by AS_D and AS_L. From those numbers, we could estimate the induced maximal differential/linear probability depending on the maximal differential/linear probability of the S-box denoted by DP/LP. Here we have $DP = 2^{-2}$ and $LP = 2^{-1}$ because our S-box acts at nibble level.

Moreover, the best differential/linear attack against the cipher has a complexity of about DP^{AS_D} (respectively LP^{AS_L}) operations. Thus, a cipher is supposed to be secure against differential/linear cryptanalysis as soon as $1/(DP^{AS_D})$ (respectively $1/(LP^{AS_L})$) is greater than the codebook. In Table 2, we evaluate the minimal number of active S-boxes up to 10 rounds for our schemes.

Table 2. Minimal number of active S-boxes for every round for CUBEAES and CUBE

Round	1	2	3	4	5	6	7	8	9	10
CUBEAES AS_D	1	5	9	25	26	30	34	50	51	55
AS_L	1	5	9	25	26	30	34	50	51	55

Round	1	2	3	4	5	6	7	8	9	10
CUBE AS_D	1	5	9	13	20	21	25	29	33	40
AS_L	1	5	9	12	19	20	24	28	31	38

As CUBEAES verifies the wide trail strategy conditions, the number of active S-boxes for both linear and differential behaviors reaches the maximal possible bounds (25 active S-boxes for 4 rounds). To estimate the number of active S-boxes for CUBE, we compute the possible differential/linear trails using 2^{40}

plaintexts and deduce the results for 5 rounds. Then, beyond, we estimate the corresponding number of active S-boxes by iterating together the different results. We also test using a branch and bound method that there is no elementary differential/linear paths with a low weight. This particular property does not occur for CUBE. We think that this fact comes from the MDS diffusion layer.

Thus, the best differential/linear cryptanalysis that could be mounted against CUBEAES is on 6 rounds. Beyond this number, the required number of plaintexts is greater than the entire codebook. For CUBE, the best differential/linear cryptanalysis could be mounted on 8 rounds.

Impossible Differential Attack. The impossible differential attack is a structural attack introduced by E. Biham et al. in [4] in 1998. Impossible differential cryptanalysis, contrary to differential cryptanalysis, exploits differences with probability 0 at some intermediate state of a cipher. The idea is to test from well chosen plaintext/ciphertext pairs some keybits and to discard keybits that verify the impossible path. We found the following impossible differential attacks: For CUBEAES, as expected and due to the wide trail strategy, we found that the best impossible differential attack could be mounted on 7 rounds using a 4 rounds impossible differential surrounded by one round at the top and 2 rounds at the end. For CUBE, the best impossible differential attack we found is on 8 rounds and uses a 5 rounds impossible differential.

Integral Attack. Integral cryptanalysis was first introduced against the Square block cipher in [10]. In [18], L. Knudsen and D. Wagner analyze integral cryptanalysis as a dual to differential attacks particularly applicable to block ciphers with bijective components. A first order integral cryptanalysis considers a particular collection of m words in the plaintexts and ciphertexts that differ on a particular word. The aim of this attack is thus to predict the values in the sums (i.e. the integral) of the chosen words after a certain number of rounds. The same authors generalize this approach to d-th order integrals: the original set to consider becomes a set of m^d vectors which differ in d components and where the sum of this set is predictable after a certain number of rounds.

For both ciphers CUBE and CUBEAES, we are able to mount a first order integral property on 3 rounds saturating one nibble and to mount 4th order integral property on 4 rounds saturating a plane (i.e. 4 nibbles). We think that for CUBEAES this 4th order integral property could not be extended at the beginning due to the MDS property whereas we conjecture that this 4th order integral could be extended by one round at the beginning in the case of CUBE, but we could not test it due to the huge induced complexity.

Against CUBEAES, the 4 rounds property could be extended by 2 rounds at the end guessing 80 subkey bits leading to attack 6 rounds with an overall complexity of about 2^{75} encryptions. For CUBE, the 5 rounds property could be extended by the same number of rounds with about the same complexity leading to an attack on 7 rounds.

Related Key and Chosen Key Attacks. The related key attacks introduced by E. Biham in [3] in 1993 allow an attacker to know some relations between different keys without knowing the keys themselves and to cipher under those keys some plaintexts. From those pairs of plaintext/ciphertext, the aim of the attacker is to recover the key. In the related key settings, we first evaluate the related key pairs that activate the lowest number of S-boxes in the key schedule.

In the case of the CUBE and the CUBEAES cipher families, the best related key attack allows to gain 2 rounds at the beginning of a classical differential attack considering a master key pair with a single bit difference placed somewhere on K_0. Under those conditions, the difference coming from the subkey addition with K_0 could be canceled using well chosen plaintext pairs, then the subkey addition with K_1 is for free and K_2 adds only one bit difference to state pairs without difference. Thus, a classical differential attack could be extended by three rounds at the beginning using a related key attack.

However, to try to improve the number of rounds gained using a related or a chosen key attack, we have implemented a branch and bound algorithm that tries to cancel the differences coming from the subkeys using differences coming from the internal state. We do not find a simple way to cancel these differences. This is mainly due to the fact that the ciphering process and the key schedule does not act at the same word size.

Resistance to Side Channel Analysis. As said before, the S-box has been chosen to offer resistance to side channel analysis at a reasonable cost. Indeed and using the algorithm proposed in [17], as the cost for boolean masking is quadratic when considering non linear operations and is linear for the XOR operation, we have chosen an S-box that is easy to mask when considering 3 shares in the algorithm proposed in [17]. Thus, our block ciphers are resistant to side channel analysis at low cost.

In summary, we conjecture that there is no attack against 8 rounds of CUBEAES and against 9 rounds of CUBE in the single key settings more efficient than the exhaustive key search leaving respectively 7 and 6 rounds of security margin. We also conjecture that there is no attack against 11 rounds of CUBEAES and against 12 rounds of CUBE in the related, known and chosen key settings more efficient than the exhaustive key search leaving respectively 4 and 3 rounds of security margin.

4   Implementation Aspects

In this Section, we sum up our implementation results concerning CUBE cipher with $n = 4$ and we compare these results with the implementation of other lightweight block ciphers using three criteria: area, power consumption and throughput (or latency). Note that we have only implemented CUBE cipher because the cost for implementing CUBEAES is about the same (the two ciphers only differ in the Permutation layer which consists in exchanging bit positions).

The majority of lightweight ciphers has been optimized in priority with chip area minimization in mind. This metric can be expressed in μm^2 but this value is dependent of the used technology and standard cells technology. To ease comparisons between implementations, the circuit area is measured in Gate Equivalences (GEs). A GE is the area of a 2-input NAND gate in the used standard cell technology. So the area of the circuit expressed in GEs is the surface of the circuit in μm^2 divided by the surface of a NAND gate.

Compared to other SPN ciphers like PRESENT, the quasi-involutive structure of CUBE cipher helps to implement both encryption and decryption modes[1] with a reasonable overhead. For example, we can save the implementation of Inverse S-boxes which can represent hundreds of GEs.

4.1 Theoretical Implementation Results

Basic Components of CUBE Cipher. To implement the confusion effect, CUBE cipher uses a 4×4 bit S-box which is by far smaller than 8×8 S-boxes and 6×4 ones. We have chosen the S-box of NOEKEON because it is compact (around 20 GEs) and it has been shown that it is relatively easy to mask with three shares at a reasonable cost.

The diffusion effect is made of a bit permutation (which costs no GE) and the MDSlayer which is area optimized: it only costs 16 2-input XORs per layer, so $16 \times 4 = 64$ XORs = 144 GEs.

The memory elements (flipflops) used to store the round keys and the cipher state are the most costly hardware elements to implement, and are those which consumes the most energy. We need $64 \times 5.75 = 368$ to store the 64-bit cipher. For storing the 128-bit key state, we have decided to implement one additional 64 bits register to have an involutive structure, i.e. $(128+64) \times 5.75 = 1104$ GEs. It requires also 64 2-input XORs to compute round key K_{i+2} from K_{i+1} and K_i, so $64 \times 2.25 = 144$ GEs.

For the Key Schedule, we have chosen to make it simple, but not as simple as KTANTAN or PRINTcipher to protect it against related key and slide attacks. To implement such kind of secure Key Schedule, we chose a simple Feistel structure iterated 3 times per round key that computes only rotations, shifts and 16 2-input XORs per round, i.e. $16 \times 3 = 48$ XORs $= 48 \times 2.25 = 108$ GEs.

Theoretical Implementation Results. We give hereafter the implementation results of a round-wise implementation of CUBE cipher. It processes 64 bits of plaintext with a 128-bit key in 25 clock cycles. Round keys are computed on-the-fly, in parallel of the cipher state processing. There is no resource sharing between the cipher state and the Key Schedule processes. The S-box is implemented in a Look-Up-Table way, so we let the compiler do its own optimizations. Only the encryption process is implemented.

Theoretically, our CUBE cipher implementation needs $1104+368 = 1472$ GEs to store both the round keys and the cipher state, $16 \times 20 = 320$ GEs for the S-box, $(64+64) \times 2.25 = 288$ GEs for all the XORs, 144 GEs for the MDS, 48 GEs for the XORs in the Feistel of the Key Schedule, 192 2-to-1 multiplexors $192 \times 2 = 384$ GEs, which selects between the encryption key (resp. the plaintext) or the

---

[1] Implementing both encryption and decryption modes is mandatory where 3-pass mutual authentication is required. ISO 9798-2 ("Entity Authentication – Mechanisms using Symmetric Encipherment Algorithms") specifies such protocols.

round key (the cipher state)). So, we can estimate (neglecting the implementation cost of the finite state machine) that our round-wise implementation of CUBE cipher needs at least $1472 + 320 + 288 + 144 + 48 + 384 = 2656$ GEs in total.

4.2 Implementation Results and Comparisons

We implemented CUBE cipher in VHDL and synthesized it using a Low-Power (LP) High Vt 65 nm standard-cell library. We used Synopsis Design Vision D-2010.03-SP5-2 for synthesis and power simulation. The foundry typical values (of 1.2 V for the core voltage and 25° for the temperature) were used. Non-scan Flip-Flops are used. We applied priority optimizations on area. Our round-wise low-power CUBE cipher occupies 2536 GEs and has a simulated power of 0.663 μW. A comparison with other ciphers follows in Table 3. We have only listed in this table block ciphers with 64-bit state and 128-bit key. When possible, we give results for circuits which makes both encryption and decryption (e.g., TWINE). The throughputs are given with a clock frequency equal to 100kHz.

Table 3. Comparison with other Lightweight Block Ciphers

	Key Size	Block Size	Lat. (cycles)	Area (GEs)	Logic Process
mCrypton	128	64	13	4108	0.13μm (theo.)
HIGHT	128	64	34	3048	0.25μm
TWINE-128	128	64	36	2285	90nm
Piccolo-128	128	64	27	1938	0.13μm (theo.)
PRESENT-128	128	64	32	1886	0.18μm (only enc.)
CUBE Cipher	128	64	25	2536	65 nm LP

Comparisons are usually difficult to make between implementations made on different technologies and with different experimental conditions, but we will however give hereafter some discussion elements.

Number of Rounds. First, if we look at the number of rounds needed to process one 64-bit block, CUBE cipher is faster than almost all its competitors (only mCrypton is faster, but is bigger). It means that implementing CUBE cipher is advantageous in terms of latency and energy.

Power Comparison. mCrypton, HIGHT, TWINE, and Piccolo did not give any result concerning average power consumption of their designs. The only authors which give such kind of results are the inventors of PRESENT (3.3 μW). CUBE cipher has a simulated power of 0.663 μW, so the gap between both propositions is important but it can be explained by the use of a low-power logic process.

Area Comparison. Compared to HIGHT, CUBE cipher is smaller and faster. Moreover, HIGHT is susceptible to Meet-in-the-Middle Attacks.

Compared to TWINE, the authors used Scan Flip-flops which allows us to save 1 GE per 1-bit storage. In our ASIC library, a Flip-Flop and 2-to-1 multiplexor cost 5.75 GEs and 2 GEs, and a Scan FF costs 6 GEs: hence the use of Scan flip-flops saves 1.75 GEs per 1-bit storage. So, if we have used scan flip-flops in our CUBE cipher implementation, we would hope save in total: $(128 + 64) \times 1.75 = 336$ GEs. So, the area of CUBE cipher will become: $2536 - 336 = 2200$ GEs. Thus, CUBE cipher will be smaller (85 GEs).

CUBE cipher does not compete on equal terms with Piccolo-128 mainly due to the very light nature of its lightweight Key Schedule (it needs only 32-bit wide 3-to-1 multiplexor to select the appropriate round key). We wanted for CUBE cipher a stronger Key Schedule to be more secure against *e.g.* Meet-in-the-Middle attacks. Moreover, the authors use scan flip-flops, so CUBE cipher will be a little bit bigger in that case (2200 GEs *vs.* 1938). The authors of Piccolo also infer AND-NOR gates to optimize XOR/XNOR gate count. They estimate that it allows us to save 0.25 GE per XOR gate. In CUBE cipher, in Key Schedule process, we have 48 XOR gates inferred in the Feistels and 64 for computing K_{i+2} from K_{i+1} and K_i. Moreover, in the round execution, we must implement 64 XORs for the AddRoundKey operation and 64 others for the MDS layer computation. So, there are: $48 + 64 + 64 + 64 = 240$ XORs in the circuit. So using the same optimization than the authors of Piccolo, we can save: $240 \times 0.25 = 60$ GEs. So, the area of CUBE cipher will become: $2200 - 60 = 2140$ GEs. So, the gap in terms of GEs between the two block ciphers should be reduced to around 200 GEs, while CUBE cipher is faster and has a more secure Key Schedule.

Compared to PRESENT-128, we have an unfavorable gap of 650 GEs. But, the results we have only concerns encryption. If we consider an implementation with both encryption/decryption purposes, it would be needed to implement in PRESENT both true and inverse S-boxes (the overhead is then equal to $16 \times 28 = 448$ GEs if we consider that the gate count of the S-box is the same than its inverse) and select their output by a multiplexer ($64 \times 2 = 128$ GEs). So, a PRESENT-128 with encryption/decryption modes would cost: $1886 + 448 + 128 = 2462$ GEs, so the gap between the two proposals is only 74 GEs.

In summary, CUBE cipher compares reasonably well to other lightweight ciphers when both encryption and decryption must be implemented. The price to pay to have a secure Key Schedule and avoid undesirable properties of PRESENT appears to be limited.

5 Conclusion

In this paper, we have presented two involutive families of block ciphers that are easy to mask and have reasonable hardware cost for the instantiations with $n = 4$. One of the main advantages of CUBEAES compared to the AES and of CUBE compared to PRESENT, is their involutive nature that allows a really for near-free implementation of the decryption process. Concerning CUBE, the addition of an MDS layer prevents the bad behaviors of the PRESENT block cipher concerning "linear hulls" and statistical saturations from happening.

In a future work, we plane to derive for those families of ciphers tweakable versions that lead to authentication-encryption schemes.

A   Instantiations with $n = 5$ and $n = 6$

As the Permutation layer will be the same whatever the value of n, we only give here the S-box called in the SbLayer, the matrices M and D of the MDSLayer and the matrices A and B used in the key schedule.

A.1   Instantiation for $n = 5$

The number of rounds r for $n = 5$ is equal to 17. The 5-bit to 5-bit involutive S-box is given in Table 4. For this S-box, we have $DP = 2^{-2.41}$, $LP = 2^{-2}$, an algebraic degree of 4, a non linearity equal to 3.

Table 4. S-box in hexadecimal notation

x	0	1	2	3	4	5	6	7	8	9	A	B	C	D	E	F	10	11	12	13	14	15	16	17	18	19	1A	1B	1C	1D	1E	1F
$S(x)$	1F	1D	1A	1B	12	1E	13	E	F	18	16	C	B	10	7	8	D	19	4	6	15	14	A	1C	9	11	2	3	17	1	5	0

The matrices A and D of the MDSLayer works on the finite field $\mathbb{F}_{32} = \mathbb{F}_2[X]/(X^5 + X^2 + 1)$ with a root of $X^5 + X^2 + 1$.

$$D = \begin{pmatrix} 0 & 1 & 0 & 0 & 0 \\ 0 & 0 & 1 & 0 & 0 \\ 0 & 0 & 0 & 1 & 0 \\ 0 & 0 & 0 & 0 & 1 \\ 1 & a^{30} & a & a & a^{30} \end{pmatrix}$$

with

$$M = D^5. \quad M_a = \begin{pmatrix} 0 & 1 & 0 & 0 & 0 \\ 0 & 0 & 1 & 0 & 0 \\ 0 & 0 & 0 & 1 & 0 \\ 0 & 0 & 0 & 0 & 1 \\ 1 & 0 & 1 & 0 & 0 \end{pmatrix} \quad \text{and} \quad M_{a^{30}} = \begin{pmatrix} 0 & 1 & 0 & 0 & 1 \\ 1 & 0 & 0 & 0 & 0 \\ 0 & 1 & 0 & 0 & 0 \\ 0 & 0 & 1 & 0 & 0 \\ 0 & 0 & 0 & 1 & 0 \end{pmatrix} \quad \text{in binary representation.}$$

The matrices A and B used in the key schedule acts on 5 blocks of 25 bits:

$$B = \begin{pmatrix} 0 & I & 0 & 0 & 0 \\ 0 & \lll 9 & I & 0 & 0 \\ 0 & 0 & 0 & I & 0 \\ \ggg 1 & 0 & 0 & 0 & I \\ I & 0 & 0 & 0 & 0 \end{pmatrix} \quad \text{and} \quad A = B^5.$$

A.2   Instantiation for $n = 6$

The number of rounds r for $n = 6$ is equal to 19. The 6-bit to 6-bit involutive S-box is given in Table 5. For this S-box, we have $DP = 2^{-3.41}$, $LP = 2^{-2.41}$, an algebraic degree of 5, a non linearity equal to 5.

The matrices M and D of the MDSLayer works on the finite field $\mathbb{F}_{64} = \mathbb{F}_2[X]/(X^6 + X^4 + X^3 + X + 1)$ with a root of $X^6 + X^4 + X^3 + X + 1$.

Table 5. S-box in hexadecimal notation

x	0	1	2	3	4	5	6	7	8	9	A	B	C	D	E	F		x	10	11	12	13	14	15	16	17	18	19	1A	1B	1C	1D	1E	1F
$S(x)$	17	13	35	A	C	26	B	23	1C	31	3	6	4	3D	3E	20		$S(x)$	16	18	14	1	12	29	10	0	11	2F	25	39	8	33	36	2E
x	20	21	22	23	24	25	26	27	28	29	2A	2B	2C	2D	2E	2F		x	30	31	32	33	34	35	36	37	38	39	3A	3B	3C	3D	3E	3F
$S(x)$	F	3A	37	7	2B	1A	5	38	3B	15	2C	24	2A	3C	1F	19		$S(x)$	32	9	30	1D	3F	2	1E	22	27	1B	21	28	2D	D	E	34

$$D = \begin{pmatrix} 0 & 1 & 0 & 0 & 0 & 0 \\ 0 & 0 & 1 & 0 & 0 & 0 \\ 0 & 0 & 0 & 1 & 0 & 0 \\ 0 & 0 & 0 & 0 & 1 & 0 \\ 0 & 0 & 0 & 0 & 0 & 1 \\ 1 & 1 & a^{61} & a^{49} & a & a^{49} \end{pmatrix} \text{ with } M = D^6, \; M_a = \begin{pmatrix} 0 & 1 & 0 & 0 & 0 & 0 \\ 0 & 0 & 1 & 0 & 0 & 0 \\ 0 & 0 & 0 & 1 & 0 & 0 \\ 0 & 0 & 0 & 0 & 1 & 0 \\ 0 & 0 & 0 & 0 & 0 & 1 \\ 1 & 1 & 0 & 1 & 1 & 0 \end{pmatrix} \; M_{a^{61}} = \begin{pmatrix} 1 & 1 & 0 & 1 & 1 & 1 \\ 1 & 0 & 1 & 1 & 0 & 1 \\ 1 & 0 & 0 & 0 & 0 & 0 \\ 0 & 1 & 0 & 0 & 0 & 0 \\ 0 & 0 & 1 & 0 & 0 & 0 \\ 0 & 0 & 0 & 1 & 0 & 0 \end{pmatrix}$$

and $M_{a^{49}} = \begin{pmatrix} 1 & 0 & 1 & 0 & 0 & 0 \\ 0 & 1 & 0 & 1 & 0 & 0 \\ 0 & 0 & 1 & 0 & 1 & 0 \\ 0 & 0 & 0 & 1 & 0 & 1 \\ 1 & 1 & 0 & 1 & 0 & 0 \\ 0 & 1 & 1 & 0 & 1 & 0 \end{pmatrix}$ in binary representation.

The matrices A and B used in the key schedule acts on 12 blocks of 18 bits:

$$B = \begin{pmatrix} 0 & I & 0 & 0 & 0 & 0 & 0 & 0 & 0 & \lll 2 & 0 & 0 \\ 0 & 0 & I & 0 & 0 & 0 & 0 & 0 & 0 & 0 & 0 & 0 \\ 0 & 0 & 0 & I & 0 & 0 & 0 & 0 & 0 & 0 & 0 & 0 \\ 0 & 0 & 0 & 0 & I & 0 & 0 & 0 & 0 & 0 & 0 & 0 \\ 0 & 0 & 0 & 0 & 0 & I & 0 & 0 & 0 & 0 & 0 & 0 \\ 0 & 0 & 0 & 0 & 0 & 0 & I & 0 & 0 & \lll 13 & 0 & 0 \\ \ggg 8 & 0 & 0 & 0 & 0 & 0 & \ggg 2 & I & 0 & 0 & 0 & 0 \\ 0 & 0 & 0 & 0 & 0 & 0 & 0 & 0 & I & 0 & 0 & 0 \\ 0 & 0 & 0 & 0 & 0 & 0 & \ggg 17 & 0 & 0 & I & 0 & 0 \\ 0 & 0 & 0 & 0 & 0 & 0 & 0 & 0 & 0 & 0 & I & 0 \\ 0 & 0 & 0 & 0 & 0 & 0 & 0 & 0 & 0 & 0 & 0 & I \\ I & 0 & 0 & 0 & 0 & 0 & 0 & \ggg 8 & 0 & 0 & \ggg 10 \end{pmatrix} \text{ and } A = B^9.$$

B Test Vectors

B.1 Test Vectors for CUBEAES and CUBE with $n = 4$

We provide the following test vectors given in little endian and in hexadeximal for CUBEAES and CUBE:

```
CUBEAES:
input_message = 0x6666666666666666
KEY = 0x0102030405060708090a0b0c0d0e0f
ciphertext = 0xee0cb8c023716ec7
CUBE:
input_message = 0x6666666666666666
KEY = 0x0102030405060708090a0b0c0d0e0f
ciphertext = 0x710ec4a98692ac3f
```

References

1. Beaulieu, R., Shors, D., Smith, J., Treatman-Clark, S., Weeks, B., Wingers, L.: The simon and speck families of lightweight block ciphers. Cryptology ePrint Archive, Report 2013/404 (2013), http://eprint.iacr.org/
2. Bertoni, G., Daemen, J., Peeters, M., Van Assche, G.: The keccak sha-3 submission. Submission to NIST (Round 3) (2011)
3. Biham, E.: New types of cryptoanalytic attacks using related keys (extended abstract). In: Helleseth, T. (ed.) EUROCRYPT 1993. LNCS, vol. 765, pp. 398–409. Springer, Heidelberg (1994)
4. Biham, E., Biryukov, A., Shamir, A.: Cryptanalysis of skipjack reduced to 31 rounds using impossible differentials. In: Stern, J. (ed.) EUROCRYPT 1999. LNCS, vol. 1592, pp. 12–23. Springer, Heidelberg (1999)
5. Biham, E., Shamir, A.: Differential cryptanalysis of des-like cryptosystems. In: Menezes, A., Vanstone, S.A. (eds.) CRYPTO 1990. LNCS, vol. 537, pp. 2–21. Springer, Heidelberg (1991)
6. Bogdanov, A.A., Knudsen, L.R., Leander, G., Paar, C., Poschmann, A., Robshaw, M., Seurin, Y., Vikkelsoe, C.: PRESENT: An ultra-lightweight block cipher. In: Paillier, P., Verbauwhede, I. (eds.) CHES 2007. LNCS, vol. 4727, pp. 450–466. Springer, Heidelberg (2007)
7. Borghoff, J., Canteaut, A., Güneysu, T., Kavun, E.B., Knezevic, M., Knudsen, L.R., Leander, G., Nikov, V., Paar, C., Rechberger, C., Rombouts, P., Thomsen, S.S., Yalçın, T.: PRINCE – A Low-Latency Block Cipher for Pervasive Computing Applications - Extended Abstract. In: Wang, X., Sako, K. (eds.) ASIACRYPT 2012. LNCS, vol. 7658, pp. 208–225. Springer, Heidelberg (2012)
8. Collard, B., Standaert, F.-X.: A Statistical Saturation Attack against the Block Cipher PRESENT. In: Fischlin, M. (ed.) CT-RSA 2009. LNCS, vol. 5473, pp. 195–210. Springer, Heidelberg (2009)
9. Collard, B., Standaert, F.-X.: Multi-trail statistical saturation attacks. In: Zhou, J., Yung, M. (eds.) ACNS 2010. LNCS, vol. 6123, pp. 123–138. Springer, Heidelberg (2010)
10. Daemen, J., Knudsen, L.R., Rijmen, V.: The block cipher square. In: Biham, E. (ed.) FSE 1997. LNCS, vol. 1267, pp. 149–165. Springer, Heidelberg (1997)
11. FIPS 197. Advanced Encryption Standard. Federal Information Processing Standards Publication 197, U.S. Department of Commerce/N.I.S.T. (2001)
12. Gérard, B., Grosso, V., Naya-Plasencia, M., Standaert, F.-X.: Block ciphers that are easier to mask: How far can we go? In: Bertoni, G., Coron, J.-S. (eds.) CHES 2013. LNCS, vol. 8086, pp. 383–399. Springer, Heidelberg (2013)
13. Grosso, V., Leurent, G., Standaert, F.-X., Varici, K.: LS-designs: Bitslice encryption for efficient masked software implementations. In: Cid, C., Rechberger, C. (eds.) FSE 2014. LNCS, vol. 8540, pp. 18–37. Springer, Heidelberg (2015)
14. Guo, J., Peyrin, T., Poschmann, A.: The PHOTON Family of Lightweight Hash Functions. In: Rogaway, P. (ed.) CRYPTO 2011. LNCS, vol. 6841, pp. 222–239. Springer, Heidelberg (2011)
15. Guo, J., Peyrin, T., Poschmann, A., Robshaw, M.J.B.: The LED Block Cipher. In: Preneel, B., Takagi, T. (eds.) CHES 2011. LNCS, vol. 6917, pp. 326–341. Springer, Heidelberg (2011)
16. Hirschfeld, J.W.P.: The main conjecture for mds codes. In: Boyd, C. (ed.) Cryptography and Coding 1995. LNCS, vol. 1025, pp. 44–52. Springer, Heidelberg (1995)

17. Ishai, Y., Sahai, A., Wagner, D.: Private circuits: Securing hardware against probing attacks. In: Boneh, D. (ed.) CRYPTO 2003. LNCS, vol. 2729, pp. 463–481. Springer, Heidelberg (2003)
18. Knudsen, L.R., Wagner, D.: Integral cryptanalysis. In: Daemen, J., Rijmen, V. (eds.) FSE 2002. LNCS, vol. 2365, pp. 112–127. Springer, Heidelberg (2002)
19. Matsui, M.: Linear cryptanalysis method for des cipher. In: Helleseth, T. (ed.) EUROCRYPT 1993. LNCS, vol. 765, pp. 386–397. Springer, Heidelberg (1994)
20. Piret, G., Roche, T., Carlet, C.: PICARO – A block cipher allowing efficient higher-order side-channel resistance. In: Bao, F., Samarati, P., Zhou, J. (eds.) ACNS 2012. LNCS, vol. 7341, pp. 311–328. Springer, Heidelberg (2012)
21. Sajadieh, M., Dakhilalian, M., Mala, H., Sepehrdad, P.: Recursive diffusion layers for block ciphers and hash functions. In: Canteaut, A. (ed.) FSE 2012. LNCS, vol. 7549, pp. 385–401. Springer, Heidelberg (2012)
22. Shibutani, K., Isobe, T., Hiwatari, H., Mitsuda, A., Akishita, T., Shirai, T.: Piccolo: An ultra-lightweight blockcipher. In: Preneel, B., Takagi, T. (eds.) CHES 2011. LNCS, vol. 6917, pp. 342–357. Springer, Heidelberg (2011)
23. Suzaki, T., Minematsu, K., Morioka, S., Kobayashi, E.: TWINE: A Lightweight Block Cipher for Multiple Platforms. In: Knudsen, L.R., Wu, H. (eds.) SAC 2012. LNCS, vol. 7707, pp. 339–354. Springer, Heidelberg (2013)
24. Wu, W., Zhang, L.: Lblock: A lightweight block cipher. In: Lopez, J., Tsudik, G. (eds.) ACNS 2011. LNCS, vol. 6715, pp. 327–344. Springer, Heidelberg (2011)

A Dynamic Attribute-Based Authentication Scheme

Huihui Yang[(✉)] and Vladimir A. Oleshchuk

Department of Information and Communication Technology,
University of Agder, Gimlemoen 25, 4630 Kristiansand S, Norway
{huihui.yang,vladimir.oleshchuk}@uia.no

Abstract. Attribute-based authentication (ABA) is an approach to authenticate users by their attributes, so that users can get authenticated anonymously and their privacy can be protected. In ABA schemes, required attributes are represented by attribute trees, which can be combined with signature schemes to construct ABA schemes. Most attribute trees are built from top to down and can not change with attribute requirement changes. In this paper, we propose an ABA scheme based on down-to-top built attribute trees or dynamic attribute trees, which can change when attribute requirements change. Therefore, the proposed dynamic ABA scheme is more efficient in a dynamic environment by avoiding regenerating the whole attribute tree each time attribute requirements change.

Keywords: Authentication · Attribute-based authentication · Attribute tree · Privacy

1 Introduction

Compared with traditional identity based authentication (IBA) [1], users in ABA schemes are authenticated by their attributes instead of identities. Since users can be authenticated anonymously in ABA schemes, it is more privacy-preserving and can be used widely in many applications, for example, e-commerce [2], eHealth [3], mobile applications [4], cloud services [5–7] and so on. To get served, users first send a request for service. After receiving the service request, service providers send the request to an entity that controls service policies, which is usually policy decision point (PDP). The PDP retrieves related policies and attribute requirements and sends them back to the user. The user checks whether it owns the required attributes. If so, it generates a signature and sends it back to the PDP. PDP communicates with attribute authorities, verifies the signature and sends the verification result to the service provider. If the signature is valid, the service provider grants the service request from the user, otherwise the request is rejected. In a more general case, the service provider and PDP here can be considered as a verifier and policy controller. Since only (non-identity) attributes are used in the authentication described above, ABA schemes can achieve anonymity. However, in a traceable ABA scheme [8], users'

© Springer International Publishing Switzerland 2015
S. El Hajji et al. (Eds.): C2SI 2015, LNCS 9084, pp. 106–118, 2015.
DOI: 10.1007/978-3-319-18681-8_9

identities can be tracked by a tracking authority, but the verifier itself can not open the signature.

In the authentication, two parts need to be authenticated. First of all, the user has to prove that it owns all the required attributes so that the signature is not generated by co-operation among different users. Secondly, a signature generated by all required attributes also needs to be verified, providing other security requirements of the ABA scheme, for example, traceability. There has already been some work on ABA scheme constructions. As far as we know, [8] is the most systematic work about ABA schemes, where a general framework is built so that a fully anonymous and traceable static ABA scheme can be obtained with the input of a fully anonymous and traceable group signature scheme, but it does not provide a general framework to generate dynamic ABA schemes. One thing to notice is that the "static" and "dynamic" mentioned here is different from what we mean in the title of this paper. In the title, "dynamic" is to describe the attribute tree construction, while here it means a user is involved in a "join in" protocol [8] to co-generate its secret keys. For some work such as attribute-based access control (ABAC) [9], attribute-based signatures (ABS) [10] and attribute-based encryption(ABE) [11], the ways how to build their cryptographic construction share a lot in common with ABA schemes. Although the way how to generate signature schemes are different, the way how to build attribute trees are almost the same. Attribute trees originate from access trees [12] in access control, where they are used to represent logical access control requirements and usually built from top to down. The main drawback of this approach is that it is impossible to achieve a new attribute tree from an existing one even though their related logical requirements are quite similar. As a result, the system has to build a new attribute tree each time the attribute requirements change. To our best knowledge, there is only one paper [13] in which the attribute tree is built from down to top and it allows dynamic attribute tree construction. In [13], the authors propose an attribute-based group signature based on this down-to-top built attribute trees. In this approach, a central attribute tree is built first and then different attribute trees can be obtained by simplifying the central attribute tree.

In this paper, we propose an ABA scheme based on down-to-top built attribute trees as described in [13]. We modify the group signature protocol proposed in [14] in the way that it can be combined with the down-to-top attribute trees to construct an attribute based authentication scheme. Therefore, our proposed scheme supports dynamic attribute tree generation, so that both computation and communication resources can be saved by avoiding re-generating attribute related parameters.

This paper is organized as follows. We first introduce the general structure and security requirements of the proposed ABA scheme in Section 2. Next we describe how to construct the proposed scheme in details in Section 3, including the down-to-top built attribute trees, signature generation, verification and opening. Then followed in Section 4, we carry out correct, security and efficiency analysis on the proposed scheme. The last part is a general conclusion of the work in this paper.

2   ABA Scheme Introduction

In Section 1, we briefly discussed how a user was authenticated in an ABA scheme. In this section, we give more details about ABA scheme structure, workflow and their security requirements in subsections 2.1 and 2.2 respectively.

2.1   Scheme Structure and Workflow

The structure of the proposed ABA scheme and its workflow can be illustrated in Fig. 1. There are usually three types of entities in ABA schemes, i.e., authorities, users and verifiers, where authorities can be divided into central authority, attribute authority, revocation authority and opener. The way how they interact with each other and how the authentication is carried out can be described as follows.

1) The first stage is system set up.
   a) The central authority generates system public and private parameters.
   b) Users obtain their secret keys from the central authority.
   c) The attribute authority retrieves these system parameters and generates private and public attribute key pairs.
   d) The opener communicates with the central authority and gains the tracking keys.
   e) Users communicate with the attribute authority for their private attribute keys.
   f) Revocation authority communicates with both the central authority and the attribute authority to establish a data of revocation information.
2) The second stage is signature generation, verification and possibly opening.
   a) After receiving a challenge or attribute requirements form the verifier, the signer (or the user) sends its signature to the verifier, where the signature is generated by signing a message with the signer's attribute keys.
   b) The verifier retrieves revocation information from the revocation authority. If the signer and related attribute keys are not revoked, the verifier checks the validity of the signature and sends a response to the signer.
   c) If the identity of the signer needs to be revealed, the verifier delivers the signature to the opener. The opener uses its tracking keys to open the signature and reveal the signer's identity.

2.2   Security Requirements

In this subsection, we generally introduce five security requirements about ABA schemes. Later in subsection 4.2, we will formally define them based on the proposed ABA scheme and prove them.

Anonymity. To achieve basic anonymity, identities of signers should be protected. Furthermore, even signers' attributes should be protected, and they only

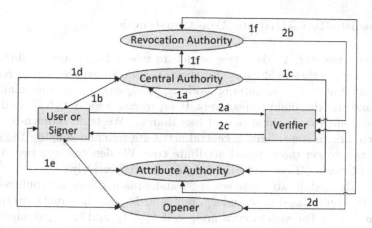

Fig. 1. Structure of the Proposed ABA Scheme

have to prove that they own the required attributes. This property is the main security requirement of an ABA scheme and is mandatory.

Unforgeability. The signer's signature should not be able to be forged by an outsider that does not belong to the system. In a system, the signature is even required unforgeable for authorities in the system. However, in a system where authorities generate all keys and secrets, the authorities obviously can forge all signatures. Therefore, "unforgeable" is defined differently in different ABA schemes. However, any system should provide at least the basic level of "unforgeability", i.e, for the outsiders.

Unlinkability. Given two signatures, if it should be impossible to decide whether they are generated by the same signer, the ABA scheme is unlinkable. If a system does not satisfy it, given enough signatures, there is a possibility to reveal the signer's identity.

Coalition Resistance. The signer can only generate the signature if he or she has all the required attributes. It should be impossible for different users to collude and generate a valid signature together if they as a whole have all the required attributes. If a system satisfies this requirement, it is coalition resistant.

Traceability. Given a valid signature, if the opener can successfully track the signer's identity, the system is traceable. It is a useful security requirement for some applications, such as obtaining evidence for legitimate issues and so on.

3   Construction of the Dynamic ABA Scheme

In this section, we present how to construct the proposed dynamic ABA scheme. We modify the group signature proposed in [14] and combine it with the down-to-top attribute tree, so that the scheme can provide attribute tree changes.

3.1 Down-to-Top Attribute Tree Construction

An attribute tree [12, 13] is a tree structure where leaves are attributes and interior nodes are threshold gates and they are used to express logical relations between attributes. For an interior node x, let l_x and k_x be the numbers of children and the threshold respectively. It represents logical "AND" and "OR" respectively when k_x is equal to and less than l_x. We build a down-to-top attribute tree in two steps: build a central attribute tree and simplify the central attribute tree to get the required attribute tree. We denote these two steps by algorithms $Create\_CTree$ and $Simplify\_CTree$ respectively.

Suppose the system attribute set is Ψ, and Ψ_i is a subset attribute set. The attribute tree built based on Ψ and Ψ_i are Γ and Γ_i, and their roots are $root$ and $root_i$ respectively. For each interior node x, if $k_x < l_x$, add $l_x - k_x$ dummy nodes as x's children, so that x can be considered as "AND". Denote the dummy leaf node set as L^{Dum}, attribute leaf node set as L^{Att} and whole leaf set as $L = L^{Dum} \cup L^{Att}$. For attribute subset Ψ_i, the node sets are L^{Dum_i}, L^{Att_i} and L_i accordingly. Each node x is indexed with a random number $ind(x)$, and each interior node x is binded to a polynomial $q_i(x)$ $(i = ind(x))$ (Refer to [8, 13] for more details.). The degree of $q_i(x)$ is $l_x - 1$. Polynomials are constructed by algorithm $Create\_CTree$ as follows.

1) Assign a secret t_j $(1 \leq i \leq |L^{Att}|, |L^{Att}| = |\Psi|)$ to each attribute leaf node.
2) Let the related polynomial $q_i(x)$ be the one passing through these points with the coordinates (j, t_j) $(y \in Child_{Att}(x), j = ind(y))$, where $Child_{Att}(x)$ is the set of child leaf nodes belonging to Ψ, so $q_i(x)$ can be computed by Lagrange's theorem [15]. Assign $q_i(0)$ to node x.
3) For x's dummy leaf child $y \in Child_{Dum}(x)$, compute d_j by $d_j = q_i(j)$ $(j = ind(y))$ and assign d_j to each dummy leaf node.
4) Repeat Steps 2) and 3) until all polynomials related to interior nodes are constructed. Then the root node is assigned with value $r = q_{root}(0)$.

After the central attribute tree Γ is built, we can simplify it to get attribute tree Γ_i based on attribute subset $\Psi_i \subset \Psi$. The simplification algorithm $Simplify\_CTree$ is as follows.

1) Delete all leaf nodes that does not belong to Ψ_i.
2) Delete an interior node together with its descendants if the number of its children is less than the threshold. The remaining part is the required attribute tree Γ_i.

For a leaf node x, let $In_{sib(x)}$ be the index set of all x's siblings including x. We define $L_x = \prod_{l \in In_{sib(x)}, l \neq ind(x)} \frac{-l}{ind(x) - l}$, $Path_x = \{x, x_1, \cdots, x_n\}$ and $\triangle_x = \prod_{z \in \{Path_x - root\}} L_z$, where $x_1 = parent(x)$, $x_{i+1} = parent(x_i)$ $(1 \leq i \leq n - 1)$ and $x_n = root$. Then for Γ_i, equation $\sum_{j \in L^{Att_i}} \triangle_j t_j + \sum_{j \in L^{Dum_i}} \triangle_{d_j} d_j = q_{root_i}(0)$ holds.

3.2   Construction Algorithms

Before introducing the scheme construction, we first briefly explain bilinear groups and q-SDH problem, on which the proposed scheme is built.

Definition 1. *([8, 13]) (**Bilinear Groups**) Let G_1, G_2 and G_3 be cyclic group of prime order p, with $g_1 \in G_1$ and $g_2 \in G_2$ as the generators. e is an efficient bilinear map if*

1) Bi-linearity: $e(g_1^a, g_2^b) = e(g_1, g_2)^{ab}$ holds for any $a, b \in \mathbb{Z}_p^*$.
2) Non-degenerate: $e(g_1, g_2) \neq 1_{G_3}$, where 1_{G_3} is the unit of G_3.

Let $\psi : G_1 \to G_2$ be an isomorphism with $\psi(g_2) = g_1$. If there is an efficient algorithm to compute it, we way ψ is computable.

Definition 2. *([8, 16]) (**q-Strong Diffie-Hellman Problem (q-SDH) in G_1 and G_2**) Let G_1 and G_2 be cyclic groups of prime order p, with $g_1 \in G_1$ and $g_2 \in G_2$ as the generators. Given a $q+2$ tuple input $(g_1, g_2, g_2^\gamma, g_2^{\gamma^2}, \cdots, g_2^{\gamma^q})$ $(\gamma \in \mathbb{Z}_p^*)$, an SDH output is a pair $(g_1^{1/(\gamma+x)}, x)$ $(x \in \mathbb{Z}_p^*)$.*

The cryptographic algorithms of the dynamic ABA scheme proposed in this paper can be constructed as follows.

- **System Setup.** Assume k is the security parameter. G_1, G_2 and G_3 are three multiplicative groups of prime order and $e : G_1 \times G_2 \to G_3$ is a bilinear map. ψ is a computable isomorphism between G_1 and G_2. $H : \{0,1\}^* \to \mathbb{Z}_q^*$ is a hash function. Randomly select a generator $g_2 \in G_2$ and $\gamma, \xi_1, \xi_2, \in \mathbb{Z}_q^*$. Then the system private parameter is $S_{pri} =< \gamma, tk >$, where $tk =< \xi_1, \xi_2 >$ is the tracking key. The public parameter is $S_{pub} =< G_1, G_2, G_3, e, H, g_1, g_2, h, u, v, w >$, where $g_1 = \psi(g_2)$, $u = h^{\xi_1}$, $v = h^{\xi_2}$ and $w = g_2^\gamma$.
- **User Key Generation.** Take γ as input and generate a private key base $bsk[i] =< A_i, x_i >$ for each user U_i, where $x_i \in \mathbb{Z}_q^*$ and $A_i = g_1^{1/(\gamma+x_i)}$ is the output of a q-SDH pair $(g_1, g_2, g_2^{r_0}, g_2^{r_0^2}, \cdots, g_2^{r_0^q})$. A_i is registered in an authority's databased for tracking if necessary.
- **Attribute Key Generation.** This algorithm is run by the attribute authority. Assume Ψ is the system attribute set. Select a number $t_j \in \mathbb{Z}_q^*$ for each attribute att_j. The private and public key pair for att_j is t_j and $bpk_{t_j} = g_2^{t_j}$. The public key related to Ψ is $gpk = \{G_1, G_2, G_3, g_1, g_2, e, H, h, u, v, w, h_1, \cdots, h_{|\Psi|}, bpk\}$, where $h_j = h^{t_j}$ and $bpk = \{bpk_{t_1}, \cdots, bpk_{t_{|\Psi|}}\}$.
- **User Attribute Key Generation.** This algorithm is run by the attribute authority. If U_i is not in the revocation list and wants to register att_j, the attribute authority calculates $T_{i,j} = A_i^{t_j}$ and sends it to U_i. Then U_i's private key is $gsk =< A_i, x_i, T_{i,1}, \cdots, T_{i,|\Psi_i|} >$, where Ψ_i is the attribute subset owned by U_i.
- **Attribute Tree Generation.** Assume Ψ is the system attribute set. Create an attribute tree Γ by algorithm *Create_CTree* described in subsection 3.1

with *root* as its root. L^{Att} ($|L^{Att}| = |\Psi|$) and L^{Dum} are the attribute leaf set and dummy leaf node set. Randomly select a $d_j \in \mathbb{Z}_q^*$ for each dummy leaf node dum_j. Calculate $gpk_{d_j} = g_2^{d_j}$. Assign gpk_{t_j} to each attribute leaf node and gpk_{d_j} to each dummy node. Make Γ and $bpk' = \{gpk_{d_1}, \cdots, gpk_{d_{|L^{Dum}|}}\}$ public.

• **Signature Generation and Verification.** Suppose M is the message to sign and Ψ_i is the selected attribute set, and then protocol runs as follows:

1) Let V be the verifier. V selects the attribute requirements and sends them to U_i.

2) After receiving Ψ_i, U_i first runs algorithm $Simplify\_CTree$ described in subsection 3.1, obtains the attribute subtree Γ_i, and computes $\triangle_j$ ($att_j \in L^{Att_i}$), $\triangle_{d_j}$ ($d_j \in L^{Dum_i}$) and its root value r_i. Next U_i calculates $g_d = \prod_{d_j \in L^{Dum_i}} gpk_{d_j}^{\triangle_{d_j}}$. Then U_i randomly selects $\zeta, \alpha, \beta, \varepsilon, r_\zeta, r_\alpha, r_\beta, r_\varepsilon, r_x, r_{\delta_1}, r_{\delta_2} \in \mathbb{Z}_p^*$ and then does the following calculations

$$C_1 = u^\zeta, C_2 = v^\beta, C_3 = A_i h^{\zeta+\beta}, C_4 = A_i w^\varepsilon, CT_j = T_{i,j} h_j^\alpha,$$
$$\delta_1 = x_i \zeta, \delta_2 = x_i \beta, R_1 = u^{r_\zeta}, R_2 = v^{r_\beta}, R_4 = C_1^{r_x} u^{-r_{\delta_1}},$$
$$R_5 = C_2^{r_x} v^{-r_{\delta_2}}, R_3 = e(C_3, g_2)^{r_x} e(h, w)^{-r_\zeta - r_\beta} e(h, g_2)^{-r_{\delta_1} - r_{\delta_2}},$$
$$R_{Att} = \frac{e(\prod_{att_j \in \Psi_i} h_j^{\triangle_j}, g_2)^{r_\alpha}}{e(w, r_i/g_d)^{r_\varepsilon}},$$
$$c = H(M, C_1, C_2, C_3, C_4, R_1, R_2, R_3, R_4, R_5, R_{Att}) \in \mathbb{Z}_p^*$$
$$s_\zeta = r_\zeta + c\zeta, s_\beta = r_\beta + c\beta, s_\alpha = r_\alpha + c\alpha, s_\varepsilon = r_\varepsilon + c\varepsilon, s_x = r_x + cx_i,$$
$$s_{\delta_1} = r_{\delta_1} + c\delta_1, s_{\delta_2} = r_{\delta_2} + c\delta_2.$$

Then the signature is $\delta = < M, C_1, C_2, C_3, C_4, c, CT_1, \cdots, CT_{|\Psi_i'|}, s_\zeta, s_\beta, s_\alpha, s_\varepsilon, s_x, s_{\delta_1}, s_{\delta_2}, \Psi_i' >$, where Ψ_i' is the attribute subset U_i decides to use.

3) The first step of V is to check whether U_i and Ψ_i' are revoked or not. If not, V continues and otherwise it rejects the signature. First of all, V runs algorithm $Simplify\_CTree$ described in subsection 3.1 to get the simplified tree Γ_i, the root value r_i and $\triangle_j$ ($att_j \in L^{Att_i}$), $\triangle_{d_j}$ ($d_j \in L^{Dum_i}$) and $g_d = \prod_{d_j \in L^{Dum_i}} gpk_{d_j}^{\triangle_{d_j}}$. Next it computes

$$R_1' = u^{s_\zeta} C_1^{-c}, R_2' = v^{s_\beta} C_2^{-c}, R_4' = u^{-s_{\delta_1}} C_1^{s_x}, R_5' = v^{-s_{\delta_2}} C_2^{s_x},$$
$$R_3' = e(C_3, g_2)^{s_x} e(h, w)^{-s_\zeta - s_\beta} e(h, g_2)^{-s_{\delta_1} - s_{\delta_2}} \left(\frac{e(C_3, w)}{e(g_1, g_2)}\right)^c,$$
$$R_{Att}' = \frac{e(\prod_{att_j \in \Psi_i} h_j^{\triangle_j}, g_2)^{s_\alpha}}{e(w, r_i/g_d)^{s_\varepsilon}} \left(\frac{e(C_4, r_i/g_d)}{e(\prod_{att_j \in \Psi_i} CT_j^{\triangle_j}, g_2)}\right)^c.$$

Finally, V checks whether $c = H(M, C_1, C_2, C_3, C_4, R_1', R_2', R_3', R_4', R_5', R_{Att}')$ holds. If valid, V accepts the signature and U_i is authenticated, and otherwise V rejects the signature.

- **Signature Open.** Before opening the signature, the opener needs to check whether the signature is valid or not. If it is valid, it computes $A_i = C_3/(C_1^{\varepsilon_1} C_2^{\varepsilon_2})$.

4 Analysis of the Dynamic ABA Scheme

In this section, we analyze the proposed ABA scheme, including their correctness, security requirements and efficiency.

4.1 Correctness Analysis

Theorem 1. *(Correctness) The construction of the proposed ABA scheme proposed in Section 3 is correct, which means:*

1) Tuple $< R_1, R_2, R_3, R_4, R_5, R_{Att} >$ equals to $< R_1', R_2', R_3', R_4', R_5', R_{Att}' >$.
2) $A_i = C_3/(C_1^{\varepsilon_1} C_2^{\varepsilon_2})$ holds.

Proof. 1) The verifier computes tuple $< R_1', R_2', R_3', R_4', R_5', R_{Att}' >$ as follows.

$$R_3' = e(C_3, g_2)^{s_x} e(h, w)^{-s_\alpha - s_\beta} e(h, g_2)^{-s_{\delta_1} - s_{\delta_2}} \left(\frac{e(C_3, w)}{e(g_1, g_2)}\right)^c$$

$$= e(C_3, g_2)^{r_x + cx_i} e(h, w)^{-(r_\alpha + c\alpha) - (r_\beta + c\beta)} e(h, g_2)^{-(r_{\delta_1} + c\delta_1) - (r_{\delta_2} + c\delta_2)} \left(\frac{e(C_3, w)}{e(g_1, g_2)}\right)^c$$

$$= R_3 \left(e(C_3, g_2)^{x_i} e(h, w)^{-(\alpha + \beta)} e(h, g_2)^{-(\delta_1 + \delta_2)} \frac{e(C_3, w)}{e(g_1, g_2)}\right)^c$$

$$= R_3 \left(e(C_3 h^{-(\alpha + \beta)}, w g_2^{x_i}) e(C_3, w)^{-1} \frac{e(C_3, w)}{e(g_1, g_2)}\right)^c$$

$$= R_3 \left(e(A_i, w g_2^{x_i})(g_1, g_2)^{-1}\right)^c = R_3$$

$$R_1' = u^{s_\varsigma} C_1^{-c} = u^{r_\varsigma + c\varsigma} (u^\varsigma)^{-c} = u^{r_\varsigma} = R_1$$

$$R_2' = v^{s_\beta} C_2^{-c} = v^{r_\beta + c\beta} (v^\beta)^{-c} = v^{r_\beta} = R_2$$

$$R_4' = u^{-s_{\delta_1}} C_1^{s_x} = u^{-r_{\delta_1} - c\delta_1} (u^\varsigma)^{r_x + cx_i} = u^{-r_{\delta_1} - c\varsigma x_i} (u^\varsigma)^{r_x + cx_i} = u^{-r_{\delta_1}} u^{\varsigma r_x}$$

$$= C_1^{r_x} u^{-r_{\delta_1}} = R_4$$

$$R_5' = v^{-s_{\delta_2}} C_2^{s_x} = v^{-r_{\delta_2} - c\delta_2} (v^\beta)^{r_x + cx_i} = v^{-r_{\delta_2} - c\beta x_i} (v^\beta)^{r_x + cx_i} = v^{-r_{\delta_2}} v^{\beta r_x}$$

$$= C_2^{r_x} v^{-r_{\delta_2}} = R_5$$

From subsections 3.1 and 3.2, we know that $\sum_{j \in L^{Att_i}} \triangle_j t_j + \sum_{j \in L^{Dum_i}} \triangle_{d_j} d_j = r_i$ and $g_d = \prod_{d_j \in L^{Dum_i}} gpk_{d_j}^{\triangle_{d_j}}$, so we have

$$e\left(\prod_{j \in L^{Att_i}} A_i^{t_j \triangle_j}, g_2\right) e\left(\prod_{j \in L^{Dum_i}} A_i^{d_j \triangle_j}, g_2\right) = e(A_i^{r_i}, g_2)$$

$$\Longleftrightarrow e\left(\prod_{j \in L^{Att_i}} A_i^{t_j \triangle_j}, g_2\right) e(A_i, g_d) = e(A_i, r_i)$$

$$\Longleftrightarrow e\left(\prod_{j \in L^{Att_i}} A_i^{t_j \triangle_j}, g_2\right) = e(A_i, r_i/g_d).$$

Based on the above equation, we have

$$
\begin{aligned}
R'_{Att} &= \frac{e(\prod_{att_j \in \Psi_i} h_j^{\triangle_j}, g_2)^{s\alpha}}{e(w, r_i/g_d)^{s\varepsilon}} \left(\frac{e(C_4, r_i/g_d)}{e(\prod_{att_j \in \Psi_i} CT_j^{\triangle_j}, g_2)} \right)^c \\
&= \frac{e(\prod_{att_j \in \Psi_i} h_j^{\triangle_j}, g_2)^{r\alpha}}{e(w, r_i/g_d)^{r\varepsilon}} \frac{e(\prod_{att_j \in \Psi_i} h_j^{\triangle_j}, g_2)^{c\alpha}}{e(w, r_i/g_d)^{c\varepsilon}} \left(\frac{e(C_4, r_i/g_d)}{e(\prod_{att_j \in \Psi_i} CT_j^{\triangle_j}, g_2)} \right)^c \\
&= R_{Att} \left(\frac{e(\prod_{att_j \in \Psi_i} h_j^{\triangle_j}, g_2)^{\alpha}}{e(w, r_i/g_d)^{\varepsilon}} \frac{e(C_4, r_i/g_d)}{e(\prod_{att_j \in \Psi_i} CT_j^{\triangle_j}, g_2)} \right)^c \\
&= R_{Att} \left(\frac{e(\prod_{att_j \in \Psi_i} h_j^{\alpha\triangle_j}, g_2)}{e(w^{\varepsilon}, r_i/g_d)} \frac{e(w^{\varepsilon}, r_i/g_d)}{e(\prod_{att_j \in \Psi_i} h_j^{\alpha\triangle_j}, g_2)} \right)^c \left(\frac{e(A_i, r_i/g_d)}{e(\prod_{att_j \in \Psi_i} T_{i,j}^{\triangle_j}, g_2)} \right)^c \\
&= R_{Att} \left(\frac{e(A_i, r_i/g_d)}{e(\prod_{att_j \in \Psi_i} A_i^{t_j\triangle_j}, g_2)} \right)^c
\end{aligned}
$$

Since $e(\prod_{j \in L^{Att_i}} A_i^{t_j\triangle_j}, g_2) = e(A_i, r_i/g_d)$, equation $R_{Att} = R'_{Att}$ holds. Thus, tuple $< R_1, R_2, R_3, R_4, R_5, R_{Att} >$ equals to $< R'_1, R'_2, R'_3, R'_4, R'_5, R'_{Att} >$.
2) $C_3/(C_1^{\varepsilon_1} C_2^{\varepsilon_2}) = A_i h^{\zeta+\beta}/((u^\zeta)^{\varepsilon_1}(v^\beta)^{\varepsilon_2}) = A_i$.

From the above proofing, we can see Theorem 1 is correct, so a user can be authenticated by the algorithm proposed in subsection 3.2.

4.2 Security Requirements Analysis

In this subsection, we prove that the proposed ABA scheme satisfies several security requirements, including anonymity, unforgeability, unlinkability, coalition resistance and traceability. Before the proof, we first introduce some definitions and assumptions based on which our proof performs.

Definition 3. *([8, 14]) (Decision Linear Diffie-Hellman Problem (DLP) in G_1) Let G be a cyclic group of prime order p, with $u, v, h \in G$ as its generators. Given $u^a, v^b, h^c \in G$ $(a, b, c \in \mathbb{Z}_p^*)$ as the input, decide whether $a + b = c$ or not.*

Definition 4. *([14]) (Decision Linear Diffie-Hellman based Encryption (DLE) in G_1) In a DLE scheme, a user's public key is $u, v, h \in G_1$ and its private key is $\varepsilon_1, \varepsilon_2 \in \mathbb{Z}_q^*$, satisfying $u^{\varepsilon_1} = v^{\varepsilon_2} = h$. To encrypt message M, the user randomly chooses $\alpha, \beta \in \mathbb{Z}_q^*$ and computes the encryption message as a triple $< C_1, C_2, C_3 >$, where $C_1 = u^\alpha$, $C_2 = v^\beta$ and $C_3 = M h^{\alpha+\beta}$. The decrypted message is calculated by $C_3/(C_1^{\varepsilon_1} C_2^{\varepsilon_2})$.*

Assumption 1. *([8]) (q-SDH Problem) For an algorithm A, if $|Pr[A(g_1, g_2, g_2^\gamma, g_2^{\gamma^2}, \cdots, g_2^{\gamma^q}) = (g_1^{1/\gamma+x}, x)] - 1/|G|| \le \varepsilon$ holds, we say that A has a negligible advantage to solve q-SDH in (G_1, G_2) and then we can assume q-SDH is hard.*

Assumption 2. (DLP) *For an algorithm A, if $|Pr[A(u^\alpha, v^\beta, h^{\alpha+\beta}) = (u^\alpha, v^\beta, h^c)] - 1/|G|| \leq \varepsilon$ holds, we say that A has a negligible advantage to solve DLP in G_1 and then we can assume DLP is hard.*

Assumption 3. (IND-CPA Security) *([16]) If DLP holds, we say that DLE is semantically secure against a chosen-plaintext attack (CPA) or IND-CPA secure.*

Assumption 4. (Forking Lemma) *([8]) Given only public data as input, if an adversary A with polynomial computation ability can find a valid signature $(M, \delta_0, c, \delta_1)$ with non-negligible probability, then there exists a replay with a different oracle, which can output new valid signatures $(M, \delta_0, c', \delta_1')$ with non-negligible probability where $c \neq c'$.*

Theorem 2. (Anonymity) *The proposed ABA scheme is fully anonymous if DLE is IND-CPA secure under the same attribute set. More specifically, given $A_{i(0)}$ and $A_{i(1)}$ and the corresponding signature δ_1 and δ_2, where $\delta_b =< M, C_1, C_2, C_{3(b)}, C_4, c, CT_1, \cdots, CT_{|\Psi_i'|}, s_\zeta, s_\beta, s_\alpha, s_\varepsilon, s_x, s_{\delta_1}, s_{\delta_2}, \Psi_i' > (b \in \{0,1\})$. Given a random toss $b \in \{0,1\}$, if the probability that an adversary A with polynomial computation ability has non-negligible advantage to guess the correct b, we say that the proposed scheme is not fully anonymous. Otherwise, the scheme is fully anonymous.*

Proof. Suppose that adversary A can break the anonymity of the proposed scheme, and then it means A has a non-negligible advantage to guess the correct b in the above statement. More precisely, given $A_{i(0)}$ and $A_{i(1)}$, the adversary A has a non-negligible advantage to distinguish the tuple $< C_1, C_2, C_{3(0)} >$ from $< C_1, C_2, C_{3(1)} >$, where $C_1 = u^\zeta$, $C_2 = v^\beta$, $C_{3(0)} = A_{i(0)} h^{\zeta+\beta}$ and $C_{3(1)} = A_{i(1)} h^{\zeta+\beta}$. From Definition 4, we know that $< C_1, C_2, C_{3(0)} >$ is a DLE tuple. If A has the ability to distinguish $< C_1, C_2, C_{3(0)} >$ from $< C_1, C_2, C_{3(1)} >$, it means A can break DLE problem, and it contradicts with Assumption 3. Thus it is impossible for A to distinguish δ_0 from δ_1 with a non-negligible probability, and the proposed ABA scheme is fully-anonymous.

Theorem 3. (Traceability) *The proposed ABA scheme is fully traceable if q-SDH is hard in G_1 and G_2. More specifically, if an adversary A with polynomial computation ability can find a valid signature $\delta =< M, C_1, C_2, C_3, C_4, c, CT_1, \cdots, CT_{|\Psi_i'|}, s_\zeta, s_\beta, s_\alpha, s_\varepsilon, s_x, s_{\delta_1}, s_{\delta_2}, \Psi_i' >$, then it can find a find a SDH pair and thus break the q-SDH problem.*

Proof. The proof is based on Forking Lemma. Suppose adversary A can forge a valid signature $\delta =< M, \delta_0, c, \delta_1, \delta_2 >$, where $\delta_0 = \{C_1, C_2, C_3, C_4\}$, $c = H(M, C_1, C_2, C_{3,4}, R_1, R_2, R_3, R_4, R_5, R_{Att})$ as computed during signature generation, $\delta_1 = \{s_\zeta, s_\beta, s_\alpha, s_\varepsilon, s_x, s_{\delta_1}, s_{\delta_2}, \Psi_i'\}$, and $\delta_2 = \{CT_1, \cdots, CT_{|\Psi_i'|}\}$. According to Forking Lemma, we can extract a tuple $< \delta_0, c', \delta_1', \delta_2 >$ from $\delta =< \delta_0, c, \delta_1, \delta_2 >$, where $c' \neq c$ and $\delta_1' \neq \delta_2'$. Then based on $< \delta_0, c', \delta_1', \delta_2 >$ and $c' \neq c$ and $\delta_1' \neq \delta_1$. Thus we can create a new SDH tuple denoted as $< A_i', x' >$, which is presented as **Theorem 5.3.5** in details in [8] and we will not repeat it here. If adversary A can create a q-SDH pair without the knowledge of γ, it can break q-SDH problem. It contradicts to Assumption 1 and thus Theorem 3 holds.

Among the above security requirements, unforgeability and coalition resistance can be promised by traceability, and unlinkability can be promised by anonymity. The general idea is as follows. Suppose S is a system that provides both anonymity and traceability. If S is not against unforgeability, it means that an adversary A can forge a valid signature δ on behalf of a valid user U. Similarly, if S is not coalition resistant, an adversary A can corrupt a few users and then use their private keys to generate a valid signature δ. In both situations, when δ is handed over to an opener, the identity revealed is U instead of the real signer A. It contradicts with "traceability" (Theorem 3) and thus both unforgeability and coalition resistance can be inferred by traceability. Anonymity promises that the system does not leak any useful information of signers given their signatures. From the description of unlinkability, we can see that it is also a kind of user identity information, meaning that anonymity is a stronger security requirement than unlinkability. Therefore, unlinkability can be deduced by anonymity.

4.3   Efficiency Analysis

We mainly focus on the computation, storage and communication costs when analyzing the efficiency of the proposed ABA scheme, and they are related to the attribute tree as well as the signature itself. From the construction of the down-to-top attribute trees, we can see that the cost is more than top-to-down built attribute trees because of the following two reasons. First of all, the attribute set Ψ for central attribute tree Γ is bigger than the required attribute subset Ψ_i, and thus attribute related keys (t_j, gpk_j), users' attribute keys $T_{i,j}$ are more than those based only on Ψ_i. Secondly, we add some dummy nodes to build the tree from down to top, which causes the computation and storage of d_j and gpk_{d_j} for dummy leaf node set L^{Dum}. At the cost of a bigger central attribute tree instead of a fixed tree and more parameters, we achieve the flexibility and dynamic of attribute trees.

For the computation and communication cost of the signature, we compare our work with two typical ABA schemes in [8] and the results are summarized in Table 1. Suppose $|G|$ is the bit length of bilinear group G_i ($i \in \{1, 2, 3\}$), k is the bit length of number in $\mathbb{Z}_q^*$ and p is the computation cost of pairing. To simplify the comparison, we only count the computation of pairing and we will not include the length of message M as well as the length of attributes to be used in the signature size. [8].1 and [8].2 (see Table 1) are the schemes proposed in Chapter 5.4 and 5.5 in [8] respectively.

Among these schemes, the signature size in [8].2 (see Table 1) is extremely short, because it does not include attribute related parameter CT_j in the signature at the cost of verifying the validity of the signer's attributes. From the result, we can see that the general signature size, computation complexity for both signing and verification are more in our scheme. The main reason is that we add parameters C_4 and R_{Att} to adjust the signature with the down-to-top attribute construction. Therefore, the ABA scheme proposed in this paper is beneficial in a dynamic environment where the attribute requirements change frequently. However, when the attribute requirements are comparatively fixed, our scheme needs more computation and communication cost, and also requires more storage space.

Table 1. Computation and Communication Efficiency Comparison of ABA Schemes

Paper	Signature Size	Sign Complexity	Verify Complexity
[8].1	$5k + (|\Psi_i| + 5)|G|$	$3p$	$5p$
[8].2	$7|G|$	$3p$	$5p$
This paper	$8k + (|\Psi_i| + 4)|G|$	$5p$	$9p$

5   Conclusions

In this paper, we have proposed a dynamic attribute tree based ABA scheme, which can adapt to a dynamic environment where attribute requirements change frequently. To gain this flexibility, the proposed ABA scheme is constructed based on down-to-top built attribute trees. In the scheme, a central attribute tree Γ is built first based on attribute set Ψ, and later attribute subtree Γ_i based on attribute subset Ψ_i can be obtained by simplifying Γ. Compared with top-to-down based ABA schemes, we add extra parameters in the signature to adjust to this new attribute tree construction approach. The scheme avoids the cost of regenerating attribute tree related parameters, but it increases costs in three aspects, storage cost for a bigger attribute tree, signature size and more computation complexity. As a results, it should be careful to chose which ABA schemes to use, i.e., down-to-top or top-to-down built attribute tree based, and it requires precise evaluation of the costs of regenerating attribute trees, signature generation and verification.

References

1. Li, H., Dai, Y., Tian, L., Yang, H.: Identity-based authentication for cloud computing. In: Jaatun, M.G., Zhao, G., Rong, C. (eds.) Cloud Computing. LNCS, vol. 5931, pp. 157–166. Springer, Heidelberg (2009)
2. Schläger, C., Sojer, M., Muschall, B., Pernul, G.: Attribute-based authentication and authorisation infrastructures for e-commerce providers. In: Bauknecht, K., Pröll, B., Werthner, H. (eds.) EC-Web 2006. LNCS, vol. 4082, pp. 132–141. Springer, Heidelberg (2006)
3. Guo, L., Zhang, C., Sun, J., Fang, Y.: Paas: A privacy-preserving attribute-based authentication system for ehealth networks. In: 2013 IEEE 33rd International Conference on Distributed Computing Systems, pp. 224–233 (2012)

4. Covington, M.J., Sastry, M.R., Manohar, D.J.: Attribute-based authentication model for dynamic mobile environments. In: Clark, J.A., Paige, R.F., Polack, F.A.C., Brooke, P.J. (eds.) SPC 2006. LNCS, vol. 3934, pp. 227–242. Springer, Heidelberg (2006)

5. Liu, X., Xia, Y., Jiang, S., Xia, F., Wang, Y.: Hierarchical attribute-based access control with authentication for outsourced data in cloud computing. In: 2013 12th IEEE International Conference on Trust, Security and Privacy in Computing and Communications (TrustCom), pp. 477–484 (July 2013)

6. Ruj, S., Stojmenovic, M., Nayak, A.: Privacy preserving access control with authentication for securing data in clouds. In: 2012 12th IEEE/ACM International Symposium on Cluster, Cloud and Grid Computing (CCGrid), pp. 556–563 (May 2012)

7. Xu, D., Luo, F., Gao, L., Tang, Z.: Fine-grained document sharing using attribute-based encryption in cloud servers. In: 2013 Third International Conference on Innovative Computing Technology (INTECH), pp. 65–70 (August 2013)

8. Khader, D.D.: Attribute-based Authentication Scheme. PhD thesis, University of Bath (2009)

9. Liu, X., Xia, Y., Jiang, S., Xia, F., Wang, Y.: Hierarchical attribute-based access control with authentication for outsourced data in cloud computing. In: 2013 12th IEEE International Conference on Trust, Security and Privacy in Computing and Communications (TrustCom), pp. 477–484 (2013)

10. Maji, H.K., Prabhakaran, M., Rosulek, M.: Attribute-based signatures. In: Kiayias, A. (ed.) CT-RSA 2011. LNCS, vol. 6558, pp. 376–392. Springer, Heidelberg (2011)

11. Yu, S., Wang, C., Ren, K., Lou, W.: Attribute based data sharing with attribute revocation. In: Proceedings of the 5th ACM Symposium on Information, Computer and Communications Security, ASIACCS 2010, pp. 261–270 (2010)

12. Goyal, V., Pandey, O., Sahai, A., Waters, B.: Attribute-based encryption for fine-grained access control of encrypted data. In: Proceedings of the 13th ACM Conference on Computer and Communications Security, CCS 2006, pp. 89–98. ACM, New York (2006)

13. Emura, K., Miyaji, A., Omote, K.: A dynamic attribute-based group signature scheme and its application in an anonymous survey for the collection of attribute statistics. In: International Conference on Availability, Reliability and Security, ARES 2009, pp. 487–492 (2009)

14. Boneh, D., Boyen, X., Shacham, H.: Short group signatures. In: Franklin, M. (ed.) CRYPTO 2004. LNCS, vol. 3152, pp. 41–55. Springer, Heidelberg (2004)

15. Armstrong, M.: Lagranges theorem. In: Groups and Symmetry, pp. 57–60. Springer, New York (1988)

16. Boneh, D., Boyen, X.: Short signatures without random oracles. In: Cachin, C., Camenisch, J.L. (eds.) EUROCRYPT 2004. LNCS, vol. 3027, pp. 56–73. Springer, Heidelberg (2004), http://dx.doi.org/10.1007/978-3-540-24676-3_4

Repeated-Root Isodual Cyclic Codes over Finite Fields

Aicha Batoul[1(✉)], Kenza Guenda[1], and T. Aaron Gulliver[2]

[1] Faculty of Mathematics USTHB,
University of Science and Technology of Algiers, Algeria
[2] Department of Electrical and Computer Engineering,
University of Victoria, BC, Canada
`a.batoul@hotmail.fr`, `kguenda@gmail.com`, `agullive@ece.uvic.ca`

Abstract. In this paper we give several constructions of cyclic codes over finite fields that are monomially equivalent to their dual, where the characteristic of the field divides the length of the code. These are called repeated-root cyclic isodual codes over finite fields. The constructions are based on the field characteristic, the generator polynomial and the length of the code.

Keywords: Repeated-root cyclic codes · Isodual codes · Multipliers · Splitting · Duadic codes

1 Introduction

An isodual code is a linear code which is equivalent to its dual. The class of isodual codes is important in coding theory because it contains the self-dual codes as a subclass. In addition, isodual codes are contained in the larger class of formally self-dual codes, and they are related to isodual lattices [1]. For some parameters, it can be shown that there are no cyclic self-dual codes over finite fields [3,5], whereas cyclic isodual codes can exist. Several types of equivalence between codes can be defined [4]. In [2] the authors gave specific constructions of self-dual and isodual codes over finite fields. Two codes C and C' are called monomially equivalent if there exists a monomial permutation, i.e. a permutation of the coordinates followed by multiplication of coordinates by nonzero field elements, which sends C to C'. Only monomial equivalence is considered here.

Jia et al. [6] considered cyclic isodual codes using multiplier equivalency. Multiplier equivalence is a monomial equivalence, but the converse is not true in general. In this paper, isodual cyclic codes over finite fields are studied. Conditions are given concerning the existence of isodual cyclic codes based on the generators polynomial, field characteristic, and length. Several constructions of isodual cyclic codes and self-dual codes are given which have good Hamming minimum distance.

The remainder of this paper is organized as follows. Some preliminary results are given in Section 2. In Section 3, the structure of the generator polynomial of cyclic codes of length $2^a m p^s$ is given using the generator polynomial of cyclic

© Springer International Publishing Switzerland 2015
S. El Hajji et al. (Eds.): C2SI 2015, LNCS 9084, pp. 119–132, 2015.
DOI: 10.1007/978-3-319-18681-8_10

codes of length m. Using the structure of cyclic codes of length $2^a m p^s$, a construction for isodual codes is given in Section 4. In Section 5, isodual codes are constructed from duadic codes over finite fields. The motivation to construct isodual codes from duadic codes is that the duadic codes are known to have good minimum distance. Examples of isodual codes are given based on the constructions presented here.

2   Preliminaries

Let $\mathbb{F}_q$ be a finite field with q a power of a prime p, and denote the corresponding group of units by $\mathbb{F}_q^*$. Let n a positive integer. A block code C of length n is called a linear code over $\mathbb{F}_q$ if it is a subspace of $\mathbb{F}_q^n$. Here, all codes are assumed to be linear. We attach the standard inner product to $\mathbb{F}_q$, i.e. $[v, w] = \sum v_i w_i$. The dual code of C is defined as

$$C^\perp = \{v \in \mathbb{F}_q^{\,n} \mid [v, w] = 0 \text{ for all } w \in C\}. \tag{1}$$

If $C \subseteq C^\perp$, the code is said to be self-orthogonal and if $C = C^\perp$, the code is self-dual.

$$x^\perp$$

A linear code C over a finite field and its dual satisfy the following

$$|C||C^\perp| = q^n, \text{ and } (C^\perp)^\perp = C. \tag{2}$$

A monomial linear transformation of $\mathbb{F}_q^n$ is an $\mathbb{F}_q$-linear transformation τ such that there exists scalars $\lambda_1, \ldots, \lambda_n$ in $\mathbb{F}_q^*$ and a permutation $\sigma \in S_n$ (the group of permutations of the set $\{1, 2, \ldots, n\}$), such that for all $(x_1, x_2, \ldots, x_n) \in \mathbb{F}_q^n$, we have

$$\tau(x_1, \ldots, x_n) = (\lambda_1 x_{\sigma(1)}, \lambda_2 x_{\sigma(2)}, \ldots, \lambda_n x_{\sigma(n)}).$$

Two linear codes C and C' of length n are called monomially equivalent if there exists a monomial transformation of $\mathbb{F}_q^n$ such that $\tau(C) = C'$. Here, whenever two codes are said to be equivalent it is meant that they are monomially equivalent. Hence in our context an isodual code is a linear code which is monomially equivalent to its dual given by (1). A linear code C of length n over $\mathbb{F}_q$ is said to be cyclic if it satisfies

$$(c_{n-1}, c_0, \ldots, c_{n-2}) \in C, \text{ whenever } (c_0, c_1, \ldots, c_{n-1}) \in C.$$

We follow the usual convention of representing vectors as polynomials. With this representation, it is well known that every cyclic code has a polynomial that generates it as an ideal of the finite ring $\mathbb{F}_q[x]/(x^n - 1)$. In general there are many generators for a given cyclic code. However, the monic generator polynomial of least degree is unique. Such a polynomial is called the generator of the code and it is a divisor of $x^n - 1$. Therefore, there is one-to-one correspondence between cyclic codes of length n over $\mathbb{F}_q$, and divisors of $x^n - 1$.

Let a be an integer such that $(a, n) = 1$. The function μ_a defined on $\mathbb{Z}_n = \{0, 1, \ldots, n-1\}$ by $\mu_a(i) \equiv ia \mod n$ is a permutation of the coordinate positions $\{0, 1, 2, \ldots, n-1\}$ and is called a multiplier. Multipliers also act on polynomials and this gives the following ring automorphism

$$\mu_a : \mathbb{F}_q[x]/(x^n - 1) \longrightarrow \mathbb{F}_q[x]/(x^n - 1) \\ f(x) \quad\quad \mapsto \mu_a(f(x)) = f(x^a). \tag{3}$$

Suppose that $f(x) = a_0 + a_1 x + \ldots + a_r x^r$ is a polynomial of degree r with $f(0) = a_0 \neq 0$. Then the monic reciprocal polynomial of $f(x)$ is

$$f^*(x) = f(0)^{-1} x^r f(x^{-1}) = f(0)^{-1} x^r (\mu_{-1}(f(x))) = a_0^{-1}(a_r + a_{r-1}x + \ldots + a_0 x^r)$$

If a polynomial is equal to its reciprocal polynomial, then it is called a self-reciprocal polynomial. If $g(x)$ is a generator polynomial of a cyclic code C of length n over $\mathbb{F}_q$, then the dual code $C^\perp$ of C is the cyclic code whose generator polynomial is $h^*(x)$ where $h^*(x)$ is the monic reciprocal polynomial of $h(x) = (x^n - 1)/g(x)$. Thus the cyclic code C is self-dual if and only if $g(x) = h^*(x)$.

3   Cyclic Codes of Length $2^a m p^s$ over $\mathbb{F}_q$

For our construction of isodual cyclic codes we need the structure of cyclic codes of length $2^a m p^s$ over $\mathbb{F}_q$ where q is a power of p and $(m, p) = 1$. We begin with the following two lemmas.

Lemma 3.1. *Let q be a power of an odd prime p, $a \geq 1$ an integer. There exists a primitive 2^a-th root of unity in $\mathbb{F}_q^*$ if and only if $q \equiv 1 \mod 2^a$.*

Proof. Suppose there exists a primitive 2^a-th root of unity $\alpha \in \mathbb{F}_q^*$. Then $\alpha^{2^a} = 1$, and since $\mathbb{F}_q^*$ is a cyclic group of order $q-1$, 2^a divides $q-1$. Conversely, if 2^a divides $q - 1$, then there exists a positive integer k such that $q - 1 = k2^a$. If β is a primitive element of $\mathbb{F}_q^*$, then $1 = \beta^{q-1} = (\beta^k)^{2^a}$. We have that in the cyclic group $\mathbb{F}_q^*$, $ord(\beta^k) = \frac{ord(\beta)}{(k, ord(\beta))} = \frac{q-1}{(k, q-1)} = \frac{k2^a}{(k, k2^a)} = 2^a$. Thus, β^k is a primitive 2^a-th root of the unity in $\mathbb{F}_q^*$. □

Lemma 3.2. *Let $a \geq 1$ be an integer and α a primitive 2^a-th root of unity in $\mathbb{F}_q^*$. Then the following holds:*

i) α^{2^i} *is a primitive 2^{a-i}-th root of unity in $\mathbb{F}_q^*$ for all i, $i \leq a$.*

ii) α^m *is a primitive 2^a-th root of unity in $\mathbb{F}_q^*$ for all odd integers m.*

iii) $\prod_{k=1}^{2^a} \alpha^k = -1$.

Proof. (i) For $i \leq a$, in the cyclic group $\mathbb{F}_q^*$ we have that $ord(\alpha^{2^i}) = \frac{ord(\alpha)}{(2^i, ord(\alpha))} = \frac{2^a}{(2^i, 2^a)} = \frac{2^a}{2^i} = 2^{a-i}$

(ii) Since $(2^a, m) = 1$, then $ord(\alpha^m) = \frac{ord(\alpha)}{(m, ord(\alpha))} = \frac{2^a}{(m, 2^a)} = 2^a$.

(iii) $(x^{2^a} - 1) = \prod_{k=1}^{2^a}(x - \alpha^k)$ so that $\prod_{k=1}^{2^a} \alpha^k = (-1)^{2^a} \frac{(-1)}{1} = -1$. □

Proposition 3.3. *Let q be a power of an odd prime p and $n = 2^a m$ a positive integer such that m is an odd integer, $(m, p) = 1$ and $a \geq 1$. Then if $\mathbb{F}_q^*$ contains a primitive 2^a-root of unity and the $f_i(x)$, $0 \leq i \leq r$ are the monic irreducible factors of $x^m - 1$ in $\mathbb{F}_q$, we have that*

$$x^{2^a m} - 1 = (x^{2^a} - 1) \prod_{i=1}^{r} f_i(\alpha^{-k} x). \tag{4}$$

Proof. Assume that $x^m - 1 = (x - 1) \prod_{i=1}^{r} f_i(x)$ (for calculation purposes we let $f_0(x) = (x - 1)$), is the factorization of $x^m - 1$ into monic factors over $\mathbb{F}_q$. This factorization is unique since it is over a unique factorization domain (UFD). Let $\alpha \in \mathbb{F}_q^*$ be a primitive 2^a-th root of unity and let $1 \leq k \leq 2^a$.

$$
\begin{aligned}
(\alpha^{-k} x)^m - 1 &= (\alpha^{-k} x - 1) \prod_{i=1}^{r} f_i(\alpha^{-k} x) \\
(\alpha^{-k})^m (x^m - (\alpha^k)^m) &= \alpha^{-k} (x - \alpha^k) \prod_{i=1}^{r} f_i(\alpha^{-k} x) \\
(x^m - \alpha^{km}) &= \alpha^{k(m-1)} (x - \alpha^k) \prod_{i=1}^{r} f_i(\alpha^{-k} x) \\
(x^m - (\alpha^m)^k) &= \alpha^{k(m-1)} (x - \alpha^k) \prod_{i=1}^{r} f_i(\alpha^{-k} x).
\end{aligned}
$$

Then by Lemma 3.2, α^m is also a primitive 2^a-th root of unity so that

$$
\begin{aligned}
\prod_{k=1}^{2^a} (x^m - (\alpha^m)^k) &= \prod_{k=1}^{2^a} \alpha^{k(m-1)} (x - \alpha^k) \prod_{i=1}^{r} f_i(\alpha^{-k} x) \\
&= \prod_{k=1}^{2^a} \alpha^{k(m-1)} \prod_{k=1}^{2^a} (x - \alpha^k) \prod_{k=1}^{2^a} \prod_{i=1}^{r} f_i(\alpha^{-k} x) \\
&= \prod_{k=1}^{2^a} \frac{\alpha^{km}}{\alpha^k} \prod_{k=1}^{2^a} (x - \alpha^k) \prod_{k=1}^{2^a} \prod_{i=1}^{r} f_i(\alpha^{-k} x) \\
&= (x^{2^a} - 1) \prod_{k=1}^{2^a} \prod_{i=1}^{r} f_i(\alpha^{-k} x).
\end{aligned}
$$

Since $(x^{2^a m} - 1) = ((x^m)^{2^a} - (\alpha^m)^{2^a}) = \prod_{k=1}^{2^a} (x^m - \alpha^{km})$, the result follows. $\square$

Corollary 3.4. *Let q be a power of an odd prime p and $n = 2^a m p^s$ a positive integer such that m is an odd integer, $(m, p) = 1$ and $a \geq 1$. Then if $\mathbb{F}_q^*$ contains a primitive 2^a-root of unity and the $f_i(x)$, $0 \leq i \leq r$ are the monic irreducible factors of $x^m - 1$ in $\mathbb{F}_q$, we have that*

$$(x^{2^a m p^s} - 1) = (x^{2^a m} - 1)^{p^s} = (x^{2^a} - 1)^{p^s} \prod_{k=1}^{2^a} \prod_{i=1}^{r} f_i^{p^s}(\alpha^{-k} x).$$

Proof. The proof is similar to that for Proposition 3.3. $\square$

In the following corollary we give the structure of cyclic codes of length $2^a m p^s$ over $\mathbb{F}_q$.

Corollary 3.5. *Let q be a power of an odd prime p, and $n = 2^a m p^s$ be a positive integer such that m is odd integer, $a \geq 1$ and $(m, p) = 1$. If $\mathbb{F}_q^*$ contains a primitive 2^a-root of unity and $(x - 1)$, $f_i(x)$, $1 \leq i \leq r$ are the monic irreducible factors of $x^m - 1$ in $\mathbb{F}_q[x]$ then any cyclic code of length $n = 2^a m p^s$ is generated by $\prod_{k=1}^{2^a} ((x - \alpha^k)^{l_k} \prod_{i=1}^{r} f_i^{j_i}(\alpha^{-k} x))$ where $0 \leq l_k, j_i \leq p^s$.*

Proof. Any cyclic code of length $n = 2^a m p^s$ is generated by a divisor of $(x^{2^a m p^s} - 1)$. By Corollary 3.4 we have that

$$(x^{2^a m p^s} - 1) = (x^{2^a m} - 1)^{p^s} = \prod_{k=1}^{2^a} ((x - \alpha^k)^{p^s} \prod_{i=1}^{r} f_i^{p^s}(\alpha^{-k} x)),$$

and the result follows. $\qquad\qquad\qquad\qquad\qquad\qquad\qquad\qquad\qquad\qquad\square$

4   Construction of Cyclic Isodual Codes of Length $2^a m p^s$ over $\mathbb{F}_q$

We first recall the following important result of Batoul et al. given in [2]

Proposition 4.1. *(Proposition 3.1 [2]) Let C be a cyclic code of length n over $\mathbb{F}_q$ generated by the polynomial $g(x)$, and $\lambda \in \mathbb{F}_q^*$ such that $\lambda^n = 1$. Then the following holds:*

(i) C is equivalent to the cyclic code generated by $g^*(x)$.
(ii) C is equivalent to the cyclic code generated by $g(\lambda x)$.

Using Proposition 4.1, we give new constructions of isodual cyclic codes of length $2^a m p^s$ over $\mathbb{F}_q$.

Theorem 4.2. *Let q be a power of an odd prime p such that $q \equiv 1 \mod 2^a$, with $a \geq 1$ an integer, n' an odd integer and $f(x)$ a polynomial in $\mathbb{F}_q[x]$ such that*

$$x^{n'} - 1 = (x - 1) f(x).$$

Then the cyclic codes of length $2^a n'$ generated by

$$(x^{2^{a-1}} - 1) \prod_{k=0}^{2^{a-1}-1} f(\alpha^{-2k-1} x),$$

and

$$(x^{2^{a-1}} + 1) \prod_{k=1}^{2^{a-1}} f(\alpha^{-2k} x).$$

are isodual codes of length $2^a n'$ over $\mathbb{F}_q$ where $\alpha \in \mathbb{F}_q^*$ is a primitive 2^a-th root of unity.

Proof. By Lemma 3.1, if $q \equiv 1 \mod 2^a$ then there exists a primitive 2^a-th root of unity $\alpha \in \mathbb{F}_q^*$ such that $\alpha^{2^a} = 1$. Suppose that $x^{n'} - 1 = (x - 1) f(x)$, so then

$$(x^{2^a n'} - 1) = (x^{2^a} - 1) \prod_{k=1}^{2^a} f(\alpha^{-k} x).$$

We have $(x^{2^a} - 1) = (x^{2^{a-1}} - 1)(x^{2^{a-1}} + 1)$, which gives that

$$(x^{2^a n'} - 1) = (x^{2^{a-1}} - 1)(x^{2^{a-1}} + 1) \prod_{k=1}^{2^a} f(\alpha^{-k} x)$$

$$= (x^{2^{a-1}} - 1)(x^{2^{a-1}} + 1) \prod_{k=1}^{2^{a-1}} f(\alpha^{-2k} x) \prod_{k=0}^{2^{a-1}-1} f(\alpha^{-2k-1} x)$$

Let

$$g(x) = (x^{2^{a-1}} - 1) \prod_{k=0}^{2^{a-1}-1} f(\alpha^{-2k-1} x),$$

so that we have

$$h(x) = (x^{2^{a-1}} + 1) \prod_{k=1}^{2^{a-1}} f(\alpha^{-2k} x),$$

and $h^*(x) = g^*(\alpha x)$. By Proposition 4.1(i), C is equivalent to the cyclic code generated by $g^*(x)$. By Proposition 4.1(ii), the cyclic code generated by $g^*(x)$ is equivalent to the cyclic code generated by $g^*(\alpha x) = h^*(x)$. As the latter code is $C^\perp$, C is isodual, so that the cyclic code generated by $g(x)$ is isodual. The same result is obtained for

$$g(x) = (x^{2^{a-1}} + 1) \prod_{k=1}^{2^{a-1}} f(\alpha^{-2k} x).$$

$\square$

Example 4.3. *Over* $\mathbb{F}_3$ *we have* $x^7 - 1 = (x+2)(x^6 + x^5 + x^4 + x^3 + x^2 + x + 1)$, *so that* $x^{14} - 1 = (x+2)(x^6 + x^5 + x^4 + x^3 + x^2 + x + 1)(x+1)(x^6 - x^5 + x^4 - x^3 + x^2 - x + 1)$, *and the cyclic codes generated by*

$$(x+2)(x^6 - x^5 + x^4 - x^3 + x^2 - x + 1) \text{ and } (x+1)(x^6 + x^5 + x^4 + x^3 + x^2 + x + 1)$$

are isodual.

Remark 4.4. *The codes generated by*

$$(x^{2^{a-1}n'} - 1) = (x^{2^{a-1}} - 1) \prod_{k=1}^{2^{a-1}} f(\alpha^{-2k} x),$$

and

$$(x^{2^{a-1}n'} + 1) = (x^{2^{a-1}} + 1) \prod_{k=0}^{2^{a-1}-1} f(\alpha^{-2k-1} x),$$

are the trivial isodual cyclic codes of length $2^a n'$.

Theorem 4.5. *Let q be a power of an odd prime p such that $q \equiv 1 \mod 2^a$, with $a \geq 1$ an integer, n' an odd integer and $f_1(x)$, $f_2(x)$ polynomials in $\mathbb{F}_q[x]$ such that*

$$x^{n'} - 1 = (x - 1)f_1(x)f_2(x).$$

Then the cyclic codes of length $2^a n'$ generated by

$$(x^{2^{a-1}} - 1) \prod_{k=1}^{2^{a-1}} f_i(\alpha^{-2k}x) \prod_{k=0}^{2^{a-1}-1} f_j(\alpha^{-2k-1}x),$$

and

$$(x^{2^{a-1}} + 1) \prod_{k=1}^{2^{a-1}} f_i(\alpha^{-2k}x) \prod_{k=0}^{2^{a-1}-1} f_j(\alpha^{-2k-1}x),$$

$i, j \in \{1, 2\}, i \neq j$, *respectively, are isodual codes of length $2^a n'$ over $\mathbb{F}_q$ where $\alpha \in \mathbb{F}_q^*$ is a primitive 2^a-th root of unity.*

Proof. By Lemma 3.1, if $q \equiv 1 \mod 2^a$ then there exists a primitive 2^a-th root of unity $\alpha \in \mathbb{F}_q^*$ such that $\alpha^{2^a} = 1$. Suppose $x^{n'} - 1 = (x - 1)f_1(x)f_2(x)$ so that

$$(x^{2^a n'} - 1) = (x^{2^a} - 1) \prod_{k=1}^{2^a} f_1(\alpha^{-k}x)f_2(\alpha^{-k}x).$$

We have $(x^{2^a} - 1) = (x^{2^{a-1}} - 1)(x^{2^{a-1}} + 1)$, which gives that

$$(x^{2^a n'} - 1) = (x^{2^{a-1}} - 1)(x^{2^{a-1}} + 1) \prod_{k=1}^{2^a} f_1(\alpha^{-k}x)f_2(\alpha^{-k}x)$$

$$= (x^{2^{a-1}} - 1)(x^{2^{a-1}} + 1) \prod_{k=1}^{2^{a-1}} f_1(\alpha^{-2k}x)f_2(\alpha^{-2k}x)$$

$$\prod_{k=0}^{2^{a-1}-1} f_1(\alpha^{-2k-1}x)f_2(\alpha^{-2k-1}x).$$

If

$$g(x) = (x^{2^{a-1}} - 1) \prod_{k=1}^{2^{a-1}} f_i(\alpha^{-2k}x) \prod_{k=0}^{2^{a-1}-1} f_j(\alpha^{-2k-1}x), \ i \neq j,$$

then we have

$$h(x) = (x^{2^{a-1}} + 1) \prod_{k=0}^{2^{a-1}-1} f_i(\alpha^{-2k-1}x) \prod_{k=1}^{2^{a-1}} f_j(\alpha^{-2k}x),$$

and $h^*(x) = g^*(\alpha^{-1}x)$. By Proposition 4.1(i), C is equivalent to the cyclic code generated by $g(x)^*$. By Proposition 4.1(ii), the cyclic code generated by $g(x)^*$ is equivalent to the cyclic code generated by $g^*(\alpha^{-1}x) = h^*(x)$. As the latter code is $C^\perp$, C is isodual, so the cyclic code generated by $g(x)$ is isodual. The same result is obtained for the code generated by

$$g(x) = (x^{2^{a-1}} + 1) \prod_{k=1}^{2^{a-1}} f_i(\alpha^{-2k}x) \prod_{k=0}^{2^{a-1}-1} f_j(\alpha^{-2k-1}x), \ i \neq j.$$

$\square$

Remark 4.6. *If*

$$g(x) = (x^{2^{a-1}} - 1) \prod_{k=1}^{2^{a-1}} f_i(\alpha^{-2k}x) \prod_{k=1}^{2^{a-1}} f_j^{p^s}(\alpha^{-2k}x), \ i \neq j$$

then $g(x) = (x^{2^{a-1}n'} - 1) = (x^{\frac{n}{2}} - 1)$ and the cyclic code generated by $g(x)$ is the trivial isodual code.

Corollary 4.7. *Let q be an odd prime power such that m is an odd integer and $f_1(x)$, $f_2(x)$ be polynomials in $\mathbb{F}_q[x]$ such that $x^{n'} - 1 = (x-1)f_1(x)f_2(x)$. Then the cyclic codes of length $2^a n'$ generated by*

$$(x^{2^{a-1}} - 1) \prod_{k=0}^{2^{a-1}} f_i(\alpha^{-2k-1}x) \prod_{k=0}^{2^{a-1}-1} f_j(\alpha^{-2k-1}x),$$

and

$$(x^{2^{a-1}} + 1) \prod_{k=1}^{2^{a-1}} f_i(\alpha^{-2k}x) \prod_{k=1}^{2^{a-1}} f_j(\alpha^{-2k}x),$$

$i, j \in \{1,2\}$, $i \neq j$ are isodual codes of length $2^a n'$ over $\mathbb{F}_q$.

Proof. The results follows immediately from Theorem 4.2. $\square$

Example 4.8. *For $q = 5$, from Lemma 3.1 there exists $\beta \in \mathbb{F}_5$ such that $\beta^4 = 1$, e.g. $\beta = 2$. If $m = 11$, we have*

$$(x^{11} - 1) = (x-1)(x^5 + 2x^4 + 4x^3 + x^2 + x + 4)(x^5 + 4x^4 + 4x^3 + x^2 + 3x + 4),$$

so that $(x^{11} - 1) = (x-1)f_1(x)f_2(x)$, and

$$(x^{44} - 1) = (x-1)f_1(x)f_2(x)(x+1)f_1(-x)f_2(-x)(x+2)f_1(2x)f_2(2x)(x-2)$$
$$f_1(-2x)f_2(-2x).$$

Then the cyclic codes generated by $g_1(x) = (x^2 \pm 1)f_1(x)f_1(-x)f_1(2x)f_1(-2x)$ and $g_2(x) = (x^2 \pm 1)f_2(x)f_2(-x)f_2(2x)f_2(-2x)$ are isodual cyclic codes over $\mathbb{F}_5$.

Example 4.9. *For $q = 17 \equiv 1 \mod 2^4$, from Lemma 3.1 there exists $\alpha \in \mathbb{F}_{17}$ such that $\alpha^{16} = 1$, e.g. $\alpha = 3$. If $m = 9$, we have*

$$(x^9 - 1) = (x-1)(x^2+x+1)(x^2+3x+1)(x^2+4x+1)(x^2+10x+1) = (x-1)f_1(x)f_2(x).$$

With $f_1(x) = (x^2 + x + 1)(x^2 + 3x + 1)$ and $f_2(x) = (x^2 + 4x + 1)(x^2 + 10x + 1)$, the cyclic codes generated by

$$(x^8 - 1)^{17^s} \prod_{k=1}^{8} f_i^{17^s}(3^{-2k}x) \prod_{k=0}^{7} f_j^{17^s}(3^{-2k-1}x),$$

and

$$(x^8 + 1)^{17^s} \prod_{k=1}^{8} f_i^{17^s}(3^{-2k}x) \prod_{k=0}^{7} f_j^{17^s}(3^{-2k-1}x),$$

$i \neq j$, are isodual cyclic codes over $\mathbb{F}_{17}$.

5   Cyclic Isodual Codes of Length $2^a m p^s$ over $\mathbb{F}_q$ from Duadic Codes

The results of Section 4 provide constructions of isodual cyclic codes over finite fields. However, a more straightforward means of finding these codes is desirable. Further, determining codes with good minimum distance is very important. In this section, infinite families of cyclic isodual codes over finite fields are constructed from duadic codes. The motivation is that duadic codes are known to have good minimum distance. Before giving our constructions of isodual cyclic codes, we recall some results about duadic codes which be used in this section. Of course isodual codes cannot be duadic since their length is even. Let q be a power of a prime p and let m be a positive odd integer such that $(m, q) = 1$. Then if $0 \le i < m$, the q-cyclotomic coset of i (mod m) is defined as

$$Cl(i) = \{iq^l \pmod{m} | l \in \mathbb{N}\}.$$

Let α be a primitive m-th root of unity in an extension field of $\mathbb{F}_q$, and C be a cyclic code over $\mathbb{F}_q$ of length m generated by a polynomial $f(x)$. C is uniquely determined by its defining set $T = \{0 \le i < m \,|\, f(\alpha^i) = 0\}$. Hence the defining set of a cyclic code over $\mathbb{F}_q$ is the union of some q-cyclotomic cosets.

Let S_1 and S_2 be unions of cyclotomic cosets modulo m such that $S_1 \cap S_2 = \emptyset$, $S_1 \cup S_2 = \mathbb{Z}_m \setminus \{0\}$, and $\mu_a S_i \mod n = S_{(i+1) \mod 2}$. Then the triple μ_a, S_1, S_2 is called a splitting modulo m. The odd-like duadic codes D_1 and D_2 are the cyclic codes over $\mathbb{F}_q$ with defining sets S_1 and S_2 and generator polynomials $f_1(x) = \Pi_{i \in S_1}(x - \alpha^i)$ and $f_2(x) = \Pi_{i \in S_2}(x - \alpha^i)$, respectively. The even-like duadic codes C_1 and C_2 are the cyclic codes over $\mathbb{F}_q$ with defining sets $\{0\} \cup S_1$ and $\{0\} \cup S_2$, respectively.

For the remainder of the paper, the notation $q = \square \mod n$ means that q is a quadratic residue modulo n. For a prime power q and integer n such that

$\gcd(q, n) = 1$, denote by $ord_n(q)$ the multiplicative order of q modulo n. This is the smallest integer l such that $q^l \equiv 1 \mod n$.

The multiplier μ_{-1} plays a special role in determining the duals of duadic codes just as it does for duals in general cyclic codes. In the following we give some important results concerning μ_{-1}.

Lemma 5.1. *(Proposition 4.4 [2]) Let $\mathbb{F}_q$ be a finite field and m a positive odd integer such that $(m, q) = 1$ and $q = \square \mod m$. Then there exists a pair of odd-like duadic codes over $\mathbb{F}_q$, D_1 and D_2, generated by $f_1(x)$ and $f_2(x)$, respectively, such that $x^n - 1 = (x - 1)f_1(x)f_2(x)$. We have the following results:*

i) *If the splitting is given by μ_{-1} then $f_1^*(x) = f_2(x)$ and $f_2^*(x) = f_1(x)$.*
ii) *If the splitting is not given by μ_{-1} then $f_1^*(x) = f_1(x)$ and $f_2^*(x) = f_2(x)$.*

We now consider when a splitting is given by μ_{-1}, and also when a splitting is left invariant by μ_{-1}.

Theorem 5.2. *[7] Let $\mathbb{F}_q$ be a finite field and $m = p_1^{s_1} p_2^{s_2} \cdots p_k^{s_k}$ be the prime factorization of an odd integer m such that $q \equiv \square \mod m$.*

i) *If $p_i \equiv -1 \mod 4$, $i = 1, 2, \ldots, k$, then all splittings mod m are given by μ_{-1}.*
ii) *If there is at least one p_i such that $p_i \equiv 1(\mod 4)$, $i \in \{1, 2, \ldots, k\}$, then there is a splitting mod m which is not given by μ_{-1}.*

Remark 5.3. *In general, the same splitting modulo m an odd integer can be given by different multipliers. For more details see [4, p. 214]. When we consider the multiplier μ_{-1}, we mean any multiplier which gives the same splitting as μ_{-1}.*

In the following we give several constructions of isodual cyclic codes over $\mathbb{F}_q$ of length $2^a m p^s$, $a \geq 1$ using generators of odd-like duadic codes over $\mathbb{F}_q$ of length m an odd integer. The construction of repeated-root isodual cyclic codes over fields with even characteristic was given in [2]. Here we give constructions of repeated-root isodual cyclic code over fields with odd characteristic.

Theorem 5.4. *Let q be a power of an odd prime p. Suppose there exists a pair of odd-like Duadic codes $D_i = \langle f_i(x) \rangle$ of odd length m, and $\alpha \in \mathbb{F}_q^*$ is a primitive 2^a-th root of unity. We then have the following:*

i) *The cyclic codes C_{ij} and C'_{ij} of length $2^a m p^s$ over $\mathbb{F}_q$ generated by*

$$(x^{2^{a-1}} - 1)^{p^s} \prod_{k=1}^{2^{a-1}} f_i^{p^s}(\alpha^{-2k}x) \prod_{k=0}^{2^{a-1}-1} f_j^{p^s}(\alpha^{-2k-1}x),$$

and

$$(x^{2^{a-1}} + 1)^{p^s} \prod_{k=1}^{2^{a-1}} f_i^{p^s}(\alpha^{-2k}x) \prod_{k=0}^{2^{a-1}-1} f_j^{p^s}(\alpha^{-2k-1}x),$$

$i, j \in \{1, 2\}, i \neq j$, *respectively, are isodual codes of length $2^a m p^s$ over $\mathbb{F}_q$.*

ii) If the splitting modulo m is given by μ_{-1}, the cyclic codes C_i and C_i' of length $2^a m p^s$ generated by

$$(x^{2^{a-1}} - 1)^{p^s} \prod_{k=1}^{2^a} f_i^{p^s}(\alpha^{-k}x),$$

and

$$(x^{2^{a-1}} + 1)^{p^s} \prod_{k=1}^{2^a} f_i^{p^s}(\alpha^{-k}x),$$

respectively, are isodual over $\mathbb{F}_q$.

iii) If the splitting modulo m is not given by μ_{-1}, then the dual of the cyclic code of length $2^a m p^s$ over $\mathbb{F}_q$ generated by

$$(x^{2^{a-1}} - 1)^{p^s} \prod_{k=1}^{2^a} f_i^{p^s}(\alpha^{-k}x),$$

is equivalent to the cyclic code generated by

$$(x^{2^{a-1}} + 1)^{p^s} \prod_{k=1}^{2^a} f_j^{p^s}(\alpha^{-k}x).$$

Proof.

i) Follows from Theorem 4.5.

ii) Let

$$C_i = \langle g_i(x) \rangle$$

$$= \langle (x^{2^{a-1}} - 1)^{p^s} \prod_{k=1}^{2^a} f_i^{p^s}(\alpha^{-k}x) \rangle. \tag{5}$$

If the splitting modulo m is given by μ_{-1} then $f_1^*(x) = f_2(x)$ and $f_2^*(x) = f_1(x)$, so that

$$C_i^{\perp} = \langle h_i^*(x) \rangle$$

$$= \langle (x^{2^{a-1}} + 1)^{*p^s} \prod_{k=1}^{2^a} f_j^{p^s}(\alpha^{-k}x)^* \rangle$$

$$= \langle (x^{2^{a-1}} + 1)^{p^s} \prod_{k=1}^{2^a} f_i^{p^s}(\alpha^{-k}x) \rangle$$

$$= \langle -g_i(\alpha^{-1}x) \rangle. \tag{6}$$

By Proposition 4.1(ii), C_i is equivalent to the cyclic code generated by $g_i(\alpha^{-1}x)$. As the latter code is $C_i^{\perp}$, C_{ii} is isodual. The same proof is used for codes generated by $g_i(x) = (x^{2^{a-1}} + 1)^{p^s} \prod_{k=1}^{2^a} f_i^{p^s}(\alpha^{-k}x)$.

iii) If the splitting modulo m is not given by μ_{-1}, then $f_1^*(x) = f_1(x)$ and $f_2^*(x) = f_2(x)$, so that

$$C_i^{\perp} = \langle h_i^*(x) \rangle$$

$$= \langle (x^{2^{a-1}} + 1)^{*p^s} \prod_{k=1}^{2^a} f_j^{p^s}(\alpha^{-k}x)^* \rangle$$

$$= \langle (x^{2^{a-1}} - 1)^{p^s} \prod_{k=1}^{2^a} f_j^{p^s}(\alpha^{-k}x) \rangle$$

$$= \langle -g_j(\alpha^{-1}x) \rangle \tag{7}$$

By Proposition 4.1(ii) $C_j \simeq C_i^{\perp}$. The same proof is used for codes generated by $g_i(x) = (x^{2^{a-1}} + 1)^{p^s} \prod_{k=1}^{2^a} f_i^{p^s}(\alpha^{-k}x)$. $\qquad\square$

Example 5.5. *For $q = 3$ and $m = 13$, $3 \equiv 16 \mod 13$, so there exist duadic codes generated by f_i $1 \le i \le 2$. Since $13 \equiv 1 \mod 4$, by Theorem 5.2 there is a splitting modulo 13 which is not given by μ_{-1} so that*

$$x^m - 1 = (x-1)(x^3 + 2x + 2)(x^3 + x^2 + x + 2)(x^3 + x^2 + 2)(x^3 + 2x^2 + 2x + 2)$$
$$= (x-1)u(x)u^*(x)v(x)v^*(x).$$

We have the following results:

(i) *If $f_1(x) = u(x)u^*(x))$ and $f_2(x) = v(x)v^*(x))$, then $f_i^*(x) = f_i(x)$, and the cyclic code of length 26 over $\mathbb{F}_3$ generated by $g(x) = (x-1)f_i(x)f_j(-x)$ $(i \ne j)$ is an isodual code with minimum distance 6.*

(ii) *If $f_1(x) = u(x)v(x))$ and $f_2(x) = u^*(x)v^*(x))$, then $f_i^*(x) = f_j(x)$, and the cyclic code of length 26 over $\mathbb{F}_3$ generated by $g(x) = (x-1)f_i(x)f_i(-x)$ is isodual with minimum distance 6.*

Example 5.6. *For $q = 5$ and $m = 11$, $5 \equiv 16 \mod 11$, so there exist duadic codes generated by f_i $1 \le i \le 2$. Since $11 \equiv -1 \mod 4$, by Theorem 5.2 all splittings are given by μ_{-1} and we have*

$$(x^{11} - 1) = (x-1)(x^5 + 2x^4 + 4x^3 + x^2 + x + 4)(x^5 + 4x^4 + 4x^3 + x^2 + 3x + 4),$$

so that $(x^{11} - 1) = (x-1)f_1(x)f_2(x) = -(x-1)f_1(x)f_1^*(x)$. Then the code of length 22 over $\mathbb{F}_5$ generated by $g_1(x) = (x-1)f_i(x)f_i^*(-x)$ is an isodual cyclic code with minimum distance 8, and the code of length 22 over $\mathbb{F}_5$ generated by $g_2(x) = (x+1)f_i(x)f_i(-x)$ is an isodual cyclic code with minimum distance 6.

Example 5.7. *For $q = 7$ and $m = 9$, $7 \equiv 1 \mod 3$, so there exist duadic codes generated by f_i $1 \le i \le 2$. Since $3 \equiv -1 \mod 4$, by Theorem 5.2 all splittings are given by μ_{-1}. From $(x^9 - 1) = (x-1)(x+3)(x+5)(x^3+3)(x^3+5)$, we have $f_1(x) = (x+3)(x^3+3)$ and $f_2(x) = (x+5)(x^3+5)$ so that $(x^9-1) = (x-1)f_1(x)f_2(x) = (x-1)f_1(x)f_1^*(x)$. Then the cyclic codes of length 18 over $\mathbb{F}_7$ generated by $(x-1)f_i(x)f_i(-x)$ and $(x-1)f_i(x)f_i(-x)$ are isodual with minimum distance 4.*

Example 5.8. *For $q = 5 \equiv 1 \mod 4$, from Lemma 3.1 there exists $\gamma \in \mathbb{F}_5$ such that $\gamma^2 = -1$, e.g. $\gamma = 2$. If $m = 11$, we have*

$$(x^{11} - 1) = (x - 1)(x^5 + 2x^4 + 4x^3 + x^2 + x + 4)(x^5 + 4x^4 + 4x^3 + x^2 + 3x + 4),$$

and therefore $(x^{11} - 1) = (x - 1)f_1(x)f_2(x)$ and

$$(x^{44} - 1) = (x - 1)f_1(x)f_2(x)(x + 1)f_1(-x)f_2(-x)(x + 2)f_1(2x)f_2(2x)(x - 2)$$
$$f_1(-2x)f_2(-2x).$$

Let $g_1(x) = (x^2 \pm 1)f_1(x)f_1(-x)f_1(2x)f_1(-2x)$ and $g_2(x) = (x^2 \pm 1)f_2(x)f_2(-x)$ $f_2(2x)f_2(-2x)$. Then the cyclic codes generated by $g_1(x)$ and $g_2(x)$ are isodual cyclic codes over $\mathbb{F}_5$.

Example 5.9. *For $q = 17 \equiv 1 \mod 2^4$, from Lemma 3.1 there exists $\alpha \in \mathbb{F}_{17}$ such that $\alpha^{16} = 1$, e.g. $\alpha = 3$. If $m = 13 \equiv 1 \mod 4$, there exist a pair of odd like duadic codes of length 13 generated by $f_1(x)$ and $f_2(x)$, such that*

$$(x^{13} - 1) = (x - 1)(x^6 + 5x^5 + 2x^4 + 4x^3 + 2x^2 + 5x + 1)$$
$$(x^6 + 13x^5 + 2x^4 + 12x^3 + 2x^2 + 13x + 1) = (x - 1)f_1(x)f_2(x).$$

Then the cyclic codes generated by

$$(x^8 - 1)^{17^s} \prod_{k=1}^{8} f_i^{p^s}(3^{-2k}x) \prod_{k=0}^{7} f_j^{17^s}(3^{-2k-1}x),$$

and

$$(x^8 + 1)^{17^s} \prod_{k=1}^{8} f_i^{17^s}(3^{-2k}x) \prod_{k=0}^{7} f_j^{17^s}(3^{-2k-1}x),$$

$i \neq j$, are isodual cyclic codes of length $2^{4\cdot13}(17)^s$, $s \geq 1$, over $\mathbb{F}_{17}$.

Example 5.10. *For $q = 17 \equiv 1 \mod 2^4$, from Lemma 3.1 there exists $\alpha \in \mathbb{F}_{17}$ such that $\alpha^{16} = 1$, e.g. $\alpha = 3$. If $m = 19 \equiv -1 \mod 4$, then there exist a pair of odd like duadic codes of length 19 generated by $f_1(x)$ and $f_2(x)$ such that*

$$(x^{13} - 1) = (x - 1)(x^9 + 4x^8 + 15x^7 + 15x^6 + 6x^5 + x^4 + 12x^3 + 2x^2 + 3x + 16)$$
$$(x^9 + 14x^8 + 15x^7 + 5x^6 + 16x^5 + 11x^4 + 2x^3 + 2x^2 + 13x + 16),$$

so that $x^{13} - 1) = (x - 1)f_1(x)f_2(x)$. Since $19 \equiv -1 \mod 4$, the splitting μ_{-1} gives codes C_i and C_i' generated by

$$(x^8 - 1)^{17^s} \prod_{k=1}^{16} f_i^{17^s}(3^{-k}x),$$

and

$$(x^8 + 1)^{17^s} \prod_{k=1}^{16} f_i^{17^s}(3^{-k}x)$$

$1 \leq i \leq 2$, respectively, are isodual codes of length $2^{4\cdot19}(17)^s$, $s \geq 1$, over $\mathbb{F}_{17}$.

References

1. Bachoc, C., Gulliver, T.A., Harada, M.: Isodual codes over Z_{2k} and isodual lattices. J. Algebra. Combin. 12, 223–240 (2000)
2. Batoul, A., Guenda, K., Gulliver, T.A.: On isodual cyclic codes over finite fields and finite chain rings: Monomial equivalence. CoRR abs/1303.1870 (2013)
3. Guenda, K.: New MDS self-dual codes over finite fields. Des., Codes, Crypt. 62(1), 31–42 (2012)
4. Huffman, W.C., Pless, V.: Fundamentals of Error-Correcting Codes. Cambridge Univ. Press, New York (2003)
5. Jia, Y., Ling, S., Xing, C.: On self-dual cyclic codes over finite fields. IEEE Trans. Inform. Theory 57(4), 2243–2251 (2011)
6. Jia, Y., Ling, S., Solé, P.: On isodual of cyclic codes over finite fields: Multiplier equivalence. Preprint
7. Smid, M.H.M.: Duadic codes. IEEE. Trans. Inform. Theory 33(3), 432–433 (1987)

Formal Enforcement of Security Policies on Parallel Systems with Risk Integration

Marwa Ziadia[✉] and Mohamed Mejri

Department of Computer Science,
Laval University, Quebec, Canada
Marwa.ziadia.1@ulaval.ca, Mohamed.Mejri@ift.ulaval.ca

Abstract. In this paper, we survey the problem of mobile security. Therefore, we introduce a formal technique allowing the enforcement of security policy on this parallel system. The main idea was to give the end-user the possibility to choose his mobile security level and to control it by choosing a risk level. So we adapted this notion to the syntax as well as the semantic of the used languages. We use an extended version of process algebra ACP (Algebra of Communicating Process) to specify the program and we define a logic that goes well with this language, to specify security policy. An example is given at the end to illustrate the approach and apply it with a real Android application from Google Play.

Keywords: Mobile security · Security policy · Enforcement · Process algebra · Risk

1 Introduction

Nowadays, securing our mobile and protecting our private life, requires intellectual effort from user. User has to inquire about applications that he wants to install, because a set of permissions are displayed each time that he requests to set up an application. These permissions corresponds to the application's potential behavior. At this level, he has no choice, he must accept all the permissions or to deny it, and in this case the installation will be aborted. Knowing that there is nothing between this two possibilities without using a third party software, which is too technical for a classic user.

Usually, when user clicks on the "Accept" button, which is the only permission decision most he ever get to make, it is for him to evaluate the risk that he will take, because nothing can guarantee if it is a malicious or benign application. With the acceptance of installation, user grants to the installed application authorities that it maybe will never need. According to [1], analytic results showed that from $141,372$ Android applications, $76,366$ $(54,01\%)$ required more permission(s) than that it really need.

This can cause negative impacts, such as leaking of private information, accessing the system tools, recording audio & video, or surreptitiously calling expensive phone numbers, etc. Even more, allowing for example an application to modify or delete the contents of the SD card means that we can give it the

© Springer International Publishing Switzerland 2015
S. El Hajji et al. (Eds.): C2SI 2015, LNCS 9084, pp. 133–148, 2015.
DOI: 10.1007/978-3-319-18681-8_11

authority to read and write any element of the card, such permission gives malicious applications the ability to replace existing files on the card. To address this, the idea was to give the end-user the possibility to control and specify his mobile security level, and this is by choosing a risk level.

Concurrent system consists of programs running in parallel. As can be seen, in our work, we mean with parallel system; the mobile platform, because concurrency or parallelism aspect is inherently existent in this system, that we could also called it, system of parallel processors. It is through the process algebra that we are able to specify such system. This mathematical framework studies the behavior of parallel or distributed systems by algebraic means [2]. This algebraic rules allow process to be simply described, manipulated and analyzed.

In this paper, we propose an algebraic approach that enforces a security policy on a given parallel system, based upon a risk evaluation. Our method relies on some assumptions, that allow us to include the risk notion and adapt it in the syntax as well as semantics of the used language.

The inputs of our problem are, a security policy φ, a process P and a risk value α (threshold) which is fixed according to the risk level chosen by the mobile user.

The output is a new process P' that respects the security policy and the user choice.

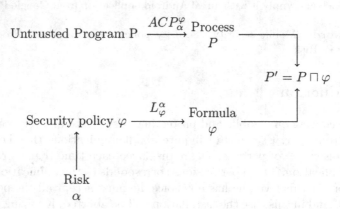

This new process has the following characteristics :

- $P'| \sim \varphi : P'$ must respect φ;
- $P' \subseteq P$: the traces of P' are also the traces of P;
- $\forall Q : ((Q| \sim \varphi) \wedge (Q \subseteq P)) \Rightarrow Q \subseteq P'$: all P traces that respect φ are also P' possible traces.

This paper is organized as follows: Section 2 is devoted to the definition of the risk notion and the used languages, with a brief description of related works. In Section 3, we present the syntax and the semantic of the logic used to specify security policies. In Section 4, we present as well the syntax and semantic used to specify processors. Section 5 is dedicated to the description of enforcement approach and how we integrated the risk to it; whether in security policy or in the enforced program. In Section 6, we exemplify the approach with an Android application and an invented security policy . Our conclusion is given in Section 7.

2   State of the Art

Even though there are diverse promising works that have treated the subject of security enforcement, the idea of introducing the risk notion in this type of problem remains original. From these works we can cite ([3], [4], [5], [7], [6], [8]).

[9] was one of the first works that has been oriented to the formalization of enforcement notion. In his paper, Schneider discusses the enforcement mechanism that works by monitoring, i.e. a monitor that supervises an untrusted program and blocks its execution when a property of security policy is violated. In [5], a new model of monitor is introduced, it's based on rewriting in which the program can be modified at run time. The monitor in this work can correct the execution of the program that violates the security policy rather than stops its execution.

Recently in [10], a program rewriting approach has been adopted to automatically enforce a security policy on an untrusted sequential program. The security enforcement is transformed to a resolution of linear systems, extracted from a computation of intersection between a process presenting the policy security and another capturing the sequential program. The basic idea of this approach is taken from a previous work [11], trying to ameliorate it formally and to prove its main results and finally to implement it in an environment denoted by FASER.

In [4], Langar and al. proposed an algebraic and automatic approach that generates from a given program and a security policy a new secure version of the original program. They defined a process algebra ACP^φ offering an algebraic framework for the enforcement and a logic that allows specifying a formal monitor. The results provide an elegant technique allowing automatically to enforce security policy on concurrent systems. This technique is restrictive, since it doesn't give the final user the possibility to choose his own level of security and to decide if such action can be executed. An action that violates a property of the security policy is simply aborted.

In our paper, we start from this work, and we adapt the technique used to our system (mobile) and to our problem (risk evaluation). So we have relied on a modified version of ACP; ACP^φ used in [4], this version is enhanced with enforcement operation, in order to modify it to cover the aspect of mobile system as well as the risk concept. So the reader may refer for instance to [4] for more details.

Starting by the risk, by definition, the risk is closely related to uncertainty. It is defined as a combination of the probability of an event and its consequences.

$$Risk = Probability * Impact \tag{1}$$

Note firstly that in our case, the risk can be considered equivalent to the impact; as we said the risk is a combination of an event (this event is the action executed by a process in our case) and its consequences (the impact of the executions of action(s)). So this cannot be related any more to the future, because the occurrence of actions is certain, which means that the probability of occurrence is equal to 1, and it's the impact of the execution that is in question. As a result, the risk is equivalent to the impact.

$$Risk \equiv Impact \tag{2}$$

So each action executed has an impact in the target program. Therefore, we can classify the risk as "low", "medium", "high" and "very high" to be compared after.

As mentioned above, our proposal includes a risk evaluation in order to add necessary information that allows the control (deny or permit) of actions executed by an application. Each application declare a list of permissions that would be not risky for a user who lack visibility into how applications use his private data, especially applications coming from unknown or unsafe sources, that could hide unauthorized access and perform sensitive operations without user consent. In [12], a new framework that analyzes smartphone application activity is proposed. It detects anomalous behavior of known applications. They show that the actions like; open(), read(), access() and chmod() are the most used system calls by malware. So we can consider these actions as risky and which need more control.

From this definition, we have adapted this concept and integrated the risk semantically into our algebraic framework.

3 L_φ^α: The Specification Logic of Security Policy

In this section, we introduce syntax, semantics and basic properties of our Logic used to define security policy.

To properly include the risk, we define a logic L_φ^α inspired from extended regular expressions. This logic expresses specific properties for enforcement of security policy. It is a linear logic which expresses the regular language class with the possibility of expressing infinite properties. In addition, it allows us to express the temporal aspect (temporal evolution of a process) and especially we can express the risk notion in a formula by this logic, besides, it fits well with process algebra syntax.

Notations
We first present notations adopted to express the syntax and semantics of each formula. See Table 1.

3.1   Syntax of a Logic L_φ^α

We started out from a formula of a simple atomic action and then we used the basic operations to compose it into more complicated formulas. The syntax of the resulted Logic L_φ^α is presented in Table 2.

Table 1. Notations

tt	Constant boolean that represents true
ff	Constant boolean that represents false
$\mathcal{A}$	Set of atomic actions
a	Action that belongs to the set $\mathcal{A}$
$\mathcal{T}$	Set of possible traces constructed from the action of $\mathcal{A}$
ε	Empty trace
$\xi_1.\xi_2$	Concatenation of two traces ξ and ξ'
α	Impact value
$Risk()$	Function that calculates the impact of a process (action(s)).

Table 2. Syntax of L_φ^α

$\varphi ::=$	$< tt >_\alpha$	(boolean constant)
$\mid$	$< ff >_\alpha$	(boolean constant)
$\mid$	$< a >_\alpha$	(atomic action)
$\mid$	$< \varphi_1 \wedge \varphi_2 >_\alpha$	(conjunction of two formulas)
$\mid$	$< \varphi_1 \vee \varphi_2 >_\alpha$	(dis-junction of two formulas)
$\mid$	$< \varphi_1.\varphi_2 >_\alpha$	(sequential composition)
$\mid$	$< \varphi_1^*\varphi_2 >_\alpha$	(iteration operator)
$\mid$	$< \neg\varphi >_\alpha$	(negation of a formula)

3.2 Semantics of L_φ^α

The semantic of L_φ^α is defined with the function :

$$[\![-]\!] : L_\varphi^\alpha \longrightarrow \mathcal{T}$$

In the following, we present the semantic followed by the intuitive sense of each formula:

- $[\![< tt >_\alpha]\!] = \mathcal{T}$:
Every trace, with a risk that does not exceed α satisfies tt;

- $[\![< ff >_\alpha]\!] = \emptyset$:
No trace satisfies ff, whatever its impact.

- $[\![< a >_\alpha]\!] = \{< a >_\alpha\}$:
  Only the trace formed from the atomic action a, and which has an impact that does not exceed α, can satisfy this formula.

- $[\![< \varphi_1 \wedge \varphi_2 >_\alpha]\!] = [\![< \varphi_1 >_{\alpha_1}]\!] \cap [\![< \varphi_2 >_{\alpha-\alpha_1}]\!]$:
  It is equal to the intersection of two sets of traces, such that the sum of impact caused by φ_1 and φ_2, i.e. $(\alpha_1 + \alpha_2)$ does not exceed α.

- $[\![< \varphi_1 \vee \varphi_2 >_\alpha]\!] = [\![< \varphi_1 >_\alpha]\!] \cup [\![< \varphi_2 >_\alpha]\!]$:
  It is equal to the union of two sets of traces φ_1 and φ_2, such that the risk produced by φ_1 or by φ_2 does not exceed α.

- $[\![< \varphi_1.\varphi_2 >_\alpha]\!] = \{\xi_1.\xi_2 | \xi_1 \in [\![< \varphi_1 >_{\alpha_1}]\!]$ and $\xi_2 \in [\![< \varphi_2 >_{\alpha-\alpha_1}]\!]\}$:
  It is a composition of a prefix ξ_1 that belongs to the semantic of $< \varphi_1 >_{\alpha_1}$ and which has an impact α_1 and a suffix ξ_2 that belongs to the $< \varphi_2 >_\alpha$ semantic and also with an impact that doesn't exceed $\alpha - \alpha_1$.

-

$$[\![< \varphi_1^* \varphi_2 >_\alpha]\!] = \begin{cases} [\![< \varphi_1 >_{\alpha_1}]\!]^* \cup \{\xi_1.\xi_2\} | \; \xi_1 \in [\![< \varphi_1 >_{\alpha_1}]\!] \; and \\ \xi_2 \in [\![< \varphi_2 >_{\alpha-\alpha_1}]\!] \quad if \; [\![< \varphi_2 >]\!] \neq \emptyset \\ \\ [\![< \varphi_1 >_{\alpha_1}]\!]^* \qquad\qquad else \end{cases} \quad (3)$$

There are two cases for the formula $< \varphi_1^* \varphi_2 >_\alpha$ that depend on φ_2 semantic:

- If there are traces that satisfy $< \varphi_2 >_{\alpha_2}$ ($[\![< \varphi_2 >_{\alpha_2}]\!] \neq \emptyset$, the semantic will be defined by the composition of a number of traces $\xi_{i\in 1...n}$ belonging to $[\![< \varphi_1 >_{\alpha_1}]\!]$ with a risk lower than α concatenated to a trace ξ_2 that belongs to $[\![< \varphi_2 >_{\alpha-\alpha_1}]\!]$.
- If there is no trace that satisfies $< \varphi_2 >_{\alpha_2}$ ($[\![< \varphi_2 >_{\alpha_2}]\!] = \emptyset$), then a trace ε that satisfies $< \varphi_1^* \varphi_2 >_\alpha$, is equal to a set possibly infinite of traces $\xi_{i\in 1...n}$ belonging to $[\![< \varphi_1 >_{\alpha_1}]\!]$ and with a risk α_1 that doesn't exceed α.

- $[\![< \neg\varphi >_\alpha]\!] = \mathcal{T} \backslash < \varphi >_{1-\alpha}$: the semantics of a formula negation is the complement of its semantic. This complement of traces must not exceed a risk equal to $1 - \alpha$.

4 ACP_α^φ: The Specification Language of Program

In this section, we introduce syntax, semantics and basic properties of our process specification. It is a modified version of ACP (Algebra of Communicating Processes). Our choice is motivated with the power of this language to describe interactions, communications and synchronizations between a collection of processes.

4.1   Syntax

The notable difference between ACP_α^φ presented in the previous section and the ACP^φ used in [4], is the integration of a condition denoted by c, that controls actions executed by a process. Therefore, a new condition c (considered as a process) is added to the syntax, to evaluate a comparison of two constants. The result of this condition is either true or false, depending of the evaluation between this constants (see Table 3).

Table 3. Syntax of ACP_α^φ

$P ::=$	1	(Constant representing successful termination)
$\mid$	δ	(Constant representing deadlock)
$\mid$	a	(Atomic action)
$\mid$	$P + Q$	(Alternative composition)
$\mid$	$P.Q$	(Sequential composition)
$\mid$	$P \parallel_\gamma Q$	(Parallel composition)
$\mid$	$P \mid_\gamma Q$	(Communication merge)
$\mid$	P^*Q	(Iteration operator)
$\mid$	$\partial_H(P)$	(Encapsulation operator, $H \subseteq A$)
$\mid$	$\tau_I(P)$	(Abstraction operator, $H \subseteq A$)
$\mid$	c	(Boolean constant)

Note that the merge operator $\parallel_\gamma$ and the communication operator $\mid_\gamma$ are parameterized by a communication function γ defined as follows:
A communication function is any commutative and associative function form $A \times A \longrightarrow A$, if :

(1) $\forall a, b \in A : \gamma(a, b) = \gamma(b, a)$ and
(2) $\forall a, b, c \in A : \gamma((a, b), c) = \gamma(a, \gamma(b, c))$.

We use the restriction and abstraction operators (respectively: ∂_H and τ_I) defined in [4], and we will recall their functionality thereafter.

4.2   Semantic

The operational semantics of ACP_α^φ extended from ACP^φ language is illustrated in Table 4. We added one new rule (R^c) that treats the risk.

- R^a : this rule indicates that a process P formed by the action a can evolve by executing this action and finish successfully.

- R_+ : This rule indicates that the process $P + Q$ can execute the action a and becomes $P' + Q$, if and only if P is able to execute the same action and become P'.

Table 4. Operational semantics of ACP_α^φ

$$(R^a)\frac{\Box}{a \xrightarrow{a} 1} \qquad\qquad (R_+)\frac{P \xrightarrow{a} P'}{P + Q \xrightarrow{a} P' + Q}$$

$$(R.)\frac{P \xrightarrow{a} P'}{P.Q \xrightarrow{a} P'.Q} \qquad\qquad (R_*)\frac{P \xrightarrow{a} P'}{P * Q \xrightarrow{a} P'.(P * Q)}$$

$$(R_{\|_\gamma})\frac{P \xrightarrow{a} P'}{P \|_\gamma Q \xrightarrow{a} P' \|_\gamma Q} \qquad (R_{|_\gamma})\frac{P \xrightarrow{a} P' \quad Q \xrightarrow{b} Q'}{P|_\gamma Q \xrightarrow{\gamma(a,b)} P' \|_\gamma Q'}\gamma(a,b) \neq \delta$$

$$(R_\tau^\varphi)\frac{P \xrightarrow{a} P'}{\tau_I(P) \xrightarrow{\tau} \tau_I(P')}a \notin I \qquad (R_\tau)\frac{P \xrightarrow{a} P'}{\tau_I(P) \xrightarrow{\tau} \tau_I(P')}a \in I$$

$$(R_{\partial_H})\frac{P \xrightarrow{a} P'}{\partial_H(P) \xrightarrow{a} \partial_H(P')}a \notin H \qquad (R^c)\frac{[\![c]\!] \equiv tt \quad P \xrightarrow{a} P'}{c.P \xrightarrow{a} P'}$$

- $R.$: The sequential composition of two processes P and Q can evolve only if P can advance.

- R^*: $P*Q$ can choose to evolve with P or Q, when P finishes, it still has the same choice.

- $R_{\|_\gamma}$: A process $P \|_\gamma Q$ can advance by executing the action a to become $P' \|_\gamma Q$ if and only if P is capable to advance with the same action and becomes P'.

- $R_{|_\gamma}$: It is the synchronization operator. In this case, a process in the form $P|_\gamma Q$ can only advance if there are two actions a and b, such that P advance by executing the action and becomes P' and Q advance with the action b and become $Q'$and the function $\gamma(a, b)$ is defined .

- R_τ^φ : This rule masks internal actions through the abstraction operator τ_I. A process of the form $\tau_I(P)$ can advance by executing the silent action τ and becomes $\tau_I(P')$, where I is any set of atomic actions called internal actions: it abstracts all output action in I by the silent action τ.

- R_{∂_H} : This rule permits with the restriction operator to prohibit the execution of certain actions by a process. A process of the form $\partial_H(P)$ can advance with the action a and become $\partial_H(P')$, if and only if P is able to advance with the same action a.

- R^c : This rule permits a process P to evolve to P', if and only if the condition c is equivalent to true, and P is able to evolve to P'.

Table 5. L_{φ}^{α} Logic translation function

$$\| - \|L_{\varphi}^{\alpha} \times \mathbb{N} \longrightarrow ACP_{\alpha}^{\varphi}$$

$$\|tt\|_i = \sum_{b \in \mathcal{A}} (\overline{b}_d^i.\overline{b}_f^i)^* \sum_{b \in \mathcal{A}} \overline{b}_d^i.\overline{b}_f^i + 1$$

$$\|ff\|_i = \delta$$

$$\|1\|_i = 1$$

$$\|\delta\|_i = \delta$$

$$\| < a >_{\alpha} \|_i = (Risk(a) \leqslant \alpha).\overline{a}_d^i.\overline{a}_f^i$$

$$\| < \varphi_1.\varphi_2 >_{\alpha} \|_i = c_1.\|\varphi_1\|_i.\, c_2.\|\varphi_2\|_i$$

$$\| < \varphi_1 > \vee < \varphi_2 >_{\alpha} \|_i = c_1.\|\varphi_1\|_i + c_2.\|\varphi_2\|_i$$

$$\| < \varphi_1^* \varphi_2 >_{\alpha} \|_i = c_1.\|\varphi_1\|_i^*.c_2.\|\varphi_2\|_i$$

$$\| < \neg a >_{\alpha} \|_i = (Risk(a^c) < \alpha).\overline{a}_d^{i\,c}.\overline{a}_f^{i\,c}.((\sum_{b \in \mathcal{A}} \overline{b}_d^i.\overline{b}_f^i)^* \sum_{b \in \mathcal{A}} \overline{b}_d^i.\overline{b}_f^i + 1)$$

5   Formal Enforcement of Security Policies with Risk Integration

As we mentioned in Section 2, to respect the user choice, the security policy will depend on a security level chosen by the mobile user. So for each level, we attribute a threshold (α) that limits and controls the program behavior; if there is an action that exceeds this value, it will be blocked. This threshold will be "consumed" by the execution of each action in the process.

We recall the reader, that the current paper borrows the main methodology of enforcement from [4], adding the necessary modifications to reflect its new concepts.

The idea of control according to the user decision is applied as follows; while the threshold is not completely consumed by the execution of each action, the process can advance. Otherwise, if this value is crossed process cannot advance, that means that actions are blocked.

To formalize this idea, we need first to introduce the following notations:

- $\mathcal{A}_c$: synchronization set, $\mathcal{C}(\mathcal{A}) = \bigcup_{a \in \mathcal{A}} \{a_d, a_f, \overline{a}_d, \overline{a}_f\}$;
- γ is a communication function, defined as follows:

$$\gamma(a|\overline{a}) = \begin{cases} a|\overline{a} & if \quad a \in \mathcal{A} \bigcup \mathcal{C}(\mathcal{A}) \\ \delta\ Else \end{cases}.$$

- $H = \mathcal{A}_c$
- $I = \bigcup_{\mu \in \mathcal{A}_c} \{\mu|\overline{\mu}\}$

First, we modify the controlled process by adding the synchronization actions that mark the start and the end of the execution of actions by the function $\lceil - \rceil$ defined in Table 6, where:

- i is an integer used to ensure the freshness of synchronization actions.
- H is a set of trusted functions in $\mathcal{A}$, representing trusted actions introduced to avoid the translation of synchronization actions [4].

Then, we transform the security policy into a controller process by adding synchronization actions and we add the condition c that monitors the process, allowing only the execution of actions that respect the security policy, i.e. which the risk does not exceed the threshold fixed by the security policy (and taken from user).

On the other side, we transform the risk to a condition and we insert it to the monitor denoted by φ_α. This condition will permit or deny the synchronization of two actions that belong to I.

Then synchronization actions are added also to the controller process via the function $\| - \|$ defined in Table 5.

Here, we explain the translation of the logic L_φ^α to a controller process illustrated in Table 5.

- The formula tt is translated to a process which can synchronize with any other process.

- The formula ff is transformed to δ (deadlock), view that no process satisfies ff.

- The formula $< a >_\alpha$ is transformed to a monitor φ_α composed from a boolean condition that control the risk followed by synchronization actions $(\overline{a}_d.\overline{a}_f)$.

- The formula $< \varphi_1.\varphi_2 >_\alpha$ is transformed to two conditions c_1 and c_2 that controls respectively the risk of φ_1 and φ_2 traces. Notice that c_1 and c_2 are calculated with the semantic presented in Section 3.2.

- $< \neg a >_\alpha$: as defined in the semantic of a formula negation, $\| < \neg a >_\alpha \| = \mathcal{T} \setminus \{< a >_{1-\alpha}\}$, this means that this formula is transformed to a process that synchronizes with any process that begins with any action different to a (except the action a), and with a risk that does not exceed α ($Risk(a^c) < \alpha$), with $a^c \in \mathcal{T} \setminus \{a\}$.

Finally to enforce φ on P, the program must be executed in parallel with the monitor. In order to explain our approach, we give the following simple examples, without and with risk integration to highlight the difference.

We start with an example without the risk integration. Thus, we suppose the following inputs:

$$P = a.b$$

Table 6. $L_\varphi^\alpha \longrightarrow ACP_\alpha^\varphi$ Process translation function

$$\lceil - \rceil ACP_\alpha^\varphi \times \mathbb{N} \times 2^{\mathcal{A}} \longrightarrow ACP_\alpha^\varphi$$

$$\lceil 1 \rceil_i^H = 1$$
$$\lceil \delta \rceil_i^H = \delta$$
$$\lceil a \rceil_i^H = \begin{cases} a & if\, a \in H \cup \{\tau\} \\ a_d^i.a.a_f^i & Else \end{cases}$$
$$\lceil P_1.P_2 \rceil_i^H = \lceil P_1 \rceil_i^H . \lceil P_2 \rceil_i^H$$
$$\lceil P_1 + P_2 \rceil_i^H = \lceil P_1 \rceil_i^H + \lceil P_2 \rceil_i$$
$$\lceil P_1^* P_2 \rceil_i^H = \lceil P_1 \rceil_i^{H_i} {}^* \lceil P_2 \rceil_i^H$$
$$\lceil P_1 \parallel_\gamma P_2 \rceil_i^H = \lceil P_1 \rceil_i^H \parallel_\gamma \lceil P_2 \rceil_i^H$$
$$\lceil \partial_{H'}(P) \rceil_i^H = \partial_{H'}(\lceil P \rceil_i^{H \cup H'})$$
$$\lceil \tau_I(P) \rceil_i^H = \tau_I(\lceil P \rceil_i^{H \cup I})$$
$$\lceil \partial_{P_\varphi}(P) \rceil_i^H = \partial_{H_i}(\tau_{I_i}(\lceil P \rceil_i^H \parallel_\gamma \parallel P_\varphi \parallel_i))$$

which is a composition of two actions a and b.

$$\varphi = a$$

which mean that only the action a is permitted.

We apply the different steps to enforce φ on P:

First step, we transform the program to a process by adding the synchronization actions (using Table 6):

$$\lceil P \rceil_1 = a_d^1.a.a_f^1.b_d^1.b.b_f^1$$

Secondly, we transform the policy security to a monitor (process):

$$\|\varphi\|_1 = \overline{a_d^1}.\overline{a_f^1}$$

The last step is to execute this two processors on parallel:

$$\partial_H(\tau_I(a_d^1.a.a_f^1.b_d^1.b.b_f^1 \parallel_\gamma \overline{a_d^1}.\overline{a_f^1}))$$

with $H = \{a_d, a_f, b_d, b_f, \overline{a_d}, \overline{a_f}\}$ and $I = \{\gamma(a_d, \overline{a_d}), \gamma(a_f, \overline{a_f})\}$

$$\partial_H(\tau_I(a_d^1.a.a_f^1.b_d^1.b.b_f^1 \parallel_\gamma \overline{a_d^1}.\overline{a_f^1}))$$

$\xrightarrow{\tau}$ $\langle Rules\ R_{\parallel\gamma},\ R_\tau^\varphi\ et\ R_{\partial_H}\ avec\ \gamma(a_d, \overline{a_d}) = a_d | \overline{a_d} \in I \rangle$

$$\partial_H(\tau_I(a.a_f^1.b_d^1.b.b_f^1 \parallel_\gamma \overline{a_f^1}))$$

$\xrightarrow{a}$ $\langle Rules\ R_{\parallel\gamma},\ R_\tau^\varphi\ et\ R_{\partial_H}\ avec\ a \notin H \rangle$

$$\partial_H(\tau_I(a_f^1.b_d^1.b.b_f^1 \parallel_\gamma \overline{a_f^1}))$$

$$\xrightarrow{\tau} \quad \langle Rules\ R_{\parallel\gamma},\ R_\tau^\varphi\ et\ R_{\partial_H}\ avec\ \gamma(a_f, \overline{a_f}) = a_f | \overline{a_f} \in I \rangle$$

$$\partial_H(\tau_I(b_d^1.b.b_f^1))$$

Here, we see that the process cannot execute the action b, view that the action b_d cannot synchronize with its complement $\overline{b_d}$, so the sub-process $b_d^1.b.b_f^1$ is blocked and as consequence, the policy security is respected.

Now, we present the second example, with risk integration. The same process as above:

$$P = a.b$$

A security policy with a risk value equal to 0.6.

$$\varphi = <a.b>_{0.6}$$

Which mean that the impact of the execution of the action a followed by the action b should not pass a risk value equal to 0.6, i.e, the sequential composition of these two actions is limited by this risk value. Given that we suppose that 0.6 present the value assigned to security level chosen by the user.

First step is to modify the process by limiting each action with synchronization actions that mark the start and the end of each action

$$\lceil P \rceil = a_d.a.a_f.b_d.b.b_f$$

On the other side, the security policy is transformed to a controller process φ_α. So we add the complement of synchronization actions used:

$$\|\varphi_\alpha\| = \overline{a_d}.\overline{a_f}.\overline{b_d}.\overline{b_f}.$$

Then, based on the semantics presented, we transform the risk into two conditions :

$$(Risk(a) \leq 0.6)\ and\ (Risk(b) \leq 0.6 - Risk(a)).$$

Now to enforce the security policy, both processors must be executed in parallel.

$$\partial_H(\tau_I(a_d.a.a_f.b_d.b.b_f |_\gamma (Risk(a \leq 0.6).\overline{a_d}.\overline{a_f}.(Risk(b) \leq 0.6 - Risk(a)).\overline{b_d}.\overline{b_f}))$$

where $I = \{\gamma(a_d, \overline{a_d}), \gamma(a_f, \overline{a_f}), \gamma(b_d, \overline{b_d}), \gamma(b_f, \overline{b_f})\}$ and $H = \{a_d, a_f, b_d, b_f, \overline{a_d}, \overline{a_f}, \overline{b_d}, \overline{b_f}\}$.

As indicated in the the rule R^c, the action a_d can synchronize with $\overline{a_d}$ only and only if the condition $(Risk(a) \leq 0.6)$ is true. If this condition is true, the process will advance and the two actions can synchronize (which is the case). We suppose that $Risk(a) = 0.4$ and $Risk(b) = 0.3$.

$$\partial_H(\tau_I(a_d).a.a_f.b_d.b.b_f|_\gamma ((Risk(a) \le 0.6).\overline{a_d}.\overline{a_f}.(Risk(b) \le$$
$$0.6 - Risk(a)).\overline{b_d}.\overline{b_f})))$$

$$\xrightarrow{\tau} \quad \langle Rules\, R_{\|\gamma}, R^c, R^\varphi_\tau\, R_{\partial_H} \text{ where } \gamma(a_d, \overline{a_d}) = a_d|\overline{a_d} \in I\rangle$$

$$\partial_H(\tau_I(a.a_f.b_d.b.b_f|_\gamma (Risk(a) \le 0.6).\overline{a_f}.(Risk(b) \le 0.6 - Risk(a)).\overline{b_d}.\overline{b_f})))$$

$$\xrightarrow{a} \quad \langle Rules\, R_{\|\gamma}, R_\tau\, R_{\partial_H} \text{ where } a \notin I\rangle$$

$$\partial_H(\tau_I(a_f.b_d.b.b_f|_\gamma (Risk(a) \le 0.6).\overline{a_f}.(Risk(b) \le 0.6 - Risk(a)).\overline{b_d}.\overline{b_f})).$$

$$\xrightarrow{\tau} \quad \langle Rules\, R_{\|\gamma}, R^c, R^\varphi_\tau\, R_{\partial_H} \text{ where } \gamma(a_f, \overline{a_f}) = a_f|\overline{a_f} \in I\rangle$$

$$\partial_H(\tau_I(b_d.b.b_f|_\gamma (Risk(a) \le 0.6).(Risk(b) \le 0.6 - Risk(a)).\overline{b_d}.\overline{b_f})).$$

To ensure that b_d can synchronize with $\overline{b_d}$, it is necessary that the condition $(Risk(b) \le 0.6 - Risk(a))$ is equivalent to true, or that it is not the case.

As can be seen, this two examples (with and without the inclusion of risk) leads to the same result (blocking the action b), except that in the second example (with the risk integration), dangerous actions wrer blocked according to the user choice.

6 Example

In this section, we present an example closer to reality, with a famous Android application called Linked In. This application allows user to make connections, access professional papers, be informed with personalized news, to view and save recommended jobs, etc.

In this example, we based on a risk classification that allows to quantify the risk of each action. This idea is inspired from the classification of Android permissions. According to [13], we see that Google classifies an application permissions to categories, that we can base on two main ones:

- *Normal permissions:* with a minimal risk, granted automatically without user's explicit approval;
- *Dangerous permissions:* with a higher-risk that could provide negative effects, this type of permissions belongs to permissions requested by the application.

From this classification, we can classify actions, and for each action we assigns a meaningful numerical value ($0 \le \alpha \le 1$) that presents the potential impact of an action, and as result, this allow us to control those actions.

Concerning this application, the following permissions are those requested by the application.

- read contacts r_c ;
- write contacts w_c ;

- read calendar r_{calend} ;
- precise location (GPS and network-based) r_{GPS} ;
- read call log r_{cl};
- write call log w_{cl};
- modify the content of USB storage m_{USB} ;
- delete the content of USB storage d_{USB} ;
- open connection o_c ;
- receive data from Internet rec_I ;
- read phone status and identity r_{id}.

We can consider the previous list of actions in the range of normal permissions, because this application comes from a known source (Google play store), so we can assign to it a certain level of confidence, but nothing prevents us from controlling the hidden actions behind this application. We assigned to this class a risk value $0 \le \alpha \le 0.2$.

The second class with a high risk level (dangerous permissions), it must be carefully controlled. Actions that can be placed in this class are more sensible than others and require a very high degree of control. Here, we can think about permissions related to phone category as making calls without user intervention. Also permissions related to SMS are very sensible too, as editing and sending SMS that user can accept without realizing it. As risk value, we assigned to this class $0, 3 \le \alpha \le 0, 5$.

- read SMS r_{SMS};
- edit SMS e_{SMS};
- send SMS s_{SMS};
- make calls *call*.

A typical example of where leakage can occur on Android, is an app that allowed to access personal information (SMS, Video, Contacts, etc) and at the same time is allowed to access the Internet. In this case, to ensure that confidential information could not be published or diffused via Internet, we could enforce a security policy that prevents access to the Internet after reading such information from the phone internal memory, and this will depend on the security level chosen.

Consider the following inputs:

First the process:

$$P = r_{SMS}.o_c + r_c.m_c$$

which mean that the process can read SMS and then open connection or read contacts and then modify it.

P must satisfy the following property:

$$\varphi = r_{SMS}.(\neg o_c)$$

which mean after the action of reading from SMS, the process cannot open connection.

As threshold ,we suppose that user chooses a medium level of security.

$$\alpha = 0,5$$

We modify the policy security by inserting the risk :

$$\varphi_\alpha = < r_{SMS}.o_c >_{0,5}$$

As we mentioned, we suppose that o_c (open connection) belongs to a class whose risk is lower than $0,2$, because it's among the actions requested by the application, and r_{SMS} belong to the second class. So, we suppose that $Risk(r_{SMS}) = 0,4$ and $Risk(o_c) = 0,2$.

The security policy formula is transformed to the monitor via the logic translation function in Table 5.

$$\overline{r_{SMSf}}.(Risk(o_c) \leq 0,5 - Risk(r_{SMS})).\overline{o_{cd}}.\overline{o_{cf}}$$

The process is transformed as well by the Process translation function to:

$$\lceil P \rceil = r_{SMSd}.r_{SMS}.r_{SMSf}.o_{cd}.o_c.o_{cf}$$

The enforcement step requires that this two processors run on parallel.

$$\partial_H(\tau_I(r_{SMSd}.r_{SMS}.r_{SMSf}.o_{cd}.o_c.o_{cf} \parallel_\gamma (Risk(r_{SMS}) \leq 0,5).\overline{r_{SMSd}}.\overline{r_{SMSf}}.(Risk(o_c) \leq 0,5 - Risk(r_{SMS})).\overline{o_{cd}}.\overline{o_{cf}}))$$

where
$$I = \{\gamma(r_{SMSd}, \overline{r_{SMSd}}), \gamma(r_{SMSf}, \overline{r_{SMSf}}), \gamma(o_{cd}, \overline{o_{cd}}), \gamma(o_{cf}, \overline{o_{cf}})\}$$
and
$$H = \{r_{SMSd}, r_{SMSf}, \overline{r_{SMSd}}, \overline{r_{SMSf}}, o_{cd}, o_{cf}, \overline{o_{cd}}, \overline{o_{cf}}\}$$

Note that this sequence violates the property φ_α, and the program should be blocked before executing the action open connection.

$$\partial_H(\tau_I(r_{SMS_d}.r_{SMS}.r_{SMSf}.o_{cd}.o_c.o_{cf} \parallel_\gamma (Risk(r_{SMS}) \leq 0,5).\overline{r_{SMSd}}.\overline{r_{SMSf}}.(Risk(o_c) \leq 0,5 - Risk(r_{SMS})).\overline{o_{cd}}.\overline{o_{cf}}))$$

$\xrightarrow{\tau}$ $\langle Rules\ R^c,\ R_{\parallel\gamma},\ R^\varphi_\tau\ and\ R_{\partial_H},\ \gamma(r_{SMSd}, \overline{r_{SMSd}}) = r_{SMSd} | \overline{r_{SMSd}} \in I \rangle$

$$\partial_H(\tau_I(r_{SMS}.r_{SMSf}.o_{cd}.o_c.o_{cf} \parallel_\gamma .\overline{o_{cf}}.(Risk(r_{SMS}) \leq 0,5 - Risk(o_c)).\overline{o_{cd}}.\overline{o_{cf}}))$$

$\xrightarrow{r_{SMS}}$ $\langle Rules\ R_{\parallel\gamma},\ R_\tau\ and\ R_{\partial_H}\ where\ r_{SMS} \notin I \rangle$

$$\partial_H(\tau_I(r_{SMSf}.o_{cd}.o_c.o_{cf} \parallel_\gamma \overline{r_{SMSf}}.(Risk(o_c) \leq 0,5 - Risk(r_{SMS})).\overline{o_{cd}}.\overline{o_{cf}}))$$

$\xrightarrow{\tau}$ $\langle Rules\ R^c,\ R_{\parallel\gamma},\ R^\varphi_\tau\ and\ R_{\partial_H}, \gamma(r_{SMSf}, \overline{r_{SMSf}}) = r_{SMSf} | \overline{r_{SMSf}} \in I \rangle$

$$\partial_H(\tau_I(o_{cd}.o_c.o_{cf} \parallel_\gamma .(Risk(o_c) \leq 0,5 - Risk(r_{SMS})).\overline{o_{cd}}.\overline{o_{cf}}))$$

As can be seen, the condition: $(Risk(o_c) \leq 0,5 - Risk(r_{SMS})) \equiv false$. The sub-process $o_{cd}.o_c.o_{cf}$ cannot execute the action open connection after reading sms, and as well the action o_{cd} cannot synchronize with $\overline{o_{cd}}$.

7   Conclusion and Future Work

In this paper, we present an idea that appears with the need to secure mobiles. The notion of risk has been crucial throughout this work, it gives a chance to the end-user to decide and choose its own level of security and to trust more applications that he installed. So we adopted a formal approach for the enforcement of security policies in parallel systems, based on the risk notion to monitor the process behavior. The results provides a technique that allows controlling applications and applying it to real language such as Java.

As future work, we want to implement the proposal approach on Android device and evaluate the results. It will also be interesting to add the Sandboxing concept to improve security by isolating applications, which could be based on a risk classification.

References

1. Johnson, R., Wang, Z., Gagnon, C., Stavrou, A.: Analysis of Android Applications' Permissions, Software Security and Reliability Companion (SERE-C). In: Software Security and Reliability Companion (SERE-C), pp. 45–46 (2012)
2. Baeten, J.C.M.: A brief history of process algebra. Theoretical Computer Science 335, 131–146 (2005)
3. Langar, M., Mejri, M.: Optimized enforcement of security policies. Foundations of Computer Security, 37–42 (2005)
4. Langar, M., Mejri, M., Adi, K.: Formal enforcement of security policies on concurrent systems. Journal of Symbolic Computation 46, 997–1016 (2011)
5. Jay, L., Lujo, B., David, W.: Edit automata: Enforcement mechanisms for run-time security policies. International Journal of Information Security 4, 2–16 (2011)
6. Khoury, R., Tawbi, N.: Corrective enforcement of security policies. In: Degano, P., Etalle, S., Guttman, J. (eds.) FAST 2010. LNCS, vol. 6561, pp. 176–190. Springer, Heidelberg (2011)
7. Ould-Slimane, H., Mejri, M., Adi, K.: Using edit automata for rewriting-based security enforcement. In: Gudes, E., Vaidya, J. (eds.) Data and Applications Security XXIII. LNCS, vol. 5645, pp. 175–190. Springer, Heidelberg (2009)
8. Chabot, H., Khoury, R., Tawbi, N.: Extending the enforcement power of truncation monitors using static analysis. Computers & Security 30, 194–207 (2011)
9. Schneider, F.B.: Enforceable Security Policies. ACM Trans. Inf. Syst. Secur. 3, 30–50 (2000)
10. Sui, G., Mejri, M.: FASER Formal and Automatic Security Enforcement by Rewriting by BPA Algebra with Test. Int. J. Grid Util. Comput. 4, 204–211 (2013)
11. Mejri, M., Fujita, H.: Enforcing Security Policies Using Algebraic Approach. New Trends in Software Methodologies, Tools and Techniques 182, 84–98 (2008)
12. Burguera, I., Zurutuza, U., Nadjm-Tehrani, S.: Crowdroid: Behavior-based Malware Detection System for Android. In: Proceedings of the 1st ACM Workshop on Security and Privacy in Smartphones and Mobile Devices, vol. 12, pp. 15–26 (2011)
13. Permission. Android Developer -API Guides- Android Manifest, http://developer.android.com/guide/topics/manifest/permission-element.html

Countermeasures Mitigation for Designing Rich Shell Code in Java Card

Noreddine El Janati El Idrissi[1]([✉]), Said El Hajji[1], and Jean-Louis Lanet[2]

[1] Laboratory of Mathematics, Computing and Applications,
Faculty of Sciences, University of Mohammed-V, Rabat, Morocco
janatinoreddine@gmail.com, elhajji@fsr.ac.ma
[2] INRIA, LHS PEC,
263 Avenue Général Leclerc, 35042 Rennes,
jean-louis.lanet@inria.fr
http://secinfo.msi.unilim.fr/lanet/

Abstract. Recently, it has been published that Java based smart cards are still exposed to logical attacks. These attacks take advantage of the lack of a full verification and dynamically use a type of confusion. Countermeasures have been introduced on recent smart card to avoid executing rich shell code and particulary dynamic bound checking of the code segment. We propose here a new attack path for performing a type confusion that leads to a Java based self modifying code. Then, to mitigate this new attack an improvement to the previous countermeasure is proposed.

Keywords: Smart Card · Shell Code · Self Modifying Code

1 Introduction

Today most of the smart cards embed a Java Card Virtual Machine (JCVM). Java Card is a type of smart card that implements the standard Java Card 3.0 [14] *Classic Edition* or *Connected Edition*. Such a smart card embeds a virtual machine, which interprets application byte codes already romized with the operating system or downloaded after issuance. Due to security reasons, the ability to download code into the card is controlled by a protocol defined by Global Platform [11]. This protocol ensures that, the code owner has the required credentials to perform the particular action.

A smart card can be viewed as a smart and secure container which stores sensitive assets. Such tokens are often the target of attacks at different levels: pure software attacks, hardware based, *i.e.* side channel of fault attacks but also mixed attacks. Security issues and risks of these attacks are increasing and continuous efforts to develop countermeasures against these attacks are sought. The main assets in a smart card are the sensitive data (*i.e.* the cryptographic keys) and the code of the program. Often attackers perform cryptanalysis using side channel to recover the keys, thus break the confidentiality of the keys. The difficulty of breaking the security properties of these assets are given bellow in decreasing order:

© Springer International Publishing Switzerland 2015
S. El Hajji et al. (Eds.): C2SI 2015, LNCS 9084, pp. 149–161, 2015.
DOI: 10.1007/978-3-319-18681-8_12

- Data confidentiality,
- Data integrity,
- Code integrity
- Code confidentiality

We have shown in our previous work [7], that it was relatively easy to break the code confidentiality then the code integrity can be broken leading to the dump of the memory. Once the memory is read, it is possible to perform memory carving to gain information on the data and in particular the key containers. Smart card manufacturers increase the security of their JCVM each in a way that the published attacks do not work anymore on recent cards. The current smart cards are now well protected against pure software attacks with program counter bound checks, typed stack and so on. For such smart cards, we propose in this paper, a new attack that mitigate the secure jump countermeasure which avoid developing rich shell code. Firstly, We demonstrate a proof of concept and then its application with the dump of a card. It is based on separating the control flow and the basic blocks of a program.

The remaining of the paper is organized as follows: the first section introduces the Java Card security. The second section presents the state of the art both in term of attacks and published countermeasures. Then, in the third section, we propose our contribution for mitigating the control flow countermeasure. Next, we propose an implementation that performs a type confusion and allows a Java based self modifying code. Finally, in the last section, we conclude.

2   Java Card Security

Smart cards security depends on the underlying hardware and the embedded software. Embedded sensors (light sensors, heat sensors, voltage sensors, *etc.*) protect the card from physical attacks. While the card detects such an attack, it has the possibility to erase quickly the content of the EEPROM preserving the confidentiality of secret data or blocking definitely the card (Card is mute). In addition to the hardware protection, softwares are designed to securely ensure that applications are syntactically and semantically correct before installation and also sometimes during execution. They also manage sensitive information and ensure that the current operation is authorized before executing it.

The Byte Code Verifier (BCV) guarantees type correctness of code, which in turn guarantees the Java properties regarding memory access. For example, it is impossible in Java to perform arithmetic on reference. Thus, it must be proved that the two elements on top of the stack are of primitive types before performing any arithmetic operation. On the Java platform, BCV is invoked at load time by the loader. Due to the fact that Java Card does not support dynamic class loading, BCV is performed at loading time, *i.e.* before installing the Converted Applet(CAP) onto the card. However, most of the Java smart cards do not have an on-card BCV as it is quite expensive in terms of memory consumption. Thus, a trusted third party performs an off-card byte code verification and sign it, and on-card its digital signature is checked.

Moreover, the Firewall performs checks at runtime to prevent applets from accessing (reading or writing) data of other applets. When an applet is installed, the system uses a unique applet identifier (AID) from which it is possible to retrieve the name of the package in which it is defined. If two applets are instances of classes coming from the same Java Card package, they are considered belonging to the same context. The firewall isolates the contexts in such a way that a method executing in one context cannot access any attribute or method of objects belonging to another context unless it explicitly exposes functionality *via* a Shareable Interface Object.

Smart card security is a complex problem with different points of view but products based on JCVM have passed successfully real-world security evaluations for major industries around the world. It is also the platform that has passed high level security evaluations for issuance by banking associations and by leading government authorities, they have also achieved compliance with FIPS 140-1 certification scheme. Nevertheless implementations have suffered severals attacks either hardware or software based. Some of them succeeded into getting access to the EEPROM (code of the downloaded applets) but as far as we know, nobody succeeded into reversing the code *i.e.* having access to the code of the VM, the operating system and the cryptographic algorithm implementations. These latter are protected by the interpretation layer which denies access to other memories than the EEPROM.

3   Embedded Countermeasures

There are three main types of attacks on a smart card. The first one is the software attack [5,9], which provides the cheapest solution to access sensitive information from the targeted cards. The second one is called side-channel or observation attack. This technique enables one either to retrieve secret cryptographic keys [8] used during a sensitive operation, or to reverse engineer the code used during a given operation [17]. The last one is the combined attack where a physical perturbation may create a logical fault which, in turn, is exploited to attack a card. We focus, in this paper, on the logical attacks which require the least knowledge for the attacker and that are the most affordable ones.

Designing a smart card attack must face several problems. The first one is the complete absence of documentation. The designer works within a black box approach. The second one is related with the embedded countermeasures. Such a product must resist to different attacks and several hardware and software fragments are dedicated to mitigate these attacks. The following section is dedicated to this class of attack and their related countermeasures.

3.1   State of the Art of Attacks Against Java Cards

Logical attacks are based on the fact that the runtime relies on the BCV to avoid costly tests. Then, once someone find an absence of a test during runtime, it is possible to perform an attack path. An attack aims to confuse the applet's control flow upon a corruption of the Java Card Program Counter or perturbation of the data.

Misleading the application's control flow purposes to execute a shell code stored somewhere in the memory. The aim of EMAN1 attack [13], explained by Iguchi-Cartigny *et al.*, is to abuse the Firewall mechanism with the unchecked static instructions (as `getstatic`, `putstatic` and `invokestatic`) to call malicious byte codes. In a malicious CAP file, the parameter of an `invokestatic` instruction may redirect the Control Flow Graph (CFG) of another installed applet in the targeted smart card. Such an attack leads for the first time to execute self modifying code in a Java Card. This attack has been mitigated through different countermeasures. EMAN2 [6] attack was related to the return address stored in the Java Card stack. They used the unchecked local variables to modify the return address, while Faugeron in [9] uses an underflow on the stack to get access to the return address.

When a BCV is embedded or if the process requires its usage, installing an ill-formed applet is impossible. To bypass an embedded BCV, new attacks exploit the idea of combining software and physical attacks. Barbu *et al.* presented and performed several combined attacks such as the attack [3] based on the Java Card 3.0 specification leading to the circumvention of the Firewall application. Another attack [2] consist on tampering the APDU that leads to access the APDU buffer array at any time. They also discussed in [1] about a way to disturb the operand stack with a combined attack wich gives the ability to alter any method regardless of its Java context or to execute any byte code sequence, even if it is ill-formed. This attack bypasses the on-card BCV [4]. In [6], Bouffard *et al.* described how to change the execution flow of an application after loading it into a Java Card. Recently, Razafindralambo *et al.* [16] introduced a combined attack based on fault enabled viruses. Such a virus is activated by hitting with a laser beam, at a precise location in the memory, where the instruction of a program (virus) is stored. Then, the targeted instruction mutates one instruction with one operand to an instruction with no operand. Then, the operand is executed by the JCVM as an instruction. They demonstrated the ability to design a code in a such way that a given instruction can change the semantics of the program. And then a well-typed application is loaded into the card but an ill-typed one is executed.

Hamadouche *et al.* [12] described various techniques used for designing efficient viruses for smart cards. The first one is to exploit the linking process by forcing it to link a token with an unauthorized instruction. The second step is to characterize the whole Java card API by designing a set of CAP files which are used to extract the addresses of the API regardless of the platform. The authors were able to develop CAP files that embed a shell code (virus). As they know all the addresses of each method of the API, they could replace instructions of any method. In [16], they abuse the on board linker in such a way that the application is only made of tokens to be resolved by the linker. Knowing the mapping between addresses to tokens thanks to the previous attacks, they have been able to use the linker to generate itself the shell code to be executed.

3.2   Mitigating the Attacks with Affordable Countermeasures

The objective of a system countermeasure is to detect an attack which occurs at linking time, run time (*e.g.* when the byte code transits on the data bus) or during the execution of another piece of code. Thus, the nature of the countermeasure is different in terms of:

- protection of variable integrity: instance field, code to be executed, evaluation stack, execution context, etc.
- protection against control flow execution modification: bypassing a test, jumping to an unauthorized data area, jumping to an argument instead of an instruction, etc.
- execution of shell code,
- type confusion, executing an instruction on an object with a given type and this object is considered in another code fragment to another type.

The integrity of application data is often used in Java Card and is called secure storage. It mainly consists of a dual storage or a checksum in order to verify whether the modification of the field is only done through the virtual machine (VM). Another integrity check concerns the VM structure and in particular the frame context. Using the EMAN 2 attack, it is possible to modify the return address in the frame using unchecked local variable indexes. Most of smart cards available on the web markets might be flooded by the modification of the CFG. Thus, it is possible to jump into an array which contains any shell code.

To prevent the execution of a shell code, there is the possibility to re-encode on the fly during the linking phase of the value of byte code. So, if someone trying to execute an arbitrary array will not be able to obtain the desired behavior. In such a method the encoded value depends on a dynamic variable, using the JPC for example as a nonce is enough to avoid any brute force attack for guessing the scrambled value.

There are lot of possibilities to protect the data and the execution of a code into the VM. Unfortunately, if all of them are activated during the execution of an application, the performance of the smart card will drastically decrease reaching an unacceptable level. For that reason, most of the smart cards available on web market implement the bound check counter measure which has been demonstrated as efficient enough to mitigate any exploitable shell code.

3.3   Checking the Jump Boundaries

An affordable countermeasure against the execution of shell code is to verify if the code is still executing within the boundaries of the current method. For each method, the system maintains several information like maxJPC. So, the domain of the JPC of a method belongs to {0..maxJPC}. An attack like the EMAN2 presented in the previous section, modifies the return address such that as it returns from method f() the control is transferred to the shell code instead of the caller. But the execution of the shell code is done within the execution context of the caller as shown Figure 1. In such a case, when the shell code ends with its own `return` instruction, it goes back to the caller of the caller of the method `f()`. The shell code can not be embedded within the method `f()` and thus is implemented as an array stored in a different area of the method.

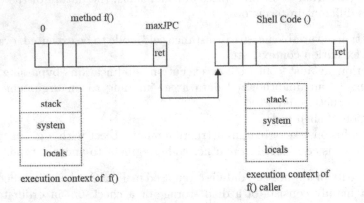

Fig. 1. Description of execution context

A naive solution should be to check if the value of JPC belongs to its domain as shown in the code fragment of the Listing 1.1.

Listing 1.1. Check the boundaries for each instruction

```
int16  BC_pop(void)
{
    vm_sp--;
    vm_pc += 1;
    if (vm_pc > maxJPC)
        return BC_SECURITY();
    return ACTION_NONE;
}
```

This increases the size of each byte code of 16 bytes on an ARM7 architecture. The original instruction requires 44 bytes. The increase for each instruction is around 36% which is too much for such a small device. The trade off is to check only the jump destination while the control flow is transferred. Thus, only the exit of a basic block will be checked, reducing the overhead. The exit instructions belong to the set {`if_xx_yy`, `goto_yy`} with yy having the value `wide` or not, depending on the domain of the offset coded on a byte or a short. The term xx has the values `type,ne`, `eq`, `lt`, `ge`, `gt`. The overhead is drastically lower impacting only a subset of the instruction set. On the Oracle Purse application it represents only 4% of overhead on the same architecture. This countermeasure is affordable and is able to detect that the control flow has been transferred to a shell code if this one requires a branch.

It does not prevent to jump to a shell code but restrict the semantics of the shell code to a linear code. In particular no loop is available, no condition evaluation and so on. As an effect, it becomes impossible to use a shell code for dumping the memory.

4   Mitigating the Control Flow Countermeasures

Two solutions are possible to bypass the countermeasure. Both of them are related to the non completeness of the countermeasure. The first one is to use the exception mechanism to transfer the control flow and data to the caller. It requires that the caller rebuilts the control flow using the catch mechanism of Java. Thus, the exception object is propagated to the caller if a handler is present, it can take decision using the **reason** embedded in the exception object. The second possibility is to simply use the return mechanism of Java if correctly handled. We have chosen the second but any avatar using the exception can get the same result.

The first step consist in implementing an EMAN2 as described by Bouffard *et al.* [5]. This attack abuses the instructions that access the local stack area[1] in order to write outside the domain of the locals. The authors succeeded in modifying the return address. When the **return** instruction is executed, it leads to a controlled execution flow modification.

A fragment of the EMAN2 exploit is shown in the Listing 1.2. The described function contains two parameters (the class instance, **this**, and the **address** parameter) and no local variable. In this function, the **sload** 1 operation pushes the value of **address** parameter onto the Java Card stack. The following operation, **sstore** 4, stores the last pushed short value into the local variable 4.

Listing 1.2. Stack overflow into a Java Card.

```
public void updateReturnAddress (short address) {
02 // flags:0 max_stack:2
20 // nargs:2 max_locals:0
16 01   sload  1 // push address from the local 1
29 03   sstore 4 // STACK OVERFLOW!
7A      return   // Jump to the shellcode
}
```

As the function's stack contains only two elements into the locals part, the authors made a stack overflow from the local variable area to set up the return address[2] by a specific value. The state of the Java Card stack is presented into

---

[1] As defined in the Java Card specification [14], accessing to the local variable is done by the **aload**, **astore**, **sload** and **sstore** instructions.

[2] On the evaluated smart cards, the references are encoded on 2-byte as short values.

the Figure 2 at left. For the current frame, we find first the arguments of the method and then the locals variables. Often, a system area is used to be able to retrieve the state of the caller. We have found some cards where the system area is not contiguous with the locals and the stack as shown in the Figure 2 right.

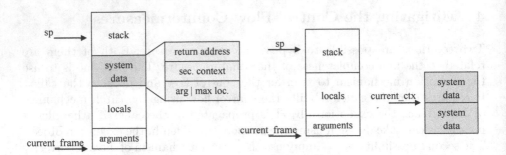

Fig. 2. Stack

4.1   Principle of the Control Flow Extraction

The objective of the attack is to split the original code fragment that we want to execute even in presence of the countermeasure into several basic blocks. Then, an instruction **sspush value** is inserted and the value is the variable that is evaluated at the beginning of the next basic block. An instruction **sreturn** finishes each of the basic block. All these basic blocks are stored consecutively into an array. The control flow is then assumed by a specific method **controlFlow()**. The **CFG** is implemented into this method which contains only decision and call to the **dummy()** method. This method plays only the role to be the context execution of the shell code and just invokes the method **shellCodeLauncher()**. This latter is the one patched thanks to the EMAN2 attack.

Once the **shellCodeLauncher()** ends its execution, it transfers the control flow to one of the basic block stored into the array. At the end of the shell code the **return** instruction is executed leading to transfer the control flow to the method **controlFlow()** as shown in Figure 3. It is important to notice that the execution context of the shell code is the **dummy** method and not the method **shellCodeLauncher()**.

With such an architecture, illegal code is executed in the **shellCode** method while the **CFG** is managed by the **controlFlow** method.

4.2   Parameters Exchange between the Controller and the Shell Code

We have seen how retrieving data from the shell code using simply the value pushed on top of the stack and send back to the caller. To provide input data to any of the basic blocks stored into the array, we can use the caller context

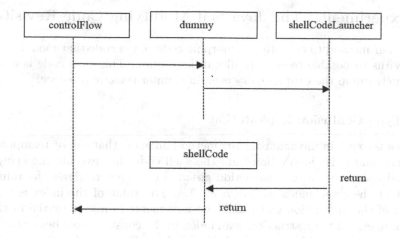

Fig. 3. Control flow derivation

i.e. the argument of the **dummy** method. The number of argument of the dummy method must be the max argument of all the basic blocks for each type of data.

For example, if the shell code is made of three basic blocks requiring the following data: $BB_1 = short, byte, BB_2 = byte, byte, BB_3 = ref, short, byte$ the maximum of generic argument of **dummy** method is 4 defined as $L_0 = short, L_1 = byte, L_2 = byte, L_3 = ref$. Note that BB_3 will be called with a reordering of its arguments: $BB_3 = short, byte, ref$. Then the argument used by each basic blocks will be the following $BB_1 = L_0, L_1, BB_2 = L_1, L_2, BB_3 = L_0, L_1, L_3$. For each basic block, the unused variables are set to their default Java value.

The first parameter of the dummy method is the offset to jump into the shell code array. In the code fragment given in the Listing 1.3, the first call in the evaluation condition is for the first segment of the shell code with the related parameters, the size of the first segment (n) is then added at the first parameter of the second segment leading to a call to BB_2. The size of the first and the second (m) is then used to call the third segment.

Listing 1.3. Calling convention of the basic blocks

```
...
if (dummy (arAdd,L_0,L_1,0,null)) // implicit call to BB1
  dummy (arAdd+n,L_0,L_1,L_2,null); //implicit call to BB2
else
  dummy (arAdd+m,L_0,0,L_2,L3); //implicit call to BB3
}
```

The only constraint is that the order of the parameters of the **dummy** method must be strictly the same as the method **shellCodeLaucher**.

5   Experiments: The Java Self Modifying Code Revisited

We use our method to execute polymorphic code, *i.e.* a code that modifies itself like a virus to be able to execute illegal instruction. This shell code is able to completely dump the card memory even in countermeasure presence.

5.1   Type Confusion Exploitation

The idea is to use in the `controlFlow` method an array that can be manipulated with read and write instructions and the shell code that execute the array. In the shell code, we use the instruction `getstatic_s` that retrieves the value of a short at the given index as shown in 1.4. The value of the index is an argument of the instruction and cannot be incremented directly by the method. The parameter of the instruction is an index in the constant pool before the link resolves the token and becomes inside the card an offset, or reference depending to the implementation.

Listing 1.4. Simple shell code to dump the memory

```
7   public void getMyAddress(){
8      // flags: 0   max_stack : 1
9      // nargs: 0   max_locals: 0
10     7D 00 00   getstatic_s 2
11     78         sreturn
12  }
```

The corresponding value in the shell code array is [7D, 00, 00, 78]. Executing this shell code retrieves the content of the memory at the address 0x0000. The `controlFlow` method has to manage the value of the address. In this basic example, the input data are only the offset in the array and the return value of the basic block must be stored in input-output buffer to be sent to the reader. The address to be modified is the content of the shell code array at offset 1 for the high byte of the address and 2 for the low one. The aim is to write in the input output buffer 128 bytes of memory.

Listing 1.5. Calling the shell code with parameters and retrieving return value

```
1   public void controlFlow(APDU apdu, byte[] buffer, short arAdd){
2     short boff=0x00;
3     for(short i=0;i<=0x7F; i++){
4     short x=dummy(arAdd);
5     Util.setShort(buffer, boff, x);
6      if(shellcode[2]==(byte)0xFF){
7        shellcode[2]=(byte)0x00;
8        shellcode[1]++);}
9      else{shellcode[2]+=2;}
10       boff=(short)(boff+2);
11     }
12    apdu.setOutgoingAndSend((short) 0x00,boff);
13  }
```

In the Listing 1.5 of the `controlFlow` method at line 4 we get the content of the memory and at line 5, we store it in the buffer. At line 9, we increase the value of the address to be dumped, and from line 6 to 8, we propagate the carry.

5.2   Completeness of the Countermeasure

We have demonstrated that such a counter measure is inefficient due to its incompleteness. The objective of the initial countermeasure was to detect the execution of a shell code outside its original position by checking the destination branch. Thus the current counter measure encompasses only the set of intra procedure instructions (*i.e.* `goto, if, jsr`). It must be extended to the set of intra procedure instructions which is more complicated. The VM has the information about the `minJPC` and the `maxJPC` which is enough to check destination branch within the boundaries.

For intra procedure instructions the VM needs to know while building or destroying the frame if the JPC belongs to a valid method. A valid method JPC depends on how methods are stored within the class. One can suggest to define the boundaries of the method pool but if the method is inherited, the check must be done with the mother class and not the current one. Moreover the method must be allowed to be called according to the current instance. This is threaten naturally by the `invoke` instruction while building the frame, no new check is required. The `return` instruction is more difficult to handle but one invariant at least must hold: at the destination the previous instruction must be an `invoke` instruction.

Listing 1.6. Deleting the frame

```
1    bool releaseFrame( value_t *ret_val )
2    {   /* mark this frame as free */
3        thr_active->curr_frame->method = NULL;
4        if ( thr_active->curr_frame->prev == NULL)
5            return false ;
6        /* update link pointers */
7        thr_active->curr_frame = thr_active->curr_frame->prev ;
8        thr_active->curr_frame->next = NULL;
9        /* copy return value in case it exists */
10       *ret_val = *(--vm_sp );
11       /* update SP and PC */
12       vm_sp = thr_active->curr_frame->sp ;
13       if ( thr_active->curr_frame->pc - 3 == BC_invoke ){
14           vm_pc = thr_active->curr_frame->pc ;
15           return true ;}
16       else return false ;
17   }
```

The check of the invariant can be done by the method that restores the previous frame as shown in Listing 1.6. At line 13, we verify whether the generic `invoke` returns true, otherwise, the caller must handle the security problem.

The overhead occurs only while retrieving the previous instruction and it ensures the completeness of the countermeasure.

6   Conclusion and Future Works

In this paper, we have demonstrated that a well known countermeasure against shell code execution can be bypassed if not all the instructions are covered by the dynamic checks. We have shown the possibility to extract the control flow and to generate a shell code that corresponds to any executable program. We use the method parameter *i.e.* its signature to provide input and recover data from the shell code. The control program can use a type confusion to execute a rich shell code, using self modifying code. As a proof of concept, we developed a program with its controller that fills an array that is executed by the shell code. We have been able to dump entirely the memory.

In the future works, we will develop a program to automatically extract the controller and the shell code for any program. then, we expect to be able to reverse the content of the dumped memory by using a memory carving program which is under development.

References

1. Barbu, G., Duc, G., Hoogvorst, P.: Java Card Operand Stack: Fault Attacks, Combined Attacks and Countermeasures. In: Prouff [15], pp. 297–313
2. Barbu, G., Giraud, C., Guerin, V.: Embedded Eavesdropping on Java Card. In: Gritzalis, D., Furnell, S., Theoharidou, M. (eds.) SEC 2012. IFIP AICT, vol. 376, pp. 37–48. Springer, Heidelberg (2012)
3. Barbu, G., Hoogvorst, P., Duc, G.: Application-Replay Attack on Java Cards: When the Garbage Collector Gets Confused. In: Barthe, G., Livshits, B., Scandariato, R. (eds.) ESSoS 2012. LNCS, vol. 7159, pp. 1–13. Springer, Heidelberg (2012)
4. Barbu, G., Thiebeauld, H., Guerin, V.: Attacks on Java Card 3.0 Combining Fault and Logical Attacks. In: Gollmann, D., Lanet, J.-L., Iguchi-Cartigny, J. (eds.) CARDIS 2010. LNCS, vol. 6035, pp. 148–163. Springer, Heidelberg (2010)
5. Bouffard, G., Iguchi-Cartigny, J., Lanet, J.L.: Combined Software and Hardware Attacks on the Java Card Control Flow. In: Prouff, E. (ed.) CARDIS 2011. LNCS, vol. 7079, pp. 283–296. Springer, Heidelberg (2011)
6. Bouffard, G., Iguchi-Cartigny, J., Lanet, J.L.: Combined Software and Hardware Attacks on the Java Card Control Flow. In: Prouff [15], pp. 283–296
7. Bouffard, G., Lanet, J.L.: Reversing the operating system of a java based smart card. Journal of Computer Virology and Hacking Techniques 10(4), 239–253 (2014), http://dx.doi.org/10.1007/s11416-014-0218-7
8. Carlier, V., Chabanne, H., Dottax, E., Pelletier, H.: Electromagnetic Side Channels of an FPGA Implementation of AES. IACR Cryptology ePrint Archive 2004, 145 (2004)
9. Faugeron, E.: Manipulating the frame information with an underflow attack. In: Francillon and Rohatgi [15], pp. 140–151

10. Francillon, A., Rohatgi, P. (eds.): CARDIS 2013. LNCS, vol. 8419. Springer, Heidelberg (2014)
11. GlobalPlatform: Card Specification. GlobalPlatform Inc., 2.2.1 edn. (January 2011)
12. Hamadouche, S., Lanet, J.L.: Virus in a smart card: Myth or reality? In: Cheng, L., Wong, K. (eds.) Journal of Information Security and Applications, vol. 18(2-3), pp. 130–137. Elsevier (2013)
13. Iguchi-Cartigny, J., Lanet, J.L.: Developing a Trojan applets in a smart card. Journal in Computer Virology 6(4), 343–351 (2010)
14. Oracle: Java Card 3 Platform, Virtual Machine Specification, Classic Edition. No. Version 3.0.4, Oracle, Oracle America, Inc., 500 Oracle Parkway, Redwood City, CA 94065 (2011)
15. Prouff, E. (ed.): CARDIS 2011. LNCS, vol. 7079. Springer, Heidelberg (2011)
16. Razafindralambo, T., Bouffard, G., Lanet, J.-L.: A Friendly Framework for Hidding fault enabled virus for Java Based Smartcard. In: Cuppens-Boulahia, N., Cuppens, F., Garcia-Alfaro, J. (eds.) DBSec 2012. LNCS, vol. 7371, pp. 122–128. Springer, Heidelberg (2012)
17. Vermoen, D., Witteman, M.F., Gaydadjiev, G.: Reverse Engineering Java Card Applets Using Power Analysis. In: Sauveron, D., Markantonakis, K., Bilas, A., Quisquater, J.-J. (eds.) WISTP 2007. LNCS, vol. 4462, pp. 138–149. Springer, Heidelberg (2007)

Weaknesses in Two RFID Authentication Protocols

Noureddine Chikouche[1]([✉]), Foudil Cherif[2], Pierre-Louis Cayrel[3],
and Mohamed Benmohammed[4]

[1] Computer Science Department, University of M'sila, Algeria
[2] Computer Science Department, LESIA Laboratory, University of Biskra, Algeria
[3] Laboratoire Hubert Curien, UMR CNRS 5516, France
[4] LIRE Laboratory, University of Constantine 2, Algeria
chiknour28@yahoo.fr

Abstract. One of the most important challenges related to Radio Frequency Identification (RFID) systems is security. In this paper, we analyze the security and performance of two recent RFID authentication protocols based on two different code-based cryptography schemes. The first one, proposed by Malek and Miri, is based on randomized McEliece cryptosystem. The second one, proposed by Li et al., is based on Quasi Cyclic-Moderate Density Parity Check (QC-MDPC) McEliece cryptosystem. We provide enough evidence to prove that these two RFID authentication protocols are not secure. Furthermore, we propose an improved protocol that eliminates existing weaknesses in studied protocols.

Keywords: McEliece cryptosystem · RFID · Authentication protocol · Desynchronization attack · Traceability attack

1 Introduction

Among the systems which were developed quickly during the last years, we can note those of Radio Frequency Identification (RFID), these are used in various domains (e.g. access control, e-health,...). RFID is a technology without contact and it makes possible to identify an object. The typical RFID systems are comprised of three main components:

1. The **RFID tag** consists of a microchip that stores data and a coupling element, such as an antenna, to communicate via radio frequency.
2. The **RFID reader** is a device which communicates with tags via radio waves.
3. The **server** (or back-end) is a centralized place that hosts all data regarding access permissions and may be consulted by the reader.

The security is one of the most important challenges related to RFID systems. The communication channel between the tag and the reader in RFID technology is insecure, which makes it open in front of active and passive attacks. In order to have secure authentication protocols, it is important that a RFID authentication protocol requires security and privacy properties, such as:

© Springer International Publishing Switzerland 2015
S. El Hajji et al. (Eds.): C2SI 2015, LNCS 9084, pp. 162–172, 2015.
DOI: 10.1007/978-3-319-18681-8_13

- **Mutual Authentication:** A RFID authentication protocol achieves mutual authentication, that is to say, it achieves tag's authentication and the reader's authentication.
- **Untraceability:** The tag is untraceable if an intruder cannot tell whether he has seen the same tag twice or two different tags [4].
- **Desynchronization Resilience:** This property specifies for RFID protocols that update a shared secret before terminating the protocol. We can define this property as follows: at session (i), the intruder can block or modify the exchanged messages between the reader and the tag. In the next session, if the authentication process fails, then the tag and the reader are not correlated and this protocol does not achieve desynchronization resilience.

In a survey of RFID authentication protocols, we can find various protocols developed using different schemes of error-correcting codes, such as [2,13,3,12,6,5]. This work is articulated around the security analysis of two recent RFID authentication protocols. The first one, proposed by Malek and Miri [6], is based on randomized McEliece cryptosystem. The second one, proposed by Li et al. [5], is based on Quasi Cyclic-Moderate Density Parity Check (QC-MDPC) McEliece cryptosystem.

The rest of this paper is structured as follows: section 2 presents the basic concepts of code-based cryptography. Section 3 analyzes the Malek and Miri's protocol. We analyze the Li et al. protocol in section 4. In section 5, we give an improved version of Malek and Miri protocol. Finally, the paper ends with a general conclusion.

2 Preliminaries

2.1 Code-Based Cryptography

Code-based cryptography allows the construction of different schemes (like public-key encryption scheme, identification scheme, etc.). The encryption and decryption are high-speed and do not require any crypto-processor. Despite those advantages, the major problem was the size of public key (for more information see [11]). Let $\mathcal{C}(n, k, t)$ be a binary linear code, where n is length, k is dimension (k and n are positive integers and $k < n$). $\mathcal{C}$ is a t-error correcting linear code, where $t = \lfloor \frac{d-1}{2} \rfloor$. The minimum distance d is the smallest weight of any non-zero codeword in the code. An example of parameters $(n, k, t) = (2048, 1278, 70)$, the public key size was about 2.5 Megabits. Recently, code-based cryptosystems were presented with small key sizes, for example [1] and [8].

2.2 Randomized McEliece Cryptosystem

The McEliece cryptosystem [7] is the first public key cryptosystem using algebraic coding theory. The security of this cryptosystem is based on two standard computational assumptions: the syndrome decoding (SD) problem is hard, and the public-key is indistinguishable.

Nojima et al. in [10] prove that padding the plaintext with a random bit-string provides the semantic security against chosen plaintext attack (IND-CPA) for the McEliece cryptosystem with the standard assumptions. A McEliece cryptosystem has the following components:

- **Key Generation:** Randomly generates a $k \times n$ generator matrix $\mathcal{G}'$ of a binary Goppa code $\mathcal{C}$. Randomly generates a $n \times n$ binary permutation matrix P and a $k \times k$ binary invertible matrix S', then computes $\mathcal{G} = S'\mathcal{G}'P$, which is another valid generator matrix. The private key is $(S', \mathcal{G}', P, \mathcal{A}(.))$, where $\mathcal{A}(.)$ is a polynomial-time decoding algorithm. The public key is $(\mathcal{G}, t)$.
- **Encryption:** Randomly generates an error vector $e \in \mathbb{F}_2^n$ of weight $\mathsf{wt}(e) \leq t$, computes the codeword $[r \parallel m]\mathcal{G}$, where $r \in \mathbb{F}_2^{k_1}$ is a random string and $m \in \mathbb{F}_2^{k_2}$ is the plaintext. The dimension k is equal to $k_1 + k_2$, with $k_1 < bk$ and $b < 1$. The ciphertext $c' \in \mathbb{F}_2^n$ is $c' = [r \parallel m]\mathcal{G} \oplus e$.
- **Decryption:** Given a ciphertext c', computes $z = c'P^{-1}$, and then applies the polynomial-time decoding algorithm $y = \mathcal{A}(z)$ and outputs $[r \parallel m] = yS'^{-1}$. The plaintext m is the last k_2 bits of the decrypted string.

2.3 McEliece Cryptography Based on QC-MDPC Codes

Quasi Cyclic-Moderate Density Parity Check (QC-MDPC) code is a linear block code with quasi-cyclic construction (see [9]) which permits to reduce the public key size.

- **Quasi-cyclic code:** An $\mathcal{C}(n, k)$-code of length $n = \ell n_0$ is a quasi-cyclic code of order ℓ (and index n_0) if $\mathcal{C}$ is generated by a parity-check matrix $H = [H_{i,j}]$ where each $H_{i,j}$ is an $\ell \times \ell$ circulant matrix.
- **MDPC codes:** An $\mathcal{C}(n, k, w)$-MDPC code is a linear code of length n and co-dimension k which stands a parity-check matrix of row weight w.

The McEliece cryptosystem based on QC-MDPC codes works as follows:

- **Key Generation:** generate $\mathcal{C}(n, k, w)$-QC-MDPC code. Select a vector $h \in \mathbb{F}_2^n$, of row weight w uniformly at random, as the initialization factor of generating $H \in \mathbb{F}_2^{k \times n}$. The parity check matrix H is obtained from $k - 1$ cyclic shifts by h. The matrix has the form $H = [H_0 | H_1 | ... | H_{n_0-1}]$, where row weight of H_i is w_i and $w = \sum_{i=0}^{n_0-1} w_i$. A generator matrix $\mathcal{G} = (I|Q)$ can be derived from the H. Note that the public key for encryption is $\mathcal{G} \in \mathbb{F}_2^{(n-k) \times n}$ and the private key is H.

$$
Q = \begin{pmatrix} (H_{n_0-1}^{-1}.H_0)^T \\ (H_{n_0-1}^{-1}.H_1)^T \\ ... \\ (H_{n_0-1}^{-1}.H_{n_0-2})^T \end{pmatrix}
$$

- **Encryption:** To encrypt the message $m \in \mathbb{F}_2^{n-k}$, randomly generate $e \in \mathbb{F}_2^n$ of $\mathsf{wt}(e) \leq t$. The ciphertext $c' \in \mathbb{F}_2^n$ is $c' = m\mathcal{G} \oplus e$.
- **Decryption:** Let $\mathcal{A}_H$ a decoding algorithm equipped with the sparse parity check matrix H. To decrypt c' into m, compute $m\mathcal{G} = \mathcal{A}_H(m\mathcal{G} \oplus e)$, and extract the plaintext m from the first $n - k$ positions of $m\mathcal{G}$.

2.4   Notations

To describe informally many authentication protocols, We, afterward, use the following notations:

T, R	the tag and the reader, respectively
A	the adversary
$h(.)$	hash function
t	integer numbers
e	error vector of length n and weight $\text{wt}(e) \le t$
id	the tag's identifier
id_R	the reader's identifier
r, r'	random numbers with length k_1
r_{old}, r_{new}	two secret synchronization values
p	random vector with length n
v	random vector with length k
A_p	circulant matrix generated from p

3   Malek and Miri's Protocol

3.1   Review of the Malek and Miri's Protocol

Malek and Miri proposed in [6] a lightweight mutual RFID authentication protocol based on randomized McEliece public-key cryptosystem. Let's note $\mathcal{G} = \begin{bmatrix} \mathcal{G}_1 \\ \mathcal{G}_2 \end{bmatrix}$, with $\mathcal{G}_1$ and $\mathcal{G}_2$ two matrices with $k_1 \times n$ and $k_2 \times n$, respectively. This protocol uses the following principle:

$$c' = [r \parallel m]\,\mathcal{G} \oplus e = r\mathcal{G}_1 \oplus m\mathcal{G}_2 \oplus e \tag{1}$$

In the initialization phase, the trusted center (e.g. server) selects a binary string id. Then it generates a random string r that uniquely identifies the tag with id. The trusted center encrypts $[r \parallel id]$ using the randomized McEliece cryptosystem. The trusted center outputs $r\mathcal{G}_1 \oplus id\mathcal{G}_2$. Then it stores $\{r\mathcal{G}_1 \oplus id\mathcal{G}_2, id\}$ in the tag's memory. The data stored in the reader are private matrices and a database composed of $\{id_R, r, id\}$, where id_R is the reader's identifier. We note that in this protocol, the tag can communicate with a set of authorized readers. So, it is possible that different parameters for different readers can be stored in the tag's memory.

The authentication phase is depicted as follows (see Fig. 1):

- The reader R sends the query message with id_R to the tag T.
- T searches the values $\{r\mathcal{G}_1 \oplus id\mathcal{G}_2, id\}$ corresponding to id_R. If T finds the corresponding values, it generates a random error vector e. T computes $y = r\mathcal{G}_1 \oplus id\mathcal{G}_2 \oplus e$ and sends it to R.
- R decrypts y to retrieve (r, id) and e and verifies the received values with id, r stored in the database. If the tag's authentication is successful, R generates a new random vector $p \in \mathbb{F}_2^n$ and computes a circular matrix A_p from p. It sends the response set $\{d_0, d_1\}$ to T, where $d_0 = r\mathcal{G}_1 \oplus id\mathcal{G}_2 \oplus p$ and $d_1 = id \oplus h(eA_p)$, where $h(.) \in \mathbb{F}_2^{k_2}$ is an hash function.

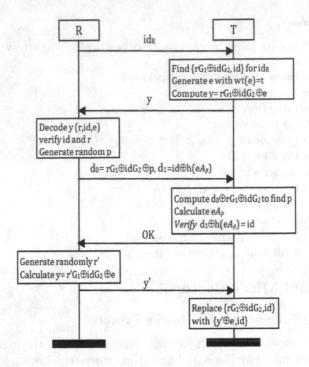

Fig. 1. Malek and Miri's Protocol [6]

- T computes $d_0 \oplus r\mathcal{G}_1 \oplus id\mathcal{G}_2 = p$ and uses its value to generate a circulant matrix A_p, in order to compute eA_p. It then, verifies that $d_1 \oplus h(eA_p) = id$. When the reader's authentication is successful, the tag requests OK to R.
- R generates a new random r' and computes $y' = r'\mathcal{G}_1 \oplus id\mathcal{G}_2 \oplus e$. It sends it to T.
- T refreshes its memory content by replacing $\{r\mathcal{G}_1 \oplus id\mathcal{G}_2, id\}$ with $\{y' \oplus e, id\}$ and terminates this session.

3.2   Desynchronization Attack

We assume that the adversary A has a complete control over the channel of communication between the reader R and the tag T. It can intercept any message passing through the network, modify or block messages, and it can also create new messages from its initial knowledge.

Fig.2 shows the message transmission of the desynchronization attack, and the following is a detailed description of each step:

1. We suppose that the system is processing normally, steps of the tag's authentication and the reader's authentication are successful. T requests OK to R and the adversary intercepts the messages transmitted between R and T.

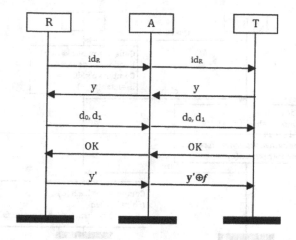

Fig. 2. Desynchronisation attack on Malek and Miri's protocol

2. R generates a new random r', computes $y' = r'\mathcal{G}_1 \oplus id\mathcal{G}_2 \oplus e$, and sends it. R updates the value of r by r'.
3. A blocks the message y', generates a vector $f \in \mathbb{F}_2^n$, and computes $y' \oplus f$. It sends it to T.
4. T updates the stored data $\{r\mathcal{G}_1 \oplus id\mathcal{G}_2, id\}$ by $\{y' \oplus f \oplus e, id\}$ and terminates the session. The new data stored is $\{r'\mathcal{G}_1 \oplus id\mathcal{G}_2 \oplus f, id\}$.
5. In the next run of the protocol, R sends the query message with id_R to T.
6. T searches $\{r'\mathcal{G}_1 \oplus id\mathcal{G}_2 \oplus f, id\}$ corresponding to id_R. T generates a random error vector e and computes $y = r'\mathcal{G}_1 \oplus id\mathcal{G}_2 \oplus f \oplus e$ and sends it to R.
7. After decrypting y, the received id'', r'' is different from id, r' (stored in the database). Thus, the tag's authentication has failed.

There is another scenario to realize the attack on desynchronization. When the intruder blocks the last message, the random value is updated in back-end and not modified in the tag. Consequentially, the tag and the reader are not correlated and this protocol does not achieve the desynchronization resilience property.

4   Li et al.'s Protocol

4.1   Review of the Li et al.'s Protocol

Li et al. proposed in [5] a mutual RFID authentication based on the QC-MDPC McEliece cryptosystem.

In the initialization phase, the trusted center (e.g. server) generates the initialization vector h', saved it in T and the database of R with identifier id. The steps of authentication phase are as follows (see Fig. 3):

– R generates a random vector v and queries T.

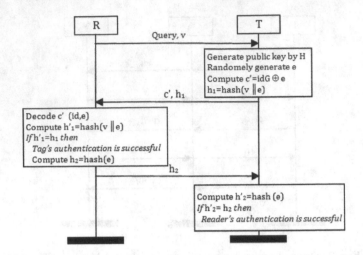

Fig. 3. Li et al.'s Protocol [5]

- After receiving v, T generates a randomly a error vector e, and then utilizes the vector h' to create public-key matrix $\mathcal{G}$ for encryption. Compute $c' = id\mathcal{G} \oplus e$ and $h_1 = h(p \parallel e)$, then sends c' and h_1 back to the reader.
- After receiving authentication message from R and transmitting them to back-end database, R performs a decoding algorithm with private key matrices and identifies the error vector e as well as id. From id, the server retrieves the corresponding value of id. It computes $h(p \parallel e)$ and compares it with h_1. If they are equal, R computes $h_2 = h(e)$ and sends it to T.
- T would compute $h(e)$, if $h(e) = h_2$, then the object of mutual authentication is achieved, authentication is successful, otherwise, the reader's authentication has failed.

4.2 Traceability Attack

In the McEliece cryptosystem, the parameters (n, k, t) are public. With these information, and particularly, the minimum distance d and the Hamming weight t; the adversary can attempt to trace the tag with the following scenario:

At session (i), the adversary intercepts $(c'_i = id\mathcal{G} \oplus e_i)$ and saves it. At session (j), it intercepts $(c'_j = id\mathcal{G} \oplus e_j)$. The intruder computes: $c'_i \oplus c'_j = id\mathcal{G} \oplus e_i \oplus id\mathcal{G} \oplus e_j$

We have $e_i \neq e_j$ and the identifier of the tag id is static in all sessions, this implicates: $c'_i \oplus c'_j = e_i \oplus e_j$. The Hamming weight of $(c'_i \oplus c'_j)$ is less than $2t + 1$, and the codeword $id\mathcal{G}$ is fixed for all sessions leads to message-resend attack, and implicates, that this protocol does not provide untraceability.

5   Improved Protocol

In the protocol of [5], the tag requires n bits in the space memory to store the vector h. In each session, the tag generates check parity matrix H, then computes the public matrix $\mathcal{G}$ from H. In the majority of RFID authentication protocols, the tag does not require to compute codeword in each process of mutual authentication, but it stores the codeword in tag, such as, the protocol of Malek-Miri [6].

In the protocol of [6] there are two major weaknesses: this protocol cannot resist desynchronization attack, and it requires an important space in volatile memory $n \times n$ bits to compute eA_p.

5.1   Algorithm of Compute eA_p

We propose the Algorithm 1 to reduce the space required from volatile memory. We symbolize eA_p by s and $\sigma(., q)$ is a circular permutation function on q positions. We present two examples of functionality of $\sigma(., q)$ with $q = 1$ and $q = 2$, respectively, as follows (2):

$$\sigma(p_1 p_2...p_n, 1) = p_n p_1...p_{n-1}$$
$$\sigma(p_1 p_2...p_n, 2) = p_{n-1} p_n...p_{n-2} \tag{2}$$

Algorithm 1. compute eA_p

1: **Input** $e = e_1 e_2...e_n$ and $p = p_1 p_2...p_n$
2: **Output** $s = s_1 s_2...s_n$
3: Initialize the vector s by values 0, ($s = 00...0$)
4: $q \leftarrow 0$
5: $i \leftarrow 0$
6: $j \leftarrow 0$
7: **while** $i < t$ **do**
8:    **if** $e_j = 0$ **then**
9:        $q \leftarrow q+1$
10:   **else**
11:       $p = \sigma(p_1 p_2...p_n, q)$
12:       $s = s \oplus p$
13:       $p = \sigma(p_1 p_2...p_n, 1)$
14:       $q \leftarrow 0$
15:       $i \leftarrow i+1$
16:   **end if**
17:   $j \leftarrow j+1$
18: **end while**
19: **return** s

Thus, using our proposed algorithm, we can reduce the size of memory required to compute eA_p from $n \times n$ into $2n$.

5.2   Description of Improved Protocol

The improved version of Malek and Miri's protocol is shown in Fig.4 and also shown as follows:

- The reader R sends the query message with id_R to the tag T.
- T searches the values $\{r\mathcal{G}_1 \oplus id\mathcal{G}_2, id\}$ corresponding to id_R. If T finds the corresponding values, it generates a random error vector e. T computes $y = r\mathcal{G}_1 \oplus id\mathcal{G}_2 \oplus e$ and sends it to R.
- R decrypts y to retrieve (id, r) and e and verifies the received values with (id, r_{old}) or (id, r_{new}) stored in the database. If the tag's authentication is successful, R generates a new random vector $p \in \mathbb{F}_2^n$ and computes eA_p by Algorithm 1, where A_p is circulant matrix of vector p. It sends the response set $\{d_0, d_1\}$ to T, where $d_0 = r\mathcal{G}_1 \oplus id\mathcal{G}_2 \oplus p$ and $d_1 = id \oplus h(eA_p)$, where $h(.) \in \mathbb{F}_2^{k_2}$ is an hash function.
- T computes $d_0 \oplus r\mathcal{G}_1 \oplus id\mathcal{G}_2 = p$ and uses its value to compute eA_p with Algorithm 1. It then, verifies that $d_1 \oplus h(eA_p) = id$. When the reader's authentication is successful, the tag requests OK to R.
- R generates a new random r', computes $y' = r'\mathcal{G}_1 \oplus id\mathcal{G}_2 \oplus e$ and $H_R = h(r'\mathcal{G}_1 \oplus id\mathcal{G}_2 \parallel e)$. It updates $r_{old} \leftarrow r_{new}$ and $r_{new} \leftarrow r'$, only in case the matched r is r_{new}.
- It sends y' and H_R to T.

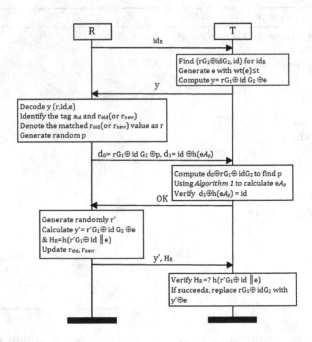

Fig. 4. Improved version of Malek and Miri's protocol

Table 1. Security comparison

	[3]	[12]	[6]	[5]	Our protocol
Mutual Authentication	N	N	Y	Y	Y
Untraceability	Y	Y	Y	N	Y
Desynchronization resilience	Y	Y	N	Y	Y
Forward secrecy	N	N	Y	N	Y

- T calculates $(y' \oplus e$ to obtaining $r'\mathcal{G}_1 \oplus id\mathcal{G}_2$. Then, it computes $h(r'\mathcal{G}_1 \oplus id\mathcal{G}_2 \parallel e)$ and compare it with received H_R. If they equal, the tag refreshes its memory content by replacing $\{r\mathcal{G}_1 \oplus id\mathcal{G}_2, id\}$ with $\{y' \oplus e, id\}$ and terminates this session.

The Table 1 presents the security comparison between the existing protocols and the proposed protocol. So, we can conclude that the improved protocol is more efficient against different attacks.

6   Conclusion

Recently, Malek-Miri and Li et al. proposed two mutual RFID authentication protocols based on error-correcting codes. In this paper, we have analyzed these two protocols in terms of security and performance. The results of security analysis show that Malek-Miri authentication protocol is vulnerable to desynchronization attack, and Li et al.'s protocol cannot resist traceability attack.

In this paper, we proposed the improved version protocol to prevent the described attacks. At the improved protocol, we used secret synchronisation values in back-end. We also proposed an algorithm to reduce the space required in volatile memory.

Acknowledgments. The third author is supported in part by NATO's Public Diplomacy Division in the framework of "Science for Peace", SPS Project 984520.

References

1. Berger, T.P., Cayrel, P.-L., Gaborit, P., Otmani, A.: Reducing key length of the mcEliece cryptosystem. In: Preneel, B. (ed.) AFRICACRYPT 2009. LNCS, vol. 5580, pp. 77–97. Springer, Heidelberg (2009)
2. Chien, H., Laih, C.: Ecc-based lightweight authentication protocol with untraceability for low-cost RFID. J. Parallel Distrib. Comput. 69(10), 848–853 (2009)
3. Cui, Y., Kobara, K., Matsuura, K., Imai, H.: Lightweight asymmetric privacy-preserving authentication protocols secure against active attack. IEICE Transactions 91-D(5), 1457–1465 (2008)
4. van Deursen, T., Mauw, S., Radomirović, S.: Untraceability of RFID protocols. In: Onieva, J.A., Sauveron, D., Chaumette, S., Gollmann, D., Markantonakis, K. (eds.) WISTP 2008. LNCS, vol. 5019, pp. 1–15. Springer, Heidelberg (2008)

5. Li, Z., Zhang, R., Yang, Y., Li, Z.: A provable secure mutual RFID authentication protocol based on error-correct code. In: Proceedings of 2014 International Conference on Cyber-Enabled Distributed Computing and Knowledge Discovery, pp. 73–78. IEEE (2014)
6. Malek, B., Miri, A.: Lightweight mutual RFID authentication. In: Proceedings of IEEE ICC 2012, pp. 868–872. IEEE (2012)
7. McEliece, R.J.: A public-key system based on algebraic coding theory. Tech. Rep. 44, Jet Propulsion Lab, DSN Progress Report (1978)
8. Misoczki, R., Barreto, P.M.: Compact McEliece keys from Goppa codes. In: Jacobson Jr., M.J., Rijmen, V., Safavi-Naini, R. (eds.) SAC 2009. LNCS, vol. 5867, pp. 376–392. Springer, Heidelberg (2009)
9. Misoczki, R., Tillich, J.P., Sendrier, N., Barreto, P.S.L.M.: MDPC-McEliece: New McEliece Variants from Moderate Density Parity-Check Codes. In: Cryptology ePrint Archive, Report 2012/409 (2012)
10. Nojima, R., Imai, H., Kobara, K., Morozov, K.: Semantic security for the McEliece cryptosystem without random oracles. Designs, Codes and Cryptography 49(1-3), 289–305 (2008)
11. Overbeck, R., Sendrier, N.: Code-based cryptography. In: Post-Quantum Cryptography, pp. 95–145. Springer, Heidelberg (2009)
12. Sekino, T., Cui, Y., Kobara, K., Imai, H.: Privacy enhanced RFID using Quasi-Dyadic fix domain shrinking. In: Proceedings of Global Telecommunications Conference, GLOBECOM 2010, pp. 1–5. IEEE (2010)
13. Suzuki, M., Kobara, K., Imai, H.: Privacy enhanced and light weight RFID system without tag synchronization and exhaustive search. In: Proceedings of the IEEE International Conference on Systems, Man and Cybernetics, pp. 1250–1255. IEEE (2006)

Square Code Attack on a Modified Sidelnikov Cryptosystem

Ayoub Otmani[1(✉)] and Hervé Talé Kalachi[2]

[1] LITIS, University of Rouen, 76821 Mont-Saint-Aignan, France
ayoub.otmani@univ-rouen.fr
[2] Department of Mathematics, University of Yaounde 1, ERAL, Cameroon
hervekalachi@gmail.com

Abstract. This paper presents a cryptanalysis of a modified version of the Sidelnikov cryptosystem which is based on Reed-Muller codes. This modified scheme consists in inserting random columns in the secret generating matrix or parity check matrix. The cryptanalysis relies on the computation of the squares of the public code. The particular nature of Reed-Muller which are defined by means of multivariate binary polynomials, permits to predicate the value of dimension of the square codes and then to fully recover in polynomial time the secret positions of the random columns. Our work shows that the insertion of random columns in the Sidelnikov scheme does not bring any security improvement.

Keywords: Sidelnikov scheme · Component-wise product · Cryptanalysis · Distinguisher

1 Introduction

Contrary to the cryptosystems based on number theory, the safety of cryptosystems based on error correcting codes appear to be resistant to the emergence of quantum computers [22]. Its other advantage is that the encryption and decryption are very fast, about five times faster for encryption, and 10 to 100 times faster for decryption compared to RSA cryptosystem. The most important representative of this cryptography is the McEliece cryptosystem [17] which is also one of the oldest public key cryptosystems. Its security is based on two problems: the difficulty of decoding a random linear code and the difficulty of recovering a decoding algorithm from a public matrix representation of a binary Goppa code. The second assumption was reformulated in a more formal way by stating there is no polynomial-time algorithm that distinguishes between a random matrix and a generating matrix of a binary Goppa code [4,21].

Although efficient, the main drawback of this scheme is the enormous size of the public key. During these last years, several authors have proposed to consider more structured codes. The common idea is to focus on codes equipped with a non-trivial permutation group.[1] This is the case for example of Misoczki and

---

[1] The permutation group of a code is the set of permutations leaving globally invariant the code.

© Springer International Publishing Switzerland 2015
S. El Hajji et al. (Eds.): C2SI 2015, LNCS 9084, pp. 173–183, 2015.
DOI: 10.1007/978-3-319-18681-8_14

Barreto [19] who proposed quasi-dyadic Goppa codes. Their worked followed Gaborit's idea to use quasi-cyclic BCH codes [11] and Berger, Cayrel, Gaborit and Otmani's paper [1] which used quasi-cyclic alternant codes. The algebraic attack given in [10] succeeds in breaking most of the parameters of [1,19]. It makes use of the fact that the underlying codes which are alternant codes come with an algebraic structure. It allows a cryptanalysis consisting in setting up a polynomial system and then solving it with Gröbner bases techniques. In the very specific the case of [1,19], the quasi-cyclic and quasi-dyadic structures allow a huge reduction of the number of variables. Recently, the attack was further improved for against [19] by exploiting more efficiently the underlying Goppa structure [8,9].

The apparition of algebraic attacks [10], although it does not undermine the security of the McEliece scheme, shows however the importance of finding a better hiding of the structure of the codes. A possible solution would be to change the description of the scheme by inserting some randomness. Probably, the first attempt towards this objective, is Berger-Loidreau's paper [2]. The authors suggest to add random rows to the description of the codes. They applied this to Niederreiter encryption scheme [20] instantiated with generalised Reed-Solomon codes. The goal is to come up with a protection against Sidelnikov and Shestakov [24]. But Wieshebrink's paper shows that component-wise product of codes [27] enables to break Berger-Loidreau's proposal.

Another simple example would be to insert random columns in the secret matrix. Several authors [25,14] have indeed proposed this technique to avoid structural attacks on similar versions of the McEliece cryptosystem. This kind modification was proposed for the first time by Wieschebrink in [25]. Its primary goal was to avoid the Sidelnikov-Shestakov attack [24] on the McEliece cryptosystem using generalized Reed-Solomon codes. Although this proposal had effectively avoided the original attack, recent studies have shown that in that case of generalized Reed-Solomon codes, the random columns can be found through considerations of the dimensions of component-wise product of codes [12,13,5]. This operation turns out to be a powerful tool. Thanks to [16], it has also helped in understanding the distinguisher of Goppa code derived in [6,7] which challenged the validity of the Goppa code indistinguishability assumption introduced in [4,21]. The paper [16] proves that the distinguisher in [6,7] has an equivalent but simpler description in terms of the component-wise product of codes. This distinguisher is even more powerful in the case of Reed-Solomon codes than for Goppa codes. Indeed, whereas for Goppa codes it is only successful for rates close to 1 [6,7], it can distinguish Reed-Solomon codes of any rate from random codes.

This paper develops a cryptanalysis of the modified version given in [14] of the Sidelnikov encryption scheme [23] which is a McEliece-type public key encryption scheme [17] based on Reed-Muller codes. The idea of [14] is to add random columns to prevent sub-exponential time key-recovery attacks of [18,3]. But, like Reed-Solomon codes, Reed-Muller codes are evaluation codes and because of this, they can be distinguished from random codes. These two families of codes share very similar properties which facilitates the recovering of the random columns. Our key-recovery attack is divided into two steps. The first one is an adaptation to Reed-Muller codes of the attacks presented in [12,13,5] in order to find the secret random columns. This is achieved in $O(n^5)$ operations in the binary field where n is the block length of the codes. The second step applies [18,3] to recover the secret permutation that hides the structure of the Reed-Muller codes. The rest of the paper is devoted to the description of the first step of the attack.

2   Preliminary Facts

We present in this section definitions and properties from coding theory we need in the paper.

Let $\mathbb{F}_q$ be the finite field of q elements, n and k be two non-zero integers such that $k \geqslant n$. A *linear code* of length n and dimension k over $\mathbb{F}_q$ is a linear subspace $\mathscr{C}$ of $\mathbb{F}_q^n$ of dimension k over $\mathbb{F}_q$. A *generating matrix* of $\mathscr{C}$ is a $k \times n$ matrix whose rows form a basis of $\mathscr{C}$. The *dual* of $\mathscr{C}$, generally denoted by $\mathscr{C}^\perp$, is the set of vectors $v \in \mathbb{F}_q^n$ such that for all $c \in \mathscr{C}$ the inner product $c \cdot v \stackrel{\text{def}}{=} \sum_i c_i v_i = 0$. A generating matrix for $\mathscr{C}^\perp$ is also called a *parity-check matrix*.

Definition 1 (Generalised Reed-Solomon). *Let* $x = (x_1, \ldots, x_n)$ *where* x_i *are distinct elements of* $\mathbb{F}_{q^m}^n$ *and let* y *be the vector* $(y_1, \ldots, y_n)$ *where* y_i *are non-zero elements of* $\mathbb{F}_{q^m}$. *The generalised Reed-Solomon code (GRS) of length* n *and dimension* k *over* $\mathbb{F}_{q^m}$ *is given by:*

$$\mathbf{GRS}_k\,(x, y) \stackrel{\text{def}}{=} \left\{ \left((y_1 f(x_1), \ldots, y_n f(x_n)) \right) \;:\; f \in \mathbb{F}_{q^m}[X],\ \deg(f) < k \right\}$$

Definition 2 (Component-Wise Product). *Given two vectors* $a = (a_1, \ldots, a_n)$ *and* $b = (b_1, \ldots, b_n)$ *in* $\mathbb{F}^n$ *where* $\mathbb{F}$ *is field, we denote by* $a \star b$ *the component-wise product:*

$$a \star b \stackrel{\text{def}}{=} (a_1 b_1, \ldots, a_n b_n).$$

Definition 3 (Product of codes). *Let* $\mathscr{A}$ *and* $\mathscr{B}$ *be two linear codes of length* n. *The star product code denoted by* $\mathscr{A} \star \mathscr{B}$ *of* $\mathscr{A}$ *and* $\mathscr{B}$ *is the vector space spanned by all products* $a \star b$ *where* a *and* b *range over* A *and* B *respectively.*

When $\mathscr{B} = \mathscr{A}$ *then* $\mathscr{A} \star \mathscr{A}$ *is called the square code of* $\mathscr{A}$ *and is rather denoted by* $\mathscr{A}^2$.

The importance of the square code construction becomes clear when we compare the dimensions of a code $\mathscr{A}$ with the dimension of its square code $\mathscr{A}^2$ and one major question is to know what one should expect. This comparison has already been made in [12,13,5] in the case of generalized Reed-Solomon codes which allowed to mount efficient attacks on several different schemes based on generalised Reed-Solomon codes [26,12,13,5]. The results of this paper are based on these comparisons in the case of Reed-Muller codes.

We recall here important facts about the product of codes.

Proposition 1. *For any linear subspaces $F \subseteq E$ and $G \subseteq E$:*

$$\dim F \star G \leqslant \dim F \dim G - \binom{\dim F \cap G}{2}. \tag{1}$$

Proof. Assume $d \overset{\text{def}}{=} \dim F \cap G$ and let $\mathcal{B} = \{b_1, \ldots, b_d\}$ be a basis of $F \cap G$. We complete $\mathcal{B}$ with vectors $\mathcal{F} = \{f_1, \ldots, f_t\}$ so that $\mathcal{B} \cup \mathcal{F}$ is a basis of F. We do the same for G by completing $\mathcal{B}$ with $\mathcal{G} = \{g_1, \ldots, g_m\}$ so that $\mathcal{B} \cup \mathcal{G}$ is a basis of G. A generating set of $F \star G$ is the union of the four sets $\{b_i \star b_j : 1 \leqslant i \leqslant j \leqslant d\}$, $\{b_i \star f_j : 1 \leqslant i \leqslant d, 1 \leqslant j \leqslant t\}$, $\{b_i \star g_j : 1 \leqslant j \leqslant d, 1 \leqslant j \leqslant m\}$ and $\{f_i \star g_j : 1 \leqslant j \leqslant t, 1 \leqslant j \leqslant m\}$. The proof is terminated by observing the equality:

$$dt + dm + tm + \binom{d+1}{2} = (t+d)(d+m) - \frac{1}{2}d(d-1).$$

Corollary 1. *For any linear subspace $F \subseteq E$:*

$$\dim F \star E \leqslant \dim F \dim E - \binom{\dim F}{2}.$$

In particular $\dim E^2 \leqslant \binom{\dim E + 1}{2}.$

3   Code-Based Public-Key Encryption Schemes

3.1   McEliece Encryption Scheme

In this section we give the basic notion about the McEliece [17] and Niederreiter [20] cryptosystems . Let $\mathcal{G}$ be a family of (n, k)-linear codes over $\mathbb{F}_q$ for which a polynomial-time algorithm to decode t-error is available. The general version of the McEliece cryptosystem is described as follows but McEliece proposed to use binary Goppa codes.

Key Generation

1. Let $\boldsymbol{G}' \in \mathcal{M}_{k \times n}(\mathbb{F}_q)$, be a generating matrix of a t-error correcting code $\mathscr{C}' \in \mathcal{G}$
2. Pick an $n \times n$ permutation matrix $\boldsymbol{P}$ and a $k \times k$ invertible matrix $\boldsymbol{S}$ at random over $\mathbb{F}_q$.
3. Compute $\boldsymbol{G} = \boldsymbol{S}^{-1}\boldsymbol{G}'\boldsymbol{P}^{-1}$ which is another generating matrix.

The public key is $(\boldsymbol{G}, t)$ and the private key is $(\boldsymbol{S}, \boldsymbol{G}, \boldsymbol{P})$.

Encryption. To encrypt the message $\boldsymbol{m} \in \mathbb{F}_q^k$, one randomly generates $\boldsymbol{e} \in \mathbb{F}_q^n$ of Hamming weight $\leqslant t$. The ciphertext is then the vector $\boldsymbol{c} = \boldsymbol{m}\boldsymbol{G} + \boldsymbol{e}$.

Decryption. The vector cP^{-1} is at a distance at most t of $\mathscr{C}$. The decoding algorithm thus allows to find the vector $y \overset{\text{def}}{=} mS^{-1}$. The plaintext is deduced by computing yS.

3.2 Niederreiter Encryption Scheme

A version of the McEliece cryptosystem that uses the parity-check matrix instead of the generating matrix has been proposed by Niederreiter [20], and has been proved to be completely equivalent in term of security. The Niederreiter cryptosystem is generally describes as follows. In the following, the transpose of matrix is denoted by ^{T}.

Key Generation

1. Let $H' \in \mathscr{M}_{(n-k) \times n}(\mathbb{F}_q)$, be a parity check matrix of a t-error correcting code $\mathscr{C}' \in \mathcal{G}$
2. Pick at random an $n \times n$ permutation matrix P and a $(n-k) \times (n-k)$ non singular matrix S over $\mathbb{F}_q$.
3. Compute $H = S^{-1}H'P^{-1}$.

   The public key is (H, t) and the private key is (S, H', P).

Encryption. For a message $m \in \mathbb{F}_q^n$ of Hamming weight $\leqslant t$. The cipher text is given by $c = Hm^T$.

Decryption. Since $c = S^{-1}H'P^{-1}m^T = S^{-1}H'(mP)^T$ and mP is a word of weight less than or equal to t, the receiver decodes Sc to get the word y. The associated plaintext is then yP.

4   Wieschebrink's Masking Technique

Here we present a masking technique first developed in [25] and then proposed several times with different families of codes. It consists in inserting random columns in the secret matrix. This technique can be used both in the McEliece cryptosystem and the Niederreiter version.

4.1 Modified McEliece Scheme

Key Generation

1. Choose three integers n_0, k, ℓ with $\ell \ll n$ and set $n \overset{\text{def}}{=} n_0 + \ell$. Pick a random a generating matrix G_0 of an (n_0, k)-code $\mathscr{C}$ that is able to decode t errors.
2. Pick randomly a $k \times \ell$ matrix R, a $k \times k$ invertible matrix S over $\mathbb{F}_q$ and an $n \times n$ permutation matrix P.
3. Set $G' = (G_0 \mid R)$ and compute $G = S^{-1}G'P^{-1}$.

   The public key is (G, t) and the private key is (S, P, G').

Encryption. To encrypt a plaintext $m \in \mathbb{F}_q^k$, one randomly generates $e \in \mathbb{F}_q^n$ of weight $\leqslant t$ and computes the ciphertext $c = mG + e$.

Decryption. To decrypt c, one computes $y = cP$ and let y' be the n_1 first columns of y. The vector y' is located within distance t from $\mathscr{C}$. The decoding of y' provides the plaintext.

4.2   Modified Niederreiter Scheme

Here one can apply the same principle as in the case of McEliece cryptosystem, but here the insertion of random columns is done in the parity check matrix.

Key Generation

1. Choose three integers n_0, k, t, ℓ with $\ell \ll n$ and set $n \overset{\text{def}}{=} n_0 + \ell$. Pick a random parity-check matrix H_0 of an (n_0, k)-code $\mathscr{C}$ that is able to decode t errors.
2. Pick randomly an $(n_0 - k) \times \ell$ matrix R and a $(n_0 - k) \times (n_0 - k)$ non singular matrix S over $\mathbb{F}_q$, and an $n \times n$ permutation matrix P.
3. Set $H' = (H_0 \mid R)$ and compute $H = S^{-1}H'P^{-1}$

The public key is (H, t) and the private key is (S, H', P).

Encryption. For a plaintext $m \in \mathbb{F}_q^n$ of Hamming weight $\leqslant t$, the corresponding ciphertext is given by $c = Hm^T$.

Decryption. Let $\mathrm{dec}(\cdot)$ be the decoding algorithm of $\mathscr{C}$. The symbol $\perp$ stands for a decoding failure[2]. The decryption of a ciphertext c is described in Algorithm 1.

Algorithm 1. Decryption of Niederreiter scheme with Wieschebrink's masking.

$u = \perp$
for all $z \in \mathbb{F}_q^\ell$ do
   $y = \mathrm{dec}\left(Sc + Rz^T\right)$
   if $y \neq \perp$ then
      $u = (y, z)P$
      return u
   end if
end for
return u

Note that it is possible for the word u to be different from the transmitted message m. But an analysis of the meaning of the received message can eliminate

---

[2] This may happen when fro instance the number of errors is greater than t.

these cases and consider them as failures decoding. The complexity of this algorithm is of order $q^{\ell}T(\mathsf{dec})$ where $T(\mathsf{dec})$ is the time complexity of the decoding algorithm $\mathsf{dec}(\cdot)$.

Although the public code seems to be random in this description, a major problem rests on the choice of the code family to use and how to reduce the size of the keys. Wieschebrink had proposed the use of Reed-Solomon codes but in [12,13,5] an attack is presented that can recover the random secret matrix $\boldsymbol{R}$.

5   Recovering the Random Columns in Polynomial Time

Recently, the paper [14] suggested the use of Reed-Muller codes along with Wieschebrink's masking technique to propose a McEliece-type encryption scheme. In the next section, we describe how to find the random columns of $\boldsymbol{R}$ in this case. Our attack uses the same technique as the one presented in [12,13,5] for the case of Reed-Solomon codes.

5.1   Reed-Muller Based Encryption Scheme

In this section, we draw inspiration from [12,13,5] to mount an attack on the version presented in [14]. But before doing so, we present some properties of Reed-Muller codes.

Definition 4 (Reed-Muller Code). *Let $\mathbb{F}_2[x_1,\ldots,x_m]$ be the set of boolean polynomials with m variables. Let us set $\{a_1,\ldots,a_n\} \overset{def}{=} \mathbb{F}_2^m$ and $n \overset{def}{=} 2^m$. The Reed-Muller code denoted by $\mathcal{RM}(r,m)$ with $0 \leqslant r \leqslant m$ is the linear space defined by:*

$$\mathcal{RM}(r,m) \overset{def}{=} \left\{ (f(a_1),\ldots,f(a_n)) \; : \; f \in \mathbb{F}_2[x_1,\ldots,x_m], \deg f \leqslant r \right\}$$

We recall an immediate fact about the dimension of Reed-Muller codes.

Fact 1. *The dimension of $\mathcal{RM}(r,m)$ is equal to $\displaystyle\sum_{i=0}^{r} \binom{m}{i}$.*

Theorem 2 ([15] Chapter 13).

$$\mathcal{RM}(r,m)^{\perp} = \mathcal{RM}(m-r-1,m)$$

Proposition 2.

$$\mathcal{RM}(r,m)^2 = \mathcal{RM}(2r,m)$$

Proof. Let $c_1 = (f(a_1),\ldots,f(a_n))$ and $c_2 = (g(a_1),\ldots,g(a_n))$ be elements of $\mathcal{RM}(r,m)$ with $\deg f \leqslant r$ and $\deg g \leqslant r$. Hence, $c_1 \star c_2$ is the vector of evaluation $(fg(a_1),\ldots,fg(a_n))$ which corresponds to polynomial fg. This means $c_1 \star c_2 \in \mathcal{RM}(2r,m)$.

Conversely, each monomial $x_1^{e_1},\ldots,x_m^{e_m}$ with $e_i \geqslant 0$ and $\sum_i e_i \leqslant 2r$ is the product of two polynomials of degree $\leqslant r$. This proves that a basis of $\mathcal{RM}(2r,m)$ is contained in $\mathcal{RM}(r,m)^2$.

Proposition 3. *Let G be a $k \times (n + \ell)$ matrix obtained by inserting ℓ random columns in the generating matrix of a Reed-Muller code $\mathcal{RM}(r, m)$ and let $\mathscr{C}$ be the code spanned by the rows of G. Assume that $\ell \leqslant \binom{k}{2}$ and $\sum_{i=0}^{2r} \binom{m}{i} \leqslant n$. Then we have:*

$$\sum_{i=0}^{2r} \binom{m}{i} \leqslant \dim \mathscr{C}^2 \leqslant \sum_{i=0}^{2r} \binom{m}{i} + \ell \qquad (2)$$

Proof. Let $\mathscr{D}_1$ be the code with generating matrix G_1 obtained from G by replacing the last ℓ columns by all-zero columns and let $\mathscr{D}_2$ be the code with generating matrix G_2 obtained by replacing in G the first n columns by zero columns. Hence $G = G_1 + G_2$ which implies $\mathscr{D}_1 \subseteq \mathscr{C} \subseteq \mathscr{D}_1 + \mathscr{D}_2$. We have $\mathscr{D}_1 \star \mathscr{D}_2 = 0$ and the following inclusion:

$$\mathscr{D}_1^2 \subseteq \mathscr{C}^2 \subseteq \mathscr{D}_1^2 + \mathscr{D}_2^2 + \mathscr{D}_1 \star \mathscr{D}_2.$$

Observe we have $\mathscr{D}_1 \star \mathscr{D}_2 = 0$. By also remarking $\dim \mathscr{D}_1^2 = \dim \mathcal{RM}(2r, m)$ and $\dim \mathscr{D}_2^2 = \min \left\{ \ell, \binom{k}{2} \right\} = \ell$, one can conclude (2) is proven.

5.2   Description of the Attack

It is easy for an adversary to use Prop. 3 to identify the random columns by computing the dimension of $\mathscr{C}^2$ where $\mathscr{C}$ is the code generated by the public matrix G as defined in Sec. 4. We recall that $\mathscr{C}$ is permuted version of a Reed-Muller code $\mathcal{RM}(r, m)$. We assume that $\sum_{i=0}^{2r} \binom{m}{i} \leqslant n_0$ where $n_0 = 2^m$ and $\ell < \binom{k}{2}$ where $k = \sum_{i=0}^{r} \binom{m}{i}$. We denote by $\mathscr{C}_i$ the code generated by the generating matrix G_i obtained by deleting the i-th column of G. We also denote by $I \subset \{1, \ldots, n\}$ the set of positions that define the random columns inserted in G. Two cases occur with high probability:

$$\dim \mathscr{C}_i^2 = \begin{cases} \dim \mathscr{C}^2 - 1 & \text{if } i \in I, \\ \dim \mathscr{C}^2 & \text{if } i \notin I. \end{cases} \qquad (3)$$

Once the set I is recovered, it is then easy to find the secret $\mathcal{RM}(r, m)$ using usual attacks on Reed-Muller code [18].

Remark 1. For the parameters in [14], we observed experimentally that (3) is always true, and the upper-bound given in (2) is always reached, that is to say:

$$\dim \mathscr{C}^2 = \sum_{i=0}^{2r} \binom{m}{i} + \ell.$$

This is way of distinguishing the random positions of the public code assumes that $\sum_{i=0}^{2r} \binom{m}{i} + \ell \geqslant n$. We will see how to deal with parameters that do not satisfy this assumption. The idea is to look at $\dim \mathscr{D}^2$ where $\mathscr{D}$ is the dual of $\mathscr{C}$. Indeed, like generalized Reed-Solomon codes, the family of Reed-Muller codes is stable under duality (Theorem 2).

Proposition 4. *Keeping with notation of Proposition 3, let $\mathscr{D}$ be the dual of $\mathscr{C}$. Assuming $\sum_{i=0}^{2r} \binom{m}{i} > n_0$, we have:*

$$\dim \mathscr{D}^2 \leqslant \frac{1}{2}\ell(\ell+1) + \ell \sum_{i=0}^{m-r-1} \binom{m}{i} + \sum_{i=0}^{2(m-r-1)} \binom{m}{i}. \qquad (4)$$

Proof. Let us set $k = \sum_{i=0}^{r} \binom{m}{i}$. We may assume without loss of generality that a generating matrix of $\mathscr{C}$ is in systematic form: $\begin{pmatrix} I_k & A & R \end{pmatrix}$ where R form the random columns and $\begin{pmatrix} I_k & A \end{pmatrix}$ generates $\mathcal{RM}(r,m)$. A parity-check matrix for $\mathscr{C}$ is then:

$$H = \begin{pmatrix} -A^T & I_{n_0-k} & 0 \\ -R^T & 0 & I_\ell \end{pmatrix}.$$

The upper-bound (4) can be readily derived from this last matrix H.

5.3 Complexity of the Attack

Proposition 5. *Let $\mathscr{A} \subset \mathbb{F}_q^n$ be a code of dimension k. The complexity of the computation of a basis of $\mathscr{A}^2$ is $O(k^2 n^2)$ operations in $\mathbb{F}_q$.*

Proof. The computation, consists first in the computation of $\binom{k+1}{2}$ generators of $\mathscr{A}^2$. This computation costs $O(k^2 n)$ operations. Then, we have to apply a Gaussian elimination to a $\binom{k+1}{2} \times n$ matrix, which costs $O(k^2 n^2)$ operations. This second step is dominant, which yields the result.

Our attack relies on the computation of the rank of n square codes so the overall complexity for guessing the random columns is $O(n^5)$ operations in the binary field.

6   Conclusion

In this paper, we study the security of the modified version of the Sidelnikov scheme [23] given in [14]. We have presented a polynomial-time method that finds the random columns inserted in a secret matrix. This cryptanalysis uses the same approach as [12,13,5] which computes the square codes. The resulting complexity is $O(n^5)$ operations in the binary field. The last step that aims to fully break the scheme ressort to using the attacks developed in [18,3]. Our work shows that the insertion of random columns in the Sidelnikov scheme does not bring any security improvement.

References

1. Berger, T.P., Cayrel, P.-L., Gaborit, P., Otmani, A.: Reducing key length of the McEliece cryptosystem. In: Preneel, B. (ed.) AFRICACRYPT 2009. LNCS, vol. 5580, pp. 77–97. Springer, Heidelberg (2009)
2. Berger, T.P., Loidreau, P.: How to mask the structure of codes for a cryptographic use. Des. Codes Cryptogr. 35(1), 63–79 (2005)
3. Chizhov, I.V., Borodin, M.A.: The failure of McEliece PKC based on Reed-Muller codes. IACR Cryptology ePrint Archive, Report 2013/287 (2013), http://eprint.iacr.org/
4. Courtois, N.T., Finiasz, M., Sendrier, N.: How to achieve a McEliece-based digital signature scheme. In: Boyd, C. (ed.) ASIACRYPT 2001. LNCS, vol. 2248, pp. 157–174. Springer, Heidelberg (2001)
5. Couvreur, A., Gaborit, P., Gauthier-Umaña, V., Otmani, A., Tillich, J.P.: Distinguisher-based attacks on public-key cryptosystems using Reed-Solomon codes. Des. Codes Cryptogr. 73(2), 641–666 (2014), http://dx.doi.org/10.1007/s10623-014-9967-z
6. Faugère, J.C., Gauthier, V., Otmani, A., Perret, L., Tillich, J.P.: A distinguisher for high rate McEliece cryptosystems. In: Proc. IEEE Inf. Theory Workshop, ITW 2011, Paraty, Brasil, pp. 282–286 (October 2011)
7. Faugère, J.C., Gauthier, V., Otmani, A., Perret, L., Tillich, J.P.: A distinguisher for high rate McEliece cryptosystems. IEEE Trans. Inf. Theory 59(10), 6830–6844 (2013)
8. Faugère, J.C., Otmani, A., Perret, L., de Portzamparc, F., Tillich, J.P.: Structural weakness of compact variants of the McEliece cryptosystem. In: Proc. IEEE Int. Symposium Inf. Theory, ISIT 2014, Honolulu, HI, USA, pp. 1717–1721 (July 2014)
9. Faugère, J.C., Otmani, A., Perret, L., de Portzamparc, F., Tillich, J.P.: Structural cryptanalysis of McEliece schemes with compact keys. Des. Codes Cryptogr. (2015), to appear, see also IACR Cryptology ePrint Archive, Report2014/210
10. Faugère, J.-C., Otmani, A., Perret, L., Tillich, J.-P.: Algebraic cryptanalysis of McEliece variants with compact keys. In: Gilbert, H. (ed.) EUROCRYPT 2010. LNCS, vol. 6110, pp. 279–298. Springer, Heidelberg (2010)
11. Gaborit, P.: Shorter keys for code based cryptography. In: Proceedings of the 2005 International Workshop on Coding and Cryptography (WCC 2005), Bergen, Norway, pp. 81–91 (March 2005)
12. Gauthier, V., Otmani, A., Tillich, J.P.: A distinguisher-based attack of a homomorphic encryption scheme relying on Reed-Solomon codes. CoRR abs/1203.6686 (2012)
13. Gauthier, V., Otmani, A., Tillich, J.P.: A distinguisher-based attack on a variant of McEliece's cryptosystem based on Reed-Solomon codes. CoRR abs/1204.6459 (2012)
14. Gueye, C.T., Mboup, E.H.M.: Secure cryptographic scheme based on modified Reed Muller codes. International Journal of Security and its Applications 7(3), 55–64 (2013)
15. MacWilliams, F.J., Sloane, N.J.A.: The Theory of Error-Correcting Codes, 4th edn. North–Holland, Amsterdam (1986)
16. Márquez-Corbella, I., Pellikaan, R.: Error-correcting pairs for a public-key cryptosystem. preprint (2012) (preprint)
17. McEliece, R.J.: A Public-Key System Based on Algebraic Coding Theory, pp. 114–116. Jet Propulsion Lab (1978), dSN Progress Report 44

18. Minder, L., Shokrollahi, M.A.: Cryptanalysis of the Sidelnikov cryptosystem. In: Naor, M. (ed.) EUROCRYPT 2007. LNCS, vol. 4515, pp. 347–360. Springer, Heidelberg (2007)

19. Misoczki, R., Barreto, P.: Compact McEliece keys from Goppa codes. In: Selected Areas in Cryptography, Calgary, Canada (August 13-14, 2009)

20. Niederreiter, H.: Knapsack-type cryptosystems and algebraic coding theory. Problems of Control and Information Theory 15(2), 159–166 (1986)

21. Sendrier, N.: Cryptosystèmes à clé publique basés sur les codes correcteurs d'erreurs. Ph.D. thesis, Université Paris 6, France (2002)

22. Shor, P.W.: Polynomial-time algorithms for prime factorization and discrete logarithms on a quantum computer. SIAM J. Comput. 26(5), 1484–1509 (1997)

23. Sidelnikov, V.M.: A public-key cryptosytem based on Reed-Muller codes. Discrete Mathematics and Applications 4(3), 191–207 (1994)

24. Sidelnikov, V.M., Shestakov, S.: On the insecurity of cryptosystems based on generalized Reed-Solomon codes. Discrete Mathematics and Applications 1(4), 439–444 (1992)

25. Wieschebrink, C.: Two NP-complete problems in coding theory with an application in code based cryptography. In: Proc. IEEE Int. Symposium Inf. Theory, ISIT 2006, pp. 1733–1737 (2006)

26. Wieschebrink, C.: Cryptanalysis of the Niederreiter public key scheme based on GRS subcodes. IACR Cryptology ePrint Archive, Report 2009/452 (2009), http://eprint.iacr.org/2009/452.pdf

27. Wieschebrink, C.: Cryptanalysis of the Niederreiter public key scheme based on GRS subcodes. In: Post-Quantum Cryptography 2010, pp. 61–72 (2010)

A Family of Six-Weight Reducible Cyclic Codes and their Weight Distribution

Gerardo Vega[✉]

Dirección General de Cómputo y de Tecnologías de Información y Comunicación, Universidad Nacional Autónoma de México, 04510 México D.F., Mexico
gerardov@unam.mx

Abstract. Reducible cyclic codes with exactly two nonzero weights were first studied by T. Helleseth [4] and J. Wolfmann [15]. Later on, G. Vega [11], set forth the sufficient numerical conditions in order that a cyclic code, constructed as the direct sum of two one-weight cyclic codes, has exactly two nonzero weights, and conjectured that there are no other reducible two-weight cyclic codes of this type. In this paper we present a new class of cyclic codes constructed as the direct sum of two one-weight cyclic codes. As will be shown, this new class of cyclic codes is in accordance with the previous conjecture, since its codes have exactly six nonzero weights. In fact, for these codes, we will also give their full weight distribution, showing that none of them can be projective. On the other hand, recently, a family of six-weight reducible cyclic codes and their weight distribution, was presented by Y. Liu, *et al.* [7]; however it is worth noting that, unlike what is done here, the codes in such family are constructed as the direct sum of three irreducible cyclic codes.

Keywords: Weight distribution · Reducible cyclic codes and Gaussian periods

1 Introduction

It is said that a cyclic code is reducible if its parity-check polynomial is factorizable in two or more irreducible factors. Each one of these irreducible factors can be seen as the parity-check polynomial of an irreducible cyclic code. Therefore, a reducible cyclic code is, basically, the direct sum of these irreducible cyclic codes. Reducible cyclic codes, whose parity-check polynomials are factorizable in exactly two different irreducible factors have been extensively studied (see, for example, [4], [15], [11], [8], [3], [14], [16] and [12]). Now, each one of these two irreducible factors might or might not correspond to the parity-check polynomial of a one-weight irreducible cyclic code. Most of the recent efforts along this line of research have been focused on the study of reducible cyclic codes constructed as the direct sum of two two-weight irreducible cyclic codes. In fact, through the easy-to-apply characterization for all semiprimitive two-weight irreducible cyclic codes over any finite field, that was recently presented in [13], it

---

Partially supported by PAPIIT-UNAM IN107515.

S. El Hajji et al. (Eds.): C2SI 2015, LNCS 9084, pp. 184–196, 2015.
DOI: 10.1007/978-3-319-18681-8_15

is interesting to note that most of the families of reducible cyclic codes studied in [8], [3], [14], [16] and [12], are constructed as a direct sum of two different semiprimitive two-weight irreducible cyclic codes of the same dimension. In this paper, we present a new class of reducible cyclic codes constructed as the direct sum of two one-weight irreducible cyclic codes. This new class of cyclic codes is different to the class of codes studied in [4], [15] and [11], and this is so because the codes presented here have, as we will see later, exactly six nonzero weights. In fact, for the codes in this new class we will explicitly give their full weight distribution, and show that none of them can be projective. Recently, on the other hand, a family of six-weight reducible cyclic codes and their weight distribution, was presented in [7], however it is worth noting that, unlike what is done here, the codes in such family are constructed as the direct sum of three different irreducible cyclic codes of the same dimension.

In order to give a detailed explanation of what is the main result of this work, let p, t, q, k and Δ be five positive integers, such that p is a prime number, $q = p^t$, and $\Delta = (q^k - 1)/(q - 1)$. In addition, let γ be a fixed primitive element of $\mathbb{F}_{q^k}$ and, for any integer a, denote by $h_a(x) \in \mathbb{F}_q[x]$ the minimal polynomial of γ^{-a}. With this notation in mind, the following result gives a description of the weight distribution of a new class of reducible cyclic codes constructed as the direct sum of two one-weight cyclic codes of the same length and dimension:

Table 1. Weight distribution of $\mathcal{C}_{(a_1,a_2)}$

Weight	Frequency
0	1
$\frac{2\lambda}{3}(q^{k-1} + 2(-1)^s q^{k/2-1})$	$q^k - 1$
$\frac{2\lambda}{3}(q^{k-1} - (-1)^s q^{k/2-1})$	$2(q^k - 1)$
$\lambda(q^{k-1} + 2(-1)^s q^{k/2-1})$	$\frac{1}{27}(q^k - 1)(q^k - 2(-1)^s q^{k/2} - 8)$
$\lambda(q^{k-1} + (-1)^s q^{k/2-1})$	$\frac{2}{9}(q^k - 1)(q^k + (-1)^s q^{k/2} - 2)$
λq^{k-1}	$\frac{2}{9}(q^k - 1)(2q^k - (-1)^s q^{k/2} - 1)$
$\lambda(q^{k-1} - (-1)^s q^{k/2-1})$	$\frac{2}{27}(q^k - 1)(4q^k + (-1)^s q^{k/2} - 14)$

Theorem 1. *Suppose that $q = p^t$ is odd, whereas tk is even. Also suppose that $3|(q^k - 1)$ and $\gcd(\Delta, \frac{q^k - 1}{3} - 1) = 3$. Let d, s, a_1 and a_2 be any four integers such that $tk = 2sd$, $a_2 = a_1 + \frac{q^k - 1}{3}$ and $\gcd(\Delta, a_1) = \gcd(\Delta, a_2) = 1$. In addition, let λ be the divisor of $q - 1$, satisfying $\gcd(q - 1, a_1) = \frac{q-1}{\lambda}$ and let $\mathcal{C}_{(a_1,a_2)}$ be the cyclic code with parity-check polynomial $h_{a_1}(x)h_{a_2}(x)$. Fix $n = \lambda\Delta$. If $3|(p^d + 1)$ then*

(A) *$h_{a_1}(x)$ and $h_{a_2}(x)$ are the parity-check polynomials of two different one-weight cyclic codes of length n and dimension k. The nonzero weight of these two irreducible cyclic codes is λq^{k-1}.*

(B) $\mathcal{C}_{(a_1,a_2)}$ is an $[n, 2k]$ cyclic code over $\mathbb{F}_q$, with the weight distribution given in Table I.

(C) If B_1 and B_2 are, respectively, the number of words of weight 1 and 2 in the dual code of $\mathcal{C}_{(a_1,a_2)}$, then $B_1 = 0$ and $B_2 = \frac{n(q-1)(\lambda-1)}{2}$. Therefore $\mathcal{C}_{(a_1,a_2)}$ is a non-projective cyclic code.

(D) $\mathcal{C}_{(a_1,a_2)}$ has, exactly, six nonzero weights.

For any integer a_1, the kind of reducible cyclic codes whose parity-check polynomials are given by the products of the form $h_{a_1}(x)h_{a_1 \pm \frac{q^k-1}{2}}(x)$, were studied in [8] and [3]. Later, a general description of this kind of codes was given in [12]. Thus, for this work, it is clear that we are dealing with the kind of reducible cyclic codes whose parity-check polynomials are given by the products of the form $h_{a_1}(x)h_{a_1 + \frac{q^k-1}{3}}(x)$. Recently, a class of this kind of reducible cyclic codes was the main subject of study in [14]. However, through the aforementioned easy-to-apply characterization in [13], it is no difficult to see that each code in this class is always constructed as the direct sum of two different semiprimitive two-weight irreducible cyclic codes of the same dimension. Conversely, the class of cyclic codes that are studied in Theorem 1 are the outcome of the direct sum of two one-weight cyclic codes, and, as will be shown later, this is so thanks to the condition $\gcd(\Delta, a_1) = \gcd(\Delta, a_1 + \frac{q^k-1}{3}) = 1$. Therefore, it is important to keep in mind that the class of codes studied in [14], and those studied here, are two different classes of codes of the same kind.

This work is organized as follows: In Section 2 we establish some notation, recall some definitions and establish our main assumption that will be considered throughout this work. We also recall, for this section, some results already known. In particular, we present the evaluation of a specific exponential sum which can be derived from a general result originally presented in [9]. Section 3 is devoted to presenting some preliminary and general results. In Section 4 we use these results in order to present a formal proof of Theorem 1 and give some examples for this result. Finally, Section 5 will be devoted to conclusions.

2   Definitions, Notation and Main Assumption

First of all, we set, for this section and for the rest of this work, the following:

Notation. By using p, t, q, k and Δ, we will denote five positive integers such that p is a prime number, $q = p^t$ and $\Delta = (q^k - 1)/(q - 1)$. From now on, γ will denote a fixed primitive element of $\mathbb{F}_{q^k}$. For any integer a, the polynomial $h_a(x) \in \mathbb{F}_q[x]$ will denote the minimal polynomial of γ^{-a}. Furthermore, we will denote by "Tr", the absolute trace mapping from $\mathbb{F}_{q^k}$ to the prime field $\mathbb{F}_p$, and by "$\mathrm{Tr}_{\mathbb{F}_{q^k}/\mathbb{F}_q}$" the trace mapping from $\mathbb{F}_{q^k}$ to $\mathbb{F}_q$. For any positive divisor m of $q^k - 1$ and for any $0 \le i \le m-1$, we define $\mathcal{D}_i^{(m)} := \gamma^i \langle \gamma^m \rangle$, where $\langle \gamma^m \rangle$ denotes the subgroup of $\mathbb{F}_{q^k}^*$ generated by γ^m. The cosets $\mathcal{D}_i^{(m)}$ are called the *cyclotomic* *classes* of order m in $\mathbb{F}_{q^k}$. In connection with these cyclotomic classes, we recall the *cyclotomic numbers* of order m. Such cyclotomic numbers are defined by

$$(i,j)^{(m,q^k)} := |(\mathcal{D}_i^{(m)} + 1) \cap \mathcal{D}_j^{(m)}|,$$

where $(\mathcal{D}_i^{(m)} + 1) = \{x + 1 \mid x \in \mathcal{D}_i^{(m)}\}$, and $0 \leq i, j \leq m - 1$.

The following definitions are important:

An *N-weight code* is a code such that the cardinality of the set of nonzero weights is N. It is important to recall that one-weight irreducible cyclic codes are also known as subfield codes.

A *projective code* is a linear code such that the minimum weight of its dual code is at least three (or, equivalently, if any two columns of its generator matrix are linearly independent).

A cyclic code is *irreducible* if its parity-check polynomial is irreducible (that is, its polynomial representation is a minimal ideal).

For this work, we are particularly interested in reducible cyclic codes, whose parity-check polynomials are factorizable in exactly two different irreducible factors. That is, we are interested in cyclic codes whose dual codes have two non conjugated zeros.

We continue with this section by recalling the definition, and a basic property of the character sums (see, for example, [6]). In order to do this, let p, q, k and γ be as before; then, the canonical additive character χ, of $\mathbb{F}_{q^k}$, is defined as

$$\chi(y) := \zeta_p^{\mathrm{Tr}(y)}, \qquad \text{for all } y \in \mathbb{F}_{q^k},$$

where $\zeta_p := \exp(\frac{2\pi\sqrt{-1}}{p})$. For the canonical additive character χ', of $\mathbb{F}_q$, the following orthogonal property will be useful:

$$\sum_{y \in \mathbb{F}_q} \chi'(y) = 0. \tag{1}$$

Now, we set, for this section and for the rest of this work, the following:

Main Assumption. From now on, we are going to suppose that q is an odd integer, whereas the product tk is an even integer. Therefore, throughout this work, we will reserve the letters s and d to represent any two positive integers that satisfy: $tk = 2sd$. In addition, we will also suppose that $3 \mid (q^k - 1)$ and $\gcd(\Delta, \frac{q^k-1}{3} - 1) = 3$. Therefore, in what follows, we will reserve the Greek letter τ in order to fix $\tau = \gamma^{\frac{q^k-1}{3}}$.

Remark 1. As a consequence of our main assumption, observe that $\mathbb{F}_q^* \subseteq \mathcal{D}_0^{(3)}$, and that the finite field element τ is a primitive three-root of unity satisfying $\tau^2 + \tau + 1 = 0$, therefore, $\tau^{1/2} = -\tau^{-1} = \tau + 1$. In addition, observe that $\tau \in \mathcal{D}_1^{(3)}$ and since $3 \mid \Delta$ we also have that $k > 1$ and, necessarily, the prime number p must be greater than 3.

Now, let χ be the canonical additive character of $\mathbb{F}_{q^k}$, and let i be any integer. Since $3|(q^k-1)$, it follows that:

$$\sum_{x\in\mathbb{F}_{q^k}}\chi(\gamma^i x^3)=1+3\sum_{z\in\mathcal{D}_i^{(3)}}\chi(z)\,.$$

We are particularly interested in the kind of exponential sums that appear in the RHS of the previous equality. These exponential sums are known as *Gaussian period* of order 3. Fortunately, the following result, which is an instance of the main result in [9, Theorem 1], gives us useful information about such Gaussian period.

Theorem 2. *With our notation and main assumption, suppose that $3|(p^d+1)$ and let η_0 and η_1 be the two integers given by:*

$$\eta_0=\frac{-2(-1)^s q^{\frac{k}{2}}-1}{3}\,,$$

$$\eta_1=\frac{(-1)^s q^{\frac{k}{2}}-1}{3}\,. \tag{2}$$

Then, for any integer i, the Gaussian period of order 3 is:

$$\sum_{z\in\mathcal{D}_i^{(3)}}\chi(z)=\begin{cases}\eta_0 & if\ i\equiv 0\pmod 3\,,\\[2mm]\eta_1 & otherwise\end{cases}.$$

Since we will be dealing with the Gaussian period of order 3, we will also need the cyclotomic numbers of order 3. The following lemma gives us important information about such cyclotomic numbers (see [1] for the general result).

Lemma 1. *Consider the same notation and hypotheses as in the previous theorem. Then*

$$(0,0)^{(3,q^k)}=\frac{(p^{sd}-(-1)^s)^2}{9}-1\,,$$

$$(i,0)^{(3,q^k)}=(0,i)^{(3,q^k)}=(i,i)^{(3,q^k)}$$

$$=\frac{(p^{sd}-(-1)^s)(p^{sd}+2(-1)^s)}{9}\,,\ for\ i=1,2,$$

$$(1,2)^{(3,q^k)}=(2,1)^{(3,q^k)}=\frac{(p^{sd}-(-1)^s)^2}{9}\,.$$

Remark 2. Observe that if $3|(p^d+1)$, which is the central hypotheses in Theorem 2, then, for any positive integer s, we have that 3 is a common divisor of both $(p^{sd}-(-1)^s)$ and $(p^{sd}+2(-1)^s)$.

3 Some Preliminary Results

The following lemma will be useful in order to show that all the cyclic codes given by Theorem 1 have exactly six nonzero weights.

Lemma 2. *Consider the same notation and hypotheses as Theorem 2. Then*
$$\eta_0 - \eta_1 = -(-1)^s q^{\frac{k}{2}}, \quad 2\eta_1 - \eta_0 = \frac{4(-1)^s q^{\frac{k}{2}} - 1}{3}, \quad 3\eta_1 - 2\eta_0 = \frac{7(-1)^s q^{\frac{k}{2}} - 1}{3}, \quad 3\eta_0 - 2\eta_1 =$$
$\frac{-8(-1)^s q^{\frac{k}{2}} - 1}{3}$ *and* $2\eta_0 - \eta_1 = \frac{-5(-1)^s q^{\frac{k}{2}} - 1}{3}$. *Therefore, none of these five previous values, as well as the values of η_0 and η_1, is equal to $\frac{q^k - 1}{3}$. Furthermore, $\eta_0 \neq \eta_1$.*

Proof. The first and third assertions come directly from (2). Since q is an odd integer, it will be enough to prove, for the second assertion, that none of the following two conditions hold: $3\eta_1 - 2\eta_0 = \frac{q^k - 1}{3}$ or $2\eta_0 - \eta_1 = \frac{q^k - 1}{3}$. Thus, supposing that $\frac{7(-1)^s q^{\frac{k}{2}} - 1}{3} = \frac{q^k - 1}{3}$, implies that $p^{sd} = 7$, where s must be a positive even integer. But clearly this conclusion is impossible, therefore $3\eta_1 - 2\eta_0 \neq \frac{q^k - 1}{3}$. On the other hand, supposing that $\frac{-5(-1)^s q^{\frac{k}{2}} - 1}{3} = \frac{q^k - 1}{3}$, implies that $p^{sd} = 5$, where $s = d = 1$. Now, since $k > 1$ (recall Remark 1) and $tk = 2sd$, we have $t = 1$, $k = 2$ and $p = q = 5$. But, under these circumstances, clearly $\gcd(\Delta, \frac{q^k - 1}{3} - 1) \neq 3$. Therefore, $2\eta_0 - \eta_1 \neq \frac{q^k - 1}{3}$.   □

The following result gives us information about the multiplicity of the elements in a very particular multiset.

Lemma 3. *With our notation and main assumption, let λ be a divisor of $q - 1$. Also let i be any integer. If $\gcd(\Delta, \frac{(q-1)}{\lambda}) = 1$, then*
$$\{xy \mid x \in \mathcal{D}_i^{(\frac{3(q-1)}{\lambda})} \text{ and } y \in \mathbb{F}_q^*\} = \lambda * \mathcal{D}_i^{(3)},$$
where $\lambda * \mathcal{D}_i^{(3)}$ is the multiset in which each element of $\mathcal{D}_i^{(3)}$ appears with multiplicity λ.

Proof. Since $\mathbb{F}_q^* \subseteq \mathcal{D}_0^{(3)}$, $(\frac{3(q-1)}{\lambda}) \mid (q^k - 1)$, and $\gcd(\Delta, \frac{3(q-1)}{\lambda}) = 3$, the result comes from the fact that $|\mathcal{D}_i^{(\frac{3(q-1)}{\lambda})}| |\mathbb{F}_q^*|/|\mathcal{D}_i^{(3)}| = \lambda$, and $\mathcal{D}_i^{(m)} = \gamma^i \mathcal{D}_0^{(m)}$, for any integer i and for any divisor m of $q^k - 1$.   □

In what follows, we will always assume that $3 \mid (p^d + 1)$. Thus, by using the finite field element τ and the cyclotomic class of order 3 in $\mathbb{F}_{q^k}$, we define the following ten sets:

$$\mathcal{E}_{i,j} - \{(\alpha, -\tau^j \alpha) \mid \tau^{1-j}(\alpha \quad \tau \alpha) \subset \mathcal{D}_i^{(3)}\}, \quad \text{for } i, j - 0, 1, 2, \text{ and}$$

$$\mathcal{G} = \{(\alpha, -\beta) \in \mathbb{F}_{q^k} \times \mathbb{F}_{q^k} \mid (\alpha - \tau^j \beta) \neq 0, \text{ for } j = 0, 1, 2\}.$$

Remark 3. By the previous definition, note that $(\alpha, -\tau^j\alpha) \in \mathcal{E}_{i,j}$ if and only if $\tau^{1-j}(1-\tau)\alpha \in \mathcal{D}_i^{(3)}$. In consequence, and since $\tau^{1-j}(1-\tau)$ is a fixed nonzero field element, we have that these ten sets are pairwise disjoint, and their cardinalities are $|\mathcal{E}_{i,j}| = |\mathcal{D}_0^{(3)}| = \frac{q^k-1}{3}$, for all $i, j = 0, 1, 2$, and $|\mathcal{G}| = q^{2k} - 1 - 9|\mathcal{E}_{0,0}| = (q^k - 1)(q^k - 2)$. Furthermore, due to Remark 1, observe that if $\tau^{1-j}(\alpha - \tau\alpha) \in \mathcal{D}_i^{(3)}$, then $\tau^{2-j}(\alpha - \tau^2\alpha) = \tau^{2-j}(\tau + 1)(\alpha - \tau\alpha) = -\tau^{1-j}(\alpha - \tau\alpha) \in \mathcal{D}_i^{(3)}$, for any $i, j = 0, 1, 2$. Therefore, the important conclusion here is that if we have $\tau^{1-j}(\alpha - \tau\alpha) \in \mathcal{D}_i^{(3)}$, then we necessarily have also that one of the following three finite field elements, $(\alpha - \tau^j\alpha)$, $\tau(\alpha - \tau^{j+1}\alpha)$ or $\tau^2(\alpha - \tau^{j+2}\alpha)$, is equal to zero and the other two elements belong to $\mathcal{D}_i^{(3)}$.

Now, for each $(\alpha, -\beta) \in \mathcal{G}$, we define the function $f_{\alpha,\beta} : \{0, 1, 2\} \to \{0, 1, 2\}$, given by the rule $f_{\alpha,\beta}(i) = j$ if and only if $\tau^i(\alpha - \tau^i\beta) \in \mathcal{D}_j^{(3)}$. With the help of these functions we induce a partition of the set $\mathcal{G}$ into the following four disjoint subsets:

$$\mathcal{S}_l = \{(\alpha, -\beta) \in \mathcal{G} \mid W_h(f_{\alpha,\beta}(0), f_{\alpha,\beta}(1), f_{\alpha,\beta}(2)) = l\}, \quad \text{for } l = 0, 1, 2, 3,$$

where $W_h(\cdot)$ stands for the usual Hamming weight function.

Remark 4. For any $\alpha, \beta \in \mathbb{F}_{q^k}$, we define $u_i = \tau^i(\alpha - \tau^i\beta)$, for $i = 0, 1, 2$. It is not difficult to see that these u values satisfy: $u_0 + u_1 + u_2 = 0$. In addition, observe that if we arbitrarily choose the values of, say, u_0 and u_2 then there must exist a unique vector $(\alpha, \beta) \in \mathbb{F}_{q^k}^2$, such that $u_0 = (\alpha - \beta)$, $u_2 = \tau^2(\alpha - \tau^2\beta)$ and $u_1 = -(u_0 + u_2)$. Therefore, if we want to calculate, for example, $|\mathcal{S}_0|$ then we can assume, without loss of generality, that u_2 can take any value in $\mathcal{D}_0^{(3)}$. This leads us to $\frac{q^k-1}{3}$ possible choices for u_2. But $u_1 = -u_2(\frac{u_0}{u_2} + 1)$ and $-1 \in \mathcal{D}_0^{(3)}$ (in fact, recall that $\mathbb{F}_q^* \in \mathcal{D}_0^{(3)}$), thus, in order that u_1 and u_0 also belong to $\mathcal{D}_0^{(3)}$, it is necessary that $(\frac{u_0}{u_2} + 1) \in \mathcal{D}_0^{(3)}$, and due to Lemma 1, the number of such instances is given by the cyclotomic number $(0,0)^{(3,q^k)}$. Consequently, we have $|\mathcal{S}_0| = \frac{q^k-1}{3}(0,0)^{(3,q^k)}$. In a quite similar way, one can obtain $|\mathcal{S}_1|$, $|\mathcal{S}_2|$ and $|\mathcal{S}_3|$.

Keeping in mind the previous definitions and observations, we now present the following result, which will be important in order to determine the weight distribution of the class of non-irreducible cyclic codes that we are interested in.

Lemma 4. *With our current notation and main assumption, we have that*

$$|\mathcal{S}_0| = \frac{q^k - 1}{3}(0,0)^{(3,q^k)}$$

$$|\mathcal{S}_1| = (q^k - 1)((0,1)^{(3,q^k)} + (0,2)^{(3,q^k)})$$

$$|\mathcal{S}_2| = (q^k - 1)((1,1)^{(3,q^k)} + (1,2)^{(3,q^k)} + (2,1)^{(3,q^k)} + (2,2)^{(3,q^k)})$$

$$|\mathcal{S}_3| = \frac{q^k - 1}{3}((0,0)^{(3,q^k)} + (0,1)^{(3,q^k)} + (1,0)^{(3,q^k)} + (1,1)^{(3,q^k)} +$$
$$(2,0)^{(3,q^k)} + (2,2)^{(3,q^k)} + (0,0)^{(3,q^k)} + (0,2)^{(3,q^k)}).$$

Furthermore, if χ denotes the canonical additive character of $\mathbb{F}_{q^k}$, and if η_0 and η_1 are as in Theorem 2, then, for any $\alpha, \beta \in \mathbb{F}_{q^k}$, we also have:

$$
\sum_{z \in \mathcal{D}_0^{(3)}} \sum_{i=0}^{2} \chi(z\tau^i(\alpha + \tau^i\beta)) =
\begin{cases}
q^k - 1 & \text{if } (\alpha, \beta) = (0,0) \\
\frac{q^k-1}{3} + 2\eta_0 & \text{if } (\alpha, \beta) \in \cup_{j=0}^{2} \mathcal{E}_{0,j} \\
\frac{q^k-1}{3} + 2\eta_1 & \text{if } (\alpha, \beta) \in \cup_{i=1}^{2} \cup_{j=0}^{2} \mathcal{E}_{i,j} \\
3\eta_0 & \text{if } (\alpha, \beta) \in \mathcal{S}_0 \\
2\eta_0 + \eta_1 & \text{if } (\alpha, \beta) \in \mathcal{S}_1 \\
\eta_0 + 2\eta_1 & \text{if } (\alpha, \beta) \in \mathcal{S}_2 \\
3\eta_1 & \text{if } (\alpha, \beta) \in \mathcal{S}_3 .
\end{cases}
$$

Proof. The first assertion comes from Remark 4. On the other hand, since $\sum_{z \in \mathcal{D}_0^{(3)}} \chi(0) = |\mathcal{D}_0^{(3)}| = \frac{q^k-1}{3}$, the second assertion comes directly from Theorem 2, Remark 3, and from the definitions of the sets $\mathcal{E}_{i,j}$ and $\mathcal{S}_l$, with $i, j = 0, 1, 2$ and $l = 0, 1, 2, 3$.   □

Considering the actual values of the cyclotomic numbers in Lemma 1, the following result is an important consequence of the previous lemma.

Table 2. Value distribution of $\displaystyle\sum_{z \in \mathcal{D}_0^{(3)}} \sum_{i=0}^{2} \chi(z\tau^i(\alpha + \tau^i\beta))$

Value	Frequency
$q^k - 1$	1
$\frac{q^k-1}{3} + 2\eta_0$	$q^k - 1$
$\frac{q^k-1}{3} + 2\eta_1$	$2(q^k - 1)$
$3\eta_0$	$\frac{1}{27}(q^k - 1)(q^k - 2(-1)^s q^{k/2} - 8)$
$2\eta_0 + \eta_1$	$\frac{2}{9}(q^k - 1)(q^k + (-1)^s q^{k/2} - 2)$
$\eta_0 + 2\eta_1$	$\frac{2}{9}(q^k - 1)(2q^k - (-1)^s q^{k/2} - 1)$
$3\eta_1$	$\frac{2}{27}(q^k - 1)(4q^k + (-1)^s q^{k/2} - 14)$

Corollary 1. *Consider the same hypotheses as in the previous lemma. Then the value distribution of the character sum $\sum_{z \in \mathcal{D}_0^{(3)}} \sum_{i=0}^{2} \chi(z\tau^i(\alpha + \tau^i\beta))$ is given in Table II. In addition, each value in Table II is different to any other value, and its corresponding frequency is different from zero.*

Proof. The first assertion comes directly from the previous lemma. Now, observe that:

$$
(q^k - 2(-1)^s q^{k/2} - 8) = (q^{k/2} + 2(-1)^s)(q^{k/2} - 4(-1)^s) ,
$$
$$
(q^k + (-1)^s q^{k/2} - 2) = (q^{k/2} + 2(-1)^s)(q^{k/2} - (-1)^s) ,
$$

$$(2q^k - (-1)^s q^{k/2} - 1) = (q^{k/2} - (-1)^s)(2q^{k/2} + (-1)^s),$$
$$(4q^k + (-1)^s q^{k/2} - 14) = (q^{k/2} + 2(-1)^s)(4q^{k/2} - 7(-1)^s).$$

Thus, by recalling that q is an odd number greater than 1, we have that the proof of the second assertion comes now from Lemma 2, and from the fact that the roots of the previous polynomials are not odd integers values (greater than 1). □

4   A Formal Proof of Theorem 1

We begin this section by recalling the following already known identity:

Let $\mathcal{C}$ be an N-weight linear code, over $\mathbb{F}_q$, of length n and dimension $2k$. Suppose that $w_1, w_2, \cdots, w_N$ are the nonzero weights of $\mathcal{C}$. For $1 \leq i \leq N$, let A_i be the number of words of weight w_i in $\mathcal{C}$ and let B_j be the number of words of weight j in $\mathcal{C}^\perp$ (the dual code of $\mathcal{C}$). Then, the third identity of Pless (see [5, p. 259] for the general result), for $\mathcal{C}$, is

$$\sum_{i=1}^{N} w_i^2 A_i = [n(q-1)(n(q-1)+1) + 2B_2 - B_1(q + 2(n-1)(q-1))] q^{2k-2}. \quad (3)$$

Remark 5. In the context of the previous identity, observe that a linear code is projective if and only if B_1 and B_2 are zero in (3).

By keeping in mind the previous notation and identity, we now proceed to present a formal proof of Theorem 1.

Proof. Part *(A)*: Suppose that $h_{a_1}(x) = h_{a_2}(x)$. Then, there must exist an integer $0 \leq v < k$ such that $a_1 q^v \equiv a_2 \pmod{q^k - 1}$. But $a_2 = a_1 + \frac{q^k-1}{3}$, thus, the last congruence implies that $a_1(q^v - 1) \equiv \frac{q^k-1}{3} \pmod{q^k - 1}$, which in turn implies that $a_1(q^v - 1) \equiv 0 \pmod{\frac{q^k-1}{3}}$. That is, $\Delta | 3a_1(\frac{q^v-1}{q-1})$. But $\gcd(\Delta, a_1) = 1$, thus $\Delta | 3(\frac{q^v-1}{q-1})$. Nevertheless, it easy to see that $\Delta > 3(\frac{q^v-1}{q-1})$, if $q \geq 3$ and $0 \leq v < k$, therefore, the condition $\Delta | 3(\frac{q^v-1}{q-1})$ is impossible. Hence $h_{a_1}(x) \neq h_{a_2}(x)$. On the other hand, $\gcd(q^k - 1, a_1) = \gcd(\Delta(q-1), a_1) = \gcd(q-1, a_1) = \frac{q-1}{\lambda}$ and $\gcd(q^k-1, a_2) = \gcd(q-1, a_1+\frac{\Delta}{3}(q-1)) = \gcd(q-1, a_1)$, that is $\gcd(q^k - 1, a_1) = \gcd(q^k - 1, a_2) = \frac{q-1}{\lambda}$. Therefore $h_{a_1}(x)$ and $h_{a_2}(x)$ are the parity-check polynomials of two different cyclic codes of the same length $n = \frac{q^k-1}{(q-1)/\lambda} = \lambda\Delta$. Due that $\gcd(\Delta, a_1) = \gcd(\Delta, a_2) = 1$, the remaining proof of this part comes now from the set of characterizations, for the one-weight irreducible cyclic codes, that was introduced in [10, Theorem 11].

Before beginning with the proof of Part *(B)*, it is important to observe that since $\gcd(\Delta, \frac{q^k-1}{3} - 1) = 3$, $a_2 = a_1 + \frac{q^k-1}{3}$ and $\gcd(\Delta, a_1) = \gcd(\Delta, a_2) = 1$, we necessary have that $\tau \in \mathcal{D}_1^{(3)}$ and $a_1 \equiv 1 \pmod 3$. In addition, since $\gcd(q - 1, a_1) = \frac{q-1}{\lambda}$ and $\gcd(\Delta, a_1) = 1$, we also have that $\gcd(\Delta, \frac{q-1}{\lambda}) = 1$.

Part *(B)*: Clearly, the code $\mathcal{C}_{(a_1,a_2)}$ has length n and its dimension is $2k$, due to Part *(A)*.

Now, for each $\alpha, \beta \in \mathbb{F}_{q^k}$, we define $c(n, a_1, a_2, \alpha, \beta)$ as the vector of length n over $\mathbb{F}_q$, which is given by:

$$(\mathrm{Tr}_{\mathbb{F}_{q^k}/\mathbb{F}_q}(\alpha(\gamma^{a_1})^0 + \beta(\gamma^{a_2})^0), \cdots,$$
$$\mathrm{Tr}_{\mathbb{F}_{q^k}/\mathbb{F}_q}(\alpha(\gamma^{a_1})^{n-1} + \beta(\gamma^{a_2})^{n-1})).$$

Thanks to Delsarte's Theorem (see, for example, [2]), it is well known that

$$\mathcal{C}_{(a_1,a_2)} = \{c(n, a_1, a_2, \alpha, \beta) \mid \alpha, \beta \in \mathbb{F}_{q^k}\}.$$

Thus the Hamming weight of any codeword $c(n, a_1, a_2, \alpha, \beta) \in \mathcal{C}_{(a_1,a_2)}$ is equal to $n - Z(\alpha, \beta)$, where

$$Z(\alpha, \beta) = \#\{i \mid 0 \le i < n, \text{ and}$$
$$\mathrm{Tr}_{\mathbb{F}_{q^k}/\mathbb{F}_q}(\alpha\gamma^{a_1 i} + \beta\gamma^{a_2 i}) = 0\}.$$

Now, if χ' is the canonical additive character of $\mathbb{F}_q$, then, by the orthogonal property in (1), we know that for each $c \in \mathbb{F}_q$ we have

$$\sum_{y \in \mathbb{F}_q} \chi'(yc) = \begin{cases} q & \text{if } c = 0 \\ 0 & \text{if } c \neq 0 \end{cases},$$

thus

$$Z(\alpha, \beta) = \frac{1}{q} \sum_{i=0}^{n-1} \sum_{y \in \mathbb{F}_q} \chi'(\mathrm{Tr}_{\mathbb{F}_{q^k}/\mathbb{F}_q}(y(\alpha\gamma^{a_1 i} + \beta\gamma^{a_2 i}))).$$

If χ denotes the canonical additive character of $\mathbb{F}_{q^k}$, then χ' and χ are related by $\chi'(\mathrm{Tr}_{\mathbb{F}_{q^k}/\mathbb{F}_q}(\varepsilon)) = \chi(\varepsilon)$ for all $\varepsilon \in \mathbb{F}_{q^k}$. Therefore, we have

$$Z(\alpha, \beta) = \frac{n}{q} + \frac{1}{q} \sum_{i=0}^{n-1} \sum_{y \in \mathbb{F}_q^*} \chi(y(\alpha\gamma^{a_1 i} + \beta\gamma^{a_2 i}))$$
$$= \frac{n}{q} + \frac{1}{q} \sum_{i=0}^{n-1} \sum_{y \in \mathbb{F}_q^*} \chi(\gamma^{a_1 i} y(\alpha + \tau^i \beta)),$$

where the last equality arises because $a_2 = a_1 + \frac{q^k-1}{3}$ and $\tau = \gamma^{\frac{q^k-1}{3}}$. Now, since $a_1 \equiv 1 \pmod 3$ and $\gcd(q^k - 1, a_1) = \frac{q-1}{\lambda}$, we have

$$
Z(\alpha, \beta) = \frac{n}{q} + \frac{1}{q} \sum_{i=0}^{n-1} \sum_{y \in \mathbb{F}_q^*} \chi(\gamma^{a_1 i} y (\alpha + \tau^{a_1 i} \beta))
$$

$$
= \frac{n}{q} + \frac{1}{q} \sum_{i=0}^{n-1} \sum_{y \in \mathbb{F}_q^*} \chi(\gamma^{\frac{q-1}{\lambda} i} y (\alpha + \tau^{\frac{q-1}{\lambda} i} \beta)) \, .
$$

But, clearly $3|n$, thus

$$
\{ \gamma^{\frac{q-1}{\lambda} i} \mid 0 \le i < n \} = \mathcal{D}_0^{(\frac{(q-1)}{\lambda})} = \mathcal{D}_0^{(\frac{3(q-1)}{\lambda})} \cup \mathcal{D}_{\frac{q-1}{\lambda}}^{(\frac{3(q-1)}{\lambda})} \cup \mathcal{D}_{\frac{2(q-1)}{\lambda}}^{(\frac{3(q-1)}{\lambda})} \, .
$$

Therefore,

$$
Z(\alpha, \beta) = \frac{n}{q} + \frac{1}{q} \sum_{i=0}^{2} \sum_{x \in \mathcal{D}_{\frac{q-1}{\lambda} i}^{(\frac{3(q-1)}{\lambda})}} \sum_{y \in \mathbb{F}_q^*} \chi(xy(\alpha + \tau^{\frac{q-1}{\lambda} i} \beta))
$$

$$
= \frac{n}{q} + \frac{1}{q} \sum_{i=0}^{2} \sum_{x \in \mathcal{D}_i^{(\frac{3(q-1)}{\lambda})}} \sum_{y \in \mathbb{F}_q^*} \chi(xy(\alpha + \tau^i \beta)) \, , \tag{4}
$$

where the last equality arises because $\mathbb{F}_q^* = \{ \gamma^{\Delta j} \mid 0 \le j < q \}$ and $3 \nmid \frac{(q-1)}{\lambda}$. Now, we already said that $\gcd(\Delta, \frac{q-1}{\lambda}) = 1$, thus after applying Lemma 3 to (4) we obtain

$$
Z(\alpha, \beta) = \frac{n}{q} + \frac{\lambda}{q} \sum_{i=0}^{2} \sum_{z \in \mathcal{D}_i^{(3)}} \chi(z(\alpha + \tau^i \beta)) \, .
$$

But, $\tau \in \mathcal{D}_1^{(3)}$, thus

$$
Z(\alpha, \beta) = \frac{n}{q} + \frac{\lambda}{q} \sum_{z \in \mathcal{D}_0^{(3)}} \sum_{i=0}^{2} \chi(z \tau^i (\alpha + \tau^i \beta)) \, .
$$

The result comes thus as a consequence of the first assertion in Corollary 1, and from the fact that the Hamming weight of any codeword of the form $c(n, a_1, a_2, \alpha, \beta)$, in $\mathcal{C}_{(a_1, a_2)}$, is equal to $n - Z(\alpha, \beta)$.

Part (C): It is well known that there are no one-weight words in the dual of any cyclic code (see for example [15]), therefore $B_1 = 0$ in identity (3). But if $B_1 = 0$ in such identity, then, with the help of Table I, it is not difficult to see that $B_2 = \frac{n(q-1)(\lambda-1)}{2}$. Thus, $B_2 = 0$ if and only if $\lambda = 1$, but recall that λ is the divisor of $q - 1$, satisfying $\gcd(q - 1, a_1) = \frac{q-1}{\lambda}$, therefore $B_2 = 0$ if and only if

$(q-1)|a_1$. We will now prove that this condition is impossible and, in order to do so, we will consider the following two possible scenarios: k is an even integer or k is an odd integer. If k is even, then clearly Δ and $q-1$ are also even integers, and consequently, if $(q-1)|a_1$ then $\gcd(\Delta, a_1) \geq 2$, which is a contradiction. Now if k is odd, then t is even (recall $tk = 2sd$) and since $p \neq 3$ (see Remark 1), we have $p^t \equiv 1 \pmod{3}$, that is $3|(q-1)$. But $3|\Delta$, thus if $(q-1)|a_1$ then $\gcd(\Delta, a_1) \geq 3$, which is again a contradiction. Therefore $(q-1) \nmid a_1$ and $B_2 > 0$. The result now follows from Remark 5.

Part (D): It is a direct consequence of the second assertion in Corollary 1. $\square$

The following are direct applications of Theorem 1.

Example 1. With our notation, let $p = q = 11$, $k = 2$, $a_1 = 25$ and $a_2 = 65$. Then $d = s = 1$, $\Delta = 12$, $\lambda = 2$ and $n = 24$. Clearly, $3|(q^k - 1)$ and $\gcd(\Delta, \frac{q^k-1}{3} - 1) = 3$, and because $3|(p^d + 1)$, we can be sure that $\mathcal{C}_{(25,65)}$ is a 6-weight cyclic code over $\mathbb{F}_{11}$, of length 24, dimension 4 and weight enumerator polynomial $A(z) = 1 + 120z^{12} + 240z^{16} + 600z^{18} + 2880z^{20} + 6720z^{22} + 4080z^{24}$. In addition, $B_2 = 120$.

Example 2. With our notation, let $p = q = 5$, $k = 4$, $a_1 = 1$ and $a_2 = 209$. Then $d = 1$, $s = 2$, $\Delta = 156$, $\lambda = 4$ and $n = 624$. Clearly, $3|(q^k - 1)$ and $\gcd(\Delta, \frac{q^k-1}{3} - 1) = 3$, and because $3|(p^d + 1)$, we can be sure that $\mathcal{C}_{(1,209)}$ is a 6-weight cyclic code over $\mathbb{F}_5$, of length 624, dimension 8 and weight enumerator polynomial $A(z) = 1 + 1248z^{320} + 624z^{360} + 116064z^{480} + 169728z^{500} + 89856z^{520} + 13104z^{540}$. In addition, $B_2 = 3744$.

5   Conclusion

A recent topic of interest has been the kind of reducible cyclic codes whose parity-check polynomials are given by products of the form $h_{a_1}(x)h_{a_1+\frac{q^k-1}{3}}(x)$, where a_1 is an integer. A class of this kind of cyclic codes was the main subject of study in [14]. In this work we presented the full weight distribution of a new class cyclic codes belonging to this kind, and we showed that no code in this class can be projective. As we already said, the class of codes studied in [14], and those studied here are two different classes of codes of the same kind. Thus, following the same idea as in [12], perhaps it is possible to develop a more general theory that allows us to present a unified explanation for these two classes of codes, and also for other classes of cyclic codes of the same kind.

References

1. Baumert, L., Mills, W., Ward, R.: Uniform cyclotomy. J. Number Theory 14(1), 67–82 (1982)
2. Delsarte, P.: On subfield subcodes of Reed-Solomon codes. IEEE Trans. Inform. Theory 5, 575–576 (1975)

3. Ding, C., Liu, Y., Ma, C., Zeng, L.: The weight distributions of the duals of cyclic codes with two zeros. IEEE Trans. Inform. Theory 57(12), 8000–8006 (2011)
4. Helleseth, T.: Some two-weight codes with composite parity-check polynomials. IEEE Trans. Inform. Theory 22, 631–632 (1976)
5. Huffman, W.C., Pless, V.S.: Fundamental of Error-Correcting Codes. Cambridge Univ. Press, Cambridge (2003)
6. Lidl, R., Niederreiter, H.: Finite Fields. Cambridge Univ. Press, Cambridge (1983)
7. Liu, Y., Yan, H., Liu, C.: A class of six-weight cyclic codes and their weight distribution. Des. Codes Cryptogr. (published online June 4, 2014) doi:10.1007/s10623-014-9984-y
8. Ma, C., Zeng, L., Liu, Y., Feng, D., Ding, C.: The Weight Enumerator of a Class of Cyclic Codes. IEEE Trans. Inf. Theory 57(1), 397–402 (2011)
9. Moisio, M.: A note on evaluations of some exponential sums. Acta Arith 93, 117–119 (2000)
10. Vega, G.: Determining the number of one-weight cyclic codes when length and dimension are given. In: Carlet, C., Sunar, B. (eds.) WAIFI 2007. LNCS, vol. 4547, pp. 284–293. Springer, Heidelberg (2007)
11. Vega, G.: Two-weight cyclic codes constructed as the direct sum of two one-weight cyclic codes. Finite Fields Appl. 14(3), 785–797 (2008)
12. Vega, G.: A General Description for the Weight Distribution of Some Reducible Cyclic Codes. IEEE Trans. Inform. Theory 59(9), 5994–6001 (2013)
13. Vega, G.: A critical review and some remarks about one- and two-weight irreducible cyclic codes. Finite Fields Appl. 33, 1–13 (2015)
14. Wang, B., Tang, C., Qi, Y., Yang, Y., Xu, M.: The weight distributions of cyclic codes and elliptic curves. IEEE Trans. Inform. Theory 58(12), 7253–7259 (2012)
15. Wolfmann, J.: Are 2-Weight Projective Cyclic Codes Irreducible? IEEE Trans. Inform. Theory 51(2), 733–737 (2005)
16. Xiong, M.: The weight distributions of a class of cyclic codes. Finite Fields Appl. 18(5), 933–945 (2012)

Codes over $\mathcal{L}(GF(2)^m, GF(2)^m)$, MDS Diffusion Matrices and Cryptographic Applications

Thierry P. Berger[1] and Nora El Amrani[1,2(✉)]

[1] XLIM (UMR CNRS 7252), University of Limoges, France
[2] Laboratory of Mathematics, Computing and Applications,
Faculty of sciences University of Mohammed V - Agdal, Rabat, Morocco
elamrani.nora@gmail.com

Abstract. The aim of this paper is to provide a general framework in the study of binary block codes. The main objective is to present a general approach in order to explore MDS diffusion matrices used for example in the design of block ciphers with a Substitution Permutation Network design (the so-called SPN block-ciphers).

In order to analyze these codes, we consider additive block codes over binary m-tuples. We are interested in the distance properties related to the block structure. To do this, we introduce a notion of $\mathcal{L}$-codes that are codes over the non-commutative ring of linear endomorphisms of $GF(2)^m$. We study the main properties of these codes, especially the notion of duality in this context. We show how most of the known families of block codes can be interpreted in this context. Finally, we conclude by practical examples that allow to derive MDS diffusion matrices over $GF(2)^m$ from MDS matrices constructed over smaller blocks.

Keywords: MDS matrices · Diffusion layers · Additive block codes · Symmetric cryptography

Introduction

Section 1 presents the notion of additive block codes and explore how most of the known results on linear codes over a finite field can be extended to additive block codes. In order to avoid some degenerated cases, we limit our study to systematic block codes. We characterize the isometries of block codes over $E = GF(2)^m$ and deduce a notion of equivalence of codes in this context. We explain the notion of MDS block codes and the link between these MDS block codes and optimal diffusion matrices used in block ciphers design. Finally, we conclude this section by presenting some previous works on codes over polynomial rings that are in fact related to our approach.

In Section 2, we introduce the notion of $\mathcal{L}$-codes that are codes over the ring $\mathcal{L} = \mathcal{L}(E, E)$. Such a ring is not commutative, in this context $\mathcal{L}$-codes are left submodules of $\mathcal{L}^r$, but are not necessarily right submodules. In this context, we clarify the notion of duality over $\mathcal{L}$ and the underlying binary duality. In Section 3, we consider some commutative subrings of $\mathcal{L}$ and explain why we obtain some classical families of codes defined on polynomial rings of the form

© Springer International Publishing Switzerland 2015
S. El Hajji et al. (Eds.): C2SI 2015, LNCS 9084, pp. 197–214, 2015.
DOI: 10.1007/978-3-319-18681-8_16

$GF(2)[x]/f(x)$. Finally in Section 4, we present some examples of application of previously presented results. The aim of this section is not to obtain optimized matrices, but to give some hints for an effective search of optimal matrices for targeted applications. Indeed, even for practical values a full exhaustive search of MDS matrices over $\mathcal{L}$ is not possible.

Notation

- $E = GF(2)^m$ is the $GF(2)$-vector space of binary m-tuples.
- $\mathcal{L} = \mathcal{L}(E, E)$ is the ring of $GF(2)$-linear endomorphisms of E.
- $\mathcal{M}_{s,t}(\mathcal{R})$ is the $\mathcal{R}$-module of matrices of size $s \times t$ over the ring $\mathcal{R}$.
- $\mathcal{M}_k(\mathcal{R}) = \mathcal{M}_{k,k}(\mathcal{R})$ is the matrix algebra of square matrices over the ring $\mathcal{R}$.
- If $\varphi \in \mathcal{L}$ is a linear endomorphism of E, $M_\varphi \in \mathcal{M}_m(GF(2))$ is its associated binary matrix with the convention: if $x = (x_1, ..., x_m) \in E$ then $\varphi(x) = xM$.

1   Additive Block Codes over $GF(2)^m$ and MDS Diffusion Matrices

The results presented in this section are mainly known. They are presented in order to easily introduce our later point of view. For more details on Error Correcting Codes and basic properties, the reader can refer to [8].

1.1   Codes over a Finite Alphabet

A code C of length r over an alphabet $\mathcal{A}$ is a subset of $\mathcal{A}^r$. Let $a = (a_1, ..., a_r)$ be an element of $\mathcal{A}^r$. As usual, the Hamming weight $w(a)$ of a is the number of non-zero coefficients a_i. The Hamming distance between two elements a and a' is the number of distinct coefficients of a and a': $d(a, a') = {}^\#\{i \mid a_i \neq a'_i\}$. The minimum distance d_C of a code C is then the minimum of the distance between two distinct elements of C.

The following proposition recalls the well-known Singleton bound that links the size of the code and its minimum distance in its more general form, *i.e.* for non-linear codes without structure.

Proposition 1. *Let C be a code of length r and minimum distance d over an alphabet $\mathcal{A}$ of size q, then :* ${}^\#C \leq q^{r-d+1}$.

Most of the times, the alphabet $\mathcal{A}$ is provided with a mathematical structure. In the following, we will focus on three of them:

- $(\mathcal{A}, +)$ is an additive group. An additive code C of length r over $\mathcal{A}$ is then a subgroup of $(\mathcal{A}, +)^r$.
- $\mathcal{A} = \mathbb{F}$ is a finite field. A linear code over $\mathbb{F}$ is an $\mathbb{F}$-subspace of the vector space $\mathbb{F}^r$.
- $\mathcal{A} = \mathcal{R}$ is a ring $\mathcal{R}$. A linear code over $\mathcal{R}$ is an $\mathcal{R}$-submodule of $\mathcal{R}^r$. Note that the multiplication in $\mathcal{R}$ is not necessary commutative, however, we are only interested in unitary rings.

In all cases, the alphabet $\mathcal{A}$ possesses an additive commutative group structure. In this situation, the minimum distance of a code becomes the minimum weight of its non-zero elements.

Definition 1. *A code C is a Maximum Distance Separable Code (an MDS code for short) if it meets the Singleton bound, i.e. $\#C = q^{r-d+1}$.*

1.2   Block Codes over E

In this paper, we are concerned with additive codes defined over E, *i.e.* the alphabet is the additive group constituted by the binary m-tuples equipped of the componentwise addition. Our motivation comes from the construction of MDS diffusion matrices for cryptographic applications. It will be explained in Section 1.6.

Definition 2. *A block code C of length r over E is an additive code over the alphabet $(E, +)$.*

From the $GF(2)$-vector space isomorphism $E^r \simeq GF(2)^{mr}$, a block code C of length r is also a binary linear code of length $n = mr$ over $GF(2)$. However, we are not interested in its binary properties, but in its block properties. In particular we do not look at the binary weight of codewords, but at block weight of codewords. In the rest of this paper, unless it is explicitly stated, $w(c)$ denotes the block weight of an element $c \in E^r$ and d_C denotes the minimum distance of the block code C.

1.3   Systematic Block Codes

To avoid some degenerated cases, we define the notion of systematic block code.

Definition 3. *Suppose that the size of C is of the form 2^{km} for some integer k. A code C is a narrow sense systematic block code of pseudo-dimension k if there exists a systematic linear encoding function Φ from E^k to E^r such that $\Phi(x) = (x_1, ..., x_k, \Phi_1(x), ..., \Phi_{r-k}(x))$, $\Phi_i \in \mathcal{L}(E^k, E)$, $x = (x_1, ..., x_k) \in E^k$ and $\Phi(E^k) = C$.*

In other words, there exists a linear encoding for C such that the first k blocks of a codeword are equal to the message x to be encoded.

This definition is equivalent to the fact that the binary image of C is of dimension $k_2 = mk$ and admits a systematic binary generator matrix $G = (I|M)$ where I is the identity matrix of size mk and M is a binary matrix of size $mk \times m(r-k)$.

Definition 4. *A (general) systematic block code C is a code which is equivalent to a narrow sense systematic code by blocks permutation (i.e. by permutation of coordinates acting on E^r).*

One can remark that the Singleton bound for systematic block codes becomes $k + d \leq r + 1$ as usual for linear codes.

1.4 $\mathcal{L}$-generator Matrix of a Systematic Block Code

Following the notations of Section 1.3, the linear applications Φ_i from E^k into E can be decomposed in k elements of $\mathcal{L}$: $\Phi_i(x) = \sum_{j=1}^k \varphi_{i,j}(x_j)$, $\varphi_{i,j} \in \mathcal{L}$.

If $M_{i,j} \in \mathcal{M}_m(GF(2))$ denotes the $m \times m$ binary matrix corresponding to the endomorphism $\varphi_{i,j}$ (i.e. $\varphi_{i,j}(a) = aM_{i,j}$ for $a \in E$), the binary systematic generator matrix of C is

$$G = \begin{pmatrix} I_m & 0_m & \cdots & 0_m & M_{1,1} & \cdots & M_{1,r-k} \\ 0_m & \ddots & \ddots & \vdots & \vdots & & \vdots \\ \vdots & \ddots & \ddots & 0_m & \vdots & & \vdots \\ 0_m & \cdots & 0_m & I_m & M_{k,1} & \cdots & M_{k,r-k} \end{pmatrix},$$

where I_m and 0_m are respectively the identity and the zero binary square matrices of size m.

We can construct an $\mathcal{L}$-generator matrix $\mathcal{G} \in \mathcal{M}_{k,r}(\mathcal{L})$ in the following way:

$$\mathcal{G} = \begin{pmatrix} Id & 0 & \cdots & 0 & \varphi_{1,1} & \cdots & \varphi_{1,r-k} \\ 0 & \ddots & \ddots & \vdots & \vdots & & \vdots \\ \vdots & \ddots & \ddots & 0 & \vdots & & \vdots \\ 0 & \cdots & 0 & Id & \varphi_{k,1} & \cdots & \varphi_{k,r-k} \end{pmatrix},$$

where Id and 0 are respectively the identity map and the zero map of $\mathcal{L}$.

The codewords c of C are then $c = x\mathcal{G}$, $x = (x_1, ..., x_k) \in E^k$. By convention, and to be consistent with the matrix notations, for $a \in E$ and $\varphi \in \mathcal{L}$, we have $a\varphi = \varphi(a)$. In addition, for φ and ψ in $\mathcal{L}$, $\varphi\psi$ denotes $\psi \circ \varphi$.

This construction can be generalized to any general systematic block code and to any binary generator matrix of such a code.

Definition 5. *Let C be a general systematic block code over E. An $\mathcal{L}$-generator matrix for C is a $k \times r$ matrix $\mathcal{G} = (\varphi_{i,j})$ over $\mathcal{L}$ such that the matrix $G = (M_{\varphi_{i,j}})$ of size $km \times rm$ is a binary generator matrix of the code C considered as $GF(2)$-linear code of dimension km and length rm.*

Following this definition, if $\mathcal{G}$ is an $\mathcal{L}$-generator matrix of a code C, the $\mathcal{L}$-generator matrices of C are those of the form $\mathcal{G}' = \mathcal{S}\mathcal{G}$ there $\mathcal{S} \in \mathcal{M}_k(\mathcal{L})$ is an invertible matrix.

1.5   Equivalence of Systematic Block Codes

In the classical situation of linear codes, two codes C and C' are equivalent if there exists an isometry Ψ (i.e. a linear endomorphism of $\mathbb{F}^n$ preserving the Hamming distance) such that $C' = \Psi(C)$. A major result on isometries in the context of linear codes over a finite field is the fact that isometries correspond to monomial matrices, i.e. the $n \times n$ matrices with one and only one non-zero element by row and by column (see [5] Ch.17 §1.5). In practice, such an isometry consists in multiplying each coordinate of a codeword by a non-zero scalar, and then by permuting these coordinates.

These properties can be easily generalized to the case of block codes. First, we will characterize some isometries on E^r.

Let $Sym(r)$ be the group of permutations acting on the set of indices $[1; r]$. A permutation $\sigma \in Sym(r)$ acts on E^r in a natural way: if $x = (x_1, ..., x_r) \in E^r$, we define $x' = \sigma(x) = (x_{\sigma^{-1}(1)}, ..., x_{\sigma^{-1}(r)})$. Clearly, σ is an isometry of E^r for Hamming block distance. If P_σ is the permutation matrix associated to σ, one have $\sigma(x) = xP_\sigma$.

The scalar multiplication in the case of linear codes is replaced by the action of invertible elements of $\mathcal{L}$, i.e. elements of the linear group $GL(m, 2)$. If $\lambda = (\lambda_1, \lambda_2, ..., \lambda_r)$ is an r-tuple of elements of $GL(m, 2)$, it acts on E^r as follows: $\lambda(x) = (\lambda_1(x_1), ..., \lambda_r(x_r)) = (x_1\lambda_1, ..., x_r\lambda_r)$. Such an application is clearly an isometry for the Hamming block distance over E^r. Moreover, the diagonal matrix D_λ with diagonal elements λ_i, is the matrix of this isometry: $\lambda(x) = xD_\lambda$.

The following proposition gives the characterization of isometries of E^r.

Proposition 2. *The isometry group of E^r for the Hamming block distance is the monomial group constituted of square matrices of size r with one and only one nonzero invertible elements on each row and each column.*

Proof. The proof is similar to that in the case of linear codes. The monomial group $Mon(GL(m, 2))$ is generated by the permutations and the diagonal invertible matrices. The elements of this group are then isometries of E^r.

Reciprocally, we look at the images of elements of E^r of weight 1 by an isometry ξ of E^r. Let $e^{(i)} \in E^r$ be the element of weight 1 such that $e_i^{(i)} = 1$. The image of $e^{(i)}$ by ξ is of the form $\lambda_j e^{(j)}$ for a fixed index j and an element $\lambda_j \in E$. The underlying permutation is entirely defined by $\sigma(j) = i$ and the element λ_j is necessarily invertible, otherwise we can construct a word of weight 1 having $0 \in E^r$ for image.

If C is a code with $\mathcal{L}$-generator matrix $\mathcal{G}$ and $\mathcal{M}$ is a monomial matrix in $Mon(GL(m, 2))$, the matrix $\mathcal{G}' = \mathcal{G}\mathcal{M}$ is an $\mathcal{L}$-generator matrix of the image C' of C by $\mathcal{M}$.

Definition 6. *Two block codes C and C' are equivalent if there exists a monomial transformation $\mathcal{M} \in Mon(GL(m, 2))$ such that C' is the image of C by $\mathcal{M}$.*

1.6   MDS Systematic Block Codes and MDS Matrices

From Definition 1, if an additive block code C is MDS, then $^{\#}C = 2^{m(r-d+1)}$, so its size is necessary a power of 2^m. Moreover, following results of [8] Ch. 11 §2, it can be shown that C admits a systematic $\mathcal{L}$-generator matrix $\mathcal{G}$. So $^{\#}C = 2^{mk}$, where k is the pseudo-dimension of C and the MDS condition becomes $k + d = r + 1$.

In addition, C is MDS if and only if the restriction of $\mathcal{G}$ to any k columns leads to an invertible matrix in $\mathcal{M}_k(\mathcal{L})$. For cryptographic applications, we are particularly interested in the redundancy part of systematic $\mathcal{L}$-generator matrices of MDS block codes.

The following theorem is a generalization of the well-known Theorem 8 ([8] Ch. 11 §4), in the case of codes over finite fields, or [1] Proposition 1 in the case of commutative ring.

Theorem 1. *An additive block code C is MDS if and only if it admits a systematic $\mathcal{L}$-generator matrix $\mathcal{G} = (I_k \mid M)$ such that every $e \times e$ square submatrix of M is a matrix of an automorphism of E^e.*

Note that, since $\mathcal{L}$ is not a commutative ring we do not use the notion of determinant for the square submatrices of M. However, the invertibility of an $e \times e$ square matrix is directly related to the invertibility of the corresponding $me \times me$ binary matrix obtained by substituting to each entry $\varphi_{i,j}$ the $m \times m$ binary matrix $M_{\varphi_{i,j}}$.

Definition 7. *A matrix $M \in \mathcal{M}_{k,s}(\mathcal{L})$ is MDS if the systematic block code C with $\mathcal{L}$-generator matrix $\mathcal{G} = (I_k \mid M)$ is an MDS block code of length $r = k + s$.*

The MDS matrices are those satisfying the conditions of Theorem 1.

In Section 1.5, we studied the action of isometries on additive block codes. Following this approach, we are able to deduce a notion of equivalence for MDS matrices.

Set $s = r - k$. Let C be a systematic block code with $\mathcal{L}$-generator matrix $\mathcal{G} = (I_k \mid M)$. In order to preserve the systematic structure of this matrix, we apply to C a permutation which separately acts on the first k positions and on the s last positions. Let $\sigma = (\sigma_1, \sigma_2) \in Sym(r)$ such that $\sigma_1 \in Sym(k)$ and $\sigma_2 \in Sym(s)$. Let $C' = \sigma(C)$ be the image of C by C'. If the systematic $\mathcal{L}$-generator matrix of C' is $\mathcal{G}' = (I_k \mid M')$, one have $M' = \Pi_{\sigma_1^{-1}} M \Pi_{\sigma_2}$, where $\Pi_{\sigma_1^{-1}}$ and Π_{σ_2} are respectively the $k \times k$ and $(s) \times (s)$ permutation matrices associated to σ_1^{-1} and σ_2.

Of this reasoning we deduce the following proposition:

Proposition 3. *Let $M \in \mathcal{M}_{k,s}(\mathcal{L})$ be an MDS matrix. A matrix M' obtained by any permutation of the rows and the columns of M is MDS.*

Similarly, suppose that $\lambda = (\lambda_1, ..., \lambda_r) \in GL(m, 2)^r$ is an r-tuple of "non-zero scalars" (*i.e.* invertible elements of $\mathcal{L}$) acting on the code C. We decompose $\lambda = (\lambda_{(1)} \mid \lambda_{(2)})$ into its first k components and its last s components. The systematic $\mathcal{L}$-generator matrix of the image C' of C by the action of λ is $\mathcal{G}' = (I_k \mid M')$ with $M' = D_{\lambda_{(1)}^{-1}} M D_{\lambda_{(2)}}$, where $D_{\lambda_{(1)}^{-1}}$ and $D_{\lambda_{(2)}}$ are respectively the $k \times k$ and $s \times s$ diagonal matrices with diagonal $\lambda_{(1)}^{-1}$ and $\lambda_{(2)}$.

So we obtain the following proposition:

Proposition 4. *Let $M \in \mathcal{M}_{k,s}(\mathcal{L})$ be an MDS matrix. A matrix M' obtained by multiplying on the left any row of M and multiplying on the right any column of M by some elements of $GL(m, 2)$ is MDS.*

Definition 8. *Two MDS matrices $\mathcal{M}$ and $\mathcal{M}'$ in $\mathfrak{M}_{k,s}(\mathcal{L})$ are equivalent if $\mathcal{M}'$ can be deduced from $\mathcal{M}$ by applying the transformations given in Propositions 3 and 4.*

1.7 MDS Diffusion Matrices for Cryptographic Applications

Classically, symmetric cryptographic algorithms alternate confusion layers and diffusion layers in their iterative cryptographic processes. The confusion layer consists in the application of a non-linear function, called an S-box which acts generally on r blocks of size m. The typical values of m are 4 or 8. The diffusion layer ensures the dissemination of any difference in input between the different r blocks. For efficiency this diffusion layer is in fact a linear application from E^r to E^r.

The goal of this diffusion layer is not to ensure a diffusion inside each block, but a diffusion between the blocks. In practice, as previously we denote by $x = (x_1, ..., x_r)$, the r input blocks and by $y = xM$ the output blocks where M can be viewed as an $r \times r$ $\mathcal{L}$-matrix or an $rm \times rm$ binary matrix.

The main example of diffusion matrix is MixColumns, the AES diffusion layer [3]. In this paper, the authors introduce the notion of branch number, which is a measure of the resistance of a diffusion matrix against linear and differential cryptanalysis in the context of SPN block ciphers.

We do not want to describe in detail the concepts of linear and differential branch numbers and their links with cryptanalysis. We just give a definition of these concepts adapted to our approach. For more details, the reader can refer to [3].

We need some notations: let $\mathcal{M}$ be an element of $\mathfrak{M}_{k,s}(\mathcal{L})$, M denotes the corresponding $km \times ks$ binary matrix. The notation $\mathcal{M}^T$ corresponds to the transpose of $\mathcal{M}$ in $\mathfrak{M}_{k,s}(\mathcal{L})$, and the matrix $\mathcal{M}^{T*}$ denotes the element of $\mathfrak{M}_{k,s}(\mathcal{L})$ associated to the binary matrix M^T. Note that $\mathcal{M}^T$ and $\mathcal{M}^{T*}$ are not equal.

Definition 9. *Let $\mathcal{M} \in \mathfrak{M}_{k,s}(\mathcal{L})$ be a diffusion matrix which takes as inputs k blocks of m bits and outputs s blocks of m bits.*
The differential branch number of $\mathcal{M}$ is the minimum distance of the additive block code generated by $(I_k \mid M)$.
The linear branch number of $\mathcal{M}$ is the minimum distance of the additive block code generated by $(I_s \mid M^{T*})$.

Note that, since we do not require that $k = s$, this definition is more general than the usual one. However, even if $k \neq s$, the differential branch number is the minimum number of blocks in input and in output that are impacted by a difference, which corresponds to a minimal word of the additive block code since it is $GF(2)$-linear. Similarly, the linear branch number is the minimum of blocks in input and in output that are impacted by a parity check equation, and corresponds to the minimum block-weight of the binary dual of the previous code. The following theorem is a direct generalization of a major result of [3] to the case where k is not necessary equal to s.

Theorem 2. *A binary diffusion matrix M of size $km \times sm$ (or equivalently a diffusion matrix $\mathcal{M} \in \mathcal{M}_{k,s}(\mathcal{L})$) has a maximal differential branch number if and only if it has a maximal linear branch number. In this situation, the both matrices $\mathcal{M}$ and $\mathcal{M}^{T*}$ are MDS.*

Suppose now that $k = s$. Let $\mathcal{C}$ be an MDS block code with systematic $\mathcal{L}$-generator matrix $\mathcal{G} = (I_k \mid \mathcal{M})$. Obviously, since $\mathcal{M}$ is a square MDS matrix, it is invertible. Moreover $(\mathcal{M}^{-1} \mid I_k)$ is another $\mathcal{L}$-generator matrix of $\mathcal{C}$. The systematic block code $\mathcal{C}'$ generated by the $\mathcal{L}$-generator matrix $\mathcal{G}' = (I_k \mid \mathcal{M}^{-1})$ is equivalent by block permutation to $\mathcal{C}$, and so it is also an MDS block code. We have proved the following proposition:

Proposition 5. *If $\mathcal{M}$ is a square MDS matrix, then $\mathcal{M}^{-1}$ is also a square MDS matrix.*

We consider now a $k \times s$ matrix $\mathcal{M}$ such that the entries $\varphi_{i,j}$ of $\mathcal{M}$ pairwise commute. This is a classical requirement for the search of MDS diffusion matrices (cf. [1,3,4]). In fact, we do not know in the literature an example of MDS diffusion matrix $\mathcal{M}$ such that the entries do not commute.

The main reason for this restriction comes from the fact that it is possible to compute the subdeterminants of $\mathcal{M}$ and to use these subdeterminants to test if the matrix is MDS. In our context, we obtain the following result:

Proposition 6. *Suppose that $\mathcal{M}$ is a (non necessary square) MDS matrix such that the entries of $\mathcal{M}$ commute pairwise. The matrix $\mathcal{M}^T$ is MDS.*

Proof. Let M be the binary $km \times sm$ matrix associated to $\mathcal{M}$. Let $\mathcal{A}$ be any square submatrix of $\mathcal{M}$ and A its associated binary submatrix. The coefficients of $\mathcal{M}$ are in fact in a commutative subring $\mathcal{R}$ of $\mathcal{L}$. Let $\delta = Det(\mathcal{A}) \in \mathcal{R}$ be the determinant of $\mathcal{A}$ computed in $\mathcal{R}$. This makes sense since $\mathcal{R}$ is commutative. Let M_δ be the binary matrix associated to $\delta \in \mathcal{L}$. From a result of Silvester [9], we have $det(A) = det(M_\delta)$, where det is the determinant of binary matrices.

To conclude our proof, we remark that $Det(\mathcal{A}) = Det(\mathcal{A}^T)$. So the submatrix $\mathcal{A}$ is invertible if and only if $\mathcal{A}^T$ is invertible.

Note that this result is false in the general case.

1.8   Ring Structures over $GF(2)^m$ and Related Additive Block Codes

A natural research direction is to explore potential mathematical structures of E to provide additive codes with additional properties, or in order to decrease an exhaustive search of good candidates for cryptographic applications.

In this section, we will look at possible ring structures over E. The most basic ring structure over E corresponds to the Hadamard product of m-tuples *i.e.* the component by component product. Most of the other ring structures over E consists in identifying E with the ring $GF(2)[x]/f(x)$ for some polynomial $f(x)$ of degree m. In both cases, we can define the notion of linear-block codes, that are not only additive block codes, but also submodules of E^r.

In these situations, it is possible to determine the minimal ideals of $\mathcal{R}$, which leads to a natural projection of such linear codes into several linear codes over finite fields.

Codes over $(GF(2)^m, +, \times)$. In this section, we consider the ring $\mathcal{R} = (GF(2)^m, +, \times) = (GF(2), +, \times)^m$, where $+$ and $\times$ are respectively the addition and multiplication of binary m-tuples coordinate by coordinate.

For i in $[0; m-1]$, we denote by $e_i \in GF(2)^m$ the element such that $e_{i,j} = 0$ except $e_{i,i} = 1$ and π_i the projection $x = (x_0, ..., x_{m-1}) \mapsto x_i$.

If C is an additive code of length r over $\mathcal{R}$, we consider the following derived codes:

$e_i C = \{e_i c = (e_i c_0, ..., e_i c_{r-1}) \mid c \in C\} \subset \mathcal{R}^r$

and $\pi_i(C) = \{\pi_i(c) = (\pi_i(c_0), ..., \pi_i(c_{r-1})) \mid c \in C\} \subset GF(2)^r$.

It is easy to verify that the minimal ideals of $\mathcal{R}$ are exactly those generated by the elements e_i.

An $\mathcal{R}$-linear code C of length r over $\mathcal{R}$ is then a submodule of $\mathcal{R}^r$. An additive code C is $\mathcal{R}$-linear if and only if the codes $e_i C$ are subcodes of C.

Proposition 7. *An additive code C of length r over $\mathcal{R}$ is an $\mathcal{R}$-linear code if and only if it is the direct sum of the codes $e_i C$, for i in $[0; m-1]$. In this situation, it is isomorphic to the direct sum of the binary codes $\pi_i(C)$.*

Corollary 1. *The minimum block-weight distance of a linear code C is the minimum of the binary non-zero minimum distances of the binary codes $\pi_i(C)$.*

In particular the only non trivial MDS linear codes over $\mathcal{R}$ are the repetition block-code and the parity check block-code, with respective generator matrices:

$$G = (I_m \, I_m \, ... \, I_m) \text{ and } G' = \begin{pmatrix} I_m & 0 & \cdots & 0 & I_m \\ 0 & \ddots & \ddots & 0 & \vdots \\ \vdots & \ddots & \ddots & 0 & \vdots \\ \vdots & & \ddots & \ddots & \vdots \\ 0 & \cdots & \cdots & 0 & I_m \end{pmatrix}$$

In addition, one can remark that if $G = (I_k \mid M)$ is the $\mathcal{L}$-generator matrix of an additive block code, the code C is linear if and only if the binary matrices of the coefficients $\varphi_{i,j}$ of M are diagonal. In this situation, the two definitions of duality given in Section 2.2 are the same.

Here is an example of such a code. We set $m = 3$, $k = 2$ and $r = 5$. The binary image of the systematic $\mathcal{L}$-generator matrix is:

$$G = \begin{pmatrix} 100 & 000 & 000 & 100 & 100 \\ 010 & 000 & 010 & 000 & 010 \\ 001 & 000 & 001 & 000 & 000 \\ \hline 000 & 100 & 100 & 000 & 100 \\ 000 & 010 & 010 & 010 & 010 \\ 000 & 001 & 000 & 001 & 001 \end{pmatrix}$$

This code is in fact the direct sums of the binary codes C_i, $1 \le i \le 3$, generated by $G_1 = \begin{pmatrix} 1 0 0 1 1 \\ 0 1 1 0 1 \end{pmatrix}$, $G_2 = \begin{pmatrix} 1 0 1 1 1 \\ 0 1 1 1 1 \end{pmatrix}$ and $G_3 = \begin{pmatrix} 1 0 1 0 0 \\ 0 1 0 1 1 \end{pmatrix}$.

Since the codes C_2 and C_3 have a minimum distance equals to 2, the minimum distance of the block code C is 2.

Codes over $GF(2)[x]/f(x)$. Let $\mathcal{R} = GF(2)[x]/f(x)$ be the ring of binary polynomials modulo a fixed polynomial $f(x)$ of degree m. In addition, we require that $f(0) = 1$ which is equivalent to say that x is invertible in $\mathcal{R}$. This condition discards some degenerated cases.

Let $f(x) = f_1^{p_1}(x) f_2^{p_2}(x) \ldots f_s^{p_s}(x)$ be the factorization of $f(x)$ into irreducible polynomials. For $i \in [1; s]$, we set $h_i(x) = f(x)/f_i(x)$. The following properties are basic polynomial algebra results: the ring $\mathcal{R}$ is a principal ring, moreover the ideals of $\mathcal{R}$ are exactly those generated by the divisors of $f(x)$. The minimal ideals of $\mathcal{R}$ are those generated by the elements $h_i(x)$.

Suppose first that $f(x)$ is irreducible, the ring $\mathcal{R} = GF(2)[x]/f(x)$ is then isomorphic to $GF(2^m)$ and the notion of linear code over $\mathcal{R}$ is exactly those of linear code of length r over $GF(2^m)$.

Suppose now that $f(x)$ is square-free, i.e. $p_i = 1$ for all $i \in [1; s]$. This condition is equivalent to the fact that $\mathcal{R}$ is a semi-simple algebra. In other words $\mathcal{R}$ is equivalent to the product of fields $\prod_{i=1}^{s} \mathbb{F}_i$ where $\mathbb{F}_i = GF(2)[x]/f_i(x)$. The minimal ideal generated by a polynomial $h_i(x)$ is simply the set of elements with all its components equal to 0 except the i-th in the product representation $\prod_{i=1}^{s} \mathbb{F}_i$.

For $i \in [1; s]$ we define the projection p_i from $\mathcal{R}$ into $\mathbb{F}_i$ by $p_i(g(x)) = g(x) \bmod f_i(x)$. We extend this projection to an application from $\mathcal{R}^r$ into $\mathbb{F}_i^r$ by applying p_i to each component. We obtain s linear codes $C_i = p_i(C)$, each of these codes is a linear code over the finite field $\mathbb{F}_i$. The following theorem is a direct consequence of the isomorphism $\mathcal{R} = \prod_{i=1}^{s} \mathbb{F}_i$.

Theorem 3. *Suppose that the polynomial $f(x)$ is square free.*
If $f(x) = f_1(x)f_2(x) \ldots f_s(x)$ is its decomposition into irreducible factors, then the linear code C over $\mathcal{R}$ is isomorphic to the direct sum $\bigoplus_{i=1}^{s} C_i$.

Note that, if C is a systematic code, all the codes C_i are distinct from $\{0\}$. As a consequence, we obtain the following corollary:

Corollary 2. *Suppose that $f(x)$ satisfies the conditions of Theorem 3. If C is a systematic linear code over $\mathcal{R}$, its minimum distance d is the minimum of the* minimum distances d_i of its projections C_i. In particular, C is MDS if and only if all the codes C_i are MDS.

The most famous example of codes over such a ring $\mathcal{R}$ is those of quasi-cyclic codes [6], where $f(x) = x^m - 1$. In that situation, $x - 1$ is always a factor of $x^m - 1$, so one of the C_i is a binary code. This particular case does not allow to build MDS block codes.

A generalization of results of [6] to any polynomial $f(x)$ can be found in [2]. We will see in Section 4 that the particular case $f(x) = f_1(x)f_2(x)$, $deg(f_1(x)) = deg(f_2(x)) = m/2$ is an interesting way in order to construct MDS block codes over E from MDS block codes over $GF(2^{m/2})$.

We do not discuss the situation where $f(x)$ is not square free. In this situation, it is possible to decompose C in a similar way, but the projections are codes defined over finite chain rings. The reader can refer to [7] for more details.

2 $\mathcal{L}$-codes

2.1   Definition of $\mathcal{L}$-codes

The set $\mathcal{L}$ is a non-commutative ring. Remember that, for consistency with our previous notations and the underlying matrix approach, the product in $\mathcal{L}$ is permuted with the composition: for $a \in E$, $a\varphi\psi = (\varphi\psi)(a) = \psi(\varphi(a)) = a(\psi \circ \varphi)$.

Following the definition of $\mathcal{L}$-generator matrix for a block code over E, we define the notion of $\mathcal{L}$-linear code.

Definition 10. *An $\mathcal{L}$-left-linear code $\mathcal{C}$ of length r over $\mathcal{L}$ is a left-submodule of $\mathcal{L}^r$.*

Note that it is possible to define in the same way an $\mathcal{L}$-right-linear code. However, our definition of $\mathcal{L}$-linear codes concerns only left-submodule since there is a one to one correspondence between (left) $\mathcal{L}$-linear codes and additive block codes over E.

Theorem 4. *Let $\mathcal{C}$ be an $\mathcal{L}$-code of length r. The set*
$C = \{a\varphi = (\varphi_1(a), ..., \varphi_r(a)) \in E^r \mid \forall a \in E \text{ and } \forall \varphi \in \mathcal{C}\}$ is an additive block code. Reciprocally, if C is an additive block code, the set $\mathcal{C} = \{\varphi = (\varphi_1, ..., \varphi_r) \in \mathcal{L}^r \mid \forall a \in E, a\varphi \in C\}$ is an $\mathcal{L}$-code. Moreover the minimum distance of C and $\mathcal{C}$ are the same.

Proof. The proof of this theorem is essentially direct verification. The only difficult point concerns the minimum distance. Clearly, we have $d_C \leq d_{\mathcal{C}}$. Indeed, if $\varphi \in \mathcal{C}$ is a non-zero element of $\mathcal{C}$ of minimum weight $d_{\mathcal{C}}$, there exists an element $a \in E$, $a \neq 0$ and $c = \varphi(a) \neq 0$, $c \in C$ and $w_E(c) \leq d_{\mathcal{C}}$, so $d_C \leq d_{\mathcal{C}}$.

Reciprocally, suppose that $c = (c_1, ..., c_r) \in C$ is a word of minimum weight d_C. We construct an element $\varphi = (\varphi_1, ..., \varphi_r) \in \mathcal{C}$ as follows: the binary matrices of the linear applications φ_i are those with the first row equals to c_i and the other rows equal to $(0, .., 0)$. Such an element satisfy the property $a\varphi = a_1 c$. In other words, the images by φ of E is the $GF(2)$-vector space of dimension 1 generated by c. This implies $\varphi \in \mathcal{C}$ and $w(\varphi) = d_C$, so $d_{\mathcal{C}} = d_C$.

Following Definition 3 and Definition 4, a narrow sense (resp. general) systematic $\mathcal{L}$-code $\mathcal{C}$ is a submodule of rank k such that C is a narrow sense (resp. general) systematic additive block code.

In this situation the $\mathcal{L}$-generator matrix of C is nothing else than a generator matrix of $\mathcal{C}$ as left submodule.

2.2 Duality of $\mathcal{L}$-codes

There is no natural notion of duality on additive block codes when they are considered as block codes and not as binary codes. The introduction of underlying $\mathcal{L}$-codes having a module structure leads us to study the concept of duality with this approach.

$\mathcal{L}$-duality. One can define a kind of scalar product on $\mathcal{L}$ in the following way: for φ and ψ in $\mathcal{L}^r$, we set $< \varphi, \psi > = \sum_{i=1}^{r} \varphi_i \psi_i = \sum_{i=1}^{r} \psi_i \circ \varphi_i \in \mathcal{L}$. Note that $\mathcal{L}^r$ is a non commutative module over $\mathcal{L}$. For $\lambda \in \mathcal{L}$ and $\varphi \in \mathcal{L}$, we denote respectively by $\lambda \varphi = (\lambda \varphi_1, ..., \lambda \varphi_r)$ and $\varphi \lambda = (\varphi_1 \lambda, ..., \varphi_r \lambda)$ the left and right product. The bilinear map is linear as left module on the left component and linear as right module on the right component. In particular $< \lambda \varphi, \psi > = \lambda < \varphi, \psi >$ and $< \varphi, \psi \lambda > = < \varphi, \psi > \lambda$. Moreover this bilinear map is non degenerated in the sense that if, for a fixed $\varphi \in \mathcal{L}^r$, $< \varphi, \psi > = 0$ for all $\psi \in \mathcal{L}^r$, then $\varphi = 0$.

We are able to define the dual of an $\mathcal{L}$-code.

Definition 11. *Let $\mathcal{C}$ be an $\mathcal{L}$ (left)-linear code. The dual $\mathcal{C}^{\perp}$ of $\mathcal{C}$ is the subset of $\mathcal{L}^r$ defined by*
$$\mathcal{C}^{\perp} = \{\psi \in \mathcal{L}^r \mid \ < \varphi, \psi > = 0, \ \forall \varphi \in \mathcal{C}\}.$$

By adapting to the particular case of our non-commutative ring $\mathcal{L}^r$ the usual demonstrations concerning the properties of the dual of a linear code, we obtain the following theorem:

Theorem 5. *Let $\mathcal{C}$ be a systematic linear code of rank k. The dual $\mathcal{C}^{\perp}$ of $\mathcal{C}$ is an $\mathcal{L}$ right-linear submodule of $\mathcal{L}^r$ of rank $r - k$.*

Note that, since $\mathcal{C}^{\perp}$ is not a left module, we cannot associate to this code an additive block code. So, this notion of dual code cannot be extended to additive block codes. However, it remains a lot of useful properties in relation with this notion of duality. We can in particular define a generator matrix of $\mathcal{C}^{\perp}$ for its right-module structure. Moreover, if $\mathcal{G}$ is a generator matrix of $\mathcal{C}$ and $\mathcal{H}$ a generator matrix of $\mathcal{C}^{\perp}$, then $\mathcal{G}\mathcal{H}^T = 0$ (but not necessary $\mathcal{H}\mathcal{G}^T = 0$).

An element $c \in E^r$ is in the additive block code C if and only if $c\mathcal{H}^T = 0$. So the matrix $\mathcal{H}$ is also called an $\mathcal{L}$-parity check matrix of the code C.

In addition, if $\mathcal{G} = (I_k \mid M)$ is a generator matrix of $\mathcal{C}$ under systematic form, then $\mathcal{H} = (M^T \mid I_{r-k})$ is a (right) generator matrix of $\mathcal{C}^{\perp}$.

A particular care must be taken to the fact that M^T denotes the transpose of M at $\mathcal{L}$ level and does not correspond to the matrix obtained by the transpose of its binary image M (*i.e.* the binary $m(r - k) \times mk$ matrix).

In addition, there exists an equivalent to Theorem 10 of [8] Ch.1 §10 which deals with the link between independence of columns of a parity check matrix and minimum distance of a code.

Theorem 6. *Let C be an additive block code and $\mathcal{H}$ be an $\mathcal{L}$-parity check matrix of C. The code C has minimum block distance d if and only if every $d-1$ columns of $\mathcal{H}$ define a linear application of rank $d - 1$ and some d columns of H define a linear application of rank strictly less than d.*

Binary Duality for $\mathcal{L}$-codes. Identifying E^r and $GF(2)^{mr}$, it is possible to define the notion of binary dual of an additive block code. This approach can be defined as in Section 2.2 using a kind of "Hermitian scalar product" on $\mathcal{L}$. We need to use the transpose of a linear application. If φ is an element of $\mathcal{L}$ with associated binary matrix M_φ, the transpose of φ is the linear application $\varphi^T \in \mathcal{L}$ with binary matrix M_φ^T.

We define a bilinear map $< \varphi, \psi >_T = \sum_{i=1}^r \varphi_i \psi_i^T \in \mathcal{L}$. One can remark that $< \lambda\varphi, \psi >_T = \lambda < \varphi, \psi >_T$ and $< \varphi, \lambda\psi >_T = < \varphi, \psi >_T \lambda^T$.

We are able to define the binary dual of an $\mathcal{L}$-code.

Definition 12. *Let C be an $\mathcal{L}$ (left)-linear code. The binary dual $C^{\perp*}$ of C is the subset of $\mathcal{L}^r$ defined by*

$$C^{\perp*} = \{\psi \in \mathcal{L}^r \mid < \varphi, \psi >_T = 0 \, \forall \varphi \in C\}.$$

Theorem 7. *Let C be a systematic linear code of rank k. The binary dual $C^{\perp*}$ of C is an $\mathcal{L}$ left-linear submodule of $\mathcal{L}^r$ of rank $r - k$, i.e. a systematic $\mathcal{L}$-code.*

Proof. The proof of this theorem comes directly from the relation $< \varphi, \lambda\psi >_T = < \varphi, \psi >_T \lambda^T$ which implies in particular that, if $< \varphi, \psi >_T = 0$, then $< \varphi, \lambda\psi >_T = 0$ for all $\lambda \in \mathcal{L}$, so $C^{\perp*}$ is a left submodule.

In order to clarify the relationship between the two types of duality, we introduce the following notation: if $\mathcal{M} = (\varphi_{i,j})$ is a matrix with entries in $\mathcal{L}$, we denote $\mathcal{M}^* = (\varphi_{i,j}^T)$ obtained by replacing each entry of the matrix by its transpose application. Note that we do not transpose the matrix itself. In particular, if $\varphi = (\varphi_1, ..., \varphi_r)$, we set $\varphi^* = (\varphi_1^T, ..., \varphi_r^T)$.

Proposition 8. *An element $\varphi \in \mathcal{L}$ is in $C^\perp$ if and only if φ^* is in $C^{\perp*}$*

Proof. It is a direct consequence of Definitions 11 and 12.

As a consequence of these results, $\mathcal{H}$ is a generator matrix of the right $\mathcal{L}$-code $C^\perp$ if and only if the matrix $\mathcal{H}^*$ is a generator matrix of the (left) $\mathcal{L}$-code $C^{\perp*}$. In particular, if $\mathcal{G} = (I_k \mid \mathcal{M})$ is a generator matrix of C under systematic form, then $\mathcal{H}^* = (\mathcal{M}^{T*} \mid I_{r-k})$ is a generator matrix of $C^{\perp*}$.

The following proposition describes the link between the $\mathcal{L}$-duality and the binary duality and justify the name of *binary dual* for $C^{\perp*}$.

Proposition 9. *Let C be a systematic block code of length r over E. Identifying E^r and $GF(2)^{mr}$, we denote by $C^\perp$ the usual binary dual of C. The $\mathcal{L}$ linear code associated to $C^\perp$ is $\mathcal{C}^{\perp *}$.*

Proof. This result comes from the remark that, for φ and $\psi \in \mathcal{L}^r$, we have $M_{<\varphi,\psi>_T} = \sum_{i=1}^{r} M_{\varphi_i} M_{\psi_i}^T$.

Note that there is no analogue to Theorem 6 for the binary $\mathcal{L}$-dual linear code.

3   Linear Codes over Subrings of $\mathcal{L}$

3.1   Notations and Remarks

In this section, we focus on systematic additive block codes having a systematic generator matrix with entries in a subring $\mathcal{R}$ of $\mathcal{L}$. We denote these codes systematic $\mathcal{R}$-codes. Following Section 2.2, we define in an obvious way the notions of $\mathcal{R}$-generator elements, $\mathcal{R}$-parity check elements, $\mathcal{R}$-generator matrices, $\mathcal{R}$-parity check matrices, and the left (resp. right) submodule $C_{\mathcal{R}}$ (resp. $C_{\mathcal{R}}^{\perp}$, $C_{\mathcal{R}}^{\perp *}$).

Suppose for instance that $\mathcal{R}$ is commutative, then the notions of left and right submodule becomes the same, so the $\mathcal{R}$-duality leads to the construction of $\mathcal{R}$-dual additive block codes. Under those hypothesis, we have two distinct notions of dual block codes.

Another possibility is the fact that $\mathcal{R}$ is include in the ring of symmetric endomorphisms, *i.e.* the elements $\varphi \in \mathcal{L}$ such that $\varphi^T = \varphi$. In that case, the $\mathcal{R}$-parity check matrix $H = (M^T | I_{n-k})$ is an $\mathcal{R}$-generator matrix of the dual code $C_{\mathcal{R}}^{\perp *}$.

3.2   Diagonal Endomorphisms

A diagonal endomorphism is an endomorphism such that its binary matrix is diagonal. We denote by $\mathcal{D}$ the ring of diagonal endomorphisms, which is isomorphic to the ring $GF(2)^m$. The $\mathcal{D}$-codes are exactly those defined in Section 1.8.

The ring $\mathcal{D}$ is commutative, moreover the elements of $\mathcal{D}$ are symmetric, so there is a single notion of duality. Moreover a code is MDS if and only if its dual is MDS. However, due to Corollary 1, the search of MDS $\mathcal{D}$-codes reduces to the search of binary MDS codes that only leads to trivial cases.

3.3   Subrings with a Single Generator

Let ψ be an invertible element of $\mathcal{L}$. Let $\mathcal{P}(\psi) = \{P(\psi) = \sum_i^{deg(P(x))} p_i \psi^i\}$ be the subring of $\mathcal{L}$ generated by ψ. The ring $\mathcal{P}(\psi)$ is commutative and was intensively studied to construct MDS matrices for cryptographic applications (cf. e.g. [1,2]). Since this ring is commutative, there exists an intrinsic notion of $\mathcal{P}(\psi)$-duality.

In that situation, if $f(x)$ is the minimal polynomial of ψ, a $\mathcal{P}(\psi)$-block code is in fact isomorphic to a block code over $GF(2)[x]/f(x)$ described in Section 1.8. If $f(x)$ is irreducible, the $\mathcal{P}(\psi)$-duality is equivalent to the duality of codes over the finite field $GF(2)[x]/f(x)$, which is distinct from the binary duality. If $f(x)$ is not irreducible, the $\mathcal{P}(\psi)$-dual of a code is obtained by taking the dual of each projection and reconstructing the codes from these projections.

A priori, as noticed in [1], in order to construct some MDS diffusion matrices for cryptographic applications, it seems preferable to limit the search to codes over finite fields. However, in order to obtain some MDS matrices suitable for efficient implementation, it remains interesting to construct by this method some MDS matrices of size m from MDS matrices from smaller size block $m' < m$ (typically $m = 2m'$).

3.4 Block-Diagonal Subrings

Suppose that $m = m' + m''$ for some non-zero integers m' and m''. An element $x = (x_1, ..., x_m)$ can be identified to the couple $x = (x', x'')$ with $x' = (x_1, ..., x_{m'}) \in E' = GF(2)^{m'}$ and $x'' = (x_{m'+1}, ..., x_m) \in E' = GF(2)^{m''}$. We denote by $\mathcal{L}'$ and $\mathcal{L}''$ the rings of linear endomorphisms of E' and E''. Using the previous identification, the ring $\mathcal{R}_{m',m''} = \mathcal{L}' \times \mathcal{L}''$ can be considered as a subring of $\mathcal{L}$. The endomorphisms of $\mathcal{R}_{m',m''}$ are those whose matrices are block diagonal matrices with a first block of size m' and a second of size m''.

In practice, an $\mathcal{R}_{m',m''}$-linear block code is constructed as a direct sum of an $\mathcal{L}'$-linear block code and an $\mathcal{L}''$-linear block code. Even if at bit level such a code is clearly not very efficient, in the context of MDS diffusion matrices, this method allows to build MDS matrices over large m from smaller MDS matrices and it may be useful for some applications. The typical values of m' and m'' are $m' = m''$, so $m = 2m'$.

4 Examples of Constructions

The aim of this section is not to present some optimized matrices for hardware or dedicated embedded software implementations, but to explain how our approach can be applied for a practical search of good candidates.

4.1 MDS Diffusion Matrices Derived from MDS Linear Codes over a Finite Field

In this example, we set $m = 3$, $k = 3$ and $r = 6$. We denote by α a primitive root of $GF(8)$. From a Reed-Solomon code of parameters $[6, 3, 4]$ over the finite field $GF(8)$, we obtain the following MDS matrix $\mathcal{M} = \begin{pmatrix} 1 & \alpha & \alpha^3 \\ 1 & \alpha^6 & \alpha^6 \\ 1 & \alpha^4 & \alpha^5 \end{pmatrix}$.

There are different ways to build a binary MDS matrix from $\mathcal{M}$.

Indifferently, we can consider α as a root of the primitive polynomial x^3+x+1, or the primitive polynomial $x^3 + x^2 + 1$. Replacing α by the companion matrix of its minimal polynomial, we obtain two distinct MDS diffusion layers:

$$M = \left(\begin{array}{ccc|ccc|ccc}
1&0&0&0&1&0&1&1&0\\
0&1&0&0&0&1&0&1&1\\
0&0&1&1&1&0&1&1&1\\
\hline
1&0&0&1&0&1&1&0&1\\
0&1&0&1&0&0&1&0&0\\
0&0&1&0&1&0&0&1&0\\
\hline
1&0&0&0&1&1&1&1&1\\
0&1&0&1&1&1&1&0&1\\
0&0&1&1&0&1&1&0&0
\end{array}\right) \quad and \quad M' = \left(\begin{array}{ccc|ccc|ccc}
1&0&0&0&1&0&1&0&1\\
0&1&0&0&0&1&1&1&1\\
0&0&1&1&0&1&1&1&0\\
\hline
1&0&0&0&1&1&0&1&1\\
0&1&0&1&0&0&1&0&0\\
0&0&1&0&1&0&0&1&0\\
\hline
1&0&0&1&1&1&1&1&0\\
0&1&0&1&1&0&0&1&1\\
0&0&1&0&1&1&1&0&0
\end{array}\right)$$

In addition, one can notice that the minimal polynomial of the $M''_\alpha = \begin{pmatrix} 1&0&1\\0&0&1\\1&1&1 \end{pmatrix}$

is $x^2 + x + 1$ and M''_α generates a ring which is isomorphic to $GF(2^3)$. So the following binary matrix is an MDS diffusion layer:

$$M'' = \left(\begin{array}{ccc|ccc|ccc}
1&0&0&1&0&1&0&0&1\\
0&1&0&0&0&1&0&1&1\\
0&0&1&1&1&1&1&1&0\\
\hline
1&0&0&1&1&0&1&1&0\\
0&1&0&1&0&1&1&0&1\\
0&0&1&0&1&0&0&1&0\\
\hline
1&0&0&1&1&1&0&1&1\\
0&1&0&1&1&0&1&0&0\\
0&0&1&1&0&0&1&0&1
\end{array}\right)$$

Note that, since this matrix M''_α is symmetric, all the blocks in M'' are also symmetric matrices.

4.2   An Example of Symmetric Automorphisms

In this section, $m = 3$, $k = 3$ and $r = 6$. We set $M_{1,1} = M_{2,1} = M_{3,1} = I_3$,

$$M_{1,2} = \begin{pmatrix} 0&0&1\\0&1&0\\1&0&1 \end{pmatrix}, M_{1,3} = \begin{pmatrix} 1&0&1\\0&1&1\\1&1&0 \end{pmatrix}, M_{2,2} = M_{3,3} = \begin{pmatrix} 1&0&0\\0&1&1\\0&1&0 \end{pmatrix}, M_{2,3} = \begin{pmatrix} 0&0&1\\0&1&0\\1&0&0 \end{pmatrix}$$

and $M_{3,2} = \begin{pmatrix} 1&0&1\\0&0&1\\1&1&1 \end{pmatrix}$. The following matrix M is an MDS block diffusion matrix

$$M = \begin{pmatrix} I_3 & M_{1,2} & M_{1,3}\\ I_3 & M_{2,2} & M_{2,3}\\ I_3 & M_{3,2} & M_{2,2} \end{pmatrix}$$

All the submatrices $M_{i,j}$ are symmetric. The matrix $M_{2,3}$ is of order 2, its minimal polynomial is $x^2 + 1$. The matrices $M_{1,2}$ and $M_{2,2}$ are of order 3, with the same minimal polynomial $x^3 + 1$. The matrices $M_{1,3}$ and $M_{3,2}$ are of order 7, with respective minimal polynomials $x^3 + x^2 + 1$ and $x^3 + x + 1$. None of these 5 matrices pairwise commute. The ring generated by these matrices is $\mathcal{L} \simeq M_3(GF(2))$.

Clearly, this example cannot be obtained by usual methods derived from finite fields or commutative subrings.

4.3   Iterative Constructions on m

From the examples given in Section 4.1, we are able to construct 3×3 MDS matrices over blocks of size $m = 6$.

First, we follow the block-diagonal method presented in Section 3.4. From a binary matrix $M_{\alpha^i} \in M_3(GF(2))$, we construct the matrix $I_2 \otimes M_{\alpha^i} = \begin{pmatrix} M_{\alpha^i} & 0 \\ 0 & M_{\alpha^i} \end{pmatrix}$ in $M_6(GF(2))$.

From the MDS matrix $M = (M_{i,j})$, $M_{i,j} \in M_3(GF(2))$, given in Section 4.1, we construct the MDS matrix $M^{(2)} = (I_2 \otimes M_{i,j})$ which acts on 3 blocks of size 6.

Even if this matrix acts separately on the subblocks of 3 bits inside the blocks of 6 bits, if this diffusion matrix is applied after a well-chosen Sbox over blocks of 6 bits, this property is no more a cryptographic weakness.

Another possible combination of MDS matrices from finite field of smaller size is those derived from Section 1.8. For example, the MDS matrix $\mathcal{M}$ over $GF(8)$ given in Section 4.1.

Recall that $\mathcal{M} = \begin{pmatrix} 1 & \alpha & \alpha^3 \\ 1 & \alpha^6 & \alpha^6 \\ 1 & \alpha^4 & \alpha^5 \end{pmatrix}$.

Set $f_1(x) = x^3 + x + 1$, $f_2(x) = x^3 + x^2 + 1$ and $f(x) = f_1(x)f_2(x)$. The corresponding polynomials rings are $\mathcal{R}_1 = GF(2)[x]/f_1(x)$, $\mathcal{R}_2 = GF(2)[x]/f_2(x)$ and $\mathcal{R} = GF(2)[x]/f(x)$.

Since $\mathcal{R}_1$ and $\mathcal{R}_2$ are isomorphic to $GF(8)$, the matrix $\mathcal{M}$ can be interpreted as a matrix in $\mathcal{R}_1$ and $\mathcal{R}_2$, with

$$\mathcal{M}_1 = \begin{pmatrix} 1 & x & x^3 \\ 1 & x^6 & x^6 \\ 1 & x^4 & x^5 \end{pmatrix} = \begin{pmatrix} 1 & x & x+1 \\ 1 & x^2+1 & x^2+1 \\ 1 & x^2+x & x^2+x+1 \end{pmatrix}$$

and

$$\mathcal{M}_2 = \begin{pmatrix} 1 & x & x^3 \\ 1 & x^6 & x^6 \\ 1 & x^4 & x^5 \end{pmatrix} = \begin{pmatrix} 1 & x & x+1 \\ 1 & x^2+x & x^2+x \\ 1 & x^2+x+1 & x+1 \end{pmatrix}.$$

The isomorphism $\mathcal{R} \simeq \mathcal{R}_1\mathcal{R}_2$ able us to construct an MDS matrix with entries in $\mathcal{R}$ using the Remainder Chinese Theorem.

If $a_1(x)f_1(x)+a_2(x)f_2(x) = 1$ is the Bézout identity, to $(g_1(x), g_2(x)) \in \mathcal{R}_1\mathcal{R}_2$ we associate the polynomial $g(x) = a_2 f_2 g_1(x) + a_1 f_1 g_2(x) \in \mathcal{R}$. In our example, $a_1(x) = x$ and $a_2(x) = x + 1$.

Our MDS matrix over $\mathcal{R}$ is then $\mathcal{M}_{\mathcal{R}} = x f_1(x)\mathcal{M}_1 + (x + 1)f_2(x)\mathcal{M}_2$, i.e.

$$\mathcal{M}_{\mathcal{R}} = \begin{pmatrix} 1 & x & x^6 + x^5 + x^4 + 1 \\ 1 & x^5 + x^4 + x^3 + x^2 & x^5 + x^4 + x^3 + x^2 \\ 1 & x^4 + 1 & x^6 + x^4 + x^3 + x + 1 \end{pmatrix}.$$

The binary MDS matrix is obtained by substituting the companion matrix M_x of $f(x)$ to x in the entries of $\mathcal{M}_{\mathcal{R}}$. Indeed, M_x is the matrix of the multiplication by x in $\mathcal{R}$.

5 Conclusion

The goal of this paper was not to construct in practice some optimized MDS matrices dedicated to specific applications, but to present a general framework for such a research. In order to have a generic approach, we introduced the notion of $\mathcal{L}$-linear codes and $\mathcal{R}$-linear codes for subrings $\mathcal{R}$ of $\mathcal{L}$.

This approach allows us to recover most of the known methods used for the construction of MDS diffusion matrices. We show that there exists some other non-explored directions of search, in particular with the ring of symmetric endomorphisms and any non-commutative subring of $\mathcal{L}$.

References

1. Augot, D., Finiasz, M.: Exhaustive search for small dimension recursive MDS diffusion layers for block ciphers and hash functions. In: Proceedings of the 2013 IEEE International Symposium on Information Theory, Istanbul, Turkey, July 7-12, pp. 1551–1555. IEEE (2013)
2. Berger, T.P., El Amrani, N.: Codes over finite quotients of polynomial rings. Finite Fields and Their Applications 25, 165–181 (2014)
3. Daemen, J., Rijmen, V.: The Design of Rijndael: AES - The Advanced Encryption Standard. Information Security and Cryptography. Springer (2002)
4. Guo, J., Peyrin, T., Poschmann, A.: The PHOTON family of lightweight hash functions. In: Rogaway, P. (ed.) CRYPTO 2011. LNCS, vol. 6841, pp. 222–239. Springer, Heidelberg (2011)
5. Huffman, W.C.: Codes and groups. In: Huffman, W.C., Pless, V. (eds.) Handbook of Coding Theory II, ch.17. Elsevier Science Inc., New York (1998)
6. Lally, K., Fitzpatrick, P.: Algebraic structure of quasicyclic codes. Discrete Applied Mathematics 111(1-2), 157–175 (2001)
7. Ling, S., Niederreiter, H., Solé, P.: On the algebraic structure of quasi-cyclic codes IV: repeated roots. Des. Codes Cryptography 38(3), 337–361 (2006)
8. MacWilliams, F.J., Sloane, N.J.A.: The theory of Error Correcting Codes. North-Holland, Amsterdam (1986)
9. Silvester, J.R.: Determinants of block matrices. The Mathematical Gazette 84(3), 460–467 (2000)

A Higher Order Key Partitioning Attack with Application to LBlock

Riham AlTawy, Mohamed Tolba, and Amr M. Youssef[✉]

Concordia Institute for Information Systems Engineering,
Concordia University, Montréal, Québec, Canada
youssef@ciise.concordia.ca

Abstract. In this paper, we present a higher order key partitioning meet-in-the-middle attack. Our attack is inspired by biclique cryptanalysis combined with higher order partitioning of the key. More precisely, we employ more than two equally sized disjoint sets of the key and drop the restrictions on the key partitioning process required for building the initial biclique structure. In other words, we start the recomputation phase of the attack from the input plaintext directly, which can be regarded as a Meet-in-the-Middle-attack where the tested keys have a predefined relation. Applying our approach on LBlock allows us to present a known plaintext attack on the full thirty two round cipher with time complexity of $2^{78.338}$ and negligible memory requirements. The data complexity of the attack is two plaintext-ciphertext pairs, which is the minimum theoretical data requirements attributed to the unicity distance of the cipher. Surprisingly, our results on the full LBlock are better, in terms of both computational and data complexity, than the results of its biclique cryptanalysis.

Keywords: Cryptanalysis · Meet-in-the-middle · Low data complexity · LBlock · Bicliques

1 Introduction

Bicliques are structures that provide a formal representation of the initial execution separation in MitM attacks [14]. These structures have become particularly important after they have been used to present a key recovery attack on the full round Advanced Encryption Standard (AES) [5]. Indeed, a biclique attack is an optimized exhaustive search attack where the whole key space is tested efficiently. Accordingly, this class of attacks is usually used to analyze the full round cipher unlike various other attacks which can only be applied to reduced round versions. As a result of the exhaustive search nature of biclique cryptanalysis, attacks employing these structure are characterized by their high computational complexity which can reach that of the brute force search. However, practical gain has been shown in a dedicated FPGA implementation of the attack on AES [4]. Most of the biclique attacks require high data complexity (depending on the length of the employed biclique) which can sometimes reach the entire codebook.

© Springer International Publishing Switzerland 2015
S. El Hajji et al. (Eds.): C2SI 2015, LNCS 9084, pp. 215–227, 2015.
DOI: 10.1007/978-3-319-18681-8_17

The need for efficient lightweight cryptography is on the rise due to the current popularity of lightweight devices such as RFID chips and wireless sensor networks. Indeed, these systems provide convenient affordable services on tiny resource constrained environments. On the other hand, these systems must guarantee certain security and privacy requirements. More precisely, the adopted primitives must fulfill the aggressive restrictions of the application environment and at the same time maintain acceptable security margins. PRESENT [6], KATAN and KTANTAN [13], LED [16], Zorro [15], and LBlock [26] are some examples of cipher designs that have been proposed to address the needs of lightweight cryptography. Most of the recent cryptanalytic attacks on lightweight ciphers aim to analyze how some design concepts which are proposed for this environments have weakened these ciphers and broadened the effect of certain types of attacks. [20,2,19,7,23].

Recently, there has been an increased interest in adopting low data complexity attacks for the analysis of ciphers. This motivation is backed by the fact that security bounds are better perceived in a realistic model [8,10]. More precisely, in a real life scenario, security protocols impose restrictions on the amount plaintext-ciphertext pairs that can be eavesdropped and/or the number of queries permitted under the same key. Given the fact that biclique cryptanalysis is characterized by its high data complexity, it has been implicitly avoided in the analysis of lightweight primitives.

In this work, we present a higher order key partitioning MitM attack. Our approach adopts only the recomputation phase from the biclique attack and does not require any specific initial biclique structure. Accordingly, we drop all the restrictions imposed by the bicliques on how the key is partitioned, and allow the use of related keys that would have been impossible otherwise. The absence of the biclique results in a low data complexity related key MitM attack in the single key setting in which the whole key space is searched efficiently through partial matching by recomputation [5]. To minimize the computational complexity of the recomputation, we employ a higher order number of disjoint sets of the master key [24]. More precisely, we partition the key space into more than two related keys (not necessarily independent related key differentials). Adopting this divide and conquer approach means that we have to deal with multiple small recomputed sets instead a dominating large set. We apply this attack on the lightweight block cipher LBlock which, similar to other lightweight ciphers, employs a simple key schedule with relatively slow diffusion to meet the resources constraints. Additionally, it adopts round subkeys that are shorter than the master key and a nibble-wise permutation. This fact allows our attack to achieve more gain over its biclique cryptanalysis counterpart [25]. Moreover, our attack on LBlock results in the minimum data requirements which makes it valid on RFID-like systems where the attacker can only acquire a very limited amount of plaintext-ciphertext pairs.

The rest of the paper is organized as follows. In the next section, we give a brief overview on the basic biclique attack. Afterwards, in Section 3, we give the specification of the lightweight block cipher LBlock. In Section 4, we provide the details of our approach and its application on LBlock. Specifically, we present a low-data complexity attack on the full thirty two round cipher. Finally, the paper is concluded in Section 5.

2   Biclique Cryptanalysis

Biclique cryptanalysis [5] was first used to present an accelerated exhaustive search on the full round AES. The basic idea of bicliques is to increase the number of rounds of the basic MitM attack by providing a formal representation of the initial structure and recomputing only the updated parts of the state. The key recovery attack starts by dividing the master key space into key sets where each key set K, is used to build one biclique. As depicted in Figure 1, a d-dimensional biclique is a structure of two sets of states P_i and S_j where $|P_i| = |S_j| = 2^d$ states and a key set K where $|K| = 2^{2d}$ keys which encrypt each state in P_i to each state in S_j. K is partitioned into three disjoint sets of key bits, i.e., $K = \{K_s, K_1, K_2\}$. Let $Enc_{[u,i,j]}(P_{i=0}^u)$ and $Dec_{[u,i,j]}(S_{j=0}^u)$ denote the encryption and decryption of the states $P_{i=0}^u$ and $S_{j=0}^u$ using the u, i, and j values of K_s, K_1, and K_2, respectively. These key sets are chosen such that for a given u of the $2^{|K_s|}$ values and all of i and j of the $2^{|K_1|}$, and $2^{|K_2|}$ values, respectively, $S_j^u = Enc_{[u,i,j]}(P_i^u)$, where $i, j \in \{0, .., 2^d - 1\}$.

The construction of bicliques imposes restrictions on the choices of K_1 and K_2 as they must result in independent related key differentials. In other words, K_1 and K_2 must be chosen such that the state variables between P_i^u and S_j^u that are affected by a change in the value of K_1 are different than those affected by a change in the value of K_2. In other words, a biclique can be constructed if for all u, i, and j of the $2^{|K_s|}, 2^{|K_1|}$, and $2^{|K_2|}$ values, respectively, the computation of $S_j^u = Enc_{[u,0,j]}(P_{i=0}^u)$ does not share any active nonlinear state variables with the computation $P_i^u = Dec_{[u,i,0]}(S_{j=0}^u)$.

The MitM key recovery attack using d-dimension bicliques starts with partitioning the master key space into $2^{|K|-2d}$ groups, where each group $K[u, i, j]$ has a single value u of the $2^{|K_s|}$ values and iterates over the 2^{2d} values of i and j of the $2^{|K_1|}$, and $2^{|K_2|}$ values, respectively. As depicted in Figure 1, usually the constructed biclique is placed on the plaintext side. The attack is divided into two main parts:

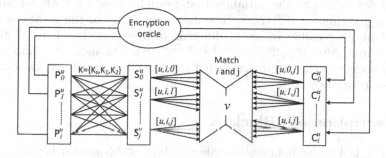

Fig. 1. Bicliques used in a MitM attack

Biclique construction: at this stage, one searches for two independent related key differentials to partition the key into key groups consisting of a given K_s and all the values of K_1 and K_2. Since we do not use any biclique structures in our attack, we refer the reader to [5] for the detailed procedure for building the bicliques.

Recomputation for MitM partial matching: during this step, partial state knowledge is computed from both the backward and forward directions to test each key group in an efficient manner. More precisely, in what follows, we give the steps performed for each key group $K[u, i, j]$.

- Choose an appropriate matching variable v between the end of the biclique and the ciphertext.
- *Forward recomputation:* for each j out of the 2^d values, do the following:
  - Compute the matching variable $\overrightarrow{v}_{0,j}^{\,u} = Enc_{[u,0,j]}(S_j^u)$ and store all the intermediate states.
  - For all the $2^d - 1$ values of i, compute the matching variable $\overrightarrow{v}_{i,j}^{\,u}$ by recomputing only those variables that differ from those previously computed using $K[u, 0, j]$ due to the effect of i.
- *Backward recomputation:* for all the 2^d values of P_i^u, ask the encryption oracle for their corresponding ciphertexts C_i^u.
- For each i of the 2^d values, do the following:
  - Compute the matching variable $\overleftarrow{v}_{i,0}^{\,u} = Dec_{[u,i,0]}(C_i^u)$ and store all the intermediate states.
  - For all $2^d - 1$ values of j, compute the matching variable $\overleftarrow{v}_{i,j}^{\,u}$ by recomputing only those variables that differ from those previously computed using $K[u, i, 0]$ due to the effect of j.

The remaining candidate keys $K[u, i, j]$ are those producing $\overrightarrow{v}_{i,j}^{\,u} = \overleftarrow{v}_{i,j}^{\,u}$. The surviving candidate keys should be further rechecked for full state matching as some of them could be false positives.

Testing each key group by the previous procedure has proved to lead to some improvements on the computational complexity. While all the 2^{2d} values of the key group are tested, we get three sets of computations. More precisely, the state variables that are affected by K_s only are computed once, those affected by either K_1 or K_2 are computed 2^d times, and the dominating large set is due to the variables that are influenced by both K_1 and K_2 which are recomputed 2^{2d} times. The data complexity of the attack is upper bounded by all possible values of different plaintext produced by all the bicliques $= min(2^{|K_s|+|K_1|}, 2^{\#\text{ of active bits in plaintext}})$. The memory complexity is upper bounded by the memory required to store the forward and backward 2^d intermediate states $\approx 2^{d+1}$.

3   Description of LBlock

LBlock [26] is a 64-bit lightweight cipher with an 80-bit master key. It employs a 32-round Feistel structure and its internal state is composed of eight 4-bit nibbles. As depicted in Figure 2, the round function adopts three nibble oriented transformations: subkey mixing, 4-bit Sboxes, and nibble permutation. The 80-bit master key, K, is stored in a key register denoted by $k = k_{79}k_{78}k_{77}\ldots\ldots k_1k_0$. The leftmost 32 bits of the register k are used as i^{th} round subkey Sk_i. The key register is updated after the extraction of each Sk_i as follows:

1. $k \lll 29$.
2. $[k_{79}k_{78}k_{77}k_{76}] = S_9[k_{79}k_{78}k_{77}k_{76}]$.
3. $[k_{75}k_{74}k_{73}k_{72}] = S_8[k_{75}k_{74}k_{73}k_{72}]$.
4. $[k_{50}k_{49}k_{48}k_{47}k_{46}] \oplus [i]_2$,

where S_8 and S_9 are two 4-bit Sboxes. For further details, the reader is referred to [26].

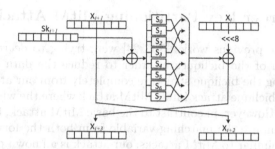

Fig. 2. The LBlock round function

LBlock [26] has been analyzed with respect to various types of attacks including impossible differential [17,18,9], integral [23,22], MitM [1], boomerang [12], and biclique cryptanalysis [25]. Particularly, the attack presented in [25] is a typical high data complexity biclique cryptanalysis where the authors presented an attack with a time complexity $= 2^{78.4}$ and a data complexity of 2^{52}. Our result for the 9th order key partitioning MitM cryptanalysis of the full round LBlock has a better time complexity and is launched with only two known plaintext-ciphertext pairs. In Table 1, we provide a summary of the current cryptanalytic results on the LBlock cipher in the single key model. In what follows, we give the notation used in our attack.

3.1   Notation

The following notation will be used throughout the remainder of the paper.

- K: The master key.
- Sk_i: i^{th} round sub key.
- X_i: The eight 4-bit nibble state at round i.
- $X_i[j]$: j^{th} nibble of the ith round state.
- $K_{[i,j]}$: i^{th} and j^{th} bits of master key K.

Table 1. Summary of the current cryptanalytic results on LBlock. DC, CP, KP, and FC stands for Differential Cryptanalysis, Chosen Plaintext, Known Plaintext, and Full Computation, respectively

Attack	#Rounds	Time	Memory	Data	Reference
Integral	22	2^{70}	2^{63}	2^{61} CP	[22]
Impossible DC	22	$2^{79.28}$	-	2^{58} CP	[17]
	23	$2^{75.36}$	2^{74}	2^{59} CP	[9]
Biclique cryptanalysis	32	$2^{78.4}$	2^{8}	2^{52} CP	[25]
Meet-in-the-middle	32	$2^{78.338}$	2^{7} FC	**2 KP**	This paper

4   Higher Order Key Partitioning MitM Attack

While most of the previous works [3,11,24] were trying to decrease the length and/or dimension of the bicliques in order to reduce the data complexity, we opted for removing the biclique structure completely from our attack. In the sequel, we turn the biclique attack into a MitM attack where the whole key space is efficiently tested. However, in contrast to the basic MitM attack, the same tested key is used to compute the matching variable from both the forward and backward directions. Similar to MitM attacks, our attack is a known plaintext attack where the number of required plaintext-ciphertext pairs is solely determined by the relationship between the block length and key lengths. Hence, given its negligible memory complexity, this approach provides an actual computational gain over exhaustive search as both of them have the same data requirements.

Our attack skips the independence requirements imposed on the choice of the related keys in the biclique attack and starts the recomputation phase from the plaintext. Thus, our approach is equivalent to a MitM attack where the tested keys have a predefined relation. Moreover, we consider a divide and conquer approach where higher order partitioning of the key space is adopted. In other words, instead of dividing the key K into three disjoint sets, we divide it into $n + 1$ sets with $n > 2$ to minimize the complexity of the recomputation in both directions. Given that the key is partitioned into a $(|K| - nd)$-bit set and n d-bit sets, adopting this higher order partitioning shares the complexity of the attack between $n + 1$ sets of recomputations where Sboxes of the ith set are recomputed $2^{(i-1)d}$ times, $i \in \{1, 2, ..., n + 1\}$.

In the sequel, we apply this technique on LBlock and present a low data complexity key recovery attack on the full round cipher. In fact, our best obtained result is a two known plaintext MitM attack where the key is partitioned into nine 4-bit sets (i.e., n=9). However, based on our trials with different values of n, we expect that further reduction in the computational complexity can be obtained with higher values of n, but given our available computational resources, the complexity of finding the optimal attack parameters as n grows is not practical. Our results are particularly interesting, because they show that removing the biclique structure from biclique-like attacks can have a good impact on both the data and computational complexities in some cases as with the case of LBlock.

4.1   A Low Data Complexity Attack on LBlock

In this section, we present a low data complexity attack on the full round LBlock. The attack exploits the weak diffusion of the key schedule. This fact enables us to partition the master key into higher order related key partitions for our MitM attack on the full cipher with some gain over the biclique attack [25]. With the aim of minimizing the computational complexity, we used an exhaustive search algorithm to test all possible 4-bit partitioning possibilities and different matching variables. Our search algorithm shows that partitioning K into K_s, K_1, K_2, K_3, K_4, K_5, K_6, K_7, K_8, and K_9 (i.e., $n = 9$) results in a MitM attack with a time complexity of $2^{78.338}$ and a memory complexity of $\approx 2^7$. The parameters of our best case for the MitM key recovery attack are as follows:

- Matching round: 22.
- Matching values: $X_{22}[2]$ and $X_{22}[7]$.
- $K_1 = K_{[76,75,74,73]}$
- $K_2 = K_{[59,58,57,56]}$
- $K_3 = K_{[38,37,36,35]}.$
- $K_4 = K_{[30,29,28,27]}.$
- $K_5 = K_{[25,24,23,22]}.$
- $K_6 = K_{[17,16,15,14]}.$
- $K_7 = K_{[13,12,11,10]}.$
- $K_8 = K_{[9,8,7,6]}.$
- $K_9 = K_{[5,4,3,2]}.$

MitM Attack with n=3. Due to the complexity of visualizing our best obtained case (i.e., $n = 9$), in what follows, we demonstrate the details of the attack in the simplest case when K is partitioned into three related partitions (n=3). The algorithm indicates that the best partitioning of the master key for the case of $n = 3$ is when $K_1 = K_{[25,24,23,22]}$, $K_2 = K_{[13,12,11,10]}$, $K_3 = K_{[3,2,1,0]}$, and $K_s = K - \{K_1, K_2, K_3\}$.

Accordingly, in our attack we test 2^{68} key groups where each group has 2^{12} keys, all of which have one value of K_S and differ in the values of K_1, K_2, and K_3. Figure 3 depicts the recomputation process adopted for our MitM attack for each key group. Our search algorithm shows that choosing the matching variable v as the two nibbles $X_{22}[2, 7]$ results in the best computational complexity for our attack. In the sequel, given one known plaintext-ciphertext pair, we evaluate the matching variable v from both the forward and backward directions with the same key and discard keys that produce $\overrightarrow{v} \neq \overleftarrow{v}$.

The computational complexity of the LBlock round function is dominated by the Sbox lookups. Consequently, to determine the gain of our approach, we use the number of Sboxes that are calculated in both directions relative to the 318 Sbox lookups used in the full thirty two rounds computation. We distinguish the color scheme used in Figure 3 as follows:

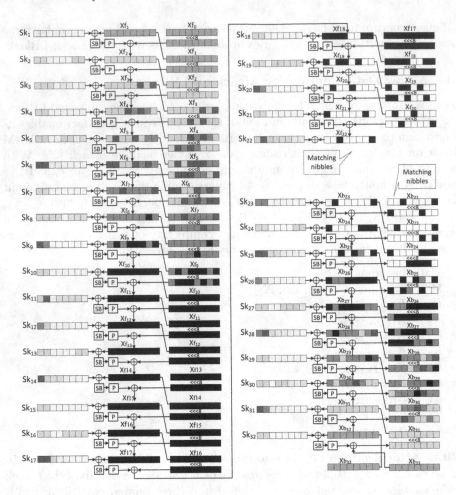

Fig. 3. Third order partitioned MitM attack on the thirty two rounds LBlock

- Gray: input nibbles which are either plaintext or ciphertext and these nibbles remain constant while testing the whole keyspace.
- Yellow: are the nibbles that are affected by changing the value of K_s only. Accordingly, when testing the 2^{12} keys within each key group, these nibbles are evaluated once.
- Red, blue, and green: are those nibbles that are influenced by a change in either K_1, K_2, or K_3, respectively. In other words, for every tested value of K_1, the values of the red nibbles are updated while the values of the blue and green nibbles remain unchanged. Same rationale applies for K_2 and K_3. Consequently, these nibbles are computed 2^4 times for each key group.
- Purple: nibbles that are affected by changing any two keys. For example, if K_2 and K_3 changed, then the values of all the blue and green nibbles, and

the values of the purple nibbles that depend on both keys should be updated. Purple nibbles are recomputed 2^8 times within one key group.
- Black: are the nibbles that are affected by a change in the values of all the three keys and these are evaluated 2^{12} times.
- White: are those nibbles whose values do not affect the value of the matching variable and thus we do not need to compute them in our attack.

Since our gain is estimated based on the number of Sbox lookups, the six right-most nibbles in the round subkeys are faded in color because only the two left-most nibbles are the ones that count in our calculations. In what follows, we give the details of the forward and backward recomputation used for testing the 2^{12} keys within a given key group. However, since each key group has one value of K_s and 2^4 values for each K_1, K_2, and K_3, we denote the tested key by $K[i,j,l]$ where each i, j, and l is one of the 2^4 values of K_1, K_2, and K_3, respectively. We also denote the matching variable that is computed using $K[i,j,l]$ as $v_{[i,j,l]}$.

Forward Recomputation: As depicted in Figure 3, the forward computation spans over states Xf_2 to Xf_{22}. The 64-bit input plaintext P is loaded in states Xf_0 and Xf_1. We now need to partially encrypt the plaintext P with all the 2^{12} keys $K[i,j,l]$ to derive 2^{12} values for the matching variable $\overrightarrow{v_{i,j,l}}$ which is the 8-bit output at $Xf_{22}[2,7]$. We first evaluate the matching variable $\overrightarrow{v_{[0,0,0]}} = Enc_{K[0,0,0]}(P)$ and store the computations. Now we do the same for all the 2^4-1 values of each key partition at a time. More precisely, we compute $\overrightarrow{v_{[i,0,0]}} = Enc_{K[i,0,0]}(P)$, $\overrightarrow{v_{[0,j,0]}} = Enc_{K[0,j,0]}(P)$, and $\overrightarrow{v_{[0,0,l]}} = Enc_{K[0,0,l]}(P)$ for all i, j, and $l \in \{1,..,2^4-1\}$, and store these computations as well. However, during these three computations, we only evaluate the nibbles that are different from the first stored computation using $K[0,0,0]$ due to the effect of either i, j, or l. These are the red, blue, green, purple, and black nibbles. Moreover, to test any two key partitions combination, e.g., $K[i,j,0]$, we only recompute the values of the purple and black nibbles that differ from that of the stored computations using $K[i,0,0]$ and $K[0,j,0]$ (i.e., where the effect of i and j overlap). Lastly, when testing any three key partitions combination, we only recompute those nibbles where the nibbles at their corresponding positions in the computations of $\overrightarrow{v_{[i,0,0]}}$, $\overrightarrow{v_{[0,j,0]}}$, and $\overrightarrow{v_{[0,0,l]}}$ overlap. As shown in Figure 3, the forward recomputation for one key group requires computing 54 Sboxes once, 25 Sboxes 2^4 times, 21 Sboxes 2^8 times, and 86 Sboxes 2^{12} times.

Backward Recomputation: The backward computation is depicted on the right side of Figure 3, where states Xb_{31} to Xb_{22} are iteratively recomputed to generate the matching variable. We use the ciphertext C corresponding to the plaintext P in states Xb_{32} and Xb_{33}. In the sequel, we proceed with partially decrypting C using the 2^{12} keys to evaluate the 2^{12} matching variable values $\overleftarrow{v_{i,j,l}}$ following the same procedure used in the forward recomputation. As depicted in Figure 3, the backward recomputation for one key group requires computing 24 Sboxes once, 14 Sboxes 2^4 times, 18 Sboxes 2^8 times, and 24 Sboxes 2^{12} times.

Surviving Candidates: As we are testing 2^{12} keys and the matching size is 8-bits, then for each key group we get 2^4 potential candidates, which need to be further retested for full state matching. This process of retesting the surviving candidates requires 2^4 full LBlock computations. Moreover, even after testing the whole key space, the relation between the key size k and the state size b determines the number of the remaining potentially right keys. If $b = k$, then only the right key remains and no further testing is required. However, in LBlock, the master key length is larger than the state size and hence, we end up with $2^{k-b} = 2^{16}$ potentially right keys. In this case, to recover the right key, the data complexity of the MitM attack is $\lceil \frac{k}{b} \rceil = 2$ plaintext-ciphertext pairs. In other words, these 2^{16} keys must be further retested with an additional plaintext-ciphertext pair.

Complexity: The computational complexity of the attack is evaluated based on the number Sbox lookups that are required in the forward and backward recomputations and testing surviving keys within a key group. The whole attack tests 2^{68} key groups, each requires the computations of $54 + 2^4(25) + 2^8(21) + 2^{12}(86) = 358086$ Sboxes in the forward direction, $24 + 2^4(14) + 2^8(18) + 2^{12}(24) = 103160$ Sboxes in the backward direction, and $2^4(318)$ Sboxes for retesting candidate keys. Accordingly, the overall computational complexity of the attack is given by $2^{68}((358086 + 103160)/318 + 2^4) \approx 2^{78.518}$. The memory complexity is upper bounded by storing 3×2^4 full LBlock computations, which is practically negligible, and the data complexity is two known plaintext-ciphertext pairs.

Generally, when adopting n key partitions of dimension d bits and m-bit matching variable, the same forward and backward recomputation procedures is applied. However, in this case, we get $n + 1$ recomputation sets. More formally, let Sb_i denote the number of Sboxes that belong to the ith recomputation set, $1 \leq i \leq n + 1$, the computational complexity of the attack is given by:

$$2^{|K|-nd}\left(\frac{\sum_{i=1}^{n+1}(2^{(i-1)d} \times Sb_i)}{318} + 2^{nd-m}\right),$$

and the memory complexity is given by $n \times 2^d$. Accordingly, given the parameters of our best obtained result when $n = 9$, the time and memory complexity of the MitM attack on the full LBlock is $2^{78.338}$ and $\approx 2^7$ full computations, respectively.

5   Conclusion

In this work, we have presented a higher order key partitioning MitM attack. Our technique adopts the recomputation phase from the biclique attack without using its initial structure. We have shown that in some cases removing the biclique structure from the MitM attack can lead to better computational and data complexity as with the case of LBlock. This fact is attributed to the restrictions imposed by the biclique on how the master key can be partitioned. On the other hand, if we search for the best key partitioning for a minimum complexity

MitM attack and begin the recomputation phase from the first round, we can get more savings. Moreover, we adopted a divide and conquer approach where the key space is divided into more than two related key sets. Thus, the computational complexity of the forward and backward recomputations is shared among multiple smaller sets and not being dominated by a large one. We applied our approach on LBlock and presented a low data complexity MitM attack on the full round cipher. Our best obtained result on the full round LBlock is a known plaintext MitM attack where the key is partitioned into nine related key sets. This attack has a time, memory, and data complexity of $2^{78.338}$, 2^7 full computations, and 2 known plaintexts, respectively, which are better than the results obtained by the biclique cryptanalysis of the cipher [25].

Inspired by biclique cryptanalysis, our attack can be described as a bruteforce-like cryptanalysis [21] which is not able to conclude that a particular primitive has some cryptanalytic weakness, as in general it covers the whole cipher. However it can help to better understand the real security provided by the primitive when no attack tweaks are adopted. Most of the applications of bruteforce-like cryptanalysis have an advantage that is sometimes much smaller than a factor of 2. Nevertheless, for lightweight ciphers with key sizes of 80 bits or less, this is very useful to know, especially when the gain compared to the optimized bruteforce search is even a factor 2. To this end, designers of lightweight symmetric primitives should be motivated to consider this class of attacks during the assessment of any new proposed design.

Acknowledgment. The authors would like to thank the anonymous reviewers for their valuable comments and suggestions that helped improve the quality of the paper. This work is supported by the Natural Sciences and Engineering Research Council of Canada (NSERC).

References

1. AlTawy, R., Youssef, A.M.: Differential sieving for 2-step matching meet-in-the-middle attack with application to LBlock. In: Eisenbarth, T., Öztürk, E. (eds.) LightSec 2014. LNCS, vol. 8898, pp. 126–139. Springer, Heidelberg (2015)
2. Bar-On, A., Dinur, I., Dunkelman, O., Lallemand, V., Tsaban, B.: Improved analysis of zorro-like ciphers. Cryptology ePrint Archive, Report 2014/228 (2014), http://eprint.iacr.org/
3. Bogdanov, A., Chang, D., Ghosh, M., Sanadhya, S.K.: Bicliques with minimal data and time complexity for AES (extended version). Cryptology ePrint Archive, Report 2014/932 (2014), http://eprint.iacr.org/
4. Bogdanov, A., Kavun, E., Paar, C., Rechberger, C., Yalcin, T.: Better than brute-force–optimized hardware architecture for efficient biclique attacks on AES-128. In: ECRYPT Workshop, SHARCS-Special Purpose Hardware for Attacking Cryptographic Systems (2012)
5. Bogdanov, A., Khovratovich, D., Rechberger, C.: Biclique cryptanalysis of the full AES. In: Lee, D.H., Wang, X. (eds.) ASIACRYPT 2011. LNCS, vol. 7073, pp. 344–371. Springer, Heidelberg (2011)

6. Bogdanov, A., Knudsen, L.R., Leander, G., Paar, C., Poschmann, A., Robshaw, M.J., Seurin, Y., Vikkelsoe, C.: PRESENT: An ultra-lightweight block cipher. In: Paillier, P., Verbauwhede, I. (eds.) CHES 2007. LNCS, vol. 4727, pp. 450–466. Springer, Heidelberg (2007)
7. Bogdanov, A., Rechberger, C.: A 3-subset meet-in-the-middle attack: Cryptanalysis of the lightweight block cipher KTANTAN. In: Biryukov, A., Gong, G., Stinson, D.R. (eds.) SAC 2010. LNCS, vol. 6544, pp. 229–240. Springer, Heidelberg (2011)
8. Bouillaguet, C., Derbez, P., Dunkelman, O., Fouque, P.-A., Keller, N., Rijmen, V.: Low-data complexity attacks on AES. IEEE Transactions on Information Theory 58(11), 7002–7017 (2012)
9. Boura, C., Minier, M., Naya-Plasencia, M., Suder, V.: Improved impossible differential attacks against round-reduced LBlock. Cryptology ePrint Archive, Report 2014/279 (2014), http://eprint.iacr.org/
10. Canteaut, A., Naya-Plasencia, M., Vayssière, B.: Sieve-in-the-middle: Improved MITM attacks. In: Canetti, R., Garay, J.A. (eds.) CRYPTO 2013, Part I. LNCS, vol. 8042, pp. 222–240. Springer, Heidelberg (2013)
11. Chang, D., Ghosh, M., Sanadhya, S.: Biclique cryptanalysis of full round AES with reduced data complexity (2013)
12. Chen, J., Miyaji, A.: Differential cryptanalysis and boomerang cryptanalysis of LBlock. In: Cuzzocrea, A., Kittl, C., Simos, D.E., Weippl, E., Xu, L. (eds.) CD-ARES 2013 Workshops. LNCS, vol. 8128, pp. 1–15. Springer, Heidelberg (2013)
13. De Cannière, C., Dunkelman, O., Knežević, M.: KATAN and KTANTAN — A family of small and efficient hardware-oriented block ciphers. In: Clavier, C., Gaj, K. (eds.) CHES 2009. LNCS, vol. 5747, pp. 272–288. Springer, Heidelberg (2009)
14. Diffie, W., Hellman, M.: Exhaustive cryptanalysis of the NBS data encryption standard. Computer 10(6), 74–84 (1977)
15. Gérard, B., Grosso, V., Naya-Plasencia, M., Standaert, F.-X.: Block ciphers that are easier to mask: How far can we go? In: Bertoni, G., Coron, J.-S. (eds.) CHES 2013. LNCS, vol. 8086, pp. 383–399. Springer, Heidelberg (2013)
16. Guo, J., Peyrin, T., Poschmann, A., Robshaw, M.: The LED block cipher. In: Preneel, B., Takagi, T. (eds.) CHES 2011. LNCS, vol. 6917, pp. 326–341. Springer, Heidelberg (2011)
17. Karakoç, F., Demirci, H., Harmancı, A.E.: Impossible differential cryptanalysis of reduced-round lBlock. In: Askoxylakis, I., Pöhls, H.C., Posegga, J. (eds.) WISTP 2012. LNCS, vol. 7322, pp. 179–188. Springer, Heidelberg (2012)
18. Liu, Y., Gu, D., Liu, Z., Li, W.: Impossible differential attacks on reduced-round LBlock. In: Ryan, M.D., Smyth, B., Wang, G. (eds.) ISPEC 2012. LNCS, vol. 7232, pp. 97–108. Springer, Heidelberg (2012)
19. Mendel, F., Rijmen, V., Toz, D., Varıcı, K.: Differential analysis of the LED block cipher. In: Wang, X., Sako, K. (eds.) ASIACRYPT 2012. LNCS, vol. 7658, pp. 190–207. Springer, Heidelberg (2012)
20. Nakahara Jr., J., Sepehrdad, P., Zhang, B., Wang, M.: Linear (Hull) and algebraic cryptanalysis of the block cipher PRESENT. In: Garay, J.A., Miyaji, A., Otsuka, A. (eds.) CANS 2009. LNCS, vol. 5888, pp. 58–75. Springer, Heidelberg (2009)
21. Rechberger, C.: On bruteforce-like cryptanalysis: New meet-in-the-middle attacks in symmetric cryptanalysis. In: Kwon, T., Lee, M.-K., Kwon, D. (eds.) ICISC 2012. LNCS, vol. 7839, pp. 33–36. Springer, Heidelberg (2013)
22. Sasaki, Y., Wang, L.: Comprehensive study of integral analysis on 22-round lblock. In: Kwon, T., Lee, M.-K., Kwon, D. (eds.) ICISC 2012. LNCS, vol. 7839, pp. 156–169. Springer, Heidelberg (2013)

23. Sasaki, Y., Wang, L.: Meet-in-the-middle technique for integral attacks against Feistel ciphers. In: Knudsen, L.R., Wu, H. (eds.) SAC 2012. LNCS, vol. 7707, pp. 234–251. Springer, Heidelberg (2013)

24. Ahmadi, S., Ahmadian, Z., Mohajeri, J., Aref, M.R.: Low data complexity biclique cryptanalysis of block ciphers with application to Piccolo and HIGHT. Cryptology ePrint Archive, Report 2013/511 (2013), http://eprint.iacr.org/

25. Wang, Y., Wu, W., Yu, X., Zhang, L.: Security on LBlock against biclique cryptanalysis. In: Lee, D.H., Yung, M. (eds.) WISA 2012. LNCS, vol. 7690, pp. 1–14. Springer, Heidelberg (2012)

26. Wu, W., Zhang, L.: LBlock: a lightweight block cipher. In: Lopez, J., Tsudik, G. (eds.) ACNS 2011. LNCS, vol. 6715, pp. 327–344. Springer, Heidelberg (2011)

A Note on the Existence of Self-Dual Skew Codes over Finite Fields

Delphine Boucher[✉]

IRMAR, CNRS, Université de Rennes 1, UMR 6625,
Université européenne de Bretagne, 5 Bd Lannec, 35 000 Rennes, France
elphine.boucher@univ-rennes1.fr

Abstract. Conditions on the existence of self-dual θ-codes defined over a finite field $\mathbb{F}_q$ are studied for θ automorphism of $\mathbb{F}_q$. When $q \equiv 1$ (mod 4) it is proven that there always exists a self-dual θ-code in any dimension and that self-dual θ-codes of a given dimension are either all θ-cyclic or all θ-negacyclic. When $q \equiv 3$ (mod 4), there does not exist a self-dual θ-cyclic code and a necessary and sufficient condition for the existence of self-dual θ-negacyclic codes is given.

1 Introduction

Conditions for the existence of self-dual cyclic and negacyclic codes have been widely studied ([4], [6]) as well as for quasi-cyclic codes ([11], [12], [7]). In [3] a formula for the number of self-dual θ-cyclic codes and self-dual θ-negacyclic codes is given over $\mathbb{F}_{p^2}$ where p is a prime number and θ is the Frobenius automorphism. The aim of this text is to give conditions for the existence of self-dual θ-cyclic codes and θ-negacyclic codes over any finite field $\mathbb{F}_q$ where θ is an automorphism of $\mathbb{F}_q$.

The text is organized as follows. In Section 2, some facts about self-dual skew codes are recalled. In Section 3, the question of the existence of self-dual skew codes generated by skew binomials (Proposition 1) is studied. One deduces from this part that for $q \equiv 1$ (mod 4) there always exists a self-dual skew code in any dimension. In Section 4, a construction of self-dual skew codes over $\mathbb{F}_q$ using least common right multiples of skew polynomials and generalizing Proposition 28 of [2] is considered (Proposition 2). This proposition is used in Section 5 when $q \equiv 3$ (mod 4) to prove that there does not exist a self-dual θ-cyclic code in any dimension and to give a necessary and sufficient condition for the existence of self-dual θ-negacyclic codes (Proposition 4). Lastly when $q \equiv 1$ (mod 4), one proves that the sufficient conditions of existence of self-dual θ-cyclic and θ-negacyclic codes given by Proposition 1 are also necessary (Proposition 5). The results of Proposition 4 and Proposition 5 are summed up in Table 1.

2 Generalities on Self-dual Skew Codes

For a finite field $\mathbb{F}_q$ and θ an automorphism of $\mathbb{F}_q$ the ring R is defined by $R = \mathbb{F}_q[X; \theta] = \{a_n X^n + \ldots + a_1 X + a_0 \,|\, a_i \in \mathbb{F}_q$ and $n \in \mathbb{N}\}$ where addition

© Springer International Publishing Switzerland 2015
S. El Hajji et al. (Eds.): C2SI 2015, LNCS 9084, pp. 228–239, 2015.
DOI: 10.1007/978-3-319-18681-8_18

is defined to be the usual addition of polynomials and where multiplication is defined by the basic rule $X \cdot a = \theta(a) X$ $(a \in \mathbb{F}_q)$ and extended to all elements of R by associativity and distributivity. The noncommutative ring R is called a **skew polynomial ring** or Ore ring (cf. [13]) and its elements are **skew polynomials**. It is a left and right Euclidean ring whose left and right ideals are principal. Left and right gcd and lcm exist in R and can be computed using the left and right Euclidean algorithm. Recall that the center of R is the commutative polynomial ring $Z(R) = \mathbb{F}_q^\theta[X^{|\theta|}]$ where $\mathbb{F}_q^\theta$ is the fixed field of θ and $|\theta|$ is the order of θ. Below, module θ-codes are defined using the skew polynomial ring R.

Definition 1 (Definition 1 of [2]). *A* **module θ-code** *(or module skew code) $\mathcal{C}$ is a left R-submodule $Rg/Rf \subset R/Rf$ in the basis $1, X, \ldots, X^{n-1}$ where $g \in R = \mathbb{F}_q[X; \theta]$ and f is a left multiple of g in R of degree n. If there exists an $a \in \mathbb{F}_q \setminus \{0\}$ such that g divides $X^n - a$ on the right, then the code $\mathcal{C}$ is (θ,a)-*constacyclic*. If $a = 1$, the code is θ-*cyclic* and if $a = -1$, it is θ-*negacyclic*. The skew polynomial g is called* **skew generator polynomial** *of $\mathcal{C}$.*

If θ is the identity then a θ-cyclic (resp. θ-negacyclic) code is a cyclic code (resp. negacyclic) code. The **(Euclidean) dual** of a linear code C of length n over $\mathbb{F}_q$ is defined with the **Euclidean scalar product** $< x, y > = \sum_{i=1}^{n} x_i y_i$ in $\mathbb{F}_q^n$ as $C^\perp = \{x \in \mathbb{F}_q^n \mid \forall y \in C, < x, y > = 0\}$. A linear code C over $\mathbb{F}_q$ is **Euclidean self-dual** or **self-dual** if $C = C^\perp$. To characterize self-dual module θ-codes, the skew reciprocal polynomial of a skew polynomial (Definition 3 of [1]) and also the left monic skew reciprocal polynomial are used :

Definition 2 ([1], Definition 3). *The* **skew reciprocal polynomial** *of $h = \sum_{i=0}^{m} h_i X^i \in R$ of degree m is $h^* = \sum_{i=0}^{m} X^{m-i} \cdot h_i = \sum_{i=0}^{m} \theta^i(h_{m-i}) X^i$. The* **left monic skew reciprocal polynomial** *of h is $h^\natural := \frac{1}{\theta^m(h_0)} \cdot h^*$.*

Since θ is an automorphism, the map $*: R \to R$ given by $h \mapsto h^*$ is a bijection. In particular for any $g \in R$ there exists a unique $h \in R$ such that $g = h^*$ and, if g is monic, there exists a unique $h \in R$ such that $g = h^\natural$.

In order to describe some properties of the skew reciprocal polynomial, the morphism of rings $\Theta: R \to R$ given by $\sum_{i=0}^{n} a_i X^i \mapsto \sum_{i=0}^{n} \theta(a_i) X^i$ ([1], Lemma 1) is useful:

Lemma 1 (See also Lemma 1 of [1]). *Let f and g be skew polynomials in R and $n = \deg(f)$. Then*

1. $(fg)^* = \Theta^n(g^*) f^*$.
2. $(f^*)^* = \Theta^n(f)$.

According to Proposition 5 of [2], a self-dual θ-code must be either θ-cyclic or θ-negacyclic. Furthermore, according to Corollary 1 of [2], a module θ-code with skew generator polynomial $g \in \mathbb{F}_q[X; \theta]$ of degree k is self-dual if and only if there exists $h \in R$ (called **skew check polynomial** of the code) such that $g = h^\natural$ and

$$h^\natural h = X^{2k} - \varepsilon \text{ with } \varepsilon \in \{-1, 1\}. \tag{1}$$

Self-dual cyclic codes exist over $\mathbb{F}_q$ if and only if the characteristic of q is 2 (Theorem 3.3 of [9] or Theorem 1 of [8]). Necessary and sufficient conditions for the existence of self-dual negacyclic codes are given in [6] when $q \equiv 1 \pmod 4$ and in [4] when the dimension is a power of the characteristic of $\mathbb{F}_q$.

According to Theorem 18 of [14], a θ-cyclic code of length n is equivalent to a quasi-cyclic code of index ℓ where $\ell = \gcd(|\theta|, n)$. Therefore, as equivalence preserves self-duality, if there exists a self-dual θ-cyclic code of length n then there exists a self-dual quasi-cyclic code of length n and index $\ell = \gcd(|\theta|, n)$. According to Lemma 2.1 of [7], for m coprime with q, self-dual quasi-cyclic codes of index ℓ with length ℓm exist over a finite field $\mathbb{F}_q$ if and only if q is of characteristic 2 and $2|\ell$ or $q \equiv 1 \pmod 4$ and $2|\ell$ or $q \equiv 3 \pmod 4$ and $4|\ell$. Therefore, if there exists a self-dual θ-cyclic code over $\mathbb{F}_q$ with $n/\gcd(|\theta|, n)$ coprime with q, then $\gcd(|\theta|, n)$ must be even if q is a power of 2 or $q \equiv 1 \pmod 4$ and it must be divisible by 4 if $q \equiv 3 \pmod 4$. If the characteristic of $\mathbb{F}_q$ is equal to 2, then for all nonnegative integer k there exists a self-dual θ-code of length $2k$. Namely, the code $(X^k + 1)_{2k}^\theta$ is such a code as the relation (1) is satisfied for $h = X^k + 1$:

$$(X^k + 1)^{\natural}(X^k + 1) = (X^k + 1)(X^k + 1) = X^{2k} + 1.$$

In next section, necessary and sufficient conditions for the existence of self-dual codes generated by skew binomials are given when *the characteristic of $\mathbb{F}_q$ is odd*.

3   Self-dual Skew Codes Generated by Skew Binomials

Over a finite field of odd characteristic, there is no self-dual cyclic code ([8]). The example below shows that it is not the case for θ-cyclic codes when θ is not the identity.

Example 1. Consider $\mathbb{F}_{3^2} = \mathbb{F}_3(a)$ with $a^2 - a - 1 = 0$, $\alpha = a^2$ and $\theta : x \mapsto x^3$. The skew polynomial $X + \alpha \in \mathbb{F}_{3^2}[X; \theta]$ is the skew check polynomial of a self-dual θ-cyclic code : $(X + \alpha)^{\natural} = \frac{1}{\alpha^3}(1 + \alpha^3 X) = X + \alpha$ and

$$(X + \alpha)^{\natural}(X + \alpha) = (X + \alpha)(X + \alpha)$$
$$= X^2 + (\alpha + \alpha^3)X + \alpha^2$$
$$= X^2 - 1.$$

According to Section VI A of [11], this code is, up to equivalence, the unique self-dual code of length 2 over $\mathbb{F}_{3^2}$, its generator matrix is $(1, \alpha)$.

The following proposition gives a necessary and sufficient condition for the existence of self-dual θ-cyclic codes and self-dual θ-negacyclic codes defined over finite fields of odd characteristic and generated by skew binomials.

Proposition 1. *Assume that $\mathbb{F}_q$ is a finite field with $q = p^e$, p odd prime number, $e \in \mathbb{N}^*$. Consider $r \in \mathbb{N}$, θ the automorphism of $\mathbb{F}_q$ defined by $\theta : x \mapsto x^{p^r}$ and k a nonnegative integer.*

1. *There exists a self-dual θ-cyclic code over $\mathbb{F}_q$ of dimension k generated by a skew binomial if and only if $p \equiv 3 \pmod 4$, e is even and $r \times k$ is odd.*
2. *There exists a self-dual θ-negacyclic code over $\mathbb{F}_q$ of dimension k generated by a skew binomial if and only if $p \equiv 1 \pmod 4$ or $p \equiv 3 \pmod 4$, e is even and $r \times k$ is even.*

Proof. — Consider $h = X^k + \alpha \in R = \mathbb{F}_q[X; \theta]$ and $\epsilon = \pm 1$. The skew binomial h is the skew reciprocal polynomial of a self-dual (θ, ϵ)-constacyclic code if, and only if, h satisfies the relation (1) i.e.

$$\left(X^k + \frac{1}{\theta^k(\alpha)} \right) \cdot (X^k + \alpha) = X^{2k} - \epsilon.$$

Developping this skew polynomial relation, one gets the equivalent conditions

$$\theta^k(\alpha) + \epsilon\alpha = \alpha^2 + 1 = 0.$$

— One then proves that there exists $\alpha \in \mathbb{F}_q$ such that $\theta^k(\alpha) + \alpha = \alpha^2 + 1 = 0$ if and only if $p \equiv 3 \pmod 4$, e is even, r and k are odd. Let us assume that $p \equiv 3 \pmod 4$, $e \equiv 0 \pmod 2$ and $r, k \equiv 1 \pmod 2$. Then -1 is a square in $\mathbb{F}_q$ and one can consider $\alpha \in \mathbb{F}_q$ such that $\alpha^2 = -1$. As r and k are odd, $p^{kr} \equiv 3 \pmod 4$ so $p^{kr} - 1 \equiv 2 \pmod 4$ and $\frac{p^{kr}-1}{2} \equiv 1 \pmod 2$. Therefore $\alpha^{p^{kr}-1} = (\alpha^2)^{\frac{p^{kr}-1}{2}} = (-1)^{\frac{p^{kr}-1}{2}} = -1$ i.e. $\theta^k(\alpha) + \alpha = 0$. Conversely, consider α in $\mathbb{F}_q$ such that $\theta^k(\alpha) + \alpha = \alpha^2 + 1 = 0$. Assume that $p \equiv 1 \pmod 4$ then -1 is a square in $\mathbb{F}_p$ so α belongs to $\mathbb{F}_p$ and α is left fixed by θ. The equality $\theta^k(\alpha) + \alpha = 0$ implies that $2\alpha = 0$, which is impossible as p is odd. Therefore $p \equiv 3 \pmod 4$ and as -1 is a square in $\mathbb{F}_q$, e must be even. As $\theta^k(\alpha) + \alpha = 0 = \alpha^2 + 1$, one gets $-1 = \alpha^{p^{kr}-1} = (\alpha^2)^{\frac{p^{kr}-1}{2}} = (-1)^{\frac{p^{kr}-1}{2}}$ so $\frac{p^{kr}-1}{2}$ is odd, and $p^{kr} - 1 \equiv 2 \pmod 4$. As $p \equiv 3 \pmod 4$, kr must be odd.
— Lastly one proves that there exists $\alpha \in \mathbb{F}_q$ such that $\theta^k(\alpha) - \alpha = \alpha^2 + 1 = 0$ if and only if $p \equiv 1 \pmod 4$ or $(p \equiv 3 \pmod 4$, e and $r \times k$ are even). First if $p \equiv 1 \pmod 4$ then -1 is a square in $\mathbb{F}_p$. Consider α in $\mathbb{F}_p$ such that $\alpha^2 = -1$, then $\theta^k(\alpha) - \alpha = 0$ because $\alpha \in \mathbb{F}_p$ is left fixed by θ. If $p \equiv 3 \pmod 4$, $e \equiv 0 \pmod 2$ and $rk \equiv 0 \pmod 2$, then -1 has a square root in $\mathbb{F}_q$ and $p^{kr} - 1 \equiv 0 \pmod 4$. Consider $\alpha \in \mathbb{F}_q$ such that $\alpha^2 = -1$. Then $\alpha^{p^{kr}-1} = (\alpha^2)^{\frac{p^{kr}-1}{2}} = 1$ because $\alpha^2 = -1$ and $(p^{kr} - 1)/2$ is even. Conversely, consider $\alpha \in \mathbb{F}_q$ such that $\theta^k(\alpha) - \alpha = \alpha^2 + 1 = 0$. Therefore -1 is a square in $\mathbb{F}_q$ and either $p \equiv 1 \pmod 4$ or $p \equiv 3 \pmod 4$ and $e \equiv 0 \pmod 2$. If $p \equiv 3 \pmod 4$ and $rk \equiv 1 \pmod 2$ then $p^{kr} - 1 \equiv 2 \pmod 4$ so $(p^{kr} - 1)/2$ would be odd and $\alpha^{p^{kr}-1}$ would be equal to -1. As $\mathbb{F}_q$ has an odd characteristic, this contradicts the hypothesis $\alpha^{p^{kr}-1} = 1$. Therefore $p \equiv 1 \pmod 4$ or $p \equiv 3 \pmod 4$, $e \equiv 0 \pmod 2$ and $rk \equiv 0 \pmod 2$.

Corollary 1. *Assume that $\mathbb{F}_q$ is a finite field with $q = p^e$, p odd prime number, $e \in \mathbb{N}^*$ and $q \equiv 1 \pmod 4$. Consider $r \in \mathbb{N}$, θ the automorphism of $\mathbb{F}_q$ defined* *by $\theta : x \mapsto x^{p^r}$ and k a nonnegative integer. Then there exists a self-dual θ-code of dimension k.*

Proof. According to Proposition 1, if $p \equiv 1 \pmod 4$ there exists a self-dual θ-negacyclic code of dimension k; if $p \equiv 3 \pmod 4$ and $e \equiv 0 \pmod 2$, then there exists a self-dual θ-cyclic code of dimension k if $r \times k$ is odd and there exists a self-dual θ-negacyclic code of dimension k if $r \times k$ is even.

Example 2. Consider $\mathbb{F}_{3^2} = \mathbb{F}_3(a)$ with $a^2 - a - 1 = 0$ and $\theta : x \mapsto x^3$. For $k \in \mathbb{N}^*$, there exists α such that $X^k + \alpha$ is the skew check polynomial of a self-dual θ-cyclic code if and only if k is odd. In this case α satisfies $\alpha^2 + 1 = 0$ and $\theta^k(\alpha) + \alpha = 0$ i.e. $\alpha^2 + 1 = 0$ and $\alpha(\alpha^2 + 1) = 0$ (because $\theta^k = \theta$ if k is odd). So α must be equal to $\pm a^2$ (see Example 1).

Remark 1. When $r = 0$ (i.e. θ is the identity), Proposition 1 gives necessary and sufficient conditions of existence of self-dual cyclic and negacyclic codes generated by binomials over finite fields of odd characteristic. If $p \equiv 1 \pmod 4$ or $p \equiv 3 \pmod 4$ and $e \equiv 0 \pmod 2$, there always exists a self-dual negacyclic code of any dimension (see also Corollary 3.3 of [4] when the dimension is p^s for $s > 0$). This seems to contradict Example 3.8 of [6] which states that a "self-dual negacyclic code of length 70 over $\mathbb{F}_5$ does not exist" and that there "is no self-dual negacyclic code of length 30 over $\mathbb{F}_9$". Namely, over $\mathbb{F}_5$, $X^{35} + 2$ generates a self-dual negacyclic code of dimension 35 whereas over $\mathbb{F}_9$, $X^{15} + a$ (with $a^2 = -1$) generates a self-dual negacyclic code of dimension 15.

4   Self-dual Skew Codes Generated by Least Common Left Multiples of Skew Polynomials

The following Lemma is inspired from Theorem 16 and Theorem 18 of [14] which state that a θ-cyclic code is either a cyclic code or a quasi-cyclic code.

Lemma 2. *Consider $\mathbb{F}_q$ a finite field, $\theta \in \mathrm{Aut}(\mathbb{F}_q)$, $R = \mathbb{F}_q[X; \theta]$, n a nonnegative integer, ℓ the greatest common divisor of n and of the order of θ, $a \in (\mathbb{F}_q)^\theta \setminus \{0\}$ and h a right divisor of $X^n - a$ in R. Then $X^\ell h = hX^\ell$ (which means that the coefficients of h belong to the fixed field of θ^ℓ, $(\mathbb{F}_q)^{\theta^\ell}$).*

Proof. Consider m the order of θ, $u, v \in \mathbb{N}$ such that $\ell = mu - nv$. Consider $\frac{1}{a^v} X^{mu} h \in Rh/R(X^n - a)$, one has $\frac{1}{a^v} X^{mu} h = hX^{mu} \times \frac{1}{a^v} = hX^\ell X^{nv} \times \frac{1}{a^v} = hX^\ell$ in $R/R(X^n - a)$, therefore $hX^\ell \in Rh/R(X^n - a)$ and there exists $Q \in R$ monic of degree ℓ such that $hX^\ell = Qh$. The constant coefficient Q_0 of Q satisfies $Q_0 h_0 = 0$, as $h_0 \neq 0$, one gets $Q_0 = 0$. Furthermore, from the coefficients of degree $1, \ldots, \ell - 1$ of $hX^\ell - Qh$, one gets that the terms of Q with degrees $\leq \ell - 1$ all cancel, therefore $hX^\ell = X^\ell h$.

The following proposition is a generalization of Proposition 28 of [2] (where θ was of order 2).

Proposition 2. *Consider* $\mathbb{F}_q$ *a finite field,* $\theta \in \mathrm{Aut}(\mathbb{F}_q)$, $R = \mathbb{F}_q[X; \theta]$, k *a nonnegative integer and* ℓ *the greatest common divisor of* $2k$ *and of the order of* θ, $\tilde{R} = (\mathbb{F}_q)^{\theta^\ell}[X; \tilde{\theta}]$ *where* $\tilde{\theta}$ *is the restriction of* θ *to* $(\mathbb{F}_q)^{\theta^\ell}$. *Consider* $s \in \mathbb{N}$ *and* $t \in \mathbb{N}$ *not multiple of* p, *such that* $2k = \ell \times p^s \times t$. *Let* $\varepsilon \in \{-1, 1\}$ *and* $Y^t - \varepsilon = f_1(Y) f_2(Y) \cdots f_m(Y) \in (\mathbb{F}_q)^\theta[Y]$, *where* $f_i(Y)$ *are monic polynomials that are pairwise coprime with the property that* $f_j^{\natural} = f_i$. *The equation* $h^{\natural} h = X^{2k} - \varepsilon \in R$ *is equivalent to* $h^{\natural} h = X^{2k} - \varepsilon \in \tilde{R}$. *Its solutions are the skew polynomials* h *defined by* $h = \mathrm{lcrm}(h_1, \ldots, h_m) \in \tilde{R}$ *where for* $i = 1, \ldots, m$, $h_i^{\natural} h_i = f_i^{p^s}(X^\ell) \in \tilde{R}$.

Proof. According to Lemma 2, the equation $h^{\natural} h = X^{2k} - \epsilon$ in $R = \mathbb{F}_q[X; \theta]$ is equivalent to $h^{\natural} h = X^{2k} - \epsilon$ in $\tilde{R} = (\mathbb{F}_q)^{\theta^\ell}[X; \tilde{\theta}]$ where $(\mathbb{F}_q)^{\theta^\ell}$ is the fixed field of θ^ℓ and $\tilde{\theta}$ is the restriction of θ to $(\mathbb{F}_q)^{\theta^\ell}$. As $\tilde{\theta}^\ell$ fixes $(\mathbb{F}_q)^{\theta^\ell}$, the order of $\tilde{\theta}$ divides ℓ and therefore it divides $2k$.

Therefore in the following, without loss of generality, one can consider the equation $h^{\natural} h = X^{2k} - \epsilon$ in $R = \mathbb{F}_q[X; \theta]$ with $\theta \in \mathrm{Aut}(\mathbb{F}_q)$ of order ℓ dividing $2k$. The proof of Proposition 28 of [2] can be adapted to this context and not all details are given.

1. ($\Leftarrow$) From $h = \mathrm{lcrm}(h_1, \ldots, h_m)$ one obtains that $h = h_i q_i$ with $q_i \in R$. Lemma 1 shows that there exists $\tilde{q}_i \in R$ such that $h^{\natural} = \tilde{q}_i h_i^{\natural}$. Therefore $h^{\natural} h = \tilde{q}_i (h_i^{\natural} h_i) q_i = \tilde{q}_i f_i^{p^s}(X^\ell) q_i = \tilde{q}_i q_i f_i^{p^s}(X^\ell)$ (because $f_i^{p^s}(X^\ell) \in (\mathbb{F}_q)^\theta[X^\ell]$ is central), showing that

   $\mathrm{lclm}((f_1)^{p^s}(X^\ell), \ldots, (f_m)^{p^s}(X^\ell)) = f_1^{p^s}(X^\ell) \cdots f_m^{p^s}(X^\ell) = X^n - \varepsilon$ is a right divisor of $h^{\natural} h$ in R. Furthermore, the degree of h is equal to the sum of the degrees of the skew polynomials h_i (because they are pairwise coprime), therefore the degree of $h^{\natural} h$ is equal to $\sum_{i=1}^m \deg((f_i)^{p^s}(X^\ell)) = 2k$ which enables to conclude that $h^{\natural} h = X^{2k} - \epsilon$.

2. ($\Rightarrow$): According to ([5], Theorem 4.1), $h^{\natural} = \mathrm{lclm}(h_1^{\natural}, \ldots, h_m^{\natural})$ where $h_i^{\natural} = \mathrm{gcrd}(f_i^{p^s}(X^\ell), h^{\natural})$ are pairwise coprime in R. In particular, according to [13], $\deg(\mathrm{lclm}(h_i^{\natural}, h_j^{\natural})) = \deg(h_i^{\natural}) + \deg(h_j^{\natural})$ for $i \neq j$ and $\deg(h^{\natural}) = \deg(\mathrm{lclm}(h_i^{\natural})) = \sum \deg(h_i^{\natural})$.

   Let us now show that h_i divides $f_i^{p^s}(X^\ell)$ and h on the left :

   - Let δ_i be the degree of $f_i^{p^s}(X^\ell)$ and d_i be the degree of h_i. Applying Lemma 1 to $f_i^{p^s}(X^\ell) = q_i h_i^*$ one obtains $(f_i^{p^s}(X^\ell))^* = \Theta^{\delta_i - d_i}(h_i^{**}) q_i^* = \Theta^{\delta_i - d_i}(\Theta^{d_i}(h_i)) q_i^* = \Theta^{\delta_i}(h_i) q_i^* = h_i q_i^*$ (because δ_i is a multiple of the order ℓ of θ). One concludes that h_i divides on the left $(f_i^{p^s}(X^\ell))^*$ and h_i divides on the left $(f_i^{p^s})^{\natural}(X^\ell) = f_i^{p^s}(X^\ell)$.
   - Since $h_i^{\natural}$ divides $h^{\natural}$ on the right, $h^* = p_i h_i^*$ for some p_i in R. Using Lemma 1, one obtains $\Theta^k(h) = h^{**} = \Theta^{k - d_i}(h_i^{**}) p_i^*$. Therefore $\Theta^k(h) = \Theta^{k - d_i}(\Theta^{d_i}(h_i)) p_i^* = \Theta^k(h_i) p_i^*$. Since Θ is a morphism of rings, h_i divides h on the left.

   Since $h_i^{\natural}$ divides $h^{\natural}$ on the right and h_i divides h on the left, there exist $g_i, \tilde{g}_i$ such that $h^{\natural} h = \tilde{g}_i h_i^{\natural} h_i g_i$. Since two factors of a decomposition of the central polynomial $h^\natural h = \tilde{g}_i h_i^\natural h_i g_i$ into two factors commute, $h_i^\natural h_i$ divides $h^\natural h = X^n - \varepsilon$ on the right. According to Theorem 4.1 of [5], $h_i^\natural h_i = \mathrm{lclm}(\mathrm{gcrd}(h_i^\natural h_i, (f_j)^{p^s}(X^\ell)), j = 1, \ldots m)$. As both $h_i^\natural$ and h_i divide the central polynomial $f_i^{p^s}(X^\ell)$, the product $h_i^\natural h_i$ divides $(f_i^{p^s})^2(X^\ell)$. For $j \neq i$, $\mathrm{gcrd}(h_i^\natural h_i, (f_j)^{p^s}(X^\ell)) = 1$ and $h_i^\natural h_i = \mathrm{gcrd}(h_i^\natural h_i, f_i^{p^s}(X^\ell))$, in particular, $h_i^\natural h_i$ divides $f_i^{p^s}(X^\ell)$.

For $i \in \{1, \ldots, m\}$ the polynomials $f_i^{p^s}(X^\ell)$ are pairwise coprime, showing that their divisors $h_i^\natural h_i$ are also pairwise coprime. Therefore

$$\deg(\mathrm{lclm}(h_i^\natural h_i)) = \sum_{i=1}^{m} \deg(h_i^\natural h_i) = 2\deg(h^\natural) = \sum_{i=1}^{m} \deg(f_i^{p^s}(X^\ell)).$$

From $\sum_{i=1}^{m} \deg(h_i^\natural h_i) = \sum_{i=1}^{m} \deg(f_i^{p^s}(X^\ell))$ and the fact that $h_i^\natural h_i$ divides $f_i^{p^s}(X^\ell)$, we obtain $h_i^\natural h_i = f_i^{p^s}(X^\ell)$.

As h_i divides h on the left, $\mathrm{lcrm}(h_i, i = 1, \ldots, m)$ also divides h on the left. Since $\mathrm{gcrd}(h_i^\natural, h_j^\natural) = 1$ implies $\mathrm{gcld}(h_i, h_j) = 1$, one gets $\deg(\mathrm{lcrm}(h_i, i = 1, \ldots, m)) = \sum \deg(h_i) = \deg(h)$. Therefore $h = \mathrm{lcrm}(h_i, i = 1, \ldots, m)$.

Corollary 2. *Consider $\mathbb{F}_q$ a finite field with odd characteristic p, $\theta \in \mathrm{Aut}(\mathbb{F}_q)$, $R = \mathbb{F}_q[X; \theta]$. Consider ℓ the greatest common divisor of $2 \times k$ and of the order of θ, $k \in \mathbb{N}^*$, $s \in \mathbb{N}$ and $t \in \mathbb{N}$ not multiple of p such that $2 \times k = \ell \times p^s \times t$.*

1. *If the order of θ is odd then there does not exist a self-dual θ-cyclic code of dimension k over $\mathbb{F}_q$.*
2. *If the order of θ is odd and if $Y^t + 1 \in (\mathbb{F}_q)^\theta[Y]$ has a self-reciprocal irreducible factor of degree > 1, then there does not exist a self-dual θ-negacyclic code of dimension k over $\mathbb{F}_q$.*

Proof. 1. Assume that there is a self-dual θ-cyclic code of dimension k, then the equation $h^\natural h = X^{2k} - 1$ has a solution in R. Furthermore $Y - 1$ divides $Y^t - 1$ and is self-reciprocal, therefore, according to Proposition 2, the intermediate equation $H^\natural H = (X^l - 1)^{p^s}$ has a solution. But the order of θ is odd so ℓ is odd, therefore the right hand side of this intermediate equation has an odd degree which is impossible as the degree of the left hand side is even.

2. Consider $f(Y) = f^\natural(Y) \in (\mathbb{F}_q)^\theta[Y]$ irreducible dividing $Y^t + 1$, then the irreducible skew factors of $f(X^\ell)$ have the same degree as $\deg(f(Y))$ and therefore a factorization of $f(X^\ell)^{p^s}$ into irreducible skew polynomials has $\ell \times p^s$ factors of degree $\deg(f(Y))$. As the order of θ is odd, ℓ is odd and $\ell \times p^s$ is odd, therefore each factorization of $f(X^\ell)^{p^s}$ into the product of irreducible factors has an odd number of irreducible factors with the same degree. Consider $H \in R$ satisfying the intermediate equation $H^\natural H = f(X^\ell)^{p^s}$. The skew polynomials H and $H^\natural$ must have the same number of irreducible factors, with the same degree and dividing $f(X^\ell)^{p^s}$. This contradicts the fact that $f(X^\ell)^{p^s}$ has an odd number of irreducible factors with the same degree. Therefore, according to Proposition 2, the equation $h^\natural h = X^{2k} + 1$ has no solution in R.

Remark 2. According to Theorem 2.2 of [6], there does not exist a self-dual nega-cyclic code of length $2k$ over $\mathbb{F}_q$, with $\mathbb{F}_q$ of odd characteristic, if the polynomial $X^{2k} + 1 \in \mathbb{F}_q[X]$ has an irreducible factor f such that $f = f^\natural$.

From Proposition 1 one deduces that there cannot exist both a self-dual θ-cyclic code generated by a binomial and a self-dual θ-negacyclic code generated by a binomial and having the same dimension. The following proposition shows that more generally there cannot exist both a self-dual θ-cyclic and a self-dual θ-negacyclic code with the same dimension.

Proposition 3. *Consider $\mathbb{F}_q$ a finite field with odd characteristic p and θ an automorphism of $\mathbb{F}_q$. There cannot exist both a self-dual θ-cyclic code and a self-dual θ-negacyclic code with the same dimension over $\mathbb{F}_q$.*

Proof. Consider $k \in \mathbb{N}$, $\epsilon \in \{-1, 1\}$. According to Lemma 2, the equation $h^\natural h = X^{2k} - \epsilon$ in $R = \mathbb{F}_q[X; \theta]$ is equivalent to $h^\natural h = X^{2k} - \epsilon$ in $\tilde{R} = (\mathbb{F}_q)^{\theta^\ell}[X; \tilde{\theta}]$ where $(\mathbb{F}_q)^{\theta^\ell}$ is the fixed field of θ^ℓ and $\tilde{\theta}$ is the restriction of θ to $(\mathbb{F}_q)^{\theta^\ell}$. As $\tilde{\theta}^\ell$ fixes $(\mathbb{F}_q)^{\theta^\ell}$, the order of $\tilde{\theta}$ divides ℓ and therefore it divides $2k$. Therefore in the following, without loss of generality, one can consider that the order ℓ of $\theta \in Aut(\mathbb{F}_q)$ divides $2 \times k$. Consider $s \in \mathbb{N}$ and $t \in \mathbb{N}$ such that $2 \times k = \ell \times p^s \times t$ where t is not a multiple of p.

1. One first considers the particular case when $t = 1$ i.e. $2 \times k = \ell \times p^s$. Assume that there exists a self-dual θ-cyclic code of dimension k. Consider $h \in R$ monic such that $h^\natural h = X^{2k} - 1$ and α the constant coefficient of h. The skew polynomial $X^\ell - \epsilon$ belongs to $(\mathbb{F}_q)^\theta[X^\ell]$ therefore it is central of degree 1 in X^ℓ and the skew factors of any of its factorizations are all of degree 1. The skew polynomial $X^{2k} - \epsilon = (X^\ell - \epsilon)^{p^s}$ shares the same property. As h divides $X^{2k} - 1$ and as $X^{2k} - 1$ factors as a product of linear skew polynomials, a factorization of h is $h = (X - \alpha_1) \cdots (X - \alpha_k)$ where $\alpha_i \in \mathbb{F}_q$ and $X - \alpha_i$ divides on the right $X^{2k} - 1$ (because $X^{2k} - 1$ is central). According to Equation (11) of [10], one has $N_{2k}(\alpha_i) = 1$ where for $m \in \mathbb{N}^*$ and $u \in \mathbb{F}_q$, $N_m(u) := u\theta(u) \cdots \theta^{m-1}(u)$ is the norm of u. As $\alpha = (-1)^k \prod_{i=1}^k \alpha_i$, one gets $N_{2k}(\alpha) = 1$. Furthermore the constant term of $h^\natural h$ is equal to $\alpha/\theta^k(\alpha)$ therefore, $\theta^k(\alpha) = -\alpha$ and $1 = N_{2k}(\alpha) = (-1)^k N_k(\alpha)^2$. Similarly if there exists a self-dual θ-negacyclic code of dimension k, then there exists β in $\mathbb{F}_q$ such that $N_{2k}(\beta) = (-1)^k$, $\theta^k(\beta) = \beta$ and $N_k(\beta)^2 = (-1)^k$. If k is even then $N_k(\alpha)^2 = 1$, therefore $N_k(\alpha) = \pm 1$ so $\alpha^{\frac{p^k-1}{p-1}} = \pm 1$ and $\alpha^{p^k-1} = (\pm 1)^{p-1}$. As p is odd, one gets $\alpha^{p^k-1} = 1$, which contradicts $\theta^k(\alpha) = -\alpha$. If k is odd then $N_k(\beta)^2 = -1 = N_k(\alpha)^2$, so $N_k(\alpha) = \pm N_k(\beta)$ and $N_{2k}(\alpha) = N_k(\beta)^2 = -1$, which contradicts $N_{2k}(\alpha) = 1$. Therefore if $t = 1$, there cannot exist both a self-dual θ-cyclic code and a self-dual θ-negacyclic code with dimension k over $\mathbb{F}_q$.

2. Consider now the case when $t > 1$. If t is even, then $Y - 1$ and $Y + 1$ divides $Y^t - 1$ in $(\mathbb{F}_q)^\theta[Y]$. If there is a self-dual θ-cyclic code of dimension k then according to Proposition 2, the intermediate skew equation $h_1^\natural h_1 =$

$X^{\ell p^s} - 1$ and $h_2^\natural h_2 = X^{\ell p^s} + 1$ must have monic solutions $h_1, h_2 \in R$, which is impossible according to the first part of the proof. Therefore no self-dual θ-cyclic code of dimension k exists. If t is odd then $Y - 1$ divides $Y^t - 1$ and $Y + 1$ divides $Y^t + 1$ in $(\mathbb{F}_q)^\theta[Y]$. According to Proposition 2, if there is a self-dual θ-cyclic code of dimension k, then the skew equation $h_1^\natural h_1 = X^{\ell p^s} - 1$ must have a monic solution $h_1 \in R$. If there is a self-dual θ-negacyclic code of dimension k, then the skew equation $h_2^\natural h_2 = X^{\ell p^s} + 1$ must also have a monic solution $h_2 \in R$. This is impossible according to the first part of the proof.

5   Existence of Self-dual Skew Codes over Finite Fields with Odd Characteristic

According to Proposition 1, if $q \equiv 3 \pmod 4$, then there is no self-dual θ-code generated by skew binomials over $\mathbb{F}_q$. The following proposition gives a necessary and sufficient condition of existence of self-dual θ-codes when $q \equiv 3 \pmod 4$. The proof uses Corollary 2.

Proposition 4. *Assume that $\mathbb{F}_q$ is a finite field of characteristic p with $q \equiv 3 \pmod 4$. Consider θ an automorphism of $\mathbb{F}_q$ and $\mu \geq 2$ the biggest integer such that 2^μ divides $p + 1$ (i.e. 2^μ divides exactly $p + 1$).*

1. *There does not exist a self-dual θ-cyclic code of dimension k over $\mathbb{F}_q$.*
2. *There exists a self-dual θ-negacyclic code of dimension k over $\mathbb{F}_q$ if, and only if, $k \equiv 0 \pmod{2^{\mu-1}}$.*

Proof. Assume that $q = p^e \equiv 3 \pmod 4$ i.e. $p \equiv 3 \pmod 4$ and $e \equiv 1 \pmod 2$. Consider $r \in \mathbb{N}$ such that θ is defined by $x \mapsto x^{p^r}$.

1. The order of θ is $e / \gcd(e, r)$, therefore as e is odd, the order of θ is also odd. According to point 1. of Corollary 2, there cannot exist a self-dual θ-cyclic code of dimension k over $\mathbb{F}_q$.
2. Consider α the biggest integer such that 2^α divides k and assume that $\alpha + 1 \geq \mu$. Therefore $2k$ is divisible by 2^μ and the skew polynomial $X^{2k} + 1$ is equal to $(X^{k/2^{\mu-1}})^{2^\mu} + 1$. One proves that the polynomial $Y^{2^\mu} + 1$ factors in $\mathbb{F}_p[Y]$ as the product of two polynomials $h(Y)$ and $h^\natural(Y)$. Namely, consider w a primitive $2^{\mu+1}$-th root of unity in $\overline{\mathbb{F}_p}$. As 2^μ divides $p + 1$, $2^{\mu+1}$ divides $p^2 - 1$ and w belongs to $\mathbb{F}_{p^2} - \mathbb{F}_p$. The polynomial $Y^{2^\mu} + 1 = (Y^{2^{\mu+1}} - 1)/(Y^{2^\mu} - 1)$ factors in $\mathbb{F}_{p^2}[Y]$ as the product of $Y - w^i$ where i describes the odd numbers of $\{0, \ldots, 2^{\mu+1} - 1\}$. This polynomial can also be written as the product of the polynomials $h_i(Y)h_i^\natural(Y)$ where $h_i(Y) = Y^2 - (w^i + w^{ip})Y + w^{i(p+1)}$ is in $\mathbb{F}_p[Y]$. One concludes that $Y^{2^\mu} + 1$ factors in $\mathbb{F}_p[Y]$ as the product of two polynomials $h(Y)$ and $h^\natural(Y)$. From this factorization, one deduces that $X^{2k} + 1 = H^\natural(X)H(X) \in \mathbb{F}_p[X]$ where $H(X) = h(X^{k/2^{\mu-1}})$. So there exists a $[2k, k]_p$ self-dual negacyclic code and as $\mathbb{F}_p$ is fixed by θ, the relation $X^{2k} + 1 = H^\natural(X)H(X)$ still holds in $\mathbb{F}_q[X; \theta]$.

Conversely, assume that $\alpha < \mu - 1$. Consider ℓ the greatest common divisor of $2k$ and of the order of θ, and t, s such that $2k = \ell \times t \times p^s$ where p does not divide t. Let us prove that $Y^t + 1 \in (\mathbb{F}_q)^\theta[Y]$ has an irreducible factor $f(Y)$ such that $f^\natural(Y) = f(Y)$. Consider $e' = \gcd(e, r)$ and $q' = p^{e'}$, then $(\mathbb{F}_q)^\theta = \mathbb{F}_{q'}$. As e is odd, and as the order of θ is equal to $e/\gcd(e, r)$, the order of θ is odd and ℓ is also odd. As $p \equiv -1 \pmod{2^\mu}$ and $q' = p^{e'}$ with e' odd, $q' \equiv -1 \pmod{2^\mu}$, furthermore $\alpha \leq \mu - 2$, $q' \equiv -1 \pmod{4 \times 2^\alpha}$. Let us consider w a primitive $4 \times 2^\alpha$-th root of unity in $\mathbb{F}_{q'^2}$. Such an w does exist as $q'^2 - 1 \equiv 0 \pmod{4 \times 2^\alpha}$, furthermore $w^{q'} = w^{-1}$ because $4 \times 2^\alpha$ divides $q' + 1$. As $2^{\alpha+1}$ divides exactly $2k$ and as $2k = \ell \times t \times p^s$, $2^{\alpha+1}$ divides exactly t, therefore $4 \times 2^\alpha$ divides exactly $2t$, $w^{2t} = 1$ and $w^t = -1$. The minimal polynomial $f \in \mathbb{F}_{q'}[Y]$ of w divides $Y^{2t} - 1$ but not $Y^t - 1$, so it divides $Y^t + 1$. Furthermore $f(w^{q'}) = 0$ and $w^{q'} = w^{-1}$ therefore $f(w^{-1}) = 0$ and $f = f^\natural$. Therefore $Y^t + 1$ has an irreducible factor $f \in \mathbb{F}_{q'}[Y] = (\mathbb{F}_q)^\theta[Y]$ such that $f = f^\natural$. Furthermore, the order of θ is odd, so according to Corollary 2, there cannot exist a self-dual θ-negacyclic code of dimension k.

Remark 3. Assume that $p \equiv 3 \pmod{4}$, e is odd. Consider $\mu \geq 2$ the biggest integer such that 2^μ divides $p + 1$. Consider $k = p^s$ with $s \in \mathbb{N}$, then $k \not\equiv 0 \pmod{2^{\mu-1}}$ and according to Proposition 4, there is no negacyclic code of dimension k. This result was previously obtained in Corollary 3.3. of [4].

To conclude, it remains to decide, when $q \equiv 1 \pmod{4}$, if the existing self-dual θ-codes are θ-cyclic or θ-negacyclic. According to Theorem 1 of [3], over $\mathbb{F}_q = \mathbb{F}_{p^2}$ with $\theta : x \mapsto x^p$ and p prime number, there exists a self-dual θ-cyclic code of length $2k$ if and only if k is an odd number and $p \equiv 3 \pmod{4}$ whereas there exists a self-dual θ-negacyclic code of dimension k if and only if $p \equiv 1 \pmod{4}$ or $p \equiv 3 \pmod{4}$ and k is even. The following proposition generalizes this result and states that the sufficient conditions of existence of self-dual skew codes given in Proposition 1 for $q \equiv 1 \pmod{4}$ are also necessary. Its proofs uses Proposition 3 which states that there cannot exist simultaneously a self-dual θ-cyclic code and a self-dual θ-negacyclic code with the same dimension:

Proposition 5. *Consider a finite field $\mathbb{F}_q$ with $q = p^e$, p odd prime number, $e \in \mathbb{N}^*$ and $q \equiv 1 \pmod{4}$ (i.e. $p \equiv 3 \pmod{4}$ and e even or $p \equiv 1 \pmod{4}$). Consider $r \in \mathbb{N}$, θ the automorphism of $\mathbb{F}_q$ defined by $\theta : x \mapsto x^{p^r}$ and k a nonnegative integer.*

1. *There exists a self-dual θ-cyclic code of dimension k over $\mathbb{F}_q$ if and only if $p \equiv 3 \pmod{4}$, e is even and $r \times k$ is odd.*
2. *There exists a self-dual θ-negacyclic code of dimension k over $\mathbb{F}_q$ if and only if $p \equiv 1 \pmod{4}$ or $p \equiv 3 \pmod{4}$, e is even and $r \times k$ is even.*

Proof. 1. According to Proposition 1 point 1., if $p \equiv 3 \pmod{4}$, e is even and $r \times k$ is odd, there exists a self-dual θ-cyclic code of dimension k (generated

Table 1. Necessary and sufficient conditions for the existence of self-dual θ-cyclic and θ-negacyclic codes of dimension k over $\mathbb{F}_q$ where $\mathbb{F}_q$ has odd characteristic p, $\mu \in \mathbb{N}$ is such that 2^μ divides exactly $p+1$ and $\theta : x \mapsto x^{p^{p^r}}$

	Self-dual θ-cyclic	Self-dual θ-negacyclic
$q \equiv 1 \pmod 4$, $p \equiv 3 \pmod 4$	$r \times k \equiv 1 \pmod 2$	$r \times k \equiv 0 \pmod 2$
$q \equiv 1 \pmod 4$, $p \equiv 1 \pmod 4$	no k	$k \in \mathbb{N}^*$
$q \equiv 3 \pmod 4$	no k	$k \equiv 0 \pmod{2^{\mu-1}}$

by a skew binomial). Conversely, assume that there exists a self-dual θ-cyclic code of dimension k, then according to Proposition 3, there is no θ-negacyclic code with dimension k therefore according to Proposition 1 point 2., $p \equiv 3 \pmod 4$, $e \equiv 0 \pmod 2$ and $r \times k \equiv 1 \pmod 2$.

2. According to Proposition 1 point 2., if $p \equiv 3 \pmod 4$, $e \equiv 0 \pmod 2$ and $r \times k \equiv 0 \pmod 2$ or $p \equiv 1 \pmod 4$, there exists a self-dual θ-negacyclic code of dimension k (generated by a skew binomial). Conversely, assume that there exists a self-dual θ-negacyclic code of dimension k, then according to Proposition 3, there is no θ-cyclic code with dimension k therefore according to Proposition 1 point 1., $p \equiv 3 \pmod 4$, $e \equiv 0 \pmod 2$ and $r \times k \equiv 0 \pmod 2$ or $p \equiv 1 \pmod 4$.

To conclude, Proposition 4 ($q \equiv 3 \pmod 4$) and Proposition 5 ($q \equiv 1 \pmod 4$) are summed up in Table 1 below.

References

1. Boucher, D., Ulmer, F.: A note on the dual codes of module skew codes. In: Chen, L. (ed.) IMACC 2011. LNCS, vol. 7089, pp. 230–243. Springer, Heidelberg (2011)
2. Boucher, D., Ulmer, F.: Self-dual skew codes and factorization of skew polynomials. Journal of Symbolic Computation 60, 47–61 (2014)
3. Boucher, D.: Construction and number of self-dual skew codes over $\mathbb{F}_{p^2}$ (2014), https://hal.archives-ouvertes.fr/hal-01090922 (preprint)
4. Dinh Hai, Q.: Repeated-root constacyclic codes of length $2p^s$. Finite Fields and Their Applications 18, 133–143 (2012)
5. Giesbrecht, M.: Factoring in skew-polynomial rings over finite fields. J. Symbolic Comput. 26(4), 463–486 (1998)
6. Guenda, K., Gulliver, T.A.: Self-dual Repeated Root Cyclic and Negacyclic Codes over Finite Fields. In: 2012 IEEE International Syposium on Information Theory Proceedings (2012)

7. Han, S., Kim, J.-L., Lee, H., Lee, Y.: Construction of quasi-cyclic self-dual codes. Finite Fields and their Applications 18(3), 613–633 (2012)
8. Jia, S., Ling, S., Xing, C.: On Self-Dual Cyclic Codes Over Finite Fields. IEEE Transactions on Information Theory 57(4) (2011)
9. Kai, X., Zhu, S.: On Cyclic Self-Dual Codes. Applicable Algebra in Engineering, Communication and Computing 19(6), 509–525 (2008)
10. Lam, T.Y.: A general theory of Vandermonde matrices. Expositiones Mathematicae 4, 193–215 (1986)
11. Ling, S., Solé, P.: On the algebraic structure of quasi-cyclic codes. I. Finite fields. IEEE Trans. Inform. Theory 47(7), 2751–2760 (2001)
12. Ling, S., Solé, P.: On the algebraic structure of quasi-cyclic codes. II. Chain rings. Designs, Codes and Cryptography 30(1), 113–130 (2003)
13. Ore, O.: Theory of Non-Commutative Polynomials. The Annals of Mathematics, 2nd Ser. 34(3), 480–508 (1933)
14. Siap, I., Abualrub, T., Aydin, N., Seneviratne, P.: Skew cyclic codes of arbitrary length. Int. J. Inf. Coding Theory 2(1), 10–20 (2011)

The Weight Distribution of a Family of Lagrangian-Grassmannian Codes

Jesús Carrillo-Pacheco[1(✉)], Gerardo Vega[2], and Felipe Zaldívar[3]

[1] Academia de Matemáticas, Universidad Autónoma de la Ciudad de México,
09790, México, D.F., México
jesus.carrillo@uacm.edu.mx

[2] Dirección General de Cómputo y de Tecnologías de Información y Comunicación,
Universidad Nacional Autónoma de México,
04510 México D.F., México
gerardov@unam.mx

[3] Departamento de Matemáticas, Universidad Autónoma Metropolitana-I,
09340 México D.F., México
fz@xanum.uam.mx

Abstract. Using Plücker coordinates we construct a matrix whose columns parametrize all projective isotropic lines in a symplectic space E of dimension 4 over a finite field $\mathbb{F}_q$. As an application of this construction we explicitly obtain the smallest subfamily of algebro-geometric codes defined by the corresponding Lagrangian-Grassmannian variety. Furthermore, we show that this subfamily is a class of three-weight linear codes over $\mathbb{F}_q$ of length $(q^4 - 1)/(q - 1)$, dimension 5, and minimum Hamming distance $q^3 - q$.

Keywords: Algebraic geometry codes · Lagrangian-Grassmannian variety · $\mathbb{F}_p$-rational points · Isotropic lines · Lagrangian-Grassmannian codes · Three-weight linear codes

1 Introduction

We consider the subfamily $C_{L(2,4)(\mathbb{F}_q)}$ of the class of algebraic-geometry codes $C_{L(m,2m)(\mathbb{F}_q)}$ defined in [1] using the Lagrangian-Grassmannian projective variety $L(m, 2m)$ of maximal isotropic subspaces of a symplectic vector space of dimension $2m$ over a finite field $\mathbb{F}_q$. Thus, as a set, $L(2,4)$ is the family of 2-dimensional isotropic subspaces of a symplectic space E of dimension 4 over any finite field $\mathbb{F}_q$. By its definition, $L(2,4)$ is a subvariety of the Grassmannian $G(2,4)$ of all 2-dimensional vector subspaces of the 4-dimensional vector space E. Now, recall that the Plücker embedding $\pi : G(2,4) \to \mathbb{P}(\wedge^2 E)$ is given by sending a 2-dimensional vector subspace $W \subseteq E$ with basis u_1, u_2 to the point in $\mathbb{P}(\wedge^2 E) \simeq \mathbb{P}^5(\mathbb{F}_q)$ given by the wedge $u_1 \wedge u_2 \in \wedge^2(E)$. The map π is a closed embedding that identifies the Grassmannian $G(2,4)$ with the closed subvariety of $\mathbb{P}^5(\mathbb{F}_q)$ defined by the quadratic Plücker equations. Moreover, if $e_1, \ldots, e_4$ is

---

G. Vega – Partially supported by PAPIIT-UNAM IN107515.

a basis of E, by choosing the standard basis $\mathcal{B} = \{e_{i_1} \wedge e_{i_2} : 1 \leq i_1 < i_2 \leq 4\}$ of $\wedge^2 E$, given a 2-dimensional vector subspace $W \subseteq E$ with basis u_1, u_2, the coordinates of $\pi(W)$ in the standard basis $\mathcal{B}$ are the (homogeneous) Plücker coordinates of the point W. We label the coordinate corresponding to the vector $e_i \wedge e_j$ by X_{ij}. Thus, the coordinates of $\mathbb{P}(\wedge^2 E)$ are $X_{12}, X_{13}, X_{14}, X_{23}, X_{24}$ and X_{34}. It is well known, see for example [2], that the Grassmannian variety $G(2,4)$ corresponds to a hypersurface in $\mathbb{P}(\wedge^2 E)$, known as the Klein quadric, given by the quadratic equation

$$X_{12}X_{34} - X_{13}X_{24} + X_{14}X_{23} = 0, \tag{1}$$

where $X_{12}, X_{13}, X_{14}, X_{23}, X_{24}$ and X_{34} are the coordinates of $\mathbb{P}(\wedge^2 E)$.

Now, if ω is the non-degenerate skew-symmetric bilinear form of the symplectic 4-dimensional space E, recall that a subspace $W \subseteq E$ is *isotropic* iff $\omega(u,v) = 0$ for all $u, v \in W$. Then, the Lagrangian-Grassmannian $L(2,4)$ is given by the set

$$L(2,4) = \{W \subseteq E : W \text{ is an isotropic 2-dimensional subspace of } E\}$$

and it is a non-singular projective variety of dimension 3. Now, for the contraction map $f : \wedge^2 E \to \wedge^0 E = \mathbb{F}_q$, given by wedging with the tensor corresponding to the bilinear form, $f(v_1 \wedge v_2) := \omega(v_1, v_2)$, it is shown in [1] that, as algebraic varieties,

$$L(2,4) = G(2,4) \cap \mathbb{P}(\ker f).$$

Observe now that f is surjective and hence $\dim \ker f = 5$. As a direct application of [1, Corollary 7] we have that $\mathbb{P}(\ker f)$ is completely determined by the unique homogeneous linear equation $X_{14} + X_{23} = 0$. From Equation (1) it follows that $L(2,4) = G(2,4) \cap \mathbb{P}(\ker f)$ is the hypersurface given by the homogeneous quadric equation

$$X_{12}X_{34} - X_{13}X_{24} - X_{14}^2 = 0. \tag{2}$$

Note that each $\mathbb{F}_q$-rational solution to Equation 2 represents an isotropic line in $\mathbb{P}(E)$. So to fully determine the set of isotropic lines in the projectivization of a vector space of dimension 4 over a finite field $\mathbb{F}_q$ it is enough to give all $\mathbb{F}_q$-solutions to the quadratic equation 2. Lastly, since the symplectic group $\text{Sp}(4)(\mathbb{F}_q)$ acts transitively [7] on $L(2,4)(\mathbb{F}_q)$, then for any point $X \in L(2,4)(\mathbb{F}_q)$ we have that

$$n := |L(2,4)(\mathbb{F}_q)| = \frac{|\text{Sp}(4)(\mathbb{F}_q)|}{|\text{Stabilizer of } X|} = (1+q)(1+q^2). \tag{3}$$

In [1, Corollary 3] it is shown that $L(2,4)(\mathbb{F}_q)$ is a non-degenerate projective system in $\mathbb{P}(\ker f) \simeq \mathbb{P}^4$. Thus, following [8], see also [4,5], $L(2,4)(\mathbb{F}_q)$ defines a non-degenerate linear code $C_{L(2,4)(\mathbb{F}_q)}$ by choosing for each point of $L(2,4)(\mathbb{F}_q) \subseteq \mathbb{P}^4(\mathbb{F}_q)$ a representative in $\mathbb{F}^5$, then the generator matrix M of the code is the $5 \times n$ matrix with columns the representatives of the $n = (1+q)(1+q^2)$ points in $L(2,4)(\mathbb{F}_q)$. Thus, $C_{L(2,4)(\mathbb{F}_q)}$ is a 5-dimensional vector subspace of $\mathbb{F}^n$ and it is the image of the linear map from the dual space $(\mathbb{F}^5)^*$ to $\mathbb{F}^n$ given by evaluation of each linear functional on the points of $L(2,4)(\mathbb{F}_q)$. The length of $C_{L(2,4)(\mathbb{F}_q)}$ is $n = |L(2,4)(\mathbb{F}_q)| = (1+q)(1+q^2)$, its dimension is 5, and its minimum distance is given by the formula

$$d(C_{L(2,4)(\mathbb{F}_q)}) = n - \max\{|L(2,4)(\mathbb{F}_q) \cap H| : H \text{ is a hyperplane of } \mathbb{P}^4\}.$$

Recall now, that since the family of codes $C_{L(2,4)(\mathbb{F}_q)}$ is a particular case of quadrics codes, their weight spectrum has been completely described in [3], including the complete generalized spectrum.

Our main contribution in this paper is a different, explicit determination of the weight distribution of the codes $C_{L(2,4)(\mathbb{F}_q)}$. Specifically, we prove that these codes are three-weight linear codes (over $\mathbb{F}_q$) of length $(q^4-1)/(q-1)$, dimension 5 and minimum Hamming distance $q^3 - q$. On the other hand, our contribution can also be seen, in essence, as an algorithm to compute all $\mathbb{F}_q$-rational points in a special algebraic variety, for all finite fields $\mathbb{F}_q$.

The paper is organized as follows. In Section 2 we determine the general conditions under which the generator matrix for the code $C_{L(2,4)(\mathbb{F}_q)}$, can be constructed. In Section 3, we use this generator matrix in order to determine the weight distribution for the code $C_{L(2,4)(\mathbb{F}_q)}$. Finally, we devote Section 4 to conclusions and point to further work.

2   Projective Isotropic Lines in a Symplectic Space of Dimension 4 over any Finite Field

In this section we will give a general construction for $L(2,4)(\mathbb{F}_p)$, giving a matrix whose columns are the points of $L(2,4)(\mathbb{F}_p)$.

Let γ be a primitive element of $\mathbb{F}_q$ and let $\alpha \in \mathbb{F}_q$ be an arbitrary element. Consider the following vectors Q and $\overline{\alpha}$ in $\mathbb{F}_q^q$ given by $Q := (0, \gamma^0, \gamma^1, \cdots, \gamma^{q-2})$ and $\overline{\alpha} := (\alpha, \cdots, \alpha)$. For elements $\alpha, \beta \in \mathbb{F}_q$, and the corresponding vectors $\overline{\alpha}, \overline{\beta} \in \mathbb{F}_q^q$, we define the vector $\varepsilon_\beta^\alpha \in \mathbb{F}_q^q$ by

$$\varepsilon_\beta^\alpha = \overline{\alpha} * Q + \overline{\beta}^2 , \tag{4}$$

where $*$ denotes the componentwise product of two vectors in $\mathbb{F}_q^q$ and $\overline{\beta}^2 = \overline{\beta} * \overline{\beta}$. With this notation we now concatenate sets of q vectors in $\mathbb{F}_q^q$, in order to construct vectors in $\mathbb{F}_q^{q^2}$ in the following way:

$$\mathcal{A}_\alpha = (\overline{1}, \cdots, \overline{1}) \text{ for each } \alpha \in \mathbb{F}_q \text{ and } \mathcal{A}_\infty = (\overline{0}, \cdots, \overline{0}) ,$$

$$\mathcal{B}_\alpha = (\overline{\alpha}, \cdots, \overline{\alpha}) \text{ for each } \alpha \in \mathbb{F}_q \text{ and } \mathcal{B}_\infty = (\overline{1}, \cdots, \overline{1}) ,$$

$$\mathcal{C}_\alpha = (\overline{0}, \overline{\gamma^0}, \cdots, \overline{\gamma^{q-2}}) \text{ for each } \alpha \in \mathbb{F}_q \text{ and } \mathcal{C}_\infty = (\overline{0}, \overline{\gamma^0}, \cdots, \overline{\gamma^{q-2}}) ,$$

$$\mathcal{D}_\alpha = (Q, \cdots, Q) \text{ for each } \alpha \in \mathbb{F}_q \text{ and } \mathcal{D}_\infty = (-\overline{0}^2, -\overline{\gamma^0}^2, \cdots, -\overline{\gamma^{q-2}}^2) ,$$

$$\mathcal{E}_\alpha = (\varepsilon_0^\alpha, \varepsilon_{\gamma^0}^\alpha, \cdots, \varepsilon_{\gamma^{q-2}}^\alpha) \text{ for each } \alpha \in \mathbb{F}_q \text{ and } \mathcal{E}_\infty = (Q, \cdots, Q) .$$

Finally, we construct the row vectors $\mathcal{A}, \mathcal{B}, \mathcal{C}, \mathcal{D}$ and $\mathcal{E}$ in $\mathbb{F}_q^{(q^4-1)/(q-1)}$ by

$$\mathcal{A} = (\mathcal{A}_0, \mathcal{A}_{\gamma^0}, \cdots, \mathcal{A}_{\gamma^{q-2}}, \mathcal{A}_\infty, \overline{0}, 0),$$
$$\mathcal{B} = (\mathcal{B}_0, \mathcal{B}_{\gamma^0}, \cdots, \mathcal{B}_{\gamma^{q-2}}, \mathcal{B}_\infty, \overline{0}, 0),$$
$$\mathcal{C} = (\mathcal{C}_0, \mathcal{C}_{\gamma^0}, \cdots, \mathcal{C}_{\gamma^{q-2}}, \mathcal{C}_\infty, \overline{0}, 0),$$
$$\mathcal{D} = (\mathcal{D}_0, \mathcal{D}_{\gamma^0}, \cdots, \mathcal{D}_{\gamma^{q-2}}, \mathcal{D}_\infty, \overline{1}, 0),$$
$$\mathcal{E} = (\mathcal{E}_0, \mathcal{E}_{\gamma^0}, \cdots, \mathcal{E}_{\gamma^{q-2}}, \mathcal{E}_\infty, Q, 1).$$

Consider now the matrix G given by

$$G = \begin{pmatrix} \mathcal{A} \\ \mathcal{B} \\ \mathcal{C} \\ \mathcal{D} \\ \mathcal{E} \end{pmatrix} = \begin{pmatrix} \mathcal{A}_0 & \mathcal{A}_{\gamma^0} & \cdots & \mathcal{A}_{\gamma^{q-2}} & \mathcal{A}_\infty & \overline{0} & 0 \\ \mathcal{B}_0 & \mathcal{B}_{\gamma^0} & \cdots & \mathcal{B}_{\gamma^{q-2}} & \mathcal{B}_\infty & \overline{0} & 0 \\ \mathcal{C}_0 & \mathcal{C}_{\gamma^0} & \cdots & \mathcal{C}_{\gamma^{q-2}} & \mathcal{C}_\infty & \overline{0} & 0 \\ \mathcal{D}_0 & \mathcal{D}_{\gamma^0} & \cdots & \mathcal{D}_{\gamma^{q-2}} & \mathcal{D}_\infty & \overline{1} & 0 \\ \mathcal{E}_0 & \mathcal{E}_{\gamma^0} & \cdots & \mathcal{E}_{\gamma^{q-2}} & \mathcal{E}_\infty & Q & 1 \end{pmatrix} =$$

$$\begin{pmatrix} \overline{1}\,\overline{1} & \cdots & \overline{1} & \overline{1}\,\overline{1} & \cdots & \overline{1} & \overline{1} & \overline{1} & \cdots & \overline{1} & \overline{0} & \overline{0} & \cdots & \overline{0} & \overline{0}\,0 \\ \overline{0}\,\overline{0} & \cdots & \overline{0} & \gamma^0\,\gamma^0 & \cdots & \gamma^0 & \gamma^{q-2} & \gamma^{q-2} & \cdots & \gamma^{q-2} & \overline{1} & \overline{1} & \cdots & \overline{1} & \overline{0}\,0 \\ \overline{0}\,\gamma^0 & \cdots & \gamma^{q-2} & \overline{0}\,\gamma^0 & \cdots & \gamma^{q-2} & \cdots & \overline{0} & \gamma^0 & \cdots & \gamma^{q-2} & \overline{0} & \gamma^0 & \cdots & \gamma^{q-2} & \overline{0}\,0 \\ Q\,Q & \cdots & Q & Q\,Q & \cdots & Q & Q & Q & \cdots & Q & -\overline{0}^2 & -\overline{\gamma^0}^2 & \cdots & -\overline{\gamma^{q-2}}^2 & \overline{1}\,0 \\ \varepsilon_0^0\,\varepsilon_{\gamma^0}^0 & \cdots & \varepsilon_{\gamma^{q-2}}^0 & \varepsilon_0^{\gamma^0}\,\varepsilon_{\gamma^0}^{\gamma^0} & \cdots & \varepsilon_{\gamma^{q-2}}^{\gamma^0} & \varepsilon_0^{\gamma^{q-2}} & \varepsilon_{\gamma^0}^{\gamma^{q-2}} & \cdots & \varepsilon_{\gamma^{q-2}}^{\gamma^{q-2}} & Q & Q & \cdots & Q & Q\,1 \end{pmatrix}$$

whose columns correspond to points of the projective space $\mathbb{P}_{\mathbb{F}_q}^4$. It is easily verified that the entries of each column $c_i = (c_{i,1}, c_{i,2}, c_{i,3}, c_{i,4}, c_{i,5})$ of G satisfy $c_{i,1}c_{i,5} = c_{i,2}c_{i,4} + c_{i,3}^2$, that is they are all solutions of the quadratic equation 2. Hence, the matrix G parametrizes all projective isotropic lines on a symplectic vector space of dimension 4 over $\mathbb{F}_q$. In other words, this matrix gives a method to determine all $\mathbb{F}_q$-rational points in the projective variety $L(2,4)$, for an arbitrary finite field.

Example 1. if $q = 2$, then $\mathbb{F}_2 = \{0,1\}$ and

$$G = \begin{pmatrix} 11 & 11 & 11 & 11 & 00 & 00 & 00 & 0 \\ 00 & 00 & 11 & 11 & 11 & 11 & 00 & 0 \\ 00 & 11 & 00 & 11 & 00 & 11 & 00 & 0 \\ 01 & 01 & 01 & 01 & 00 & 11 & 11 & 0 \\ 00 & 11 & 01 & 10 & 01 & 01 & 01 & 1 \end{pmatrix}.$$

Example 2. If $q = 3$, then $\mathbb{F}_3 = \{0,1,2\}$ and

$$G = \begin{pmatrix} 111\,111\,111 & 111\,111\,111 & 111\,111\,111 & 000\,000\,000 & 000 & 0 \\ 000\,000\,000 & 111\,111\,111 & 222\,222\,222 & 111\,111\,111 & 000 & 0 \\ 000\,111\,222 & 000\,111\,222 & 000\,111\,222 & 000\,111\,222 & 000 & 0 \\ 012\,012\,012 & 012\,012\,012 & 012\,012\,012 & 000\,222\,222 & 111 & 0 \\ 000\,111\,111 & 012\,120\,120 & 021\,102\,102 & 012\,012\,012 & 012 & 1 \end{pmatrix}.$$

3 $C_{L(2,4)}(\mathbb{F}_q)$ is a Class of Three-Weight Linear Codes

If w_H is the Hamming weight of a finite-length vector, then, from Equation 4, we have that $w_H(\varepsilon_0^0) = 0$, $w_H(\varepsilon_\alpha^0) = q$ and $w_H(\varepsilon_\beta^\alpha) = q-1$, for all $\alpha \in \mathbb{F}_q^*$ and $\beta \in \mathbb{F}_q$. Therefore, $w_H(\mathcal{A}) = w_H(\mathcal{B}) = w_H(\mathcal{D}) = w_H(\mathcal{E}) = q^3$ and $w_H(\mathcal{C}) = q^3 - q$.

Since G is the generator matrix for the code $C_{L(2,4)}$, then for each $X \in C_{L(2,4)}$ there must exist scalars a, b, c, d and e in the finite field $\mathbb{F}_q$ such that $X = a\mathcal{A} + b\mathcal{B} + c\mathcal{C} + d\mathcal{D} + e\mathcal{E}$. Clearly, if for each $\alpha \in \mathbb{F}_q$ we construct the vectors $X_\alpha = a\mathcal{A}_\alpha + b\mathcal{B}_\alpha + c\mathcal{C}_\alpha + d\mathcal{D}_\alpha + e\mathcal{E}_\alpha \in \mathbb{F}_q^{q^2}$, and if we take $X_\infty = a\mathcal{A}_\infty + b\mathcal{B}_\infty + c\mathcal{C}_\infty + d\mathcal{D}_\infty + e\mathcal{E}_\infty \in \mathbb{F}_q^{q^2}$, then $X = (X_0, X_{\gamma^0}, \cdots, X_{\gamma^{q-2}}, X_\infty, \overline{d} + eQ, e)$.

Lemma 1. $C_{L(2,4)}(\mathbb{F}_q)$ *is a three-weight linear code over* $\mathbb{F}_q$ *of length* $(q^4 - 1)/(q - 1)$ *and dimension 5, with nonzero weights* q^3, $q^3 - q$ *and* $q^3 + q$.

Proof. We must prove that $w_H(X) \in \{0, q^3, q^3 - q, q^3 + q\}$. With the previous notation observe first that $w_H(a\mathcal{A} + b\mathcal{B} + c\mathcal{C}) \in \{0, q^3, q^3 - q\}$. In fact, $w_H(a\mathcal{A} + b\mathcal{B} + c\mathcal{C}) = w_H(\mathcal{C}) = q^3 - q$ if and only if $c \neq 0$. Thus, for any scalars a, b and c, we may assume that $d \neq 0$ or $e \neq 0$.

First Case: $d \neq 0$ and $e = 0$. Without loss of generality we may assume that $d = 1$. For $\alpha, \beta \in \mathbb{F}_q$, let $\delta_\beta^\alpha \in \mathbb{F}_q$ such that $\delta_\beta^\alpha = a + b\alpha + c\beta$. Then, $X_\alpha = (\overline{\delta_0^\alpha} + Q, \overline{\delta_{\gamma^0}^\alpha} + Q, \cdots, \overline{\delta_{\gamma^{q-2}}^\alpha} + Q)$. Since $Q = (0, \gamma^0, \gamma^1, \cdots, \gamma^{q-2})$, there must exist permutations $\sigma_1, \cdots, \sigma_q \in S_q$ (the *symmetric group* on q entries) such that $X_\alpha = (\sigma_1(Q), \cdots, \sigma_q(Q))$ and therefore $w_H(X_\alpha) = q(q - 1)$, for each $\alpha \in \mathbb{F}_q$. Now, let $p_1(z) \in \mathbb{F}_q[z]$ be given by $p_1(z) = b + cz - z^2$. Since $X_\infty = (\overline{b} + \overline{c0} - \overline{0}^2, \overline{b} + \overline{c\gamma^0} - \overline{\gamma^0}^2, \cdots, \overline{b} + \overline{c\gamma^{q-2}} - \overline{\gamma^{q-2}}^2)$, then

$$w_H(X_\infty) = \begin{cases} q^2, & \text{if } p_1(z) \text{ has no roots in } \mathbb{F}_q, \\ q^2 - q, & \text{if } p_1(z) \text{ has one root in } \mathbb{F}_q, \\ q^2 - 2q, & \text{if } p_1(z) \text{ has two roots in } \mathbb{F}_q. \end{cases}$$

Clearly $w_H(\overline{1}) = q$ and therefore $w_H(X) \in \{q^3, q^3 - q, q^3 + q\}$.

Second Case: $e \neq 0$. Without loss of generality we may assume that $e = 1$. By an argument similar to the previous case, we see that $w_H(X_\infty) = q(q - 1)$. For $\alpha \in \mathbb{F}_q$, let $\xi_\alpha \in \mathbb{F}_q$ such that $\xi_\alpha = a + b\alpha$. Then, from Equation (4) we have

$$X_\alpha = ((\overline{d + \alpha}) * Q + \overline{\xi_\alpha} + \overline{c0} + \overline{0}^2, (\overline{d + \alpha}) * Q + \overline{\xi_\alpha} + \overline{c\gamma^0} + \overline{\gamma^0}^2, \cdots,$$
$$(\overline{d + \alpha}) * Q + \overline{\xi_\alpha} + \overline{c\gamma^{q-2}} + \overline{\gamma^{q-2}}^2).$$

Now, if $d + \alpha \neq 0$ there must exist permutations $\sigma_1, \cdots, \sigma_q \in S_q$ such that $X_\alpha = (\sigma_1(Q), \cdots, \sigma_q(Q))$ and therefore $w_H(X_\alpha) = q(q - 1)$. On the contrary, if $d + \alpha = 0$, and if we take $p_2(z) \in \mathbb{F}_q[z]$ such that $p_2(z) = \xi_\alpha + cz + z^2$, then

$$w_H(X_\alpha) = \begin{cases} q^2, & \text{if } p_2(z) \text{ has no roots in } \mathbb{F}_q, \\ q^2 - q, & \text{if } p_2(z) \text{ has one root in } \mathbb{F}_q, \\ q^2 - 2q, & \text{if } p_2(z) \text{ has two roots in } \mathbb{F}_q. \end{cases}$$

Since there is just one $\alpha \in \mathbb{F}_q$ such that $d + \alpha = 0$, and since $w_H(\overline{d} + Q, 1) = q$, then $w_H(X) \in \{q^3, q^3 - q, q^3 + q\}$.   □

We now recall the following known identities: Let C be a linear code of length n and dimension k over $\mathbb{F}_q$. Let $w_1, w_2, \ldots, w_N$ be the nonzero Hamming weights of C, and suppose that the minimum Hamming distance of $C^\perp$ (the dual of C) is at least 3. For $1 \leq i \leq N$, let A_{w_i} be the number of words of weight w_i in C. The set $\{A_{w_i}\}_{i=1}^N$ is called the weight distribution of the linear code C. The first three identities of Pless (see [6] for the general result, alternatively see [10]) for the code C are:

1) $\sum_{i=1}^N A_{w_i} = q^k - 1$,
2) $\sum_{i=1}^N A_{w_i} w_i = n(q-1)q^{k-1}$,
3) $\sum_{i=1}^N A_{w_i} w_i^2 = n(q-1)(n(q-1)+1)q^{k-2}$.

With the previous notation and identities, our main result is:

Theorem 1. $C_{L(2,4)(\mathbb{F}_q)}$ is a three-weight linear code over $\mathbb{F}_q$ of length $(q^4 - 1)/(q-1)$ and dimension 5, whose weight distribution is as follows:

$$A_{q^3} = q^4 - 1,$$
$$A_{q^3 - q} = q^2(q^2 + 1)(q-1)/2,$$
$$A_{q^3 + q} = q^2(q^2 - 1)(q-1)/2.$$

Proof. As seen before, the generator matrix for the code $C_{L(2,4)(\mathbb{F}_q)}$ corresponds to the parity check matrix of a Hamming code of order 5 with some of its columns deleted. Therefore this code is necessarily a *projective* one, which in turn means (by definition) that the minimum Hamming distance of its dual code is at least 3. The result now follows from Lemma 1 and a direct application of Pless identities with $n = (q^4-1)/(q-1)$, $k = 5$, $N = 3$, $w_1 = q^3$, $w_2 = q^3-q$ and $w_3 = q^3+q$.   □

Example 3. If $q = 3$, then $C_{L(2,4)(\mathbb{F}_q)}$ is a three-weight linear code over $\mathbb{F}_3$ of length 40, dimension 5 and minimum Hamming distance 24, whose weight distribution is $A_{27} = 80$, $A_{24} = 90$ and $A_{30} = 72$.

Remark 1. It is important to observe that the previous code has the best known parameters according to the tables of linear codes [11] maintained by Markus Grassl.

Remark 2. As pointed out by a referee, the weights of $C_{L(2,4)}$ are the cardinalities of the complements of hyperplane sections of the non-degenerate parabolic quadric 2 in $\mathbb{P}^4$. These intersections have three possibilities: a cone, a hyperbolic quadric or an elliptic quadric, which give the three non-zero weights of $C_{L(2,4)}$.

4 Conclusion

We provide a general method to construct all $\mathbb{F}_q$-rational solutions of the Plücker equation of the parabolic quadric $L(2, 4)$. With this method, we studied the sub-family of codes $C_{L(2,4)(\mathbb{F}_q)}$ contained in the family of Lagrangian-Grassmannian codes $C_{L(m,2m)(\mathbb{F}_q)}$. We have shown that such a subfamily is a class of three-weight linear codes over $\mathbb{F}_q$ with parameters $[(q^4 - 1)/(q - 1), 5, q^3 - q]$, whose weight distribution is given by Theorem 1. An obvious open problem is to find the parameters for other subfamilies of Lagrangian-Grassmannian codes $C_{L(m,2m)(\mathbb{F}_q)}$.

Acknowledgements. We would like to thank the referees for a careful reading of the manuscript.

References

1. Carrillo-Pacheco, J., Zaldivar, F.: On Lagrangian-Grassmannian Codes. Des. Codes and Cryptogr. 60, 291–298 (2011)
2. Fulton, W.: Young Tableaux, with Applications to Representation Theory and Geometry. Cambridge University Press, Cambridge (1997)
3. Nogin, D.: Generalized Hamming Weights of Codes on Multi-Dimensional Quadrics. Problems Inform. Transmission 29(3), 218–227 (1993)
4. Nogin, D.: Codes Associated to Grassmannians. In: Arithmetic Geometry and Coding Theory (Luminy 1993), pp. 145–154. Walter de Gruyter, Berlin-New York (1996)
5. Nogin, D.: The Spectrum of Codes Associated with the Grassmannian Variety $G(3,6)$. Problems Inform. Transmission 33(3), 114–123 (1997)
6. Pless, V.: Power Moment Identities on Weight Distributions in Error-Correcting Codes. Inf. Contr. 6, 147–152 (1962)
7. Suzuki, M.: Group Theory I. Springer, Berlin (1982)
8. Tsfasman. M. A., Vladut, S.G.: Algebraic-Geometric Codes. Kluwer, Dordrecht (1991)
9. Wan, Z.X.: The Weight Hierarchies of the Projective Codes from Non-degenerate Quadrics. Des. Codes Cryptogr. 4(3), 283–300 (1994)
10. Wolfmann, J.: Are 2-weight Projective Cyclic Codes Irreducible? IEEE Trans. Inform. Theory 51, 733–737 (2005)
11. Code Tables, http://codetables.de

Algorithms of Constructing Linear and Robust Codes Based on Wavelet Decomposition and its Application

Alla Levina[✉] and Sergey Taranov

ITMO University, 49 Kronverksky Pr., St. Petersburg,
197101 Russia
alla_levina@mail.ru

Abstract. This article presents the algorithms of constructing error detecting codes using wavelet decomposition. Linear code, presented in the paper, based on the coefficients of scaling function of wavelet transformation. Constructed linear code was used for creation of robust codes that have a smaller number of undetectable errors and have an ability to detect any error with a predetermined probability. Robust codes are generated by applying a nonlinear function to the redundancy part of the linear code. The article describes comparative characteristics between the proposed wavelet code constructions and other error detecting codes. The paper proposes two constructions of robust code, first robust code base on the multiplicative inverse in a finite field, redundancy part of second code construction build as a cube in the field of the information component. The paper describes a model of application proposed code constructions in ADV612 system. Characteristics of robustness of the described model for uniform and nonuniform codeword distribution are also presented in the paper.

Keywords: Robust code · Linear code · Wavelet decomposition · Scaling function · Error masking probability

1 Introduction

Wavelet transformation has become well known and widely used for signal analysis in many fields of science. The basics of wavelet theory can be found in the works of Daubechies [9,11]. Many types of wavelets provide quick but very inaccurate compression. Methods of data compression using wavelet expansions are described in [8,10]. This article discusses the use of wavelet decompositions in coding theory and in the design of robust codes.

The paper presents an error-correcting coding scheme based on biorthogonal wavelet transform. The second section of this article contains the basic theory of wavelet decompositions. Also, this section reveals that the idea of multiresolution analysis is signal decomposition produced by the orthogonal basis formed by shifts and multiresolution copies of wavelet function. In the third part, we construct linear code based on matrix representation of wavelet decompositions.

© Springer International Publishing Switzerland 2015
S. El Hajji et al. (Eds.): C2SI 2015, LNCS 9084, pp. 247–258, 2015.
DOI: 10.1007/978-3-319-18681-8_20

Next, we derive a generator and a check matrix for the proposed construction. If we substitute the coefficients of wavelet scaling functions in the proposed matrix, we obtain specific linear codes.

Based on linear wavelet codes, we develop two constructions of robust codes. Robust codes are new nonlinear systematic error- detecting codes that provide uniform protection against all errors, whereas linear error-detecting code detects only a certain class of errors. Therefore, defence by the linear codes can be ineffective in many channels and environments when error distribution is unknown. Constructions of systematic robust codes were first introduced in [5]. Different types of robust codes, partially robust codes and minimum distance robust codes, were offered in [2], [3] and [6]. For the construction of robust codes proposed in this paper, we calculate the number of undetected errors and the error masking probability.

The presented construction of codes may be used to achieve high efficiency in the systems that use wavelet decompositions. Reuse of scaling function coefficients can simplify and accelerate the encoding process in the proposed code. We explore application of the proposed wavelet code in the ADV612 chip. We present a model of the error-coding scheme by proposed wavelet code that allows the detection of errors in the ADV612 chip even in cases of nonuniform codeword distribution. Proposed wavelet codes provide greater benefit when using the Gray mapping than the robust code from [2]. Also, without Gray mapping the proposed wavelet codes are less susceptible to downward trend of the error masking probability that are inherent to robust codes according to [2].

2   The Basic Tenets of the Wavelet Transform

The idea of the wavelet transform is a partition of the signal $s(t)$ into two components, approximating $d_i(t)$ and detailing $a_i(t)$:

$$s(t) = \sum_i d_i(t) + \sum_i a_i(t)$$

The basis of the wavelet transform is the use of two functions:
-wavelet function $\psi(t)$: this function defines the details of the signal $d_i(t)$ and generates the detail coefficients:

$$d_i(t) = \sum_k d_{ik} \psi_{ik}(t),$$

where d_{ik} are the detail coefficients.
-scaling function $\phi(t)$: this function defines the rough approximation of the signal $a_i(t)$ and generates the coefficients of the approximation:

$$a_i(t) = \sum_k a_{ik} \phi_{ik}(t),$$

where a_{ik} is the coefficient of approximation.

Function ψ is created on the basis of a particular basis function ψ_0 that determines the type of wavelet. A wavelet denoted as ψ_0 is called the mother wavelet, as it generates a certain class of wavelets. For given m and n, the function $\psi(t)$ is a wavelet:

$$\psi_{m,n}(t) = m^{-1/2}\psi_0(\frac{t-n}{m}); m, n \in \mathbf{R},$$

where $\psi_{m,n}(t)$ is a certain wavelet function; ψ_0 is a mother wavelet; m is a scaling parameter; and n is a shift parameter.

An orthogonal property significantly simplifies the analysis, allows the reconstruction of signals, and allows the implementation of fast wavelet transform algorithms.

Multiresolution analysis is a description of the space $L^2(R)$ through hierarchically nested subspaces that are disjoint; their union gives all $L^2(R)$:

$$\cdots \subset V_2 \subset V_1 \subset V_0 \subset V_{-1} \subset V_{-2} \subset \cdots$$

These space have the next properties:
1) $\cap_{m\in\mathbf{z}}V_m = 0$;
2) $\cup_{m\in\mathbf{z}}V_m = L^2(R)$;
3) Function $f(t) \in V_m$ and its compressed version $f(2t)$ must belong to subspace V_{m-1};
4) Also, there is a function $\phi \in V_0$ such that its shift $\phi_{0,n} = \phi(x-n)$, where n is a shift parameter, and $n \in \mathbf{Z}$ forms an orthonormal basis of the space V_0.

The last property implies that the functions

$$\phi_{m,n}(x) = 2^{-m/2}, \quad \phi(2^{-m}x - n)$$

form an orthonormal basis of the space V_m, where n is a shift parameter, $2^{-m/2}$ is a normalization factor, and $n, m \in \mathbf{Z}$. These basis functions are called scaling functions.

As $V_0 \subset V_{-1}$ and $\phi_{-1k}(t)$ is an orthonormal basis of space V_{-1} therefore:

$$\phi(t) = \phi_{0,0}(t) = \sqrt{2}\sum_n h_n\phi_{-1,n}(t) = 2\sum_n h_n\phi(2t-n) \qquad (1)$$

where h_n are the coefficients of scaling function $\phi(t)$, and n is a shift parameter.

We now derive the coefficient of the wavelet function. Consider the spaces W_m, which is the orthogonal addition of spaces V_m and V_{m-1}:

$$V_{m-1} = V_m \oplus W_m, \cap_{m\in\mathbf{z}}W_m = 0, \overline{\cap_{m\in\mathbf{z}}W_m} = L^2(R).$$

Denote $\psi(t) = \psi_{0,0}(t)$ as the basis function of space W_0.
Then

$$\psi(t) = \psi_{0,0}(t) = \sqrt{2}\sum_n g_n\phi_{-1,n}(t) = 2\sum_n g_n\phi(2t-n), \qquad (2)$$

where g_n are the wavelet coefficients, $\psi(t)$ is a mother wavelet, and $\phi(t)$ is a scaling function.

3   The Construction of Linear Code Based on Wavelet Transform

A *linear codes* is a separable code such that its redundancy part is a result of linear operations over its information part.

Linear codes denote (n, k). Here n is the length of the codeword; k is the number of information symbols; $r = n - k$ is the number of redundant symbols. In coding theory, the generator matrix is commonly used to find the corresponding codeword.

For linear codes, multiplying the generator matrix by any combinations of the information part obtains according to this message codeword. Let $(x_1, x_2, \cdots, x_n)$ be a codeword and $(x_1, x_2, \cdots, x_k)$ denote all possible combinations of the information part for a specific code. Then generator matrix of the linear code has the following form:

$$
\begin{bmatrix} x_1 \\ x_2 \\ \vdots \\ x_n \end{bmatrix} = \mathbf{G} \begin{bmatrix} x_1 \\ x_2 \\ \vdots \\ x_k \end{bmatrix},
$$

where $\mathbf{G}$ is the generator matrix of linear code. This generator matrix has size $k \times n$.

Check that the matrix has dimension $((n - k) \times n)$ and associate it with the linear code as follows:

$$
H\mathbf{x}^T = 0,
$$

where $\mathbf{x}$ is the received sequence, and H is the check matrix.

The check and generator matrices have a relationship. If we submit both matrices in the canonical form, we get the following constructions that have the coinciding matrix A:

$$
G = [I_k | - A^T], \quad H = [A | I_{n-k}],
$$

where I is the identity matrix.

We will represent the wavelet transform in a matrix form and prove that the wavelet transform can be represented as a matrix consisting of the scaling function coefficients.

Theorem 1. Wavelet function $\psi(t)$ and scaling functions $\phi(t)$ are completely defined by the scaling coefficients h_n.

Proof:
Find a relationship between the wavelet coefficients g_n and the scaling function coefficients h_n. Since W_m is the orthogonal complement V_m, then the functions $\psi(t)$ and $\phi(t)$ must be orthogonal. Hence, there is an inner product of these functions:

$$
\langle \psi(t), \phi(t) \rangle = 0.
$$

Substituting (1) and (2) in the last expression, we obtain:

$$2 \sum_n \sum_k h_n g_k \langle \phi_{-1n}(t), \psi_{-1p}(t) \rangle = 0,$$

$$2 \sum_n h_n g_n = 0$$

Also, comparing the equations (1) and (2), we see that:

$$g_n = (-1)^n h_{2m-1-n},$$

where m is a wavelet order.

Based on the above formula and the equations (1) and (2), it can be seen that the wavelets $\psi(t)$ are completely defined by the scaling function $\phi(t)$. In turn, the scaling function $\phi(t)$ is completely determined by its scaling coefficient h_n. ∎

For practical calculations of wavelet transform, it is not necessary to know the wavelet structure; it is sufficient to know its coefficients h_n. This set of coefficients h_n uniquely determines the scaling function $\phi(t)$ and wavelet $\psi(t)$. We will prove that wavelet transform can be represented as a multiplication of vectors; on the matrix that correspond to this transformation.

Theorem 2. Any wavelet transform can be represented as a multiplication of the input data vector s on the matrix A consisting of the scaling function coefficients h_n.

Proof:

A digital signal can be represented as a vector $s = s_1, ..., s_N$ of the length N. By definition, the wavelet transform of signal $s(t)$ has the following form:

$$s(t) = \sum_k a_{mk} \phi_{mk}(t) + \sum_m \sum_k d_{mk} \psi_{mk}(t), \tag{3}$$

where ψ and ϕ are wavelet and scaling functions, respectively, m is the decomposition level of the signal, and $a_{mk} = \langle s(t), \phi_{mk}(t) \rangle$, $d_{mk} = \langle s(t), \phi_{mk}(t) \rangle$ are the decomposition coefficients.

The wavelet transform is a linear transform, hence it can be associated with a matrix. By the definition (3), there are two mappings, $V_{m-1} \rightarrow V_m$ and $V_{m-1} \rightarrow W_m$. Denote the matrix representations of this mapping as H and G.

Let $h_1, ..., h_N$ denote the scaling function coefficients, and $g_1, ..., g_N$ the wavelet coefficients. Then matrix H is cyclic matrix with size $N/2 \times N$ and with shift that is equal to the wavelet order:

$$H = cir_d(h_1, h_2, \cdots, h_{N-1}, h_N),$$

where d - shift of matrix H.

Matrix G is obtained in the same way from the wavelet coefficients $g_1, ..., g_N$. However, this can be reduced to a form that will depend on the scaling function coefficients $h_1, ..., h_N$. Using the equation that was obtained in Theorem 1

$$g_n = (-1)^n h_{2m-1-n},$$

we can replace the coefficients $g_1, ..., g_N$ for coefficients $h_1, ..., h_N$.

As a result, the matrix representation of mapping $V_{m-1} \to V_m$ and $V_{m-1} \to W_m$ can be represented by the scaling function coefficient h_n. ∎

Now we will describe the requirements for the check and generator matrices of the proposed construction of linear wavelet code.

Denote $\mathbf{H}$ and $\mathbf{G}$ as the check and generator matrices that are used for encoding process. In turn, let $\overline{\mathbf{H}}$ and $\overline{\mathbf{G}}$ denote matrices that are necessary for decoding. The set of these matrices must satisfy the following conditions:

1) biorthogonality condition:

$$\begin{cases} \overline{\mathbf{H}}\mathbf{H}^T = I \\ \overline{\mathbf{G}}\mathbf{G}^T = I \\ \overline{\mathbf{H}}\mathbf{G}^T = O \\ \overline{\mathbf{G}}\mathbf{H}^T = O \end{cases}$$

2) condition of exact recovery:

$$\mathbf{H}^T\overline{\mathbf{H}} + \mathbf{G}^T\overline{\mathbf{G}} = I$$

We will describe the algorithm for obtaining the linear wavelet code. Error-correcting linear code is defined by a single level of space decomposition, so the dimension of the vectors can be any even number. The proposed linear code is constructed as follows. The information part $v = \{v_1, v_2, \cdots, v_{N/2}\}$ of the codeword is a sequence of field elements $GF(q)$; N is even. Hence, the redundant part has the form $w = \{w_1, w_2, \cdots, w_{N/2}\}$. The set of all codewords can be defined by the next generator matrix and check matrix:

1) generator matrix of wavelet linear code:

$$\mathbf{H}^T + a\mathbf{G}^T J$$

2) check matrix of wavelet linear code:

$$\overline{\mathbf{H}}^T + bJ^T\overline{\mathbf{G}}^T,$$

where a, b are vectors of the field $GF(q)$ with length $N/2$ and satisfying to condition $ab = (p - 1)modp, p \in GF(q)$, matrix $J = cir(0, 1, 0, ..., 0)$ of size $N/2 \times N/2$.

As can see from the above matrix that the resulting code is a linear cyclic code. Hence, principles of syndrome decoding that use cyclic codes can be applied to that proposed in this construction of linear wavelet code.

For storing or transferring all codewords, linear codes store significantly fewer of them in the memory of the encoder or decoder; more precisely, only words that form the basis of linear space. This simplifies the implementation of encoding and decoding devices and makes linear codes very attractive for practical applications. Although linear codes effectively detect and correct few, but large, error blocks, their effectiveness decreases if errors are frequent or controlled by an attacker.

4 The Construction of Robust Code Based on Wavelet Linear Code

In this section, we propose constructions of robust codes developed by wavelet cyclic codes from the previous section.

Robust codes are nonlinear systematic error-detecting codes that provide uniform protection against all errors without any (or that minimize) assumptions about the error and fault distributions, capabilities and methods of an attacker.

One of the main criteria for evaluating the effectiveness of a robust code is the *error masking probability*. The error masking probability $Q(e)$ can be defined as:

$$Q(e) = \frac{|\{x| \in C, x + e \in C\}|}{M},$$

where C is the robust code, x is a codeword that belongs to the code C, e is an error, and M is the number of codewords in the code C.

Robust codes for the same codeword length and number of redundant bits have fewer undetectable errors and lower masking error probability than the corresponding linear code.

The robustness of codes depends on the nonlinearity of functions. The function nonlinearity P_f can be measured by using the directional derivative

$$D_l f(x) = f(x + l) - f(x),$$

where l is a vector that define the direction.

Let

$$f : GF(q^k) \to GF(q^r) : l \to s = f(x).$$

then

$$P_f = max_{l \in GF(q^k)} max_{s \in GF(q^r)} Pr(D_l f(x) = s) \tag{4},$$

where $Pr(z)$ denotes the fractions of cases when z occurs, l, s - vectors of two fields between which there is mapping f.

As shown in the paper [1], robust code can be constructed on the basis of existing linear codes. Let C be a binary linear wavelet and (n, k) be a code over $GF(q)$ with generator matrix:

$$\mathbf{H}^T + a\overline{\mathbf{G}}J.$$

Code C can be converted into a nonlinear systematic robust code C_R by taking the multiplicative inverse in $GF(q^r)$ of the r redundant bits.

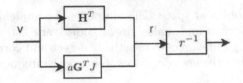

Fig. 1. The process of encoding by the wavelet robust code based on the multiplicative inverse

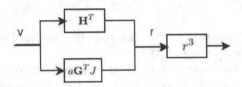

Fig. 2. The process of encoding by the cube wavelet robust code, where v - information part of codeword, r - redundancy part

Let v denote information part of codeword, r denote redundancy part, then the process of encoding for wavelet robust code based on the multiplicative inverse is shown in Figure 1.

Let x be a polynomial from $GF(q^k)$. Then x^{-1} is a polynomial from field $GF(2^{r=k})$. Then for error $e = (e_1, e_2)$, where $e_1, e_2 \in GF(2^{r=k})$, the condition of error masking equals $f(x_1 + e_1) = f(x_1) + e_2$. Respectively to (4) we get $D_{e_1} f(x_1) = e_2$. For function $f(x) = 1/x$, we have $P_f = 2^{-r}$. As a result, we obtain the next parameters of wavelet code $(2r, 2^r, 2)_2$. Therefore, this robust code construction has masking error probability 2^{-r} and number of undetectable errors 0.

Also, wavelet robust code C_R can be obtained by a calculations cube in $GF(q^r)$ of the r redundant bits. The process of encoding by the proposed wavelet robust code is shown in Figure 2.

Table 1. Comparison of the optimum robust duplication code and Hamming linear code with proposed construction. $Q(e)$ is an error masking probability, k is the number of information symbols in codewords, r is the number of redundant symbols.

Code	$Q(e)$	Undetectable errors
Hamming linear code	1	2^k
Partially robust Hamming code	1	2^{k-r}
Robust quadratic systematic code [2]	2^{-r}	0
Robust duplication code [2]	2^{-k}	0
Wavelet linear code	1	2^k
Wavelet robust code with encoding function $1/x$	2^{-k}	0
Wavelet robust code with encoding function x^3	2^{-k}	0

For function $f(x) = x^3$, we also have $P_f = 2^{-r}$. The robust wavelet code based on the cube encoding function has masking error probability of 2^{-r}, and the number of undetectable errors equals 0. Comparison of the error masking probability and the number of undetectable error for linear and robust wavelet codes are shown in Table 1.

As we can see from Table 1, the proposed wavelet codes are not inferior to the representatives of their code class. If linear codes are characterised by an efficient encoding and decoding process, robust codes provide protection against some side channel attacks. However, the speed of encoding and decoding for these codes is significantly less than for linear codes, since robust code uses laborious processes of finding multiplicative inverse elements and calculation of the cube in the field $GF(q)$.

As proposed in this paper, coding methods are useful in systems that use wavelet transforms. These include systems that implements the digital signal processing and analysis, image coding, or the construction of various filters, because since in such systems, the coefficients of scaling functions can be used in the check and generator matrices of the proposed constructions.

5   Implementation of Wavelet Robust Codes in ADV612 Chip

In particular, the proposed scheme of coding by the robust wavelet codes was tested on the ADV612 chip. The ADV612 is wavelet-based single-chip system. This system can implement the real-time compression and decompression of digital video at very high image quality. The ADV612 system consists of three main blocks: wavelet filter bank, quantizer, run length coder and Huffman coder.

Our method for modifying an ADV612 architecture is based on the method proposed in [2]. The model architecture is based on adding redundancy around an original device to create data redundancy that can be used to verify data integrity and the correct operation of the ADV612 system. The architecture is composed of three hardware components: original hardware, Encoder of wavelet code for predicting the redundancy bits of the original device, and a check mechanism that verifies the predetermined relationship of the output of the original device and the result of the Encoder work. By defining an appropriate code and implementing the suitable function in the encoder, a desired level of robustness can be guaranteed for the desired output of the ADV612 system. The model describing this system is presented in Figure 3.

In our scheme, we applied wavelet codes presented in this paper to ensure the integrity of the ADV612. The k bits of the ADV612 output and the r redundant output bits of the Encoder of wavelet code form the $n = k + r$ generalised output of the device. The generalised output forms a codeword of the systematic wavelet code that can be used to detect errors in the original hardware or in the Encoder. It is the Check mechanism that verifies that the generalised output of the ADV612 device belongs to the corresponding wavelet code. If it does not, then the Check mechanism generates an error.

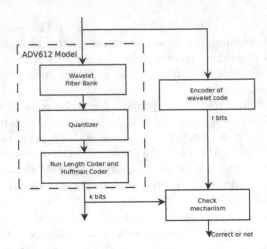

Fig. 3. General architecture for protection of the ADV612 system with the wavelet code

Table 2. Comparison of the four different classes of robust codes in cases of uniform and nonuniform distribution

Code	Distribution	Maximum of error masking probability without Gray mapping	Maximum of error masking probability with Gray mapping
(32,16) robust wavelet code with encoding function x^3	uniform	0.571	0.571
	nonuniform	0.711	0.643
(32,16) robust code based on scheme $(x,(Px)^3)$	uniform	0.571	0.571
	nonuniform	0.758	0.692
(24,16) robust quadratic code based on $x_1x_2 + x_3x_4 + ...$	uniform	0.5	0.5
	nonuniform	0.780	0.653
(32,16) robust code based on scheme $(x,(Px)^{-1})$	uniform	0.724	0.724
	nonuniform	0.945	0.825

The proposed scheme of coding by wavelet robust code yields higher gain compared with other classes of robust codes [1,2,3] in the case of nonuniform codeword distribution. The comparison of four robust code constructions is shown in Table 2.

Detailed description of the constructions of robust (32,16) code based on scheme $(x,(Px)^3)$, (24,16) robust quadratic code based on $x_1x_2 + x_3x_4 + ...$, (32,16) robust code based on scheme $(x,(Px)^{-1})$ can be find in [2]. In the case of nonuniform codeword distribution, using Grey mapping of the most probable codewords to a predefined set reduces the error masking probabilities of the robust codes. Also, this mechanism can be used for improving the characteristics of the presented robust wavelet code. As we can see from Table 2, the proposed wavelet codes provide greater benefit when using the Gray mapping than the robust codes from [2]. Also, without Gray mapping, the proposed wavelet codes are less susceptible to the downward trend of the error masking probability that is inherent to all robust codes.

6   Conclusion

In this paper, we have described a general algorithm for obtaining linear and robust codes based on wavelet decompositions. For each algorithm, we derived check and generator matrices. Interaction theory of wavelet decomposition and coding theory can help optimise the transmitted information. For example, if one plans to use robust code for transferring image and video information, one can discard part of the redundant information, and then we have the opportunity to restore the original sequence.

If the system uses wavelet decomposition, application of the codes proposed in this paper can yield effective results. The scaling function coefficients of the specific wavelet decompositions can be used to encode information by wavelet linear code or if there is a need to ensure security by robust wavelet code. In future works will be improved the proposed coding schemes and explore the error masking probability in cases of using Daubechies wavelets, spline wavelets and others.

References

1. Kulikowski, K.J., Karpovsky, M.G., Taubin, A.: Robust Codes and Robust, Fault Tolerant Architectures of the Advanced Encryption Standard. Journal of System Architecture (2007)
2. Karpovsky, M.G., Kulikowski, K., Wang, Z.: On-Line Self Error Detection with Equal Protection Against All Errors. Int. Journal of Highly Reliable Electronic System Design (2008)
3. Wang, Z., Karpovsky, M.G., Kulikowski, K.: Design of Memories with Concurrent Error Detection and Correction by Non-Linear SEC-DEC Codes. Journal of Electronic Testing (2010)
4. Shumsky, I., Keren, O., Karpovsky, M.: Robustness of Security-Oriented Binary Codes Under Non-Uniform Distribution of Codewords. In: Proc. Int. Depend Symp. (2013)
5. Karpovsky, M.G., Taubin, A.: A New Class of Nonlinear Systematic Error Detecting Codes. IEEE Trans. Info Theory (2004)
6. Karpovsky, M.G., Wang, Z.: Design of Strongly Secure Communication and Computation Channels by Nonlinear Error Detecting Codes. IEEE Trans. Computers (2013)
7. Karpovsky, M.G., Kulikowski, K.J., Taubin, A.: Robust protection against fault-injection attacks on smart cards implementing the advanced encryption standart. dsn (2004)
8. U. K. Demyanovich, Minimal splines and wavelets, Vestnik SPSU (2008)
9. Daubechies, I.: Ten lectures on wavelets, CBMS-NSF conference series in applied mathematics. SIAM Ed. (1992)
10. Guha, S., Harb, B.: Approximation algorithms for wavelet transform coding of data streams. Information Theory (2008)
11. Daubechies, I., Sweldens, W.: Factoring wavelet transforms into lifting steps. Journal of Fourier Analysis and Applications (1998)
12. Caire, G., Grossman, R.L., Poor, H.V.: Wavelet Transforms Associated with Finite Cyclic Groups. IEEE Trans. Inf. Theory (1993)

13. Fekri, F., Mersereau, R.M., Schafer, R.W.: Theory of wavelet transform over finite fields. IEEE Trans. Inform. Theory (2002)
14. Fekri, F., McLaughlin, S.W., Mersereau, R.M., Schafer, R.W.: Double circulant self-dual codes using finite-field wavelet transforms. In: Applied Algebra, Algebraic Algorithms and Error Correcting Codes Conference (1999)
15. Louis, A.K., Rieder, A.: Wavelets Theory and Applications. John Wiley Sons (1997)

Failure of the Point Blinding Countermeasure Against Fault Attack in Pairing-Based Cryptography

Nadia El Mrabet[1,2](✉) and Emmanuel Fouotsa[3,4]

[1] LIASD – Université Paris 8, France
[2] SAS - CMP Gardanne, France
elmrabet@ai.univ-paris8.fr, nadia.el-mrabet@emse.fr
[3] Dep of Mathematics, Higher Teacher's Training College,
University of Bamenda - Cameroun
[4] LMNO – Université de Caen, France
emmanuelfouotsa@yahoo.fr

Abstract. Pairings are mathematical tools that have been proven to be very useful in the construction of many cryptographic protocols. Some of these protocols are suitable for implementation on power constrained devices such as smart cards or smartphone which are subject to side channel attacks. In this paper, we analyse the efficiency of the point blinding countermeasure in pairing based cryptography against side channel attacks. In particular, we show that this countermeasure does not protect Miller's algorithm for pairing computation against fault attack. We then give recommendation for a secure implementation of a pairing based protocol using the Miller algorithm.

Keywords: Miller's algorithm · Identity Based Cryptography · Side Channel Attacks · Fault Attacks · Countermeasure

1 Introduction

Pairings are bilinear maps defined on the group of rationals points of elliptic or hyper elliptic curves [36]. Nowadays, more and more protocols using pairings are proposed in the literature [10,21,6]. Among these protocols, only those constructed on the identity based model involve a secret which is one of the argument during the computation of a pairing. The implementation of a pairing based protocol is efficient enough to allow the use of pairing based cryptography on power constrained device such as smart cards and mobile phones [31,22,19]. Smart cards are by nature sensitive to side channel attacks. Side channel attacks are powerful attacks that use the implementation of a protocol to obtain

This work was supported in part by the French ANR-12-INSE-0014 SIMPATIC Project. The second author is supported by The Simons Foundations through Pole of Research in Mathematics with applications to Information Security, Subsaharan Africa.

S. El Hajji et al. (Eds.): C2SI 2015, LNCS 9084, pp. 259–273, 2015.
DOI: 10.1007/978-3-319-18681-8_21

information on the secret. They are divided into two families: invasive and non invasive attacks. Invasive attacks are based on the model of fault attacks. The execution of a protocol is disturbed, the result is then a faulty one and the analysis of this faulty result can provide information on the secret. In non invasive attacks, the information can be leaked by the time of execution, the electric consumption or the electromagnetic emission of the device. Several works have investigated the robustness of identity based cryptography to side channel attacks. They are mainly focused on fault attacks [27,37,11,2]. Few works consider differential power analysis attack [27,13,5]. As the secret during an identity based protocol can be recovered by side channel attacks, several countermeasures were proposed. Those countermeasures are the same for invasive and non invasive attacks [14]. In [16], Ghosh, Mulhopadhyay and Chowdhury proposed an analysis of countermeasures to fault attack presented in [27]: the new point blinding method and the alliterating point blinding method. They concluded that the countermeasures are not sufficient and proposed new one. However, their explanations on the non efficiency of the countermeasure are not convincing. Later, Park et al. [28] clearly exposed the weaknesses of the point blinding technique against fault attacks described by Page and Vercauteren [27].

In this article we analyze and extend the work in [16,28] on the efficiency of the point blinding countermeasure in pairing based cryptography. Especially, we generalize the attack of Park et al. [28] and expose its failure to protect the Miller algorithm, main tool in pairing computation. As the most efficient pairings are constructed on the model of the Tate pairing, we focus on the Miller algorithm, used for the Tate pairing considering Weierstrass elliptic curve. Obviously, this analysis is the same for the (optimal) Ate, twisted Ate or pairing lattices; and for every model of elliptic curve or coordinates.

The rest of this paper is organized as follows: The Section 2 presents brief concepts on pairings that are useful to understand this work. In Section 3 we present side channel attacks with emphasis on fault attacks in pairing based cryptography. In Section 4 we explicitly demonstrate that the point blinding countermeasure fails to protect the Miller algorithm against fault attack. Finally we conclude the work in Section 5.

2   Background on Pairings

In this section, we briefly recall basics on pairings and on the Miller algorithm [25], main tool for an efficient computation of pairings. Let E be an elliptic curve defined over a finite field $\mathbb{F}_q$, with q a prime number or a power of a prime. The neutral element of the additive group law defined on the set of rational points of E is denoted P_∞. Let r be a large prime divisor of the group order $\sharp E(\mathbb{F}_q)$ and k the embedding degree of E with respect to r, i.e. the smallest integer k such that r divides $q^k - 1$. The integer k is also the smallest integer such that $E\left(\overline{\mathbb{F}_q}\right)[r] \subset E(\mathbb{F}_{q^k})$, where $E\left(\overline{\mathbb{F}_q}\right)[r] = \{P \in E\left(\overline{\mathbb{F}_q}\right) : [r]P = P_\infty\}$ with $[r]P = \underbrace{P + P + \ldots + P}_{r \text{ times}}$ and $\overline{\mathbb{F}_q}$ is the algebraic closure of $\mathbb{F}_q$.

In general, the sizes of r, q and k are dependent from the security level and the currently recommendations are at least $r > 2^{160}$ and $q^k > 2^{2024}$ [15]. The recent results for the discrete logarithm problem [20,3] imply that the number q must be a large prime number. The security recommendations allow the choice of k to be a product of power of 2 and 3. A consequence of the fact that $k \equiv 0 \mod 2$ is the use of a twist representation for the point Q. This representation using a twisted elliptic curve allow the denominator elimination optimization [23].

Definition of a Twisted Elliptic Curve. We explain here the concept of twist of elliptic curve in the context of Weierstrass elliptic curve. This will help us to understand the choice of the coordinates of points in Section 4. The quadratic twist of the elliptic curve $E : y^2 = x^3 + ax + b$ over $\mathbb{F}_{p^k}$ is the elliptic curve $\tilde{E} : \frac{1}{\nu}y^2 = x^3 + ax + b$ where $\{1, \nu\}$ is a basis of $\mathbb{F}_{q^k}$ as $\mathbb{F}_{q^{k/2}}$ vector space. The two curves are isomorphic via

$$\psi : \tilde{E}(\mathbb{F}_{q^{k/2}}) \longrightarrow E(\mathbb{F}_{q^k})$$
$$(x, y) \longmapsto (x, y\sqrt{\nu}).$$

This isomorphism is particularly useful since it enables to take the point $Q \in E(\mathbb{F}_{q^k})$ in the following manner $Q = \psi(Q')$ where $Q' = (x_Q, y_Q)$ with $x_Q, y_Q \in \mathbb{F}_{q^{k/2}}$. This ensures an efficient computation since many computations will be consequently done instead in the subfield $\mathbb{F}_{q^{k/2}}$ and more interestingly, it enables to avoid the inversions in the Miller algorithm. This elimination is the denominator elimination [23].

Indeed, if $P_1(x_1, y_1)$ and $P_2(x_2, y_2)$ are two points of the elliptic curve in Weierstrass form $E : y^2 = x^3 + ax + b$ then the function h_{P_1, P_2} with divisor

$$\text{Div}(h_{P_1, P_2}) = (P_1) + (P_2) - (P_1 + P_2) - (P_\infty),$$

is $h_{P_1, P_2} = \frac{\ell_{P_1, P_2}}{v_{P_1 + P_2}}$ where ℓ_{P_1, P_2} is the straight line defining $P_1 + P_2$ and $v_{P_1 + P_2}$ is the corresponding vertical line passing through $P_1 + P_2$. Explicitly, we have

$$h_{P_1, P_2}(x, y) = \frac{y - \lambda x - \alpha}{x - x_3},$$

where x_3 is the first coordinate of $P_1 + P_2$ and $\lambda = \frac{y_2 - y_1}{x_2 - x_1}$ if $P_1 \neq P_2$, $\lambda = \frac{3x_1^2 + a}{2y_1}$ if $P_1 = P_2$ and $\alpha = y_1 - \lambda x_1$.

In the particular case of doubling ($P_1 = P_2$), a straightforward computation gives, after changing to Jacobian coordinates ($x_1 = \frac{X_1}{Z_1^2}$, $y_1 = \frac{Y_1}{Z_1^3}$)

$$h_{P_1, P_1}(Q) = h_{P_1, P_2}(x_Q, y_Q\sqrt{\nu}) = \frac{2Y_1 Z_1^3 y_Q \sqrt{\nu} - 2Y_1^2 - (3X_1^2 + aZ_1^4)(x_Q Z_1^2 - X_1)}{2Y_1 Z_1^3 (x_Q - x_3)},$$

We then remark that the denominator of the previous expression is an element of $\mathbb{F}_{q^{k/2}}$ and consequently will be equal to 1 during the final exponentiation.

So the main expression that will be used in the Miller algorithm is:

$$h_{P_1,P_1}(Q)=h_{P_1,P_2}(x_Q,y_Q\sqrt{\nu})=2Y_1Z_1^3y_Q\sqrt{\nu} - 2Y_1^2 - (3X_1^2 + aZ_1^4)(x_QZ_1^2 - X_1) \tag{1}$$

The expression given by equation 1 is used in algorithms 1 and 2 and will be particularly useful in Section 4 to illustrate our attack.

The Tate Pairing. Consider a point $P \in E(\mathbb{F}_q)[r]$, the principal divisor $D = r(P) - r(P_\infty)$ and a function $f_{r,P}$ with divisor Div $(f_{r,P}) = D$. Let $Q \in E(\mathbb{F}_{q^k})[r]/E(\mathbb{F}_q)$ and μ_r be the group of r-th roots of unity in $\mathbb{F}_{q^k}^*$. The reduced Tate pairing e_r is a bilinear and non degenerate map defined as

$$e_r : E(\mathbb{F}_q)[m] \times E(\mathbb{F}_{q^k})[r] \to \mu_m$$
$$(P,Q) \qquad \mapsto f_{r,P}(Q)^{\frac{q^k-1}{r}}$$

The value $f_{r,P}(Q)$ can be determined efficiently using Miller's algorithm [25].

Algorithm 1. Miller's Algorithm

Input : $P \in E(\mathbb{F}_q)[r]$, $Q \in E(\mathbb{F}_{q^k})[r]$, $m = (1, m_{n-2}, ...m_1, m_0)_2$.
Output: $f_{m,P}(Q)$

1: Set $f \leftarrow 1$ and $T \leftarrow P$
2: **For** $i = n - 2$ **down to** 0 **do**
3:     $f \leftarrow f^2 \cdot h_{T,T}(Q)$, with $h_{T,T}$ the Equation (1) of the tangent to E at point T
4:     $T \leftarrow 2T$
5:    **if** $m_i = 1$ **then**
6:     $f \leftarrow f \cdot h_{T,P}(Q)$, with $h_{T,P}$ the equation of the line (PT)
7:     $T \leftarrow T + P$
8:    **end if**
9: **end for**
10: **return** f

Fig. 1. The Miller algorithm

More information on pairings can be found in [9]. In order to obtain the result of the Tate pairing, the output of Miller's algorithm must be raised to the power $\frac{q^k-1}{r}$, this operation is called the final exponentiation.

We call a Tate-like pairing any pairing constructed on the following model: an execution of the Miller algorithm followed by a final exponentiation. Every Tate-like pairing was an improvement of the previous. The ate pairing [18] was an improvement of the Tate pairing [29], the twisted ate pairing [18] an improvement of the ate pairing, the notion of optimal pairings [35] an improvement of the ate and twisted ate pairing and finally the pairing lattices [17] another way to deal with optimal pairings. The algorithmic difference between the Tate pairing and a Tate-like pairing is principally the number of iterations, sometimes it could also be the role playing by P and Q. In Algorithm 1, we describe the Miller algorithm. In order to keep our explanations general , the number of iterations in the Miller algorithm is indexed over m. The integer m would be r for the Tate pairing, or smaller than r for a Tate-like pairing. We describe the attack considering that we are computing a pairing using $f_{m,P}(Q)$, for m the integer giving the number of iterations of the pairing. Obviously, the discussion can be straightforward adapted for the computation of $f_{m,Q}(P)$.

Obviously, the system of coordinates influences the equations of the Miller algorithm, but if the attack is efficient over one model of elliptic curve for one system of coordinates, then the same attack will be efficient over any other model of elliptic curve and considering any other system of coordinates.

3 Side Channel Attacks on Pairing-Based Cryptography and Countermeasures

In this section we briefly recall and describe existing side channel attacks and countermeasures in the context of pairing-based cryptography. Especially, we analyse the point blinding countermeasure presented in [27] and its weakness exposed in [28].

3.1 Background on Side Channel Attacks

The first analysis of side channel attacks against a pairing was proposed by Page and Vercauteren [27]. They attack the Duursma and Lee algorithm used to compute a pairing over super singular elliptic curves. Page and Vercauteren described a new fault attack model and mention without development the differential power analysis against pairings. The fault model consists in the modification of the number of iterations of an algorithm. The fault attack was adapted by further works on the Miller algorithm [37,11,2]. Whelan et Scott [37] highlighted the fact that pairings without a final exponentiation are more sensitive to a sign change fault attack. They analyzed the Weil, the Tate and Eta pairing. They used a simplified version of Page and Vercauteren attack. After that, El Mrabet [11] generalized the attack of Page and Vercauteren to the Miller algorithm used to compute all the recent optimizations of pairings. El Mrabet considered only the Miller algorithm and did not take into account the final exponentiation. The target of El Mrabet's attack is the loop counter in the Miller algorithm. The final exponentiation was attacked by Lasherme et al. [24]. They used three faults to inverse the final exponentiation of the Tate pairing, which is the same for Ate and twisted ate pairing. Recently, an attack against a whole pairing, i.e. the Miller algorithm together with the final exponentiation, was published by Blömer et al. in [4]. The attack consists in modifying the clock of the device and as a consequence, the device returns intermediary results that allow to recover the secret. Few works consider differential power analysis. In [13] El Mrabet et al. highlight the fact that without protection the Miller algorithm is sensitive to a differential power analysis attack. Their work was recalled in [5]. In practice, the efficiency of side channel attacks does not lay on the choice of the characteristic, neither on the choice of the elliptic curve, nor on the choice of the coordinates. To each attack, several countermeasures were proposed. The countermeasures rely on the bilinearity of pairings, or on the homogeneity of the coordinates [14].

3.2   Description of Fault Attack

In an Identity Based Encryption scheme [6], one argument of the pairing is secret. So fault attacks can be performed to reveal the secret. We describe the attack against the Miller algorithm. As stated in the introduction, fault attack on pairing algorithm tries to corrupt the loop bound (which is $\log(m)$) of the Miller algorithm. The attacker injects fault repetitively in such a way that he can obtain two consecutive loop bounds $\log(m - s)$ and $\log(m - s) + 1$ and the corresponding pairings $e_{m-s}(P, Q)$ and $e_{m-s+1}(P, Q)$, for a certain integer s. It has been shown in [11] that it is possible to obtain such consecutive integers in a finite number of fault injections.

The clock glitch attack described in [4] highlights the fact that in practice a modification of the glitch can make the device stop and return intermediary results, such as internal results of Miller's algorithm. In order to explain how the attacker can obtain the secret point from the erroneous pairings $e_{m-s}(P, Q)$ and $e_{m-s+1}(P, Q)$ we consider the two following situations.

First Situation: Excluding the Final Exponentiation. Instead of obtaining the values $e_{m-s}(P, Q)$ and $e_{m-s+1}(P, Q)$ after the final exponentiation, the attacker tries to get the final values obtained after $\log(m - s)$ and $\log(m - s) + 1$ iterations, just before the final exponentiation. A method to obtain those intermediary values is the use of a clock glitch attack [4]. We denote these values by $f_{m-s,P}(Q)$ and $f_{m-s+1,P}(Q)$. Depending on the last bit corresponding to each iteration, we have four possibilities for the expression of $f_{m-s,P}(Q)$ and $f_{m-s+1,P}(Q)$.

Without lost of general ity, we can consider the case when

$$f_{m-s+1,P}(Q) = (f_{m-s,P}(Q))^2 \times h_{[j]P,[j]P}(Q),$$

with j the integer composed by the $log_2(m - s)$ most significant bits of m.

Consequently, the attacker knows

$$S = \frac{f_{m-s+1,P}}{f_{m-s,P}^2}(Q) = h_{[2j]P,[2j]P}(Q).$$

The trick of the attacker is now to use the representation of S and $h_{[2j]P,[2j]P}(Q)$ $\in \mathbb{F}_{q^k}$ in a basis of $\mathbb{F}_{q^k}/\mathbb{F}_q$ in order to obtain by identification, a system of linear or non-linear equations. The resolution of this system leads to the obtention of the coordinates of the secret point. A successful such attack has been mounted against the Miller algorithm [12]. We briefly recall the attack and refer to [11] for a complete description of this attack.

We recall that the point Q is public, the point P is secret and R is random in $E(\mathbb{F}_{q^k})$. For efficiency reasons, the embedding degree k is smooth and at least divisible by 2, or 4 or for the best cases by 6. A smooth integer is a number that admits a factorisation into small prime numbers. This condition on k enables efficient computation of pairings and the denominator elimination thanks to the twist of the elliptic curve. A consequence is that the points Q and R are seen as images of points belonging to the twist. The coordinates of R are composed by at most k values in $\mathbb{F}_{q^{k/d}}$, where d is the degree of the twist. The point P could be given in affine, projective or Jacobian coordinates. The choice will depend on the most efficient computation for the pairing. Whatever the choice is, the coordinates of point P will always count as 2 unknown values X_P and Y_P. This is obvious if P is given in affine coordinates. If P is given in projective or Jacobian coordinates, P would be characterized and gives improvement of the pairing computations by 3 unknown values X_P, Y_P and Z_P. But, using the homogeneity of projective and Jacobian coordinates, we could consider that the point P is in fact X'_P, Y'_P and 1. Indeed, we know that for $Z \neq 0$ in projective coordinates $(X, Y, Z) \cong (X/Z, Y/Z, 1)$ and in Jacobian coordinates $(X, Y, Z) \cong (X/Z^2, Y/Z^3, 1)$.

Putting all together one obtains a system of $k + 2$ polynomial equations in $k + 2$ unknown values. This system admits solutions as it is derived from a constructive algorithm. The points P and R are defined by construction. So, we can use the Gröbner basis [8] for instance to solve the system and find the coordinates of the point P. If the secret is the point Q, the attack is easier and successful [11].

Second Situation: Including the Final Exponentiation. In this situation we consider the values $e_{m-s}(P, Q)$ and $e_{m-s+1}(P, Q)$ obtained after the final exponentiation. Then

$$\frac{e_{m-s+1}}{e_{m-s}^2} = \left[h_{[2j]P,[2j]P}(Q) \right]^{\frac{(q^k-1)}{r}}$$

The aim here is, since it has been easy to obtain $e_{m-s}(P, Q)$ and $e_{m-s+1}(P, Q)$ contrary to situation 1, to reverse the exponent $\frac{(q^k-1)}{r}$, such that an application of the method in situation 1 may lead to the obtaining of the secret. In secured pairing based protocols, it has been shown that the exponent $\frac{(q^k-1)}{r}$ is difficult to reverse mathematically [30,24]. So the attack in this situation requires a fault model that would neutralize the final exponentiation, which is possible experimentally. One possibility can be to combine two fault models to neutralize the final exponentiation. For instance use a fault attack to reduce the number of iterations as in [11] and a fault attack to reverse the exponentiation as in [24]. Another way would be to use a fault model that modifies the time of execution as modification of the glitch or under voltage attack [4].

Remark 1. In the case of super singular elliptic curves, the final exponentiation can be reversed by mathematical considerations, the form of the exponent combined with a sparse decomposition in the basis of $\mathbb{F}_{p^k}$ allow this operation [27].

This is specific to pairings over supersingular elliptic curves and cannot be applied to ordinary elliptic curves.

3.3   The Point Blinding Countermeasure and Weaknesses

In [16], Ghosh, Mulhopadhyay and Chowdhury proposed an analysis of countermeasures to fault attack presented in [27]. They analyze what they called the new point blinding technique:

$$e(P, Q) = e([x]P, [y]Q) \text{ for random } x, y \text{ such that } xy \equiv 1 \mod (r)$$

and the altering traditional point blinding:

$$e(P, Q) = \frac{e(P, Q + R)}{e(P, R)},$$

for R a random point in $E(\mathbb{F}_q)$ such that the pairings $e(P, Q)$ and $e(P, R)$ are defined. They conclude that these two countermeasures are not sufficient against the fault attack described in [27]. However their analysis was not convincing. Concerning the new point blinding method, they claim that the intermediary steps of a pairing computation are bilinear which is not the case. The ratio obtained in the attack depends on the coordinates of the points $[x]P$ and $[y]Q$, with x and y unknown to the attacker. They do not explain how they can recover the value of the secret point used during the pairing computation. Concerning the altering traditional point blinding method, their analysis was not clear enough. In [16] the explanation did not take into account the randomness induced by the point R. We demonstrate in the next section that this countermeasure is not efficient with a precise approach and we develop the corresponding equation.

In [28] Park et al. exposed the weaknesses of the point blinding technique against fault attacks of Page and Vercauteren [27]. They presented an attack where they omit the last iteration of the Duursma and Lee algorithm. We generalize their approach to the Miller algorithm and for every iteration not only the last one.

4   Attack Against the Point Blinding Countermeasure during Miller's Algorithm

In this section, we first explain how the Miller algorithm can be implemented with the point blinding technic. As far as we know, this is the first time that an algorithm is proposed for the implementation of this counter measure. The aim of point blinding method is to add randomness to the known entry of the pairing computation. Indeed, a side channel attack is successful principally because the attacker knows the value of data combined with the secret. The point blinding countermeasure is made to blind the knowledge of the attacker. As the point R is random, the point $Q + R$ is also random. This countermeasure is considered as sufficient to prevent any side channel attack against a pairing implementation.

We then show how this countermeasure does not really protect the algorithm against fault attack.

4.1 Implementation of the Countermeasure

We discuss here the possible ways to implement the Miller algorithm using the point blinding countermeasure: $e(P, Q) = \frac{e(P,Q+R)}{e(P,R)}$.

Case 1: We consider that the secret is the point $P \in E(\mathbb{F}_q)$. The point $Q \in E(\mathbb{F}_{q^k})$ is public. The countermeasure consists in adding randomness to the point Q, expecting that it would be then impossible to perform the fault attack. The randomness is the choice of a point R such that the pairings $e(P, R)$ and $e(P, Q + R)$ are defined.

In practice, for optimization reason, k is smooth. In order to simplify the explanation, we consider that $k \equiv 0 \mod 2$. The point Q is represented as the image of a point Q' belonging to the twisted elliptic curve E' of E and defined over $\mathbb{F}_{q^{k/2}}$. The coordinates of Q are $Q = (x_Q, y_Q \sqrt{\nu})$, for a quadratic twist. If another twist is used, the scenario is the same, but the equation must be adapted in consequence.

The device is implemented to compute $\frac{e(P,Q+R)}{e(P,R)}$. For efficiency reasons, as these two pairing computations are performed during the scalar multiplication of the point P, the two computations $e(P, Q + R)$ and $e(P, R)$ would be done in parallel. In order to compute only one exponentiation on the elliptic curve. The inversion in the field $\mathbb{F}_{q^k}$ and the final exponentiation are expensive operations. So, once obtained the results $f_{m,P}(Q+R)$ and $f_{m,P}(R)$, it will be more efficient to perform the inversion followed by the final exponentiation instead of two final exponentiations followed by an inversion. In practice, the discussion about inverting the final exponentiation is the same for the altering point blinding countermeasure and the classical Miller algorithm recalled in Section 3.2. Given these efficiency considerations, the Miller algorithm that would be used for the point blinding countermeasure would likely to be as presented in Algorithm 2. For clarity of explanations, we add the inversion at the end of Miller algorithm (step 14), it could be performed outside the Miller algorithm and that would not change our discussion.

Case 2: We consider that the point $P \in E(\mathbb{F}_q)$ is public and the secret is the point $Q \in E(\mathbb{F}_{q^k})$. The randomness, considering the point blinding countermeasure would be added to the point P. The device would be implemented in order to compute $\frac{e(P+R,Q)}{e(R,Q)}$. The implementations of the two Miller algorithms would then be done either in parallel or consecutively. The choice would highly depend on the target for the implementation. On a multiple processor device the parallel solution would be preferred. On a constrained device, as a smart card, the computation would be done one after the other, or delegated to a more powerful device. Considering this hypothesis we do not try to give a general way to

---

Algorithm 1. Miller's Algorithm with the point blinding countermeasure

Input : $P \in E(\mathbb{F}_q)[r]$, $Q \in E(\mathbb{F}_{q^k})[r] \setminus E(\mathbb{F}_q)[r]$, $m = (1, m_{n-2}, \dots m_1, m_0)_2$.

Output: $\dfrac{f_{m,P}(Q+R)}{f_{m,P}(R)}$

1:    Choose R randomly in $E(\mathbb{F}_{q^k})[r] \setminus E(\mathbb{F}_q)[r]$
2:    If $R = -Q$, go to 1.
3:    Set $f \leftarrow 1$, $g \leftarrow 1$ and $T \leftarrow P$
4:    **For** $i = n - 2$ **down to** 0 **do**
5:          $f \leftarrow f^2 \cdot h_{T,T}(Q+R)$
6:          $g \leftarrow g^2 \cdot h_{T,T}(R)$
7:          $T \leftarrow 2T$
8:          **if** $m_i = 1$ **then**
9:             $f \leftarrow f \cdot h_{T,P}(Q+R)$
10:         $g \leftarrow g \cdot h_{T,P}(R)$
11:         $T \leftarrow T + P$
12:        **end if**
13:   **end for**
14:   **return** $\frac{f}{g}$

Fig. 2. The modified Miller algorithm

perform the computation. Indeed, either the same counter will be used and if it is modified once, it will be for the two computations. Either two counters will be used and then two faults would be necessary to modify them. The case of a delegation of the computation would require a whole article. We do not describe it here.

4.2 Description of the Attacks

We describe here the fault attack against the Miller algorithm implemented using the point blinding countermeasure $e(P,Q) = \frac{e(P,Q+R)}{e(P,R)}$.

Case 1: When the Secret is the Point P. We consider that the secret is the point P, we can freely choose the point Q and the randomness is the point R such that the pairings $e(P,R)$ and $e(P,Q+R)$ are defined. The device is implemented to compute $\frac{e(P,Q+R)}{e(P,R)}$ using the modified Miller algorithm described in the Algorithm 2.

The target of the fault attack is the counter given the number of iterations in the modified Miller algorithm. The aim of the fault is to reduce the number of iterations performed during the execution of the Miller algorithm. For instance, the fault can be induced by a laser [1,34] or a modification of the glitch [4]. The probability to obtain two shortened Miller algorithms with consecutive number of iterations is high enough to made this hypothesis realistic [11]. So, we suppose that we have obtained the results of the modified Miller algorithm after the m'^{th} and the $(m'+1)^{th}$ iterations, for m' an integer smaller than m the original number of iterations. We exactly know what happens during the $(m'+1)^{th}$ iteration.

Let f'_m and g'_m denote the results stored in f and g at the m'^{th} iteration, let m_i be the value of the corresponding bit. Then, in order to express $f_{m'+1}$ and $g_{m'+1}$ we must consider two possibilities, either the m_i is 0, or 1.

If $m_i = 0$, then $f_{m'+1} = f'^2_m \times h_{T,T}(Q + R)$ and $g_{m'+1} = g^2_m \times h_{T,T}(R)$, with $T = [1m_{n-1} \ldots m_{i+1}m_i]P$.

If $m_i = 1$ then $f_{m'+1} = \left(f'^2_m \times h_{T,T}(Q + R)\right) \times h_{2T,P}(Q + R)$ and $g_{m'+1} = \left(g'^2_m \times h_{T,T}(R)\right) \times h_{2T,P}(R)$. The attacker will receive the two values $\frac{f'_m}{g'_m}$ and $\frac{f_{m'+1}}{g_{m'+1}}$ in $\mathbb{F}_{q^k}$. We could be tempted to follow the scheme of the attacks described in [27,11], i.e. compute the exact value in $\mathbb{F}_{q^k}$ of the ratio $\frac{\frac{f_{m'+1}}{g_{m'+1}}}{\left(\frac{f'_m}{g'_m}\right)^2}$, use its theoretical decomposition (if $m_i = 0$ it is $\frac{h_{T,T}(Q+R)}{h_{T,T}(R)}$ or if $m_i = 1$ it is $\frac{h_{T,T}(Q+R) \times h_{2T,P}(Q+R)}{h_{T,T}(R) \times h_{2T,P}(R)}$) and after use the identification in the basis of $\mathbb{F}_{q^k}$ in order to obtain k equations depending on the coordinates or P, Q and R. The equation of the elliptic curve gives two more equations as P and R are on the curve.

But be careful! The point R is randomly chosen at each execution of the Algorithm 2. So in practice, we obtain $\frac{f'_m}{g'_m}(P, Q, R_1)$ and $\frac{f_{m'+1}}{g_{m'+1}}(P, Q, R_2)$, for R_1 and R_2 two random points in $E(\mathbb{F}_{q^k})[r] \setminus E(\mathbb{F}_q)[r]$. In this case, the theoretical decomposition of the ratio $\frac{\frac{f_{m'+1}}{g_{m'+1}}(P,Q,R_2)}{\frac{f'^2_m}{g'^2_m}(P,Q,R_1)}$ would not admit any simplification and the previous description inspired from [27,11] is no longer possible. We have to describe a more painful and awful attack.

In this attack, we need only one faulty result $\frac{f'_m}{g'_m}(P, Q, R)$, for P secret, Q chosen and R random. After one iteration of the Miller algorithm, assuming that the corresponding bits of m are 0, we have $f_1 = h_{P,P}(Q + R)$ and $g_1 = h_{P,P}(R)$. After two iterations, $f_2 = h_{[2]P,[2]P}(Q + R) \times (h_{P,P}(Q + R))^2$ and $g_2 = h_{[2]P,[2]P}(R) \times (h_{P,P}(R))^2$. We can express the equation of $h_{P,P}$ and $h_{[2]P,[2]P}$ in terms of the coordinates of P. The evaluation of these functions at the points $Q + R$ and R will give a polynomial expression in the coordinates of P and R.

The theoretical description of the coordinates of R will admit a decomposition in the basis of $\mathbb{F}_{q^k}$. If we are able to obtain the result of the Miller algorithm after m' iterations (denoted $\lambda_0 + \lambda_1\sqrt{\nu}$, with λ_0 and $\lambda_1 \in \mathbb{F}_{q^{k/2}}$), we have on one hand the theoretical description and on an other the value in $\mathbb{F}_{q^k}$ of this description:

$$\frac{f_{m'}(P, Q, R)}{g_{m'}(P, Q, R)} = \lambda_0 + \lambda_1\sqrt{\nu}. \tag{2}$$

We know the value of λ_0, λ_1 and the theoretical description of $f_{m'}(P, Q, R)$ and $g_{m'}(P, R)$. Exactly like at the end of the attack described in [11], by identification in the basis of $\mathbb{F}_{p^k}$, we obtain a system of k polynomial equations with coordinates of P and R as unknown. The degree of the polynomial depends on the number of iterations. That is why an important step of the attack is to minimize the number of iterations that are executed by the Miller algorithm.

As illustration we have for one iteration, $f_{m'}(P, Q, R) = h_{P,P}(Q + R)$ and $g_{m'}(P, R) = h_{P,P}(R)$. The equation 2 gives $h_{P,P}(Q+R) = (\lambda_0 + \lambda_1\sqrt{\nu}) \times h_{P,P}(R)$ which is a degree 3 polynomial in X_P, a degree 2 in Y_P, a degree 6 for Z_P and a degree 1 polynomial in x_R. We give the equations of $h_{P,P}(Q + R)$ and $h_{P,P}(R)$ (see section 2 for details) in order to illustrate an idea of the system.

$$P = (X_P, Y_P, Z_P), X_P, Y_P, Z_P \in \mathbb{F}_q$$
$$Q + R = (x_{Q+R}, y_{Q+R}\sqrt{\nu}), x_{Q+R}, y_{Q+R} \in \mathbb{F}_{q^{k/2}}$$
$$R = (x_R, y_R\sqrt{\nu}), x_R, y_R \in \mathbb{F}_{q^{k/2}}$$
$$h_{P,P}(Q + R) = 2Y_P Z_P^3 y_{Q+R}\sqrt{\nu} - 2Y_P^2 - (3X_P^2 + aZ_P^4)(x_{Q+R}Z_P^2 - X_P)$$
$$h_{P,P}(R) = 2Y_P Z_P^3 y_R\sqrt{\nu} - 2Y_P^2 - (3X_P^2 + aZ_P^4)(x_R Z_P^2 - X_P)$$

The equation of the elliptic curve in P and R gives us 2 more equations and we still have $k + 2$ unknown values in $\mathbb{F}_q$. To conclude the attack, we will use the Gröbner basis. In order to ensure the fact that the solution will be in $\mathbb{F}_q$, we have to add the equation $\xi^p \equiv \xi$ mod p for each unknown value. We therefore obtain a system of $2k + 2$ polynomial equations for k+2 unknown values. The Gröbner basis is the perfect tool for solving this system, that admits solutions by construction.

Obviously for a greater number of iterations, by hand it is difficult to develop the theoretical expression without any mistake. We do not describe it even for one iteration. Fortunately, we have mathematical softwares that can help us, like PariGP[33], Sage [32], Magma [7] or Maple [26].

If we consider that each iteration raise the degree of the polynomials in the system by a power of 2, than after μ iterations, the degree of the polynomial would be 2^μ in the coordinates of P. In practice, the evaluation of the degree is more complex. The degree of $h_{P,P}(R)$ is 6 in Z_P. After 2 iterations, the degree of $g_{2,P}(R)$ will be at the most $6 \times 2 + 6$ for Z_P and 3 for the coordinates of R. (The degree of g and f are the same, we choose to describe it for g for clarity. The degree of f depends on the coordinates of $Q + R$.) For n iterations, $n > 2$, we can estimate the degree of the polynomial with the formulas:

$$\deg(n, Z_P) = 2 \times \deg(n - 1, Z_P) + 6^{n-2} \times 13$$
$$\deg(n, R) = 2 \times \deg(n - 1, R) + 1,$$

where $\deg(n, Z_P)$ represents the degree of the polynomial system after n iterations in the unknown value Z_P and $\deg(n, R)$ is the degree in the coordinates of R. The degree of the polynomial for X_P and Y_P is smaller than the degree for Z_P.

The interesting question is how many iterations can we deal with? What would be the maximum degree of the polynomial system that can be solved by Gröbner basis in a reasonable time? We refer to [8] for more details on Gröbner basis.

Case 2: When the Secret is the Point Q. We consider that the secret is the point Q, we can freely choose the point P, the randomness is the point R such that the pairings $e(P+R, Q)$ and $e(R, Q)$ are defined. The device is implemented to compute $\frac{e(P+R,Q)}{e(R,Q)}$ using a modified version of the Miller algorithm. If the same counter is used to perform the computation it would be modified once and used for both computations. If two counters are used, as in [34] we need two faults to modify the counters. After that, the scheme of the attack is the same. Once we obtain the intermediate results $\frac{f'_m}{g'_m}(P, R, Q)$, for P public, R random and Q secret. The theoretical expression of R, $P + R$ and $h_{T,T}(Q)$ depending on the coordinates of P, R and Q combined with the value of $\frac{f'_m}{g'_m}(P, R, Q)$ will give a polynomial system in the unknown coordinates of R and Q. This polynomial system would be solved using the Gröbner basis.

5 Conclusion

In this paper we analysed the efficiency of the point blinding countermeasure in pairing based cryptography considering fault attacks in Miller's algorithm. We describe a theoretical fault attack. We highlighted the fact that the point blinding countermeasure alone is not a protection in the case of pairing based cryptography. Whenever the secret is the first or the second parameter, a fault attack gives the coordinates of the secret.

In our opinion, we believe that the only way to provide a secure implementation of the pairing relies on the discrete logarithm problem. The computation of $e(P, Q)$, should be $e([a]P, [b]Q)$, with a and b integers such that $ab \equiv 1 \mod r$. Of course, the computation of $[a]P$ and $[b]Q$ should be secured.

Acknowledgments. This work was supported by the French ANR-12-INSE-0014 SIM-PATIC Project financed by the Agence National de Recherche (France). We would like to thank the anonymous reviewers for their numerous suggestions and remarks which have enables us to substantially improve the paper.

References

1. Anderson, R., Kuhn, M.: Tamper resistance – a cautionary note. In: The Second USENIX Workshop on Electronic Commerce Proceedings, pp. 1–11 (1996)
2. Bae, K., Moon, S., Ha, J.: Instruction fault attack on the Miller algorithm in a pairing-based cryptosystem. In: 2013 Seventh International Conference on Innovative Mobile and Internet Services in Ubiquitous Computing (IMIS), pp. 167–174 (July 2013)
3. Barbulescu, R., Gaudry, P., Joux, A., Thomé, E.: A heuristic quasi-polynomial algorithm for discrete logarithm in finite fields of small characteristic. In: Nguyen, P.Q., Oswald, E. (eds.) EUROCRYPT 2014. LNCS, vol. 8441, pp. 1–16. Springer, Heidelberg (2014)

4. Blömer, J., da Silva, R.G., Günther, P., Krämer, J., Seifert, J.-P.: A practical second-order fault attack against a real-world pairing implementation. In: Proceedings of Fault Tolerance and Diagnosis in Cryptography (FDTC) (2014) (to appear), Updated version at http://eprint.iacr.org/2014/543

5. Blömer, J., Günther, P., Liske, G.: Improved side channel attacks on pairing based cryptography. In: Prouff, E. (ed.) COSADE 2013. LNCS, vol. 7864, pp. 154–168. Springer, Heidelberg (2013)

6. Boneh, D., Franklin, M.: Identity-Based Encryption from the Weil pairing. SIAM J. of Computing 32(3), 586–615 (2003)

7. Bosma, J., Cannon, W., Playout, C.: The Magma algebra system I. the user language. J. Symbolic Comput. 24(3-4), 235–265 (1997)

8. Buchberger, B.: An algorithm form finding the basis elements of the residue class ring of a zero dimensional polynomial ideal (phd thesis 1965). In: Elsevier (eds.) Journal of Symbolic Computation, vol. 41, pp. 475–511. Elsevier (2006)

9. Cohen, H., Frey, G. (eds.): Handbook of elliptic and hyperelliptic curve cryptography. Discrete Math. Appl. Chapman & Hall/CRC (2006)

10. Dutta, R., Barua, R., Sarkar, P.: Pairing-based cryptography: A survey. Cryptology ePrint Archive, Report 2004/064 (2004)

11. El Mrabet, N.: What about vulnerability to a fault attack of the Miller algorithm during an Identity Based Protocol? In: Park, J.H., Chen, H.-H., Atiquzzaman, M., Lee, C., Kim, T.-h., Yeo, S.-S. (eds.) ISA 2009. LNCS, vol. 5576, pp. 122–134. Springer, Heidelberg (2009)

12. El Mrabet, N: Fault attack against Miller's algorithm. IACR Cryptology ePrint Archive, 2011:709 (2011)

13. El Mrabet, N., Di Natale, G., Flottes, M.-L., Rouzeyre, B., Bajard, J.-C.: Differential Power Analysis against the Miller algorithm. Technical report. Published in Prime 2009. IEEE Xplore (August 2008)

14. El Mrabet, N., Page, D., Vercauteren, F.: Fault attacks on pairing-based cryptography. In: Joye, M., Tunstall, M. (eds.) Fault Analysis in Cryptography, Information Security and Cryptography, pp. 221–236. Springer, Heidelberg (2012)

15. Freeman, D., Scott, M., Teske, E.: A taxonomy of pairing-friendly elliptic curves. J. Cryptology 23(2), 224–280 (2010)

16. Chowdhury, D.R., Santosh, G., Debdeep, M.: Fault attack and countermeasures on pairing based cryptography. International Journal of Network Security 12(1), 21–28 (2011)

17. Hess, F.: Pairing lattices. In: Galbraith, S.D., Paterson, K.G. (eds.) Pairing 2008. LNCS, vol. 5209, pp. 18–38. Springer, Heidelberg (2008)

18. Hess, F., Smart, N., Vercauteren, F.: The Eta Pairing Revisited. IEEE Transactions on Information Theory 52, 4595–4602 (2006)

19. Iyama, T., Kiyomoto, S., Fukushima, K., Tanaka, T., Takagi, T.: Efficient implementation of pairing on brew mobile phones. In: Echizen, I., Kunihiro, N., Sasaki, R. (eds.) IWSEC 2010. LNCS, vol. 6434, pp. 326–336. Springer, Heidelberg (2010)

20. Joux, A.: A new index calculus algorithm with complexity $l(1/4 + o(1))$ in small characteristic. In: Lange, T., Lauter, K., Lisoněk, P. (eds.) SAC 2013. LNCS, vol. 8282, pp. 355–379. Springer, Heidelberg (2014)

21. Joye, M., Neven, G.: Identity-based Cryptography. Cryptology and information security series. IOS Press (2009)

22. Kawahara, Y., Takagi, T., Okamoto, E.: Efficient implementation of Tate pairing on a mobile phone using java. In: 2006 International Conference on Computational Intelligence and Security, vol. 2, pp. 1247–1252 (November 2006)

23. Koblitz, N., Menezes, A.: Pairing-based cryptography at high security levels. In: Smart, N.P. (ed.) Cryptography and Coding 2005. LNCS, vol. 3796, pp. 13–36. Springer, Heidelberg (2005)
24. Lashermes, R., Fournier, J., Goubin, L.: Inverting the final exponentiation of Tate pairings on ordinary elliptic curves using faults. In: Bertoni, G., Coron, J.-S. (eds.) CHES 2013. LNCS, vol. 8086, pp. 365–382. Springer, Heidelberg (2013)
25. Miller, V.: The Weil pairing and its efficient calculation. Journal of Cryptology 17, 235–261 (2004)
26. Monagan, M.B., Geddes, K.O., Heal, K.M., Labahn, G., Vorkoetter, S.M., McCarron, J., DeMarco, P.: Maple 10 Programming Guide. Maplesoft, Waterloo ON (2005)
27. Page, D., Vercauteren, F.: A fault attack on Pairing-Based Cryptography. IEEE Transactions on Computers 55(9), 1075–1080 (2006)
28. Park, J., Sohn, G., Moon, S.: Fault attack on a point blinding countermeasure of pairing algorithms. ETRI Journal 33(6) (2011)
29. Scott, M.: Computing the Tate pairing. In: Menezes, A. (ed.) CT-RSA 2005. LNCS, vol. 3376, pp. 293–304. Springer, Heidelberg (2005)
30. Scott, M., Benger, N., Charlemagne, M., Dominguez Perez, L.J., Kachisa, E.J.: On the Final Exponentiation for Calculating Pairings on Ordinary Elliptic Curves. In: Shacham, H., Waters, B. (eds.) Pairing 2009. LNCS, vol. 5671, pp. 78–88. Springer, Heidelberg (2009)
31. Scott, M., Costigan, N., Abdulwahab, W.: Implementing cryptographic pairings on smartcards. In: Goubin, L., Matsui, M. (eds.) CHES 2006. LNCS, vol. 4249, pp. 134–147. Springer, Heidelberg (2006)
32. Stein, W.: Sage mathematics software (version 4.8). The Sage Group (2012), http://www.sagemath.org
33. The PARI Group, Bordeaux. PARI/GP, version 2.7.0 (2014), http://pari.math.u-bordeaux.fr/.
34. Trichina, E., Korkikyan, R.: Multi fault laser attacks on protected CRT-RSA. In: 2010 Workshop on Fault Diagnosis and Tolerance in Cryptography (FDTC), pp. 75–86. IEEE (2010)
35. Vercauteren, F.: Optimal pairings. IEEE Trans. Inf. Theor. 56(1), 455–461 (2010)
36. Washington, L.C.: Elliptic curves, number theory and cryptography. Discrete Math. Aplli., Chapman and Hall (2008)
37. Whelan, C., Scott, M.: The importance of the final exponentiation in pairings when considering Fault Attacks. In: Takagi, T., Okamoto, T., Okamoto, E., Okamoto, T. (eds.) Pairing 2007. LNCS, vol. 4575, pp. 225–246. Springer, Heidelberg (2007)

Impossible Differential Properties
of Reduced Round Streebog

Ahmed Abdelkhalek, Riham AlTawy, and Amr M. Youssef[✉]

Concordia Institute for Information Systems Engineering
Concordia University, Montréal, Québec, Canada
youssef@ciise.concordia.ca

Abstract. In this paper, we investigate the impossible differential properties of the underlying block cipher and compression function of the new cryptographic hashing standard of the Russian federation Streebog . Our differential trail is constructed in such a way that allows us to recover the key of the underlying block cipher by observing input and output pairs of the compression function which utilizes the block cipher in Miyaguchi-Preneel mode. We discuss the implication of this attack when utilizing Streebog to construct a MAC using the secret-IV construction. Moreover, we present two versions of the attack with different time-data trade-offs.

Keywords: Cryptanalysis · Hash functions · MAC · Secret-IV · Miss in the middle · Impossible Differential · GOST R 34.11-2012 · Streebog

1 Introduction

In late 2012, Streebog [2] was announced as the new Russian cryptographic hashing standard GOST R 34.11-2012. It officially replaced GOST R 34.11-94 which has been theoretically broken in [25] and further analyzed in [24,23]. The output length of the Streebog hash function can be either 512 or 256-bit. Its compression function is based on a 12-rounds AES-like block cipher with 8×8-byte internal state, followed by an XOR operation with a whitening key. The compression function operates in Miyaguchi-Preneel mode and is plugged in Merkle-Damgård domain extender with a finalization step [19]. Literature related to the cryptanalysis of Streebog includes the analysis of the collision resistance of its compression function and internal cipher by AlTawy *et al.* [3], and Wang *et al.* [27]. An integral analysis of the compression function has been presented by AlTawy and Youssef where integral distinguishers for the reduced compression function was proposed [4]. Moreover, preimage attacks on the reduced hash function have been independently proposed by Altawy and Youssef [5], and Zou *et al.* [28], and later the attacks were improved by Bingka *et al.* [22]. Also, Kazymyrov and Kazymyrova presented an analysis of the algebraic aspects of the function [19], and a long second preimage attack was proposed by Guo *et al.* [17]. Finally, a malicious version of the whole hash function was presented in [7], and a differential fault analysis of the function when used in different MAC schemes was proposed [6].

© Springer International Publishing Switzerland 2015
S. El Hajji et al. (Eds.): C2SI 2015, LNCS 9084, pp. 274–286, 2015.
DOI: 10.1007/978-3-319-18681-8_22

A Message Authentication Code (MAC) [8] is a symmetric-key construction that provides mutual entity authentication and data integrity. Two common approaches are used to construct MAC schemes. The first approach employs a block cipher or a permutation, e.g., Cipher Block Chaining (CBC)-MAC [1], PELICAN-MAC [14], and ALPHA-MAC [13]. The second approach is based on hash functions where a secret key shared between the communicating parties is processed in a specific construction by the hash function which is consequently viewed as a keyed hash function. Examples of this approach include simple prefix MAC [26], secret-IV MAC [26], NMAC [8], and the internationally standardized HMAC [8]. Attacks on MAC schemes usually aim to investigate their resistance against forgery attacks and key recovery attacks. The latter attack is more devastating since it directly grants the attacker the ability to impersonate any of the communicating parties and consequently forge any given message. As a result, analyzing hash-based MACs with respect to key recovery attacks has been the main aspect of many proposed works [18,16].

When considering a hash function in a given MAC scheme, the first step is to analyze the security of the underlying primitives operating in the secret key model against key recovery attacks. Consequently, key recovery attacks on the underlying primitives has been considered as a valuable analytic model for the hash function. Such model has been adopted by Bouillaguet et al. in their analysis of the SHA-3 submission Lesamnta [11], where they presented a key recovery attack on the internal cipher reduced to 22 rounds. Additionally, the cryptanalysis of the SHA-3 submission EDON-R [21], where Laurent presented a key recovery attack on the function used in the Secret-IV MAC. One of the prospective applications of Streebog, as any other hash function, is using it in MAC schemes. Though both the simple prefix and the secret-IV MACs are vulnerable to length extension attacks, and the nested HMAC construction is internationally standardized, Streebog is by design not vulnerable to length extension attacks. This property may tempt users to adopt simpler MAC constructions such as the secret-IV setting. In this approach, the standard initial value is replaced by the secret key in the iterative construction of the hash function. More formally, $MAC(M) = H(K, M)$, where $H(K, M)$ is the hash value of the message M using the secret key K as the IV. Indeed, the designers of the NIST SHA-3 hash function, keccak [9] [12], state on their website that since keccak is not vulnerable to length extension attacks, it does not need HMAC and propose that MAC computation can be done by concatenating the key with the message [20]. It should also be noted that the proof of security of the Miyaguchi-Preneel mode assumes that the underlying block cipher is ideal and must exhibit no distinguishing property. Accordingly, the results presented in this work are also interesting from this perspective since they are relevant to these indistinguishability claims.

In 2000, Biham and Keller presented a 4-round impossible differential property of AES [10] was the basis for all the succeeding impossible differential attacks on AES. This property specifies that given an input pair at round i which has just one non zero difference byte (in the literature, this is usually referred to as an active byte), the corresponding pair at round $i + 3$ cannot be equal in any of the four columns after applying the ShiftRow transformation. This 4-round impossible differential property consists of two deterministic differentials; one 2-round forward differential and the other is a 2-round backward differential that contradict with each other.

In this work, we provide a security evaluation of the Streebog compression function when used in the secret key model where the IV is replaced by a secret key. More precisely, we present an impossible differential (ID) property of the underlying block cipher and compression function and employ it to recover the secret-IV of the compression function. We also present two versions of the attack in a time-data trade-off approach where one version uses less message queries but requires more time and memory complexity while the other needs less time and memory complexity but requires more message queries. Table 1 provides a summary of current cryptanalytic results on the Streebog hash function. The rest of the paper is organized as follows. In the next section, the specification of the Streebog hash function along with the notation used throughout the paper are provided. A brief overview of impossible differential cryptanalysis is given

Table 1. Summary of the current cryptanalytic results on Streebog

Target	#Rounds	Time	Memory	Data	Attack	Reference
Internal cipher	5	2^8	2^8	-	Free-start collision	[3]
	8	2^{64}	2^8	-		
	3.75	-	-	-	ID distinguisher	Sec. 3
Internal permutation	6.5	2^{64}	-	2^{64}	Distinguisher	[4]
	7.5	2^{120}	-	2^{120}		
Compression function	7.75	2^{184}	2^8	-	Semi free-start collision	[3]
	4.75	2^8	-	-		
	7.75	2^{72}	2^8	-	Semi free-start near collision	
	8.75	2^{128}	2^8	-		
	9.75	2^{184}	2^8	-		
	6.75	$2^{399.5}$	2^{349}	$2^{427.1}$	Secret-IV recovery	Sec. 4
	6.75	$2^{261.5}$	2^{205}	$2^{495.5}$		
	6	2^{64}	-	2^{64}	Distinguisher	[4]
	7	2^{120}	-	2^{120}		
Hash function	5	2^{122}	2^{64}	-	Collision	[28]
	6	2^{496}	2^{64}	-	Preimage	[22]
	12	-	2^{14}	-	Differential fault analysis	[6]
	12	2^{266}	–	-	Long second preimage	[17]

in Section 3. Afterwards, in Section 4, we provide detailed description of the impossible differential attack on the block cipher and the complexity of the attack. Finally, the paper is concluded in Section 5.

2   Specification of Streebog

Streebog outputs a 512 or 256-bit hash value and can process up to 2^{512}-bit message. The compression function iterates over 12 rounds of an AES-like cipher with an 8×8 byte internal state and a final round of key mixing. The compression function operates in Miyaguchi-Preneel mode and is plugged in Merkle-Damgård domain extender with a finalization step. The input message M is padded into a multiple of 512 bits by appending one followed by zeros. Given $M = m_n \| .. \| m_1 \| m_0$, the compression function g_N is fed with three inputs: the chaining value h_{i-1}, a message block m_{i-1}, and the block size counter $N_{i-1} = 512 \times i$. (see Figure 1). Let h_i be a 512-bit chaining variable. The first state is loaded with the initial value IV and assigned to h_0. The hash value of M is computed as follows:

$$h_i \leftarrow g_N(h_{i-1}, m_{i-1}, N_{i-1}) \text{ for } i = 1, 2, .., n+1$$
$$h_{n+2} \leftarrow g_0(h_{n+1}, |M|, 0)$$
$$h(M) \leftarrow g_0(h_{n+2}, \sum(m_0, .., m_n), 0),$$

where $h(M)$ is the hash value of M. As depicted in Figure 1, the compression function g_N consists of:

- K_N: a nonlinear whitening round of the chaining value. It takes a 512-bit chaining variable h_{i-1} and the block size counter N_{i-1} and outputs a 512-bit key K.
- E: an AES-based cipher that iterates over the message for 12 rounds in addition to a finalization key mixing round. The cipher E takes a 512-bit key K and a 512-bit message block m as a plaintext. As shown in Figure 2, it consists of two similar parallel flows for the state update and the key scheduling.

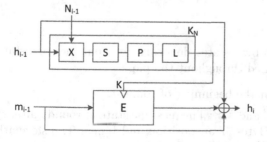

Fig. 1. Streebog's compression function g_N

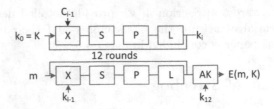

Fig. 2. The internal block cipher (E)

Both K_N and E operate on an 8×8 byte key state K. E updates an additional 8×8 byte message state M. In one round, the state is updated by the following sequence of transformations

- AddKey(X): XOR with either a round key, a constant, or a block size counter (N)
- SubBytes (S): A nonlinear byte bijective mapping.
- Transposition (P): Byte permutation.
- LinearTransformation (L): Left multiplication by an MDS matrix in GF(2).

Initially, state K is loaded with the chaining value h_{i-1} and updated by K_N as follows:

$$k_0 = L \circ P \circ S \circ X(N_{i-1})$$

Now K contains the key k_0 to be used by the cipher E. The message state M is initially loaded with the message block m and $E(k_0, m)$ runs the key scheduling function on state K to generate 12 round keys $k_1, k_2, .., k_{12}$ as follows:

$$k_i = L \circ P \circ S \circ X(C_{i-1}), \text{ for } i = 1, 2, .., 12,$$

where C_{i-1} is the ith round constant. The state M is updated as follows:

$$M_i = L \circ P \circ S \circ X(k_{i-1}), \text{ for } i = 1, 2, ..., 12.$$

The final round output is given by $E(k_0, m) = M_{12} \oplus k_{12}$. The output of g_N in the Miyaguchi-Preneel mode is $E(K_N(h_{i-1}, N_{i-1}), m_{i-1}) \oplus m_{i-1} \oplus h_{i-1}$. For further details, the reader is referred to [2].

2.1   Notation

Let M be (8×8)-byte states denoting an input message state. The following notation will be used throughout the paper:

- M_i^I: A state at the beginning of round i.
- M_i^X, M_i^S, M_i^P and M_i^O: The message state at round i after the application of AddKey, SubBytes, Transposition and Linear Transformation, respectively. intuitively, $M_{i-1}^O = M_i^I$ for $i >= 2$.

- $M_i[r, c]$: A byte at row r and column c of state M_i. Another representation of state bytes is an enumeration $0, 1, 2, 3, ..., 63$ as shown in Figure 3.
- $M_i[\text{row } r]$: Eight bytes located at row r of M_i state.
- $M_i[\text{col } c]$: Eight bytes located at column c of M_i state.

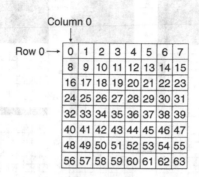

Fig. 3. The 8×8 state of Streebog

3   Impossible Differential Cryptanalysis of the Compression Function

Although the Streebog compression function employs an AES-like cipher, applying commonly used 4-round impossible differential of AES as is on the Streebog compression function would not be of value as in this case, we would recover the key of the last round of the block cipher masked by the chaining value (recall that Streebog's compression function works in Miyaguchi-Preneel mode). Therefore, we opted to reverse the impossible differential as detailed below to help recover the key of the first round, i.e., k_0. Since the key scheduling is bijective, once k_0 is recovered, we can recover the secret chaining value in the case of a secret-IV MAC construction when applied only at the level of the compression function. We note that the impossible differential property of the compression function would be limited to 3.75-rounds because, unlike the AES, in the Streebog underlying block cipher, the linear transformation in the last round is not omitted. As depicted in Figure 4, this impossible differential property states that given a pair of M_i^I with any 7 active bytes in the same arbitrary row (row 0 is chosen in the figure for illustration purposes), M_{i+3}^P cannot have only one active byte (similarly, that active byte can be any byte out of the 64 bytes state). The deterministic differentials in this property are different than those of the

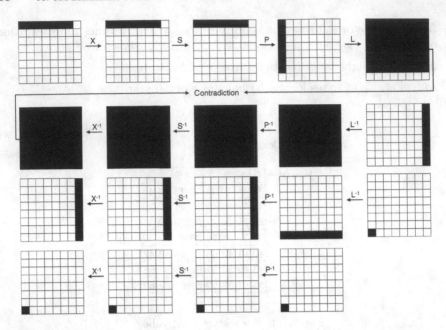

Fig. 4. Impossible differential property of the internal block cipher

AES property; on top of being swapped, the forward differential is just 1-round while the backward differential is 2.75 rounds. The property is rationalized as follows: any 7 active bytes in the same row of M_i^I give 56 active bytes by M_i^O with one entire row being equal (which row will depend on the position of the zero-difference byte in the input). On the other hand, one active byte in M_{i+3}^P leads to a full active state where all 64 bytes are active in M_{i+1}^I, which means that the middle states contradicts with each other as illustrated in Figure 4. As explained above, this impossible differential holds regardless of the row and also the positions within that row. Figure 5 gives an example for impossible input and output patterns for the compression function. When the compression function message input has specific non-zero difference at bytes 0 to 6 (δ_0 to δ_6 in the figure) and zero difference in all the other bytes, then after the feedforward the output difference cannot have the same values as the input difference at bytes 0 to 6 (δ_0 to δ_6) and a non-zero difference at byte 56 (δ_7). Such input and output difference patterns are impossible on the compression function level. It is to be note that there exists $8 \times 8 \times 255^7$ such input patterns (the input differences can be in any row and the inactive byte can be at any column of that row and each of the differences can take $2^8 - 1$ possible values. There are also $((7 \times 8) + 1) \times 255 = 57 \times 255$ contradicting output patterns (the non-zero output difference byte can be at any column of 7 rows, i.e., all but the input differences row and takes the position of the inactive byte on that row).

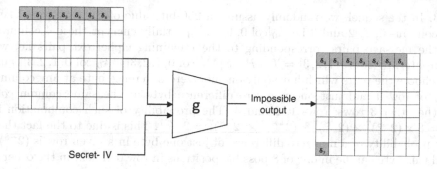

Fig. 5. Example of impossible differentials for the 3.75 round reduced compression function

4 Impossible Differential Attack on 6.75 rounds of the Compression Function

Considering the aforementioned impossible differential property, a 6.75-rounds attack on the Streebog compression function can be mounted as detailed hereafter. In our attack model, we assume access to the keyed Streebog reduced compression function oracle which allows us to query the keyed oracle with chosen messages and get the corresponding compression function outputs. Later, we show how the attack can be done in less time but at the expense of requiring more message queries using a simple time-data trade-off approach.

4.1 Attack Algorithm

The first version of the attack is illustrated in Figure 6. In what follows, we give its details.

1. The keyed compression function oracle is fed with 2^n structures where each structure consists of 2^{256} messages having the same value in columns 4, 5, 6 and 7 and assuming all possible 2^{256} values in columns 0, 1, 2 and 3. Accordingly, each structure offers $2^{256} \times 2^{256} \times 1/2 \simeq 2^{511}$ pairs of messages. Thus we have a total of 2^{n+256} messages, and 2^{n+511} message pairs for the 2^n structures.

2. Since we have access to the output of the compression function which operates in Miyaguchi-Preneel mode as depicted in Figure 1, the output h_i that we observe is $h_{i-1} \oplus m_{i-1} \oplus c_{i-1}$ where c_{i-1} is the corresponding output of the reduced variant of the block cipher when its input is m_{i-1}. Therefore, for each message query, we first XOR the compression function output with the corresponding input message, i.e., $h_i \oplus m_{i-1}$, to get $c_{i-1} \oplus h_{i-1}$ and keep only the pairs that have non-zero difference in just one column. Consequently, it is expected to have $2^{n+511} \times 2^{-448} = 2^{n+63}$ pairs.

3. In the sequel, we randomly assume a 256-bit value of the first round key at columns 0, 1, 2 and 3 i.e. $k_0[\text{col } 0, 1, 2, 3]$, partially encrypt these 4 columns of the message pairs corresponding to the remaining ciphertext pairs i.e. we compute $M_1^O[\text{row } 0, 1, 2, 3] = L \circ P \circ S[M_1^I[\text{col } 0, 1, 2, 3] \oplus k_0[\text{col } 0, 1, 2, 3]]$ and we choose the pairs which have only one non-zero difference byte at any column col_i of row 0 and just one non-zero difference byte at the same column col_i in the other 3 rows; rows 1, 2 and 3. The probability of such combination is $q_1 = 8 \times (2^{-8})^7 \times (2^{-8})^7 \times (2^{-8})^7 \times (2^{-8})^7 = 2^{-221}$. This is due to the fact that the probability of a non-zero difference of just one byte in a given row is $(2^{-8})^7$ and this byte can be in one of 8 possible positions in row 0. Then, in the other 3 rows the non-zero byte will be at a fixed position, i.e., the same position of the non-zero difference byte in row 0. Accordingly, $2^{n+63} \times 2^{-221} = 2^{n-158}$ message pairs are expected to pass after this step.

4. Afterwards, we assume a 32-bit value for the bytes 0, 1, 2 and 3 of column 0 of the key k_1 i.e. bytes 0, 8, 16, 24 as in Figure 3 (As discussed, this could have been any other column as well), partially encrypt these bytes through the second round to compute $M_2^O[\text{row } 0] = L \circ P \circ S[M_2^I[(0, 1, 2, 3), 0] \oplus k_1[(0, 1, 2, 3), 0]$. We choose the pairs which have only one zero-difference byte at any column of that row. The probability of such pairs is $q_2 = 8 \times 2^{-8} = 2^{-5}$. So after this step, we have $2^{n-158} \times 2^{-5} = 2^{n-163}$ message pairs.

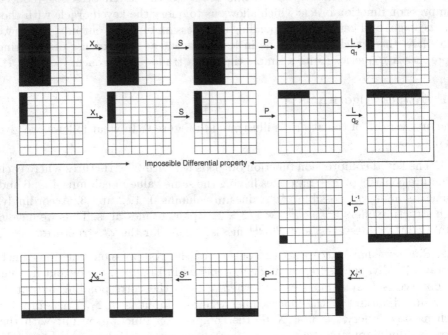

Fig. 6. 6.75-rounds impossible differential attack on the compression function

5. Following on, we assume a 64-bit value for the last column of $k_7[\text{col } 0] \oplus h_{i-1}[\text{col } 0]$ so that we end up with the block cipher output (Note: h_{i-1} is the

targeted secret-IV to be recovered and it has the same value for all the pairs we have). Specifically, we calculate $M_7^P[\text{col } 7] = (c_{i-1}[\text{col } 7] \oplus h_{i-1}[\text{col } 7]) \oplus (k_7[\text{col } 7] \oplus h_{i-1}[\text{col } 7])$ for all the pairs we have so far. The former value is the output we get from step 2, while the latter is the value we just assumed.

6. For each of the filtered pairs, we partially decrypt column 7. In other words, we compute $L^{-1} \circ ((S^{-1} \circ P^{-1}(M_7^P[\text{col } 7])) \oplus (S^{-1} \circ P^{-1}(M_7^{*P}[\text{col } 7]))$ and we choose the pairs which have only one non-zero difference byte at any position of row 7, which happens with probability $p = 8 \times (2^{-8})^7 = 2^{-53}$. This difference is impossible, hence each key (or to be exact each $k_7 \oplus h_{i-1}$ i.e., $k_7 \oplus const$) that results with such a difference is a wrong key. Therefore, after analyzing 2^{n-163} pairs, only $2^{64} \times (1 - 2^{-53})^{2^{n-163}} \simeq 2^{64} \times (e^{-1})^{2^{n-216}} \simeq 2^{64} \times 2^{-1.4 \times (2^{n-216})}$ wrong values of the last column of $k_7 \oplus h_{i-1}$ remains.

To be able to find the correct partial keys, we discard the 64-bit values for $k_7 \oplus h_{i-1}$ unless the initial guess of the 256-bit value of k_0 and the 32-bit value of k_1 is correct. The wrong values (k_0, k_1, k_7) remain with probability: $(2^8)^{(32+4)} \times 2^{64} \times 2^{-1.4 \times (2^{n-163})} = 2^{352-1.4 \times (2^{n-163})}$ which should be made as small as possible, e.g., less than 2^{-30} (that value is chosen to maximize the probability of finding the correct tuple without having a significant impact on the number of messages needed) which means $2^{352-1.4 \times (2^{n-163})} < 2^{-30}$ resulting in $n > 171.09$. Accordingly, when we start with $2^{171.1}$ structures and there remains a value of $k_7 \oplus h_{i-1}$, we consider the assumed 256-bit value for k_0 correct and the probability of wrong values (k_0, k_1, k_7) is $2^{-32.1}$.

4.2 Attack Complexity

With n set to 171.1, the attack requires $2^{n+256} = 2^{427.1}$ chosen messages. The time complexity of the attack is calculated as follows:

- In step 3, row 0 requires $2 \times 2^{64} \times 2^{n+63} \times 1/8 = 2^{n+125}$ one round encryptions, row 1 requires $2 \times 2^{64} \times 2^{64} \times 2^{n+10} \times 1/8 = 2^{n+136}$ one round encryptions, row 2 requires $2 \times 2^{64} \times 2^{64} \times 2^{64} \times 2^{n-46} \times 1/8 = 2^{n+144}$ one round encryptions and row 3 requires $2 \times 2^{64} \times 2^{64} \times 2^{64} \times 2^{64} \times 2^{n-102} \times 1/8 = 2^{n+152}$ one round encryptions.
- In step 4, row 0 requires $2 \times 2^{256} \times 2^{32} \times 2^{n-158} \times 1/16 = 2^{n+127}$ one round encryptions.
- In step 6, column 7 decryption requires $2 \times 2^{256} \times 2^{32} \times 2^{64} \times (1 + (1 - 2^{-53}) + (1 - 2^{-53})^2 + ... + (1 - 2^{-53})^{2^{n-163}}) \times 1/16 \simeq 2^{402}$ one round encryptions.
- For $n = 171.1$, the overall complexity of the attack is about $(2^{296.1} + 2^{307.1} + 2^{315.1} + 2^{323.1} + 2^{298.1} + 2^{402})/6.75 \simeq 2^{399.5}$ encryptions to recover 256 bits of k_0.

Then, the other half of k_0 can be found by an exhaustive search. Hence the whole k_0 can be recovered with time complexity of $2^{399.5} + 2^{256} \simeq 2^{399.5}$ queries. Once k_0 is recovered, we can easily recover the secret-IV h_{i-1}. Finally, the memory requirements is dominated by the memory needed to store the list of the deleted key tuples (k_0, k_1, k_7), so we need $2^{352}/2^3 = 2^{349}$ bytes.

4.3   Time-Data Trade-Off to Recover the Secret-IV

In the above variant of the attack, the attack is launched with 4 active columns in the messages. However, if as illustrated in Figure 7, the messages are chosen so that they have just 2 active columns, we will be able to launch the attack successfully using a smaller number of queries but with more data. Indeed, in such variant, we need $2^{495.5}$ messages, the time complexity drops to around $2^{261.5}$ and the memory requirement is reduced to $2^{208}/2^3 = 2^{205}$ bytes. The change in the messages will reflect on the attack algorithm details but no major changes in the attack strategy.

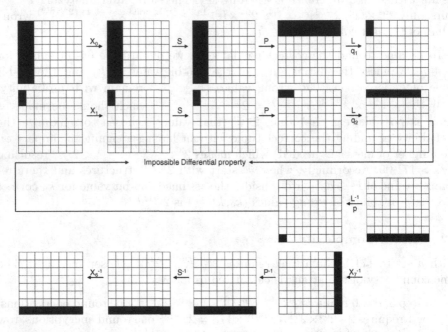

Fig. 7. Another 6.75-rounds impossible differential attack requiring less time and memory but more data

5   Conclusion

In this paper, we have analyzed the Streebog compression function in the secret key model. More precisely, we have proposed Secret-IV recovery attack, at the level of the compression function, based on impossible differential properties of the compression function. The attack has a trade-off between data and time. One variant of the attack requires $2^{427.1}$ messages, has a time complexity equivalent to $2^{399.5}$ queries to the compression function reduced to 6.75 rounds and needs

2^{349} bytes of memory. If more data is permissible, the attack can be performed with $2^{261.5}$ queries of the Streebog compression function reduced to 6.75 rounds with $2^{495.5}$ messages and 2^{205} bytes of memory.

Finally, it should be noted that this attack does not directly contradict the security claims of Streebog and does not present any immediate practical threat to its security. However, it helps as a cautionary note for using Streebog in this mode since it might be tempting to do so because the finalization stage of Streebog is strengthened against length extension attacks which are the main reasons for not using secret-IV or secret-prefix MAC constructions. It is interesting to not that, while the results in [15] show that the modular sum finalization stage weakens the function when used in HMAC construction, extending our attack to the full hash function remains a challenge and requires further investigation of the compression function one wayness properties.

Acknowledgment. The authors would like to thank the anonymous reviewers for their valuable comments and suggestions that helped improve the quality of the paper. This work is supported by the Natural Sciences and Engineering Research Council of Canada (NSERC).

References

1. ISO/IEC 9797-1. Information technology-security techniques-data integrity mechanism using a cryptographic check function employing a block cipher algorithm. international organizatoin for standards

2. The National Hash Standard of the Russian Federation GOST R 34.11-2012. Russian Federal Agency on Technical Regulation and Metrology report (2012), https://www.tc26.ru/en/GOSTR34112012/GOST_R_34_112012_eng.pdf

3. AlTawy, R., Kircanski, A., Youssef, A.M.: Rebound attacks on striibog. In: Lee, H.-S., Han, D.-G. (eds.) ICISC 2013. LNCS, vol. 8565, pp. 175–188. Springer, Heidelberg (2014)

4. Altawy, R., Youssef, A.M.: Integral distinguishers for reduced-round Stribog. Information Processing Letters 114(8), 426–431 (2014)

5. AlTawy, R., Youssef, A.M.: Preimage Attacks on Reduced-Round Stribog. In: Pointcheval, D., Vergnaud, D. (eds.) AFRICACRYPT. LNCS, vol. 8469, pp. 109–125. Springer, Heidelberg (2014)

6. AlTawy, R., Youssef, A.M.: Differential Fault Analysis of Streebog. In: Lopez, J., Wu, Y. (eds.) Information Security Practice and Experience. LNCS, vol. 9065, pp. 35–49. Springer, Heidelberg (2015)

7. Altawy, R., Youssef, A.M.: Watch your Constants: Malicious Streebog. IET Information Security (2015) (to appear)

8. Bellare, M., Canetti, R., Krawczyk, H.: Keying hash functions for message authentication. In: Koblitz, N. (ed.) CRYPTO 1996. LNCS, vol. 1109, pp. 1–15. Springer, Heidelberg (1996)

9. Bertoni, G., Daemen, J., Peeters, M., Van Assche, G.: Keccak sponge function family main document. Submission to NIST (Round 2) 3 (2009)

10. Biham, E., Keller, N.: Cryptanalysis of reduced variants of Rijndael. In: 3rd AES Conference, New York, USA (2000)

11. Bouillaguet, C., Dunkelman, O., Leurent, G., Fouque, P.-A.: Attacks on hash functions based on generalized feistel - application to reduced-round Lesamnta and SHAvite-3$_{512}$. Cryptology ePrint Archive, Report 2009/634 (2009), http://eprint.iacr.org/2009/634.pdf.
12. Chang, S.-J., Perlner, R., Burr, W.E., Turan, M.S., Kelsey, J.M., Paul, S., Bassham, L.E.: Third-round report of the SHA-3 cryptographic hash algorithm competition. Citeseer (2012)
13. Daemen, J., Rijmen, V.: A New MAC Construction ALRED and a Specific Instance ALPHA-MAC. In: Gilbert, H., Handschuh, H. (eds.) FSE 2005. LNCS, vol. 3557, pp. 1–17. Springer, Heidelberg (2005)
14. Daemen, J., Rijmen, V.: The Pelican MAC function. Cryptology ePrint Archive, Report 2005/088 (2005), http://eprint.iacr.org/2005/088.pdf
15. Dinur, I., Leurent, G.: Improved Generic Attacks against Hash-Based MACs and HAIFA. In: Garay, J.A., Gennaro, R. (eds.) CRYPTO 2014, Part I. LNCS, vol. 8616, pp. 149–168. Springer, Heidelberg (2014)
16. Fouque, P.-A., Leurent, G., Nguyen, P.Q.: Full Key-Recovery Attacks on HMAC/NMAC-MD4 and NMAC-MD5. In: Menezes, A. (ed.) CRYPTO 2007. LNCS, vol. 4622, pp. 13–30. Springer, Heidelberg (2007)
17. Guo, J., Jean, J., Leurent, G., Peyrin, T., Wang, L.: The Usage of Counter Revisited: Second-Preimage Attack on New Russian Standardized Hash Function. In: Joux, A., Youssef, A. (eds.) SAC 2014. LNCS, vol. 8781, pp. 195–211. Springer, Heidelberg (2014)
18. Handschuh, H., Preneel, B.: Key-Recovery Attacks on Universal Hash Function Based MAC Algorithms. In: Wagner, D. (ed.) CRYPTO 2008. LNCS, vol. 5157, pp. 144–161. Springer, Heidelberg (2008)
19. Kazymyrov, O., Kazymyrova, V.: Algebraic aspects of the russian hash standard GOST R 34.11-2012. In: CTCrypt, pp. 160–176 (2013), http://eprint.iacr.org/2013/556
20. Keccak team. "Strengths of Keccak - Design and security", http://keccak.noekeon.org/ (last accessed February 20, 2014)
21. Leurent, G.: Practical key recovery attack against secret-IV EDON-$\mathcal{R}$. In: Pieprzyk, J. (ed.) CT-RSA 2010. LNCS, vol. 5985, pp. 334–349. Springer, Heidelberg (2010)
22. Ma, B., Li, B., Hao, R., Li, X.: Improved Cryptanalysis on Reduced-Round GOST and Whirlpool Hash Function. In: Boureanu, I., Owesarski, P., Vaudenay, S. (eds.) ACNS 2014. LNCS, vol. 8479, pp. 289–307. Springer, Heidelberg (2014)
23. Matyukhin, D., Shishkin, V.: Some methods of hash functions analysis with application to the GOST P 34.11-94 algorithm. Mat. Vopr. Kriptogr 3, 71–89 (2012)
24. Mendel, F., Pramstaller, N., Rechberger, C.: A (Second) Preimage Attack on the GOST Hash Function. In: Nyberg, K. (ed.) FSE 2008. LNCS, vol. 5086, pp. 224–234. Springer, Heidelberg (2008)
25. Mendel, F., Pramstaller, N., Rechberger, C., Kontak, M., Szmidt, J.: Cryptanalysis of the GOST Hash Function. In: Wagner, D. (ed.) CRYPTO 2008. LNCS, vol. 5157, pp. 162–178. Springer, Heidelberg (2008)
26. Preneel, B., van Oorschot, P.C.: On the security of iterated message authentication codes. IEEE Transactions on Information Theory 45(1), 188–199 (1999)
27. Wang, Z., Yu, H., Wang, X.: Cryptanalysis of GOST R hash function. Information Processing Letters 114(12), 655–662 (2014)
28. Zou, J., Wu, W., Wu, S.: Cryptanalysis of the Round-Reduced GOST Hash Function. In: Lin, D., Xu, S., Yung, M. (eds.) Inscrypt 2013. LNCS, vol. 8567, pp. 307–320. Springer, Heidelberg (2014)

Security Issues on Inter-Domain Routing with QoS-CMS Mechanism

Hafssa Benaboud[1]([✉]), Sara Bakkali[1],
and José Johnny Randriamampionona[1,2]

[1] LRI, Performance Evaluation Team, FS Rabat, Mohammed V University
[2] Nokia Networks, Ankorondrano, Madagascar
benaboud@fsr.ac.ma, bakkalisara@gmail.com,
jose.randriamampionona@nsn.com

Abstract. Ensuring end-to-end Quality of Service for traffic that traverse multiple Autonomous Systems (AS) is today a major challenge for ISPs. the QoS requirement became inevitable with the evolution of the amount of traffic flowing over the Internet, and also of the important diversity of these traffic types. Each type of traffic requires a specific QoS parameters. To respond to this need, we proposed in a previous work a new mechanism which is mainly based on Class Manager server set in each AS and can provide the same traffic QoS guarantees even during the passage through several ASs. These Class Manager Servers collect information concerning quality of service set up within the AS, and then they exchange them. This exchange may present a serious security weakness of all the architecture, and it can be an important vulnerability of the whole network. In this paper, we discuss the main security issues of the proposed mechanism, concerning the communication between the CM server and the internal routers and also the exchange between the CM Servers. We give an architecture to avoid vulnerablity during the exchange of information.

Keywords: Inter-domain routing · QoS management · Class Manager Server · Vulnerability

1 Introduction

Internet is an interconnection of multiple IP networks. Initially, an IP network was conceived to route only the traffic data which requires a very limited QoS constraints concerning a low loss ratio. So, the Internet network does not have to provide any QoS and it's named best effort network. However, with the permanent internet's evolution due to the huge quatity and diversity of applications and services offered to clients (voice over IP, streaming, ...), IP networks have to route several types of traffic. Considering the wide diversity of traffic and the limited network ressources, the best-effort IP network cannot route these traffic in the appropriate conditions, since each type of traffic has its own QoS requirements. To solve this problem, several models and techniques have been proposed

© Springer International Publishing Switzerland 2015
S. El Hajji et al. (Eds.): C2SI 2015, LNCS 9084, pp. 287–296, 2015.
DOI: 10.1007/978-3-319-18681-8_23

that ensure QoS in an IP network, and differentiate network behavior according to each type of traffic ([3], [4], [5]). However, the problem remains unsolved for any traffic crossing multiple domains. Therefore, some research is currently focusing on how to ensure end-to-end QoS in an inter-domain environment and the corresponding challenge is growing with the current Internet infrastructure. ([13],[14], [15]).

In order to ensure the real inter-domain end-to-end QoS, we proposed in [1] a new method, which we call QoS-CMS in this paper, and is based on Class Manager servers. These Class Manager Servers collect all the information related to the QoS in each edge router in their own AS and exchange them to each other. Each CM server has to communicate with every edge router in the AS in order to fill its CT table. During these communications, routers send to the CM server all the data in clear text which could cause an important security flaw. Also, the exchange between neighboring CM Servers is transported using a regular TCP session and this could present a vulnerability of the whole mechanism.

In this paper, we discuss these security weaknesses and we propose some alternatives to secure the exchange, both between the Edge routers and the CM Server, and between the CM servers themselves. We will show also that, the vulnerability problem will be solved without degrading the performances of the QoS-CMS mechanism in term of delay and loss rate.

This paper is organized as follows. We first recall the principle of our mechanism in Section 2. Section 3 discusses the problem of security between the CM server and edge routers, as well as between CM servers. Section 4 provides some alternatives to overcome security problems. Finally, we conclude our paper in section 5.

2   Mechanism Principle

Before giving a brief description of our mechanism principle, we present some related works.

2.1   Related Works

Ensuring QoS for traffic circulating within the same AS, can be performed by several solutions and technologies that have been proposed and implemented, such as IntServ (Integrated Services) [3] model, DiffServ (Differentiated Services) [4] model or even MPLS (Multi Protocol Label Switching) [5]. However, traffic that crosses two different ASs is still facing a problem concerning routing conditions. This problem is mainly due to the fact that QoS constraints, required by the client and which the operator undertakes to provide (usually specified in the Service Level Agreement, SLA [6]), are defined in the classes of service, while the definition of the classes of service is assured by the domain administrator, they are consequently specific to each domain, and are valid only within this domain. Also, information concerning topology and available resources on the links, that are necessary for ensuring QoS, cannot be communicated between the various operators that are in competition. Thus, in this case, at the transition to another domain, the QoS constraints offered to the traffic will not be the same as in the source domain. Therefore the QoS required by the client at the beginning will not be provided from the end-to-end until its destination.

Various studies have been conducted to propose solutions that solve the problem of inter-domain routing with QoS constraints. The proposed solutions can be classified into two main classes:

- Theoretical (or analytical) solutions: mainly based on algorithms for computing a path that satisfies the various constraints imposed by the different traversed domains. Among these solutions, we cite the following. A reliable routing with QoS guarantees for multi-domain IP/MPLS networks approach presented in [14], a hybrid approach for the inter-domain multi-constraints paths computation described in [15], and a multi-constraints end-to-end path computation through several inter-domain routes presented in [16].
- Technical solutions: are mainly extensions or improvements of existing technologies. Indeed, several solutions that have been proposed are based on operational technologies like MPLS or Border Gateway Protocol (BGP) [9]. In this context, we cite: the inter-domain MPLS Traffic Engineering [11], a BGP extension presented in [10], and also other solutions that introduce new principles like in [12] and [13].

All these inter-domain solutions do not provide to clients traffic the same QoS required as in its source domain. In this context, we introduced a solution that offers to client's traffic the same QoS constraints even during the transition to another domain. In the following section we remind briefly our solution.

2.2   Brief Description of the Proposed Solution

We introduced in [1] a new method that provides a new mechanism for inter-domain traffic treatment. This mechanism ensures continuity of QoS constraints offered to the client even after the transition to other domains. In this section we remind the main points of the proposed method. The approach is based on designing in each domain a server responsible for the management of the different classes of service, named the Class Manager (CM). On this server a table named Class Table (CT) is defined, and contains all information concerning the different classes defined in this domain (such as bandwidth, loss rate, delay, etc.). Once the CM of each domain fills its CT, it sends it to the neighboring domain. In this way, each CM has all the information about its neighbors' classes of service, and then, upon receiving a packet from the neighboring domain, the router in the current domain can classify it in a class that has the same characteristics as the source class. In this manner, the client flow retains the same QoS constraints throughout its path to the destination, and receives the same treatment from end-to-end.

Our mechanism is studied in [2] using simulation of various scenarios. Simulation results show that, using this mechanism, network performances in term of delay and loss rate are improved and also show a better optimization of network resources by reducing the utilization rate of the end to end link.

However, the study carried out in [2] didn't take into account security issues, and hence, the impact of security mechanisms which could be implemented to mitigate these problems. In the following section, we discuss some security issues of our mechanism and we propose a solution which could be implemented to overcome these problems.

3 Security Issues

As we mentioned previously and according to the mechanism principle, each CM server communicates with each edge router in the AS to fill its CT table. During this communication the edge router sends to the CM server all information concerning the QoS offered to client's traffic. However, all communicated information are sent in clear text, which means that they can easily be intercepted by any unauthorized third parties. Also, the mechanism is based on the exchange of CT tables between CM neighbors. The exchange is performed via a regular TCP session without any authentication of the CM servers or any encryption of exchanged data. So, the communication between CM servers is exposed to a various security threats. Our objective is to study these security problems, and to propose alternatives that may resolve them and secure the CM server mechanism.

Because of the confidential and important information contained in CT tables, the security issues of the CM server mechanism concerns two principal points: the communication between edge routers and the CM server, and the exchange between neighboring CM servers. In this section we discuss the security issues observed at each point.

3.1 The Communication Between Edge Routers and CM Server

The vulnerabilities of the communication between edge routers and CM server concern two parts : the first one is about the identity of both edge router and the CM server, and the second one is about the integrity and authenticity of the data sent by the routers.

Identity of the Nodes. CM server mechanism has to use a method to verify the identity of the CM server and of the edge router sending the data. If not, any outsider can spoof the IP address of the CM server to impersonate it, and then, it can receive all information coming from edge routers and modify them which can affect the clients' traffic routing conditions. In the same way, any unauthorized party can spoof an edge router's IP address and send incorrect data to the CM server, and this can also badly affect the routing conditions of the clients' traffic.

Authenticity of Data. In the other hand, the proposed method has to verify the integrity and authenticity of data sent by the edge router. Indeed, an error in data sent by an edge router can modify the treatment of traffic in terms of QoS. Errors can be in any entry of the CT table corresponding to any field of this table. Actually, any syntactic error or any wrong nature of an entry of CT table may change the QoS allowed to a certain client traffic. Also, an unauthorized person can easily modify the values of CT tables entries without being detected which can damage the traffic routing conditions.

The two issues discussed previously are not so important as is the next one. Indeed, the exchange between the CM and the router takes place in the same domain. For the next issue, this exchange is between two different domains.

3.2 The Exchange Between Neighboring CM Servers

Another security problem of CM server mechanism concerns the exchange of CT tables between CM servers. In fact, the exchange between neighboring CM servers is done via a regular FTP transfer. So, the mechanism uses the protocol TCP as a transport protocol, which means that the communication between CMs is exposed to all attacks against TCP which are much more common in Internet environment. Actually, the use of TCP may be a source of multiple vulnerabilities which concerns multiple TCP parameters including: synchronization, acknowledgment and reset.

TCP Synchronization. Indeed, a synchronization message "SYN" and synchronization acknowledgment message "SYN ACK" are sent during the TCP session establishment. However, a CM server cannot verify the identity of the CM server that is requesting the establishment of the TCP session. That is why, any outsider can sent these packets to a CM server at the same time with another CM server trusted node, so the first node may reject the trusted connection and establish a session with the outsider, and this may prevent the achievement of the mechanism and affect the traffic routing conditions.

TCP Acknowledgment. Also, the TCP acknowledgment message is used to complete the establishment of the TCP session. It can also be spoofed by an outsider to be connected with a CM server and receives all exchanged data of the CT table, which must remain confidential and if its intercepted that may negatively act on routing conditions of traffic.

TCP Reset. The receipt of a TCP reset message "RST" causes immediately the TCP session closure, which means the interruption of the exchange between CM servers, and this way the CM server mechanism could not be performed. So, it may present an important threat if an outsider can spoof this message.

4 Proposed Solution

4.1 The Communication Between Edge Routers and CM Server

In order to authenticate both the router and the CM server, an authentication server could be implemented; this could be done using a simple AAA (Authentication, Authorization and Accounting) server [17] and Kerberos [18].

A direct transfer from the router to the CM server is not really recommended. In order to add more security layer, using a secure file transport such as SCP is recommended during the backup. The router configuration is then transferred to a central backup server (using a secure transfer protocol), and all required information are stripped from the configuration file and put in an encrypted class file. The transfer from the CB (Central Backup) to the CM server could be implemented using a simple secure file transport, such FTP over SSL/TLS [19] or FTP over SSH, or SFTP over SSH.

Figure 1 illustrates a solution to secure communication between edge routers and CM server.

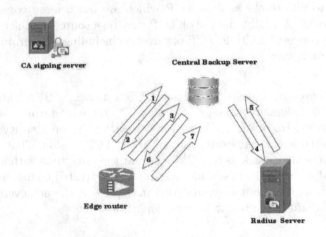

Transfer flow:
1. The edge router seeks to connect to the backup server
2. SSL/TLS handshake is performed to authenticated both ends
3. A secure channel is established and the credential
(username/password) is sent towards the backup server.
4. AAA flow begins using the credential in step 3.
5. Authentication successful: Radius-Reply Allow-access-request is sent
to the backup server.
6. and 7. The file transfer begins

Fig. 1. Securing communication between edge routers and CM server

As highlighted above, a direct transfer from the router to the CM server is not recommended from a security point of view. Assuming that a backup policy is in place, and the Edge router's configuration file is pushed regularly to a backup server. The first part of the idea is to define a high secure environment between the Edge router and the central backup server, and then work on the transfered configuration file in order to strip all QoS and Class related information. The new output file will be formated as ".CLASS" file, before it is pushed to the next CM server.

How the security between the router and backup server should be implemented? Two new servers come in the scenario: AAA server and the CA signing sever. The second one is used ONLY to sign the certificate request (private key) from the client and could/should be disconnected from the network. The first one is to authenticate the temporary user that is being used to transfer the file itself.

Router to the Backup Server

1. The private key and the signing request is generated on the client
2. The same is sent or shared to the signing server and signed - this is applicable for all nodes within the same AS.
3. Since both routers and the backup server are in the same CA, they are authorized to communicate to each other (hardware authentication - level 01)
4. During the handshake between the backup server and the router the cipher to use is selected by both ends.
5. The secure channel is established and the credentials (username,password) are sent towards the CM server and proxied to the Radius server
6. The access request response - ACCEPT is sent and the data (configuration file) is sent.

In the Backup Server

1. The file is stripped by the mean of a simple script and all the lines containing all critical information are removed.
2. The file is then encrypted using the same keys.

From the Backup Server to the CM Server

1. The scenario is same as the first one (router to the backup server) but this time the data is encrypted (doubly).

4.2   The Exchange between Neighboring CM Servers

FTP or a simple synchronization is not a solution. The identity of both server could be managed using a dual AAA server in both ends, the transfer could be secured using one of the following flows:

- FTP over SSL or SFTP or SCP (FTP over SSH is not safe as the data will not be channelized correctly inside the SSH channel)
- The data is encrypted using an OTP (One Time Password) [20] and PGP (Pretty Good Privacy) before it is transferred using the above flow.

Figure 2 illustrates a solution to secure communication between CM servers.

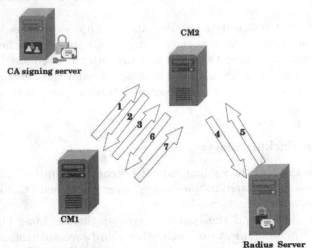

Transfer flow:
1.a. In the backup server, the .CFG file is stripped and all critical lines are removed.
  b. The resulting .CLASS file and is encrypted (SSL/ Random strong password) before it is pushed to CM1.
  c. When completely pushed to CM1, it is stored, extracted and compared to the most recent existing .CLASS file.
  d. If it is not different than the previous one, then do nothing. Else, merge the difference between both .CLASS files, and store the difference in .CLASSUPDATE file, encrypt the same and prepare it to be pushed to CM2.
  e. CM1 seeks to connect to CM2.
2. SSL Handshakes is performed to authenticate both ends.
3. A secure channel is established, and the credentials (username/password) is sent towards CM2.
4. Radius authentication flow begins.
5. An authorization is sent towards CM2
6.7. The transfer begins.

Fig. 2. Securing communication between CM servers

1. The encrypted file is then compared to the previous encrypted backup files if any.
2. If it is the first time then it will be shared right away towards the 2nd CM server (CLASS-SEND)
3. If different, the difference (plain text) is stripped from both files and encrypted again
4. The difference is then shared to the 2nd CM server if required, if not it will not be transfered (CLASS-UPDATE)

5   Conclusion

In this paper, we first reminded the main points of the QoS-CMS mechanism proposed in [1] to ensure end-to-end QoS in inter-domain. We then listed the various security issues that could affect the network if a safety mechanism is not implemented. Vulnerability risks are possible during communication between border routers and CM servers in the same domain, as well as during the exchange of CT between CM servers of different domains. To overcome this problem, we discussed proposals to follow and we gave a solution for every level of risk. Note that, by implementing the proposed security solution, the performance of our mechanism will not be affected. Indeed, the exchange of data between border routers and CM servers do not happen often, so the time factor has no impact on delay or throughput. Similarly, exchange tables between CM servers of different ASs is very rarely and happens only when there is a change of service classes in an AS domain.

References

1. Bakkali, S., Benaboud, H., Ben Mamoun, M.: On Ensuring End-to-End Quality of Service in Inter-Domain Environment. In: Gramoli, V., Guerraoui, R. (eds.) NETYS 2013. LNCS, vol. 7853, pp. 326–330. Springer, Heidelberg (2013)
2. Bakkali, S., Benaboud, H., Ben Mamoun, M.: Management of Inter-domain Quality of Service Using DiffServ Model in Intra-domain. In: Świątek, J., Grzech, A., Świątek, P., Tomczak, J.M. (eds.) Advances in Systems Science. AISC, vol. 240, pp. 727–736. Springer, Heidelberg (2014)
3. Braden, R., Clark, D., Shenker, S. : Integrated Services in the Internet Architecture: an Overview. IETF Informational, RFC 1633 (1994)
4. Blake,S., Black,D., Carlson,M., Davies,E., Wang,Z., Weiss,W.: An Architecture for Differentiated Services. IETF Informational, RFC 2475 (1998).
5. Rosen, E., Viswanathan, A., Callon, R.: Multiprotocol Label Switching Architecture. IETF Standards Track, RFC 3031 (2001).
6. Bourasa, C., Sevasti, A.: Service level agreements for DiffServ-based services' provisioning. Journal of Network and Computer Applications 28(4), 285–302 (2005)
7. Van Mieghem, P., Kuipers, F.A.: Concepts of exact QoS routing algorithms. IEEE/ACM Transaction on Networking 12(5), 851–864 (2004)
8. Korkmaz, T., Krunz, M.: Multi-constrained optimal path selection. In: INFOCOM 2001 Twentieth Annual Joint Conference of the IEEE Computer and Communications Societies, pp. 834–843 (2001)
9. Rekhter, Y., Li, T., Hares, S.: A Border Gateway Protocol 4 (BGP-4). IETF Standards Track, RFC 4271 (2006)
10. L. Xiao, Lui, K.-S., Wang, J., Nahrstedt, K.: QoS extension to BGP. In: Proceedings of the 10th IEEE International Conference on Network Protocols, pp. 100–109 (2002)
11. Farrel, A., Vasseur, J.-P., Ayyangar, A. : A Framework for Inter-Domain Multiprotocol Label Switching Traffic Engineering.IETF Informational, RFC 4726 (2006).
12. Howartha, P., Boucadairb, M., Floglmaa, P., Wanga, N., Pavloua, G., Morandb, P., Coadicb, T., Griffinc, D., Asgarid, A., Georgatsosen, P.: End-to-end quality of service provisioning through inter-provider traffic engineering. Computer Communications 29, 683–702 (2006)

13. Misseri, X., Rougier, J.-L., Moretti, S.: Auction-type framework for selling inter-domain paths. In: Proceedings of The International Conference on Network and Service Management (CNSM), pp. 284–291 (2013)
14. Sprintson, A., Yannuzzi, M., Orda, A., Masip-Bruin, X.: Reliable Routing with QoS Guarantees for Multi-Domain IP/MPLS Networks. In: 26th IEEE International Conference on Computer Communications IEEE INFOCOM 2007, pp. 1820–1828 (2007)
15. Frikha, A., Lahoud, S., Cousin, B.: A Hybrid End-to-End QoS Path Computation Algorithm for PCE-Based Multi-Domain Networks. Journal of Network and Systems Management, 1–27 (2013)
16. Djarallah, N.B., Pouyllau, H., Lahoud, S., Cousin, B.: Multi-constrained path computation for inter-domain QoS-capable services. International Journal of Communication Networks and Distributed Systems 12(4), 420–441 (2014)
17. Metz, C.: AAA protocols: authentication, authorization, and accounting for the Internet. IEEE Internet Computing 3(6), 75–79 (1999)
18. Neuman, B.C.: Kerberos: an authentication service for computer networks. IEEE Communications Magazine 32(9), 33–38 (1994)
19. Dierks, T., Allen, C.: The TLS Protocol. RFC 2246, IETF Network Working Group (January 1999)
20. Haller, N., Metz, C., Nesser, P., Straw, M.: A One-Time Password System. RFC 2289, Internet Standard (February 1998)

Uncovering Self Code Modification in Android

Faisal Nasim, Baber Aslam(✉), Waseem Ahmed(✉), and Talha Naeem(✉)

National University of Science and Technology, Islamabad, Pakistan
faisalnasimkhan445@gmail.com, ababer@mcs.edu.pk
http://www.nust.edu.pk/Pages/Home.aspx

Abstract. Android has proved itself to be the defacto standard for smart phones. Today Android claims a major share of smart phones OS market . The increasing popularity of Android has attracted attention of developers from around the globe in a very short span of time. Concerns and resultantly techniques involving code protection were evolved. The techniques were focused to hide sensitive logic of important pieces of code.These techniques were also used by malicious code writers to hide the malicious functionality of their code. This paper will analyze the techniques being employed in Android applications for code obfuscation. In addition, one of obfuscation technique i.e. runtime code modification in Android would be analyzed in detail.The major part of the paper would focus on tools and techniques for extracting dex files from the memory and analyzing them in order to recover code which has been injected in application process during runtime and is actually being executed in the memory.

Keywords: Android reverse engineering · Code modification · Obfuscation

1 Introduction

In recent years, an explosive growth for smart phones sale and usage has been observed [5]. Increased processing and memory have transformed the smart phones into a full fledged personnel computing device. Wide range of applications belonging to categories such as entertainment, lifestyle, education, business and personalization etc are available on different Android markets. New and advanced applications for smart phones are being released on every passing day. Smart phones have increased the dependence of users on them to very high degree. This increased dependence on smart phones has created some serious security concerns for the users. Despite the positive contributions which the advancing technology has imparted, some negative aspects in the form of malware have also emerged.

Modern Android malware have not only inherited many of the code hiding techniques from their x86 counterparts but have also taken tremendous advantage of obfuscation techniques which were basically meant to protect the intellectual rights of the application. Reversing these malware is important because they reveal sensitive information such as the extent of damage they can do, the

S. El Hajji et al. (Eds.): C2SI 2015, LNCS 9084, pp. 297–313, 2015.
DOI: 10.1007/978-3-319-18681-8_24

method they have chosen to hide themselves from AV tools and the strategy they have adopted to spread themselves etc. All this information contribute towards a better and resilient AV product.

Android have inherited most of the code obfuscation techniques from their x86 counterparts. Most of the techniques which were valid for x86 applications are also effective in case of Android. These techniques are meant to disrupt static and dynamic analysis and make reverse engineering as difficult as possible. Dynamic analysis is a relatively new analysis technique in which an analyst try to understand the functionality of the code during runtime. Common examples of code obfuscation techniques include string manipulation, identifier mangling and dead code injection. These all techniques modify the source code in such a way that efforts to analyze a software increases without disturbing the overall logic of the application. Self code modification is another obfuscation technique in which the source code is only modified during run time within the memory. Modification in the memory does not leave any artifacts of code alteration on the disk. Under Android, code modification during runtime is achieved by using Native code (ARM ELF) which can be executed as part Android application as dynamic library. Analyzing dynamically modified code with static analysis is extremely difficult. Traditionally analysts make use of dynamic analysis tools to understand the behavior of self code modification. This paper discusses techniques through which dynamically modified code can be captured from the memory after which it can be statically analyzed to reveal the functionality of the application. We have analyzed the use of a tool named as *Volatility* for retrieving of dex file from memory. In section 2 of our paper we will discusses different obfuscation techniques used by Android application. Section 3 will delve on self code modification under Android. Section 4 will discuss techniques and tools designed to uncover Self code modification.

2   Code Obfuscation

Code obfuscation is a technique that deliberately make such changes in the code that it becomes difficult to understand it [6]. Although code obfuscation puts a smoke screen around the code but retains the overall logic of the application. Aim of obfuscation is to make reverse engineering computationally hard. Code Obfuscation is used by the developers to protect their code from being reverse engineered. The central theme of code obfuscation is security of the application through obscurity. In Android, code obfuscation can be applied at both the source and bytecode level. Tools that can perform code obfuscation for Android bytecode include *ProGuard*, *Dalvik-obfuscator*, *APKObfuscator* and *DexGuard*. In the upcoming sections few of the most commonly used techniques will be discussed.

2.1   Identifier Mangling

Identifier Mangling involves replacing names of identifiers such as variable names, method names and class names with meaning less identifiers that does not give any meta information regarding the code associated with it [8] [18] [16]. In case of *ProGuard* identifiers are replaced with minimal lexical sorted strings i.e {a,b,c..aa,ab..ba,bc...}. Figure 1 shows the *Casting* class before obfuscation whereas figure 2 shows the same *Casting* class after obfuscation. *APKObfuscator* exploits Unix restriction that an identifier can be upto 255 characters. Since strings in a Dex files are sorted in alphabetical order therefore *APKObfuscator* appends extra data at the end of identifiers so that its position in the string table inside Dex file does not change. *Identifier Mangling* hide the meta information that can possibly reveal valuable information about the behaviour of the application. This technique not only effectively obscures the functionality associated with certain identifier but also drastically reduces the memory requirements of the application.

```
public class Casting
{
    private String ascStr="abc";
    private int sendXML(String password)
    {...}
    private String recieveXML(int identifier)
    {...}
}
```

```
public class a
{
    private String ab="abc";
    private int b(String ac)
    {......}
    private String c(int ad)
    {......}
}
```

Fig. 1. Before Obfuscation   **Fig. 2.** After Obfuscation

2.2   String Obfuscation

Strings can be very useful in process of reverse engineering. They can, not only be used to find the location of code of interest but also reveal sensitive information such as passwords and cryptographic keys. String obfuscation involves changing string to a value that can not be extracted simply by matching through regular expression in some editor [18] [16]. This can be achieved through many methods such as replacing the string with any non ASC-II combination, XORing the string with some preset value or by using any encryption algorithm for encryption. The process of string alteration must be reversible since it has to be reversed to its original value before it can be used in the application or presented to the user on some GUI. This constraint dictates the presence of decryption stub within the code. Practically string obfuscation can be performed on source code and bytecode level. Source code string obfuscation involves a custom decryption method which can decrypt and handover the string to application when required. Strings are entered in the source code in encrypted form. The disadvantage of this technique is that encrypted string are present in *string_ids* constant pool and are visible to the analyst. *APKProtect* uses this technique. Theoretically speaking bytecode string obfuscation is possible but practically it will become a huge programming task due to following three reasons.

– Ordering of strings in *string_ids* is alphanumeric
– *String_ids* table does not contain repeating entries
– Fixing all references to *string_ids* table across entire Dex file

This technique of obfuscation is effective only against static analysis and can easily be defeated through dynamic analysis.

2.3 Dead Code Injection

Dead code injection injects code in an executable that will never be executed. This is achieved by introducing non conditional and conditional branches. Conditional branches will always be true so that execution-flow always follow the valid code instead of dead code. Such conditions whose results are known in advance are called *Opaque Predicate*. This approach will add extra nodes and paths in the execution-flow diagram and thus add complexity in its analysis. Furthermore, different techniques of adding dead code can also be used to break / dodge disassemblers using different parsing algorithms. Adding non conditional branches e.g *goto*, in front of dead code, make linear disassemblers disassemble incorrect instructions but does not affect correct disassembling capability of disassemblers using recursive traversal algorithm [10] [11]. In case of Android, dead code can be introduced using pseudo instruction like *fill-array-data-payload* that should never be encountered during normal flow of execution [9] beside other traditional techniques.

2.4 Packing

Packing is the technique in which an executable is transformed in such a way that it becomes impossible for an analyst to reverse engineer it [12]. This transformation of an executable can achieved through variety of techniques. One such technique can be the encryption of the executable and decrypting it just before executing it. In this case there will be nothing available on the disk for the analyst to analyze but in the memory. The decryption stub associated with such an executable will have to perform four basic functions.

– Loading of encrypted file into the memory
– Decryption of the file
– Loading of decrypted file into the memory
– Execute the loaded file

Android packers encrypt the *dex* file in an APK which is decrypted by ARM ELF file at runtime [13]. The decrypted version is then loaded and executed using *DexclassLoader*. The packers applications usually have the capability to change the structure and overall control flow of an APK which ultimately increases the complexity to reverse engineer the application. *HoseDex2Jar* and *Bangcle* are two such packers that encrypt the original dex file and decrypt them only in memory for its execution[26]. *HoseDex2Jar* puts the encrypted *dex* in the header of packing dex file and update its header length.

2.5   Dynamic Code Loading

Dynamic code loading gives an application the ability to load code at runtime and execute [15]. This ability has been provided by the *Android Platform* itself and is extensively used by malware to evade detection of malicious functionality during off line application analysis. All applications submitted to google play store are subject to off line application analysis before being accepted to the store. This off line analysis is done by a software known as *Bouncer*. *Bouncer* analyzes the submitted applications for any malicious functionality so that undesired applications can be filtered before admission [23]. Common technique for a malware to evade detection by bouncer is to exclude malicious code altogether once it is submitted to google. Once it makes its way to google play store and installed on any user device, it download malicious piece of code at runtime and execute it. An application cannot be declared malicious just due to the presence of runtime code loading functionality since many benign applications use this capability for legitimate purposes. Android class *DexClassLoader* is used for loading classes dynamically. *DexClassLoader* does not put restrictions on the location of the code. It means that it can load and execute code from any location over the Internet. This capability can prove harmful for benign applications if they have instructions for loading code from location which can be written by any other application i.e. writable location.

2.6   Self Modifying Code

Application code that can alter itself during run time is referred to as self modifying code [4]. This implies that the code which is executed at run time may not be the same code that form part of application on the disk. This technique is used as an effective obfuscation measure to hide the functionality of the application. Self code modification has many uses e.g run time code generation, patching of subroutine address calling, prevent reverse engineering and evade detection by virus / spyware scanning software [1]. It is an effective method to evade analysis by static analysis tools. Injecting self code modification techniques in a malware code can increase cost and effort of static analysis to an undesirable value. Footprints of self code modification can be detected by observing behavior of an application during dynamic analysis. Since modern malware are capable of detecting the environment in which they are running i.e actual physical device or an emulator and can therefore decide at runtime whether to go for self code modification or not[24]. Dynamic analysis tools can therefore be rendered useless due to the fact that most of these tools make use of emulator as the base platform.

3   Android Self Code Modification

Most Android applications are written in Java and can use C/C++ code from within the Java code. Java code is compiled into Java bytecode and packed in a .*class* file format. This .*class* file is then transformed into Android specific Dalvik bytecode in a .*dex* file. Google choice for a Dalvik .*dex* format instead of Java .*class* format is based on the better performance results achieved using register based Dalvik Virtual Machine. *Dx* tool, available with Android SDK, is used for the conversion of .*class* to .*dex* format.This bytecode is interpreted by the Dalvik Virtual Machine. Unlike Java .*class* file, all the bytecode of the an Android application is packaged into a single .*dex* file. An Android application besides using Java can also take advantage of Native code for performing processor intensive tasks. Native code is the one which can be run directly by the processor i.e. ARM assembly code in case of ARM devices. Native code is compiled as a shared library and its functions can be called from within the Java code in an Android application. All the code of an Android application whether it is Dalvik bytecode or Native, lives with in the bounds of the same process and therefore share the same privileges [2]. A typical cut out section of Android process memory layout can be seen in figure 3.

```
47ec2000-47f72000 r--p 00000000 1f:01 658   /data/dalvik-cache/data@app@com.example.hellondk-1.apk@classes.dex
47f72000-47f73000 r--s 00000000 1f:01 876   /data/app/com.example.hellondk-1.apk
47f73000-47f76000 r-xp 00000000 1f:01 652   /data/data/com.example.hellondk/lib/libhello.so
47f76000-47f78000 rw-p 00002000 1f:01 652   /data/data/com.example.hellondk/lib/libhello.so
47f78000-47f7e000 rw-s 00000000 00:07 1534  /dev/ashmem/InputChannel 41559030 com.example.hellondk/com.example
Activity (deleted)
47f7e000-47f81000 r-xp 00000000 1f:00 870   /system/lib/hw/gralloc.goldfish.so
47f81000-47f82000 rw-p 00003000 1f:00 870   /system/lib/hw/gralloc.goldfish.so
```

Fig. 3. Android Process Map

In this section of Android application memory layout it can be seen that *classes.dex* file (*/data/dalvik-cache/data@app@com.example.hellondk-1.apk@ classes.dex*)which contains the Dalvik bytecode of the application has been mapped into the process memory as read-only. Dalvik virtual machine reads the instructions, it comes across in the *classes.dex* file, and interprets them. In the third line of the screen shot, data section of the native library (ARM ELF) *libhello.so* can be seen. In the immediate next line the actual executable code of the library *libhello.so* is mapped as read-execute.

Instruction set of Dalvik bytecode consists of limited number of instruction (as compared to x86 or ARM) and does not contain any instructions that can be used to modify bytecode within *classes.dex* file during execution. The *classes.dex* file therefore cannot contain any code that can be used for self modification [16]. Since, the Native library bundled together with the application is mapped into the same address space as that of Dalvik bytecode and can access any arbitrary address inside the process, it can therefore access and overwrite Dalvik instructions. Self code modification with in *classes.dex* file would therefore need the services of Native code.

Since *classes.dex* file is mapped as read-only into the application's address space therefore any attempt to alter it from within Native code would fail and result in the termination of the process. Before attempting any tampering, *classes.dex* file has to be remapped writable. Once memory pages containing mapped *classes.dex* files is remapped writable, any type of modification can be performed on it through the Native code.

Another technique for self code modification which is commonly used on x86 and ARM platform is to create a new mapping with read and execute permissions in the virtual address space of the calling process using *mmap* system call. The next step is to write the code into the newly created mapping and execute by jumping to it.

4 Inspecting Self Modified Code in Android

As already discussed in section 3 that all Android applications have *classes.dex* file mapped into the memory pages as read-only. We can confirm it from memory space layout of an Android application displayed in figure 3. If *classes.dex* file is to be changed through the Native code, all or few pages in the memory where *classes.dex* has been mapped will have to be mapped writable. The system call which is used for changing the permissions of memory pages belonging to certain application is *mprotect*. Four types of permissions which can be assigned to a memory page, as standalone or in combination, are *PROT_NONE* (The memory cannot be accessed at all), *PROT_READ* (The memory can be read), *PROT_WRITE* (The memory can be modified) and *PROT_EXEC*(The memory can be executed). Figure 4 shows the cut out section of an Android application in which code has been altered during the runtime.Here we see that, *classes.dex* has been partitioned in three sections. The middle partition is *read-write*. This is the partition where the code has been replaced with new one during execution. Note that the malware writer changed memory permissions of only those pages where he intended to perform code modification i.e from memory address 0x47ee9000 till 0x47eea000. Permissions of the rest of the pages were left unaltered. The two unaltered *read-only* sections of *classes.dex* extends from 0x47ec2000 till 0x47ee9000 and from 0x47eea000 till 0x47f28000. Since the smallest unit of memory where permission enforcement is possible is a memory page therefore *mprotect* can alter the permission of the an entire page and not some part of it. An application will always be allocated an entire page inside memory. This is the reason why the size of all three sections of *classes.dex* are the multiples of page size.

```
47ec2000-47ee9000 r--p 00000000 1f:01 914    /data/dalvik-cache/data@app@com.      .lab.poc-1.apk@classes.dex
47ee9000-47eea000 rw-p 00027000 1f:01 914    /data/dalvik-cache/data@app@com.      .lab.poc-1.apk@classes.dex
47eea000-47f28000 r--p 00028000 1f:01 914    /data/dalvik-cache/data@app@com.      .lab.poc-1.apk@classes.dex
47f28000-47f29000 r--s 00028000 1f:01 878    /data/app/com.      .lab.poc-1.apk
47f29000-47f2c000 r-xp 00000000 1f:01 905    /data/data/com.      .lab.poc/lib/lib   .so
47f2c000-47f2e000 rw-p 00002000 1f:01 90£    /data/data/com.      .lab.poc/lib/lib   .so
```

Fig. 4. Android Process Map

The fact which we will be exploiting in order to detect and inspect the self code modification is that the *classes.dex* is a memory-mapped file. A memory-mapped file is a segment of virtual memory which has been assigned a direct byte-for-byte correlation with some portion of a file or file-like resource [14]. The general strategy for detecting self code modification would be comprising of following steps.

- Obtaining memory dump of pages where classes.dex file has been mapped.
- Comparing the extracted *classes.dex* file with *classes.dex* present on the Android file system (*/data/dalvik-cache*).
- Interpreting the bytecode which has been modified to figure out modified instructions.
- Diagrammatically, the overall process can be seen in Figure 5

4.1   Obtainign Memory Dump

Traditionally memory dumps on Linux machines are obtained using */dev/mem* device. Using this method initial 896 MB of RAM can be read without the need of loading any additional code into the kernel. Android does not add the ability to expose physical memory through */dev/mem* and therefore cannot be used for physical RAM capture. Furthermore, using this technique, memory captures over and above 896 MB is not possible on Linux machines. To work around this limitation *Ivor Kollar* created *fmem* [22], a loadable kernel module, which supports memory captures on Linux machines. *fmem* uses *page_is_ram* function to check the presence of page in the memory. *fmem* loadable kernel module cannot be used on Android devices since *page_is_ram* function is absent on ARM architecture. The other choice for obtaining memory dump on an Android device is *dmd* module developed by *Joe Sylve*[25]. This is also a loadable kernel module and offer memory acquisition over TCP and device SD card. This module takes the approach of parsing the kernels *iomem_resource* structure and obtaining physical addresses of system RAM. We use *dmd* module for obtaining the memory dump of an Android device and getting the memory contents of our interest for our further analysis. One issue in using loadable kernel modules is the enforcement of stringent sanity checks before loading of any module in the kernel. The checks enforced by kernel ensures that the module has been compiled for that very version of the kernel failing which results in loading error. Android kernel is compiled with the options of loading and unloading the external kernel modules as unchecked. With these options as unchecked we cannot load the *dmd* module for obtaining Android RAM image. The general steps for obtaining memory dump of Android device using *dmd* are enlisted below [19] [17].

- The first step is to build the kernel module with options for loading and unloading the external kernel modules i.e. *CONFIG_MODULES, CONFIG_MODULES_UNLOAD, CONFIG_MODULES_FORCE_UNLOAD* as checked.
- Cross compile *dmd* against the kernel built in the first step.

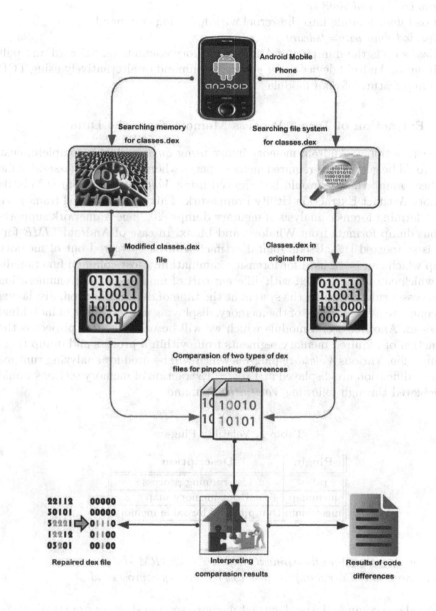

Fig. 5. Process of replacing and analysing tempered instructions

- Boot the Android device with the custom built kernel.
- Push the *dmd* module on SD card using *adb* command i.e. *adb push dmd-evo.ko /sdcard/dmd.ko*
- Load *dmd* module into the kernel with following command.
  insmod dmd path=/sdcard
- Last step is the dumping of Android memory contents on SD card and pull it out of Android device using *adb pull* command or alternatively using TCP dump feature of *dmd* module .

4.2   Extraction of Target Process Memory from the Dump

Once extraction of Android memory image using *dmd* module is complete, next step would be getting the required memory pages where our target *classes.dex* file has been mapped. This would be achieved using *Volatility*. *Volatility* is Volatile Memory Artifact Extraction Utility Framework. This is an advanced framework for performing forensic analysis of memory dumps [20]. The framework supports various dump formats from Windows and Linux. In case of Android *LIME* format is supported [17]. Useful digital artifacts can be extracted out of memory dump which can be of used for forensic examination. Most common functionalities which can be performed with different sort of modules include enumeration of processes running under the system at the time of dump collection, displaying the complete address space of the memory, displaying address space of individual processes. Another useful module which we will be using for this project is the extraction of required memory segments from within a process and dump them as single file. Various *Volatility* plug-ins which will be used for analyzing runtime code modification are displayed in Table 1. Extraction of memory sections would be achieved through following *Volatility* command.

Table 1. Volatility Plug-ins

Plugin	Description
pslist	List running process
memmap	Print memory map
memdump	Dump the addressable memory

 # python vol.py –profile=LinuxGoldfish-2_6_29ARM -f
/home/ram.lime linux_dump_map -s 0x47ec2000 -p <process_id >

The above command would extract memory section starting from 0x47ec2000 till 0x47ee9000. The other required sections of memory would also be extracted on the same lines. In our case, we have extracted the three sections of *classes.dex* file visible in figure 4. The three extracted sections can be combined through any hex editor to form a single file. This single file is *classes.dex* which was present in the memory at the time when memory dump was created by *dmd* kernel module.

We call this file as *classes.dex-M* in the rest of the paper. Similarly the *classes.dex* file present on Android file system i.e. */data/dalvik-cache* would be referred to as *classes.dex-F* in the text to follow. The third type of *classes.dex* file that would be extracted from *apk* would be refered to as *classes.dex-A*. Since *classes.dex-M* file is memory-mapped therefore there should be no difference between the two *classes.dex* files i.e *classes.dex-M* and *classes.dex-F*. The two files will be different in case where runtime code modification has been performed.

4.3   Analysis of *classes.dex-M* Files

At the beginning of this project it was visualized that the *classes.dex-M* will be compared with *classes.dex-F* for analyzing the run time code modification. But when the comparisons were done it was revealed that the similarity between two types of files was ranging between 50% - 85%. These results were in contradiction to our previous intuition that only a small percentage of the dex file should change which represents the modified bytecode where run time code modification has been done. Applications where no dynamic code modification was performed should have no difference between the two types of *classes.dex* files. Graph displayed in figure 6 depicts the similarity percentage between *classes.dex-M* and *classes.dex-F* of some commonly used applications available on Google Play. Here *classes.dex-M* was obtained using *dmd* module. *dmd* was unable to retrieve the *classes.dex-M* in case of *App Lock* and *Hot Spot Shield*.

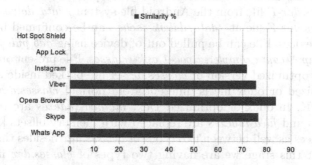

Fig. 6. Similarity %

It was not possible to go ahead with run time code modification analysis with the accuracy of results achieved using *dmd* for Android. In order to get the reliable memory contents of *classes.dex* file, a tool was designed that was capable of extracting memory contents of pages associated with the process using *ptrace* system call. The tool made use of */proc/<pid>/maps* to get the memory addresses associated with the *classes.dex* of an Android application. The *ptrace* system call provides a means by which one process (the "Tracer") may observe and control the execution of another process (the "Tracee"), and can examine and change the Tracee's memory and registers[21].The obtained addresses were further used by *ptrace* utility to read the required contents. A comparative analysis of the similarity results between *Volatility* and the designed tool can be seen in Table 2

Table 2. *: Unable to retrieve image **: Unable to attach to process

Application	Volatility	Custom Tool
Whats App	49.5745	99.9997
Skype	76.2894	99.9994
Opera	83.5615	100
Viber	75.3598	99.9999
Instagram	71.5729	100
App Lock	*	100
Hot Spot Shield	*	**

4.4   Comparison of *classes.dex* Files

In oder to explore the tampering done with the run time code modification it would be necessary to compare *classes.dex-M* with *classes.dex-F* so that we can find out what all has been tampered. Before comparing the two files we would extract *classes.dex-F* file from the Android file system */data/dalvik-cache*. Presence of *classes.dex-F* file at */data/dalvik-cache* can be confirmed by using shell on Android devices. File can be pulled out of device using *adb pull /data/dalvik-cache/data@app@com.example.ex.poc-1.apk@classes*. File present on Android file system is the optimized version of *classes.dex-A* file packed inside an APK file.

The optimized or odex file is different from normal *classes.dex* file packed inside an APK. In odex or optimized Dex, *method indexes* are replaced with *vtable indexes* and *filed indexes* are replaced with memory offsets [3]. There are other differences as well between two types of *classes.dex* besides the ones stated previously. At this stage we are having two types of *classes.dex* files available with us i.e. *classes.dex-M* with *classes.dex-F*. The next step would involve the comparison of two available files. Here we would perform analysis on a POC application developed by *Patrick Schulza* of *Bluebox Mobile Security* [2]. For this purpose we used *vbindiff* available for Linux platform. Figure 7 shows the difference of dex bytecode between the two type of *classes.dex* files of the POC application in different colour. The upper half of the screen shot shows the dex bytecode retrieved from *classes.dex-F* file whereas the lower half shows the dex bytecode of *classes.dex-M* file. Note that we see the different byte code at the same addresses of POC application. This difference in color shows the tempered bytecode inside the memory.

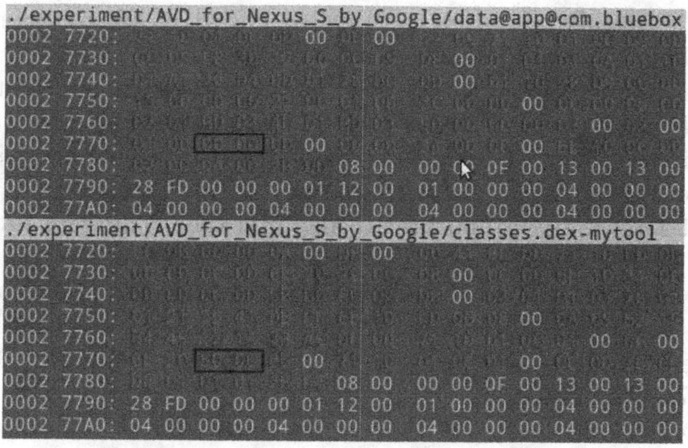

Fig. 7. The Difference between classes.dex files

4.5 Interpreting Bytecode

At this moment it can not be figured out what the tempered bytecode actually refers to in *classes.dex*. This bytecode can belong to any section of *classes.dex* file e.g data area or it can be bytecode representing the instructions of any method. There can be two approaches to figure out the bytecode which is actually being executed. One approach can be manual parsing of *dex* file. Manual parsing for figuring out the exact code being executed is practically not feasible since the size of dex file can be in MBs. One variant of this approach is to make use of available tools that parse the dex file e.g. *010 Editor*. This can be very useful as the editor would parse the entire dex file for you and we can exactly pin point the target bytecode. In this case it would be required to disassemble the dex instructions by the user himself. The other more practical approach would be to disassemble the entire *dex* file using disassemblers like *IDA Pro*, *dexdump* etc. This would straight away present you with the disassembled bytecode in *Smali* format. Both of the previously stated approaches can be effective as long as the *dex* file presented to the previously stated tools is valid. But a small invalid modification of the dex file would crash all these tools leaving the user with the only approach of manually parsing the dex file and figuring out the tampered bytecode which practically speaking is cumbersome.

Under normal circumstances, when an application is installed on an Android device its dex file is optimized and verified for the validity of the dex bytecode it contains. A verified and optimized version of the dex file will be stored on the system, protected by file system permission (you cannot change it afterwards unless you have rooted your device). Once the bytecode modification takes place during runtime inside the memory, verifier does not comes in the loop for verification of tempered bytecode. It is therefore much easier to inject bytecode that can make reverse engineering tools to crash.

The same happened in case of the POC application. When different reverse engineering tools including IDA Pro, Dexdump, 010 Editor etc were used to disassemble *classes.dex-M*, they all crashed. The only tool that hinted towards the cause of crash was IDA Pro. In case of our POC *classes.dex-M* file, the number of *tries_ size* in *code_item* of *Ljava/lang/String;->append* method was more than the permitted 128. This number was injected during runtime and verifier was unable to validate it. It is important to mention that *code_item* data can only be tempered in case where the method under consideration has not been called during the course of execution of the application. In our case *Ljava/lang/String;->append* was never called from anywhere inside the classes.dex file. The number of tries in method can be seen in figure 7 with in black rectangles. The figure 7 show the difference between two files from in *vbindiff*.

```
027670:                    |[027670] java.lang.String.add:(Ljava/lang/String;)V
027680: 1202               |0000: const/4 v2, #int 0 // #0
027682: 6e10 1f0c 0c00     |0001: invoke-virtual {v12}, Ljava/lang/String;.length:()I // method@0c1f
027688: 0a04               |0004: move-result v4
02768a: 2205 db01          |0005: new-instance v5, Ljava/util/HashMap; // type@01db
02768e: 7010 4d0c 0500     |0007: invoke-direct {v5}, Ljava/util/HashMap;.<init>:()V // method@0c4d
027694: 1a00 0000          |000a: const-string v0, "" // string@0000
027698: 5bb0 fc02          |000c: iput-object v0, v11, Ljava/lang/String;.content:Ljava/lang/String; // field@02fc
02769c: 6e10 220c 0c00     |000e: invoke-virtual {v12}, Ljava/lang/String;.toCharArray:()[C // method@0c22
0276a2: 0c06               |0011: move-result-object v6
0276a4: 2167               |0012: array-length v7, v6
0276a6: 0123               |0013: move v3, v2
0276a8: 3473 0300          |0014: if-lt v3, v7, 0017 // +0003
0276ac: 0e00               |0016: return-void
```

Fig. 8. Instructions from *classes.dex-M*

Due to the failure of current reverse engineering tools in parsing and analyzing *classes.dex-M*, an automated approach is required for the automatic analysis of dex files altered in memory. With this in mind we have developed a tool in Python which made use of *Androguard* Library. This tool is capable of parsing and analyzing the runtime modified bytecode. The algorithm used by this tool is explained below.

- Classes.dex-M *and* classes.dex-F *are input to the Python script.*
- *Both files are compared byte by byte to figure out the difference in bytes along with their address.*
- *Make* DalvikOdexVMFormat *object of* classes.dex-F. *Any such attempt on* classes.dex-M *would fail (In case where meta information of some method has been altered to unauthorized value).*
- *Get the starting and ending addresses of all the methods in* classes.dex-M.
- *Compare the addresses obtained in step 2 for presence between starting and ending address of all the methods. This would pinpoint the methods that are modified in the memory.*
- *Addresses that are not matched with addresses of any of the methods do not form part of any method.*

- *Based on starting and ending addresses of altered methods, extract the complete byte code of these methods from* classes.dex-M.
- *Extracted bytecode would be replaced with the bytecode of modified methods at the same addresses in* classes.dex-F.
- *The previous step will replace the original bytecode with modified bytecode in* classes.dex-F *while keeping the meta information of the code intact.*
- *In the last step, the* classes.dex-F *would be analyzed using any dex file disassembler to figure out the actual instructions being executed from memory.*

```
027670:                    |[027670] java.lang.String.add:(Ljava/lang/String;)V
027680: 1201               |0000: const/4 v1, #int 0 // #0
027682: ee10 0600 0c00     |0001: +execute-inline {v12}, [0006] // inline #0006
027688: 0a00               |0004: move-result v0
02768a: ee10 0600 0c00     |0005: +execute-inline {v12}, [0006] // inline #0006
027690: 0a03               |0008: move-result v3
027692: ee10 0600 0c00     |0009: +execute-inline {v12}, [0006] // inline #0006
027698: 0a04               |000c: move-result v4
02769a: ee10 0600 0c00     |000d: +execute-inline {v12}, [0006] // inline #0006
0276a0: 0a05               |0010: move-result v5
0276a2: ee10 0600 0c00     |0011: +execute-inline {v12}, [0006] // inline #0006
0276a8: 2206 db01          |0014: new-instance v6, Ljava/util/HashMap; // type@01db
0276ac: 7010 4d0c 0600     |0016: invoke-direct {v6}, Ljava/util/HashMap;.<init>:()V // method@0c4d
0276b2: f810 1a00 0c00     |0019: +invoke-virtual-quick {v12}, [001a] // vtable #001a
```

Fig. 9. Instructions from *classes.dex-F*

The modified instructions in case of the POC application can be seen in figure 8. The actual instructions which were present in *classes.dex* on disk are visible in figure 9. We can clearly see that that instructions in memory are altogether different from instructions on disk. As an example we can see that bytecode for *classes.dex-F* present at memory address 0x27698 is 0a04 whereas bytecode for the same location in case of *classes.dex-M* is 5bb0 fc02.

5   Conclusion

This paper has proposed two tools. One tool for retrieving the dex file from the memory using system call *Ptrace* and the other tool for analyzing the retrieved dex file for code modification during the runtime. The dex file analysis tool can detect the modified bytecode and disassemble it for easy comprehension of an analyst. The output *classes.dex* file can be used as an input into many of static analysis systems for further analysis. In addition, this paper has also analyzed different code obfuscation techniques which are relevant in case of Android. These techniques included static techniques which can be applied on the source code or byte code present on the disk and dynamic techniques which includes *self code modification* and *dynamic code loading*.

References

1. Uses of Self Modifying Code, http://stackoverflow.com/questions/516688/what-are-the-uses-of-self-modifying-code
2. Android Security Analysis Challenge, https://bluebox.com/technical/android-security-analysis-challenge-tampering-dalvik-bytecode-during-runtime
3. My Life with Android, http://mylifewithandroid.blogspot.com/2009/02/optimized-dex-files.html
4. Self Modifying code, http://tibasicdev.wikidot.com/selfmodify
5. Smartphone OS Market Share, Q2 2014, http://www.idc.com/prodserv/smartphone-os-market-share.jsp
6. How obfuscation helps protect Java from reverse engineering, http://www.techrepublic.com/blog/software-engineer/how-obfuscation-helps-protect-java-from-reverse-engineering/
7. Android Bytecode Obfuscation, http://www.dexlabs.org/blog/bytecode-obfuscation
8. How obfuscation helps protect Java from reverse engineering, http://www.techrepublic.com/blog/software-engineer/how-obfuscation-helps-protect-java-from-reverse-engineering/
9. Bytecode for the Dalvik VM, http://s.android.com/tech/dalvik/dalvik-bytecode.html
10. Linear Sweep vs Recursive Disassembling Algorithm, http://resources.infosecinstitute.com/linear-sweep-vs-recursive-disassembling-algorithm/
11. What is the algorithm used in Recursive Traversal disassembly? http://reverseengineering.stackexchange.com/questions/2347/what-is-the-algorithm-used-in-recursive-traversal-disassembly
12. What are Suspicious Packers? http://www.kaspersky.com/internet-security-center/threats/suspicious-packers
13. Android packer: facing the challenges, building solutions, https://www.virusbtn.com/conference/vb2014/abstracts/Yu.xml
14. Memory-Mapped files, https://msdn.microsoft.com/en-us/library/dd997372(v=vs.110).aspx
15. Custom Class Loading in Dalvik, http://android-developers.blogspot.com/2011/07/custom-class-loading-in-dalvik.html
16. Patrick, S.: Code Protection in Android. Communication and Communicating Devices (2012)
17. Joe, S.: LiME - Linux Memory Extractor, Instructions v1.1
18. Alexandrina, K.: Efficient Code Obfuscation for Android. University of Luxembourg (2013)
19. Holger, M.: Live Memory Forensics on Android with Volatility. Friedrich Alexander University (2013)
20. The Volatility Framework, https://code.google.com/p/volatility/
21. Ptrace, http://linux.die.net/man/2/ptrace
22. How to acquire memory from a running Linux system, https://gist.github.com/adulau/5094750
23. Sebastian, P., Yanick, F., Antonio, B., Christopher, K., Giovanni, V.: Execute This! Analyzing Unsafe and Malicious Dynamic Code Loading in Android Applications. In: Network and Distributed System Security Symposium (2014)
24. Thanasis, P., Giannis, V., Elias, A., Michalis, P., Sotiris, I.: Rage Against the Virtual Machine:Hindering Dynamic Analysis of Android Malware. In: European Workshop on Systems Security (2014)

25. Joe, S., Andrew, C., Lodovico, M., Golden, G.R.: Acquisition and analysis of volatile memory from android devices. Elsevier 8(3-4), 175–184 (2012)
26. Axelle, A., Ruchna, N.: Obfuscation in Android Malware and how to fight back. Virus Bulletin (2014)
27. Yury, Z., Maqsood, A., Olga, G., Bruno, C., Fabio, M.: StaDynA: Addressing the Problem of Dynamic Code. In: ACM Conference on Data and Application Security and Privacy (2015)

Performance of LDPC Decoding Algorithms with a Statistical Physics Theory Approach

Manel Abdelhedi[1]([✉]), Omessaad Hamdi[2], and Ammar Bouallegue[1]

[1] Syscom Lab, National Engineering School of Tunis, ENIT, Tunisia
abdelhedi_manel@yahoo.fr, ammar.bouallegue@enit.rnu.tn
[2] University of Cartage, Supcom, Tunisia
omessaad.hamdi@gmail.com

Abstract. In 1989, N.Sourlas used the parallel that exists between the information theory and the statistical physics to bring out that low density parity check (LDPC) codes correspond to spins glass models. Such a correspondence is contributing nowadays to the similarity between the statistical physics and the error correcting codes. Hence, the statistical physics methods have been applied to study the properties of these codes. Among these methods the Thouless-Anderson-Palmer (TAP) is an approach which is proved to be similar to the Belief Propagation (BP) algorithm. Unfortunately, there are no studies made for the other decoding algorithms.

The main purpose of this paper is to provide a statistical physics analysis of LDPC codes performance. First, we investigate the Log-Likelihood Ratios-Belief Propagation (LLR-BP) algorithm as well as its simplified versions the BP-Based algorithm and the λ-min algorithm with the TAP approach. Second, we evaluate the performances of these codes in terms of a statistical physics parameter called the magnetization on the Additive White Gaussian Noise (AWGN) channel. Simulation results obtained in terms of the magnetization show that the λ-min algorithm reduces the complexity of decoding and gets close to LLR-BP performance. Finally, we compare our LLR-BP results with those of the replica method.

Keywords: LDPC codes · Ising spin · LLR-BP algorithm · BP-Based · λ-min · TAP approach

1 Introduction

Nowadays, the Low-Density Parity-Check (LDPC) codes have become a part of essentially all new communication standards such as DSB-S2, DVB-S2X, DVB-T2 [1]. Actually, these codes can get very close to the Shannon limit by the mean of an iterative decoding algorithm called Belief-Propagation (BP) algorithm [2]. The implementation of the BP algorithm is a difficult task. In order to overcome this difficulty, all these standards use reduced-complexity decoding algorithm such as the BP-Based [3] algorithm and the λ-min algorithm [4].

In 1989, N.Sourlas established that the LDPC codes and the spin glass models are equivalent [5]. Since then, the statistical physics methods initially developed

© Springer International Publishing Switzerland 2015
S. El Hajji et al. (Eds.): C2SI 2015, LNCS 9084, pp. 314–330, 2015.
DOI: 10.1007/978-3-319-18681-8_25

for the study of disordered systems have also been applied to analyze the properties of these codes. Nowadays, this result has become a common tool to study the properties of LDPC codes in a wide range of channel such as Binary Symmetric Channel (BSC) [6,7,8], asymmetric channel [9], binary memory asymmetric channel [10], but only for the BP algorithm.

Motivated by this fact, we propose to analyze the decoding algorithms of LDPC codes by statistical physics methods. Actually, the main objective is to develop the Log-Likelihood Ratios-Belief Propagation (LLR-BP) algorithm and its simplified versions i.e. the BP-Based algorithm and the λ-min algorithm with the TAP approach. We evaluate the mentioned decoding algorithms performance in terms of a statistical physics parameter called magnetization in the case of an additive white Gaussian noise (AWGN) channel.

Our paper is organized as follows. Section 2 introduces the LDPC codes and their BP decoding algorithm. Section 3 briefly introduces some tools relative to statistical physics. Section 4 demonstrates the similarity between LDPC codes and spin glass models. In Sect. 5, we develop the LLR-BP algorithm and its simplified versions the BP-Based algorithm and the λ-min algorithm with the TAP approach. Simulation results in terms of the magnetization are presented in Sect. 6. The performance of the LLR-BP algorithm is compared to the results of the replica method. Finally, we end this paper with a conclusion and we propose some directions of future work in Sect. 7.

2   LDPC Codes and Decoding Algorithm

2.1   Low-Density Parity Check Codes

Let M and N be two integers. The binary LDPC code is a linear block code described by its parity check matrix A, with N columns denoting the codeword length and M rows denoting the parity-check equations.

In the following, we will use notations adapted from [3]. Let $L(\mu) = \{j/A_{\mu j} = 1\}$ be the set of bits j that involve in check μ. Similarly, let $M(j) = \{\mu/A_{\mu j} = 1\}$ be the set of checks in which bit j involves. Finally, $z = s + \zeta$ denotes the received word.

2.2   Belief Propagation Algorithm

The objective of the decoding algorithm is to get the most probable codeword $\hat{s}$ that have been transmitted over the channel when given the channel output. This section gives a summary of the iterative decoding of LDPC codes based on the BP algorithm according to [2]. In the following, we will denote $q_{\mu j}^a$ the probability that bit s_j of s is equal to $a = \{0, 1\}$, when the information obtained by checks rather than the μ check is known. The quantity $r_{\mu j}^a$ denotes the probability of check μ is satisfied when the bit s_j is fixed at a and the other bits have a separable distribution following the probabilities $\{q_{\mu l} : l \in L(\mu) \backslash j\}$ [3].

The following steps describe the standard iterative decoding algorithm based on the BP approach.

- **Initialization:** The variables $q^0_{\mu j}$ and $q^1_{\mu j}$ are initialized respectively to the values $p^0_j = P(s_j = 0 \backslash z_j)$ and $p^1_j = P(s_j = 1 \backslash z_j)$.
- **Iterative processing:** For each $i = 1$ to the maximum number of iterations
  1. Define $\delta q_{\mu j} = q^0_{\mu j} - q^1_{\mu j}$ and for each μ, $j \in L(\mu)$, and for $a = \{0,1\}$, compute

$$\delta r_{\mu j} = \prod_{l \in L(\mu) \backslash j} \delta q_{\mu l} ,$$

$$r^a_{\mu j} = \frac{1}{2}(1 + (-1)^a \delta r_{\mu j}) .$$

  2. For each j and $\mu \in M(j)$, and for $a = \{0,1\}$ update

$$q^a_{\mu j} = \alpha_{\mu j} p^a_j \prod_{\nu \in M(j) \backslash \mu} r^a_{\nu j} ,$$

$$q^a_j = \alpha_j p^a_j \prod_{\mu \in M(j)} r^a_{\mu j} .$$

  where $\alpha_{\mu j}$ and α_j are chosen such that

$$q^0_{\mu j} + q^1_{\mu j} = 1 \quad \text{and} \quad q^0_j + q^1_j = 1 .$$

  3. Create the detection word $\hat{s} = (\hat{s}_1 ... \hat{s}_N)$ of the transmitted codeword such that:

$$\hat{s}_j = 0 \quad \text{if} \quad \delta q_j > 0 ,$$

$$\hat{s}_j = 1 \quad \text{if} \quad \delta q_j \leq 0 .$$

This process of decoding ends if $A.\hat{s}^T = 0$. Otherwise, a next BP iteration is performed. When the maximum number of iterations is reached and $\hat{s}$ is not a valid codeword, the algorithm stops with a failure result.

3   Overview of Statistical Physics

N.Sourlas showed that there is a similarity between statistical physics and error correcting codes [5]. Our aim is to use this fact to analyze LDPC codes. In this section, we give an overview of some statistical physics tools we need in the development of this work.

Statistical physics is a part of physics devoted to understand the behavior of physical systems with a large number of components that interact together. Initially, statistical physics dealt with a solid or a liquid material having a large number of atoms, ions or molecules. Nowadays, one can consider macroscopic items such as cars on a road, matrices and graphs, etc. [7]. To understand the properties of these physical systems, the theorists have developed physical models like Ising model among others.

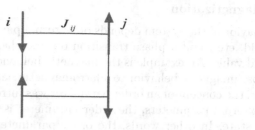

Fig. 1. A two dimensional Ising model

3.1 Ising Model

The Ising model describes the magnetic moments by Ising spins that are localized at the vertices of a certain region of the d-dimensional cubic lattice [7]. Figure 1 depicts an Ising model with $d = 2$.

There is an Ising spin σ_i on each vertex i. Each spin is modeled by an arrow that points up if $\sigma_i = +1$ and points down if $\sigma_i = -1$. The coupling J_{ij} represents interaction between pairs of spins.

- If $J_{ij} = J > 0$ then this model represents a ferromagnetic Ising model.
- If $J_{ij} > 0$ or $J_{ij} < 0$ then this model represents an Ising spin glass.

A configuration $\sigma = (\sigma_1 \ldots \sigma_N)$ of the system is obtained by assigning the values of all the spins in the system. The energy or the Hamiltonian of a configuration is defined as follows [7] :

$$H(\sigma) = -\sum_{(ij)} J_{ij}\sigma_i\sigma_j - B\sum_i \sigma_i \ . \tag{1}$$

The RHS of (1) is composed of two types of contributions :

- A term $-J_{ij}\sigma_i\sigma_j$ for each edge (ij) of the graph.
- A term $-B\sigma_i$ for each spin σ_i, due to an external magnetic field applied to the system.

The Boltzmann distribution gives the probability to find the system in the configuration σ [7]:

$$P_\beta(\sigma) = \frac{1}{Z(\beta)} \exp\left[-\beta H(\sigma)\right] \ ,$$

$$Z(\beta) = \sum_\sigma \exp\left[-\beta H(\sigma)\right] \ ,$$

where the parameter $\beta = 1/T$ is the inverse of the temperature. The normalization constant $Z(\beta)$ is called the partition function.

3.2   Magnetization

The behavior of the system depends on external parameters (temperature, magnetic field, etc.) and a phase transition occurs when this parameter achieves a threshold value. An example is the magnetic behavior transformation of a metal from a paramagnetic behavior to a ferromagnetic behavior. To describe a system behavior, the concept of an order parameter was introduced. For appropriate values of external parameters, the order parameter is nonzero and it indicates an ordered state. In other words, the order parameter is a measure of the order degree of the system [7]. The order parameter of the Ising model is the average value of the spins for all configurations, called magnetization.

- If $m = 1$ the system is in an ordered phase called ferromagnetic phase. Figure 2 shows a simple example of this behavior. All spins are in the same state.

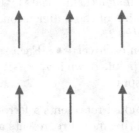

Fig. 2. A ferromagnetic phase

- If $m = 0$ the system is in a disordered phase called paramagnetic phase. Figure 3 shows a simple example of this behavior. All spins are independent.

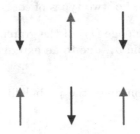

Fig. 3. A paramagnetric phase

To conclude this section, we shall remark that all these concepts are not common on the field of information theory. How these tools can be used to analyze the LDPC codes will be the subject of the next sections.

4   Decoding Problem from Statistical Physics Point of View

4.1   Statistical Physics Analogy

We have described in the previous section the LDPC codes using the additive boolean group $(\{0,1\}, +)$. For convenience, we introduce a multiplicative binary group $(\{+1, -1\}, \times)$ which is more suitable for the statistical physics methods [5].

Each code can be considered as a spin system from the statistical physics description. Each bit $S_j = (-1)^{s_j}$ is called a spin and is equal to $\{\pm 1\}$. The word $S = ((-1)^{s_1}, ..., (-1)^{s_N})$ is a collection of N spins called a configuration. The parity-check matrix performs the interactions between the spins [5].

The decoding problem depends on a posteriori probability $P(S\backslash J)$ where J is the evidence (received message or syndrome vector) and S is an estimation of the original codeword (or an estimation of the noise vector) [5]. In this section, J represents the syndrome vector and S represents an estimation of the noise vector. By applying Bayes' theorem the a posteriori probability can be written in the following form [5]

$$P(S\backslash J) = \frac{P(J\backslash S)P(S)}{\sum\limits_{S} P(J\backslash S)P(S)} \tag{2}$$

$$= \frac{1}{Z} \exp\left[\ln P(J\backslash S) + \ln P(S)\right] .$$

From a statistical physics point of view, the probability (2) can be seen as a Boltzmann distribution at inverse temperature β [11].

$$P(S\backslash J) = \frac{1}{Z} \exp\left[-\beta H(S\backslash J)\right] , \tag{3}$$

where $H(S\backslash J)$ denotes the Hamiltonian of the system. The inverse temperature in (3) don't has any relationship with the temperature of the communication device. It is simply an auxiliary parameter in the decoding process.

In this Hamiltonian, we identify two essential components in the analysis of the LDPC codes.

- A first term that ensures that all parity checks are fulfilled. It can be represented by the Kronecker's delta δ [5].

$$P(J\backslash S) = \prod_{\mu=1}^{M} P\left(J_\mu \backslash S\right)$$

$$= \prod_{\mu=1}^{M} \delta\left[J_\mu, \prod_{j \in L(\mu)} S_j\right] .$$

Note that J_μ represents the μ-th component of the syndrome vector J, therefore $J_\mu = \prod_{j \in L(\mu)} \zeta_j$. Thus,

$$P(J_\mu \backslash S) = \begin{cases} 1 & \text{if } J_\mu = \prod_{j \in L(\mu)} S_j \\ 0 & \text{otherwise .} \end{cases}$$

According to [11], the δ's can be substituted by a soft constraint

$$P(J \mid S) = \exp\left[\beta \sum_{\mu=1}^{M} J_\mu \prod_{j \in L(\mu)} S_j \right] ,$$

where $\beta \to \infty$.

- A prior term that yields some statistical information on the dynamical variables S. It can be written with the prior distribution

$$P(S) = \frac{\exp\left(F \sum_{j=1}^{N} S_j \right)}{(2 \cosh F)^N} ,$$

where F is the external field that depends on the transmission channel. For a BSC channel, characterized by the error probability p, it's equal to

$$F = \frac{1}{2} \ln \frac{1-p}{p} .$$

Therefore, the Hamiltonian $H(S \backslash J)$ can be expressed as follows:

$$H(S \backslash J) = - \sum_{\mu=1}^{M} J_\mu \prod_{j \in L(\mu)} S_j - \frac{F}{\beta} \sum_{j=1}^{N} S_j .$$

According to the final form of the Hamiltonian, we conclude that LDPC codes are similar to a spin glass system with muti-spin coupling J_μ in an external field F.

4.2 Decoding with the Statistical Physics

The decoding technique consists of finding local magnetization at inverse temperature β, $m_j = \langle S_j \rangle_\beta$. After that, the estimates of the codeword bit is calculated as in [5].

$$\hat{S}_j = \text{sign}(m_j) . \tag{4}$$

The magnetization value allows the measure of the decoding performance in the statistical physics approach [5]. This magnetization is defined by the overlap between the transmitted codeword S and its estimated word $\hat{S}$:

$$m = \frac{1}{N} \left\langle \sum_{j=1}^{N} S_j \hat{S}_j \right\rangle_{A,\zeta} . \tag{5}$$

Let us notice that the average $\langle \cdots \rangle$ is performed over the disorder defined by the noise ζ and the matrices A of a particular LDPC code. This value gives information about the performance of the code. Indeed, N.Sourlas [5] has shown that the bit error probability is given by :

$$p_b = \frac{1-m}{2} .$$

In order to compute the value of the magnetization, two main methods can be used: the replica method [11] and the TAP approach [12]. We will focus only on the TAP approach in this paper.

4.3   TAP Approach

Kabashima et al. [12] have shown that there exists a similarity between the equations obtained by the TAP approach [13] and those derived from BP. In the statistical physics language, the field $q_{\mu j}^a$ corresponds to the mean influence of sites other than the site j and the field $r_{\mu j}^a$ represents the influence of j back over the system (reaction fields) [14].

This connection can be highlighted by the proportionality observed between the likelihood $P(J_\mu \backslash S)$ and the Boltzmann weight [14] :

$$\omega_B(J_\mu \backslash \{S_j : j \in L(\mu)\}) = \exp\left(-\beta J_\mu \prod_{j \in L(\mu)} S_j\right) .$$

The conditional probability $r_{\mu j}^{S_j}$ can be thought as a normalized effective Boltzmann weight (effective Boltzmann probability) deduced when the bit S_j is kept fixed [14]

$$r_{\mu j}^{S_j} = \alpha_{\mu j}\, \omega_{eff}(J_\mu \mid S_j)$$
$$= \alpha_{\mu j} \sum_{S_l : l \in L(\mu) \backslash j} \omega_B(J_\mu \mid \{S_l : l \in L(\mu)\}) \prod_{l \in L(\mu) \mid j} q_{\mu l}^{(S_l)} .$$

Since spin variable S_j takes only two values ± 1, it is convenient to express the BP/TAP algorithm using spin averages $\sum_{S_j = \pm 1} S_j q_{\mu j}^{S_j}$ and $\sum_{S_j = \pm 1} S_j r_{\mu j}^{S_j}$ rather than the distributions $q_{\mu j}^{S_j}$ and $r_{\mu j}^{S_j}$ themselves.

We denote $\sum_{S_j=\pm1} S_j q_{\mu j}^{S_j} = \delta q_{\mu j} = m_{\mu j}$. Similarly, we denote $\delta r_{\mu j} = \hat{m}_{\mu j} = \sum_{S_j=\pm1} S_j r_{\mu j}^{S_j}$. Further, it was shown in [14] that the following statements hold

$$\hat{m}_{\mu j} = \tanh(\beta J_\mu) \prod_{l \in L(\mu)\backslash j} m_{\mu l} \ , \tag{6}$$

$$m_{\mu j} = \tanh\left(\sum_{\nu \in M(j)\backslash \mu} \tanh^{-1}(\hat{m}_{\nu j}) + \beta F \right) \ . \tag{7}$$

Then, the pseudo-posteriori might be computed

$$m_j = \tanh\left(\sum_{\mu \in M(j)} \tanh^{-1}(\hat{m}_{\mu j}) + \beta F \right) \ , \tag{8}$$

which gives a way to calculate the Bayes optimal decoding as follows:

$$\hat{S}_j = \text{sign}(\delta q_j) = \text{sign}(m_j) \ . \tag{9}$$

The equality $\delta q_j = m_j$ is proved as follows:

$$
\begin{aligned}
m_j &= \langle S_j \rangle_\beta \\
&= \frac{1}{Z} \sum_S S_j \exp\left[-\beta H(S \mid J)\right] \\
&= \sum_S S_j P(S \mid J) \\
&= \sum_{S_j} S_j \sum_{S\mid S_j} P(S \mid J) \\
&= \sum_{S_j} S_j P(S_j \mid J) \\
&= P(S_j = +1 \mid J) - P(S_j = -1 \mid J) \\
&= q_j^{+1} - q_j^{-1} \\
&= \delta q_j \ .
\end{aligned}
$$

5   LLR-BP Algorithm and Its Simplified Version with TAP Approach

5.1   LLR-BP Algorithm with TAP Approach

Instead of using probabilities as in [2], we prefer to deal with the LLRs in the LLR-BP algorithm since it yields more in implementation advantages [3].

In this section, our first contribution is the development of the LLR-BP with the TAP approach. Let $x_{\mu j}$ be the LLR of the probability sent from bit S_j to check node μ and $y_{\mu j}$ be the LLR of the probability sent from the check node μ to bit node S_j. According to the statistical physics definition, the LLR is defined as follows [7]:

$$x_{\mu j} = \frac{1}{2\beta} \ln \frac{q_{\mu j}^{+1}}{q_{\mu j}^{-1}} = \frac{x'_{\mu j}}{2\beta} \; , \tag{10}$$

and

$$y_{\mu j} = \frac{1}{2\beta} \ln \frac{r_{\mu j}^{+1}}{r_{\mu j}^{-1}} = \frac{y'_{\mu j}}{2\beta} \; . \tag{11}$$

The following result is helpful

$$\tanh\left(\beta x_{\mu j}\right) = \tanh\left(\frac{1}{2} \ln\left(\frac{q_{\mu j}^{+1}}{q_{\mu j}^{-1}}\right)\right) \tag{12}$$

$$= q_{\mu j}^{+1} - q_{\mu j}^{-1} = m_{\mu j} \; .$$

From (12), the variable $\hat{m}_{\mu j}$ can be written as

$$\hat{m}_{\mu j} = \tanh(\beta y_{\mu j}) \; . \tag{13}$$

From equations (6), (12) and (13) we get

$$x_{\mu j} = \frac{1}{\beta} \tanh^{-1} \tanh\left(\sum_{\nu \in M(j)\backslash\mu} \tanh^{-1}(\hat{m}_{\nu j}) + \beta F\right)$$

$$= \sum_{\nu \in M(j)\backslash\mu} y_{\nu j} + F \; . \tag{14}$$

Besides, the extrinsic information $y_{\mu j}$ can be written as

$$y_{\mu j} = \frac{1}{\beta} \tanh^{-1}(\hat{m}_{\mu j}) \tag{15}$$

$$= \frac{1}{\beta} \tanh^{-1}\left(\tanh(\beta J_\mu) \prod_{l \in L(\mu)\backslash j} \tanh(\beta x_{\mu l})\right) \; .$$

Finally, the a posteriori information of bit S_j can be expressed as:

$$x_j = \sum_{\mu \in M(j)} y_{\mu j} + F \; . \tag{16}$$

It's not simple to implement (15) due to the product form. Our idea is to split the extrinsic information into sign and magnitude processing as in [4]. We get from (15)

$$\tanh(\beta y_{\mu j}) = \tanh(\beta J_\mu) \prod_{l \in L(\mu)\backslash j} \tanh(\beta x_{\mu l}) \; . \tag{17}$$

Replacing (βJ_μ) by $\text{sign}\,(\beta J_\mu) \times |\beta J_\mu|$ and $(\beta x_{\mu l})$ by $\text{sign}\,(\beta x_{\mu l}) \times |\beta x_{\mu l}|$ in (17) yields

$$\text{sign}\,(\beta y_{\mu j}) = \prod_{l \in L(\mu)\backslash j} \text{sign}(J_\mu x_{\mu l}) \;, \tag{18}$$

$$\tanh |\beta y_{\mu j}| = \tanh |\beta J_\mu| \prod_{l \in L(\mu)\backslash j} \tanh |\beta x_{\mu l}|$$

$$\tanh \left| \frac{y'_{\mu j}}{2} \right| = \tanh \left| 2\frac{\beta J_\mu}{2} \right| \prod_{l \in L(\mu)\backslash j} \tanh \left| \frac{x'_{\mu l}}{2} \right| \;. \tag{19}$$

Let $f(t)$ be defined by

$$f(t) = -\ln \left(\tanh \left(\frac{t}{2} \right) \right) = \ln \frac{e^t + 1}{e^t - 1} \;. \tag{20}$$

Applying the logarithm to the inverse of both sides of (19) yields

$$-\ln \tanh \left| \frac{y'_{\mu j}}{2} \right| = -\ln \tanh \left| 2\frac{\beta J_\mu}{2} \right| + \sum_{l \in L(\mu)\backslash j} -\ln \tanh \left| \frac{x'_{\mu l}}{2} \right|$$

$$f\left(\left| y'_{\mu j} \right| \right) = f\left(|2\beta J_\mu| \right) + \sum_{l \in L(\mu)\backslash j} f\left(\left| x'_{\mu l} \right| \right) \;. \tag{21}$$

The equation (20) verifies

$$f\left(f\left(t \right) \right) = t \;. \tag{22}$$

Hence, (21) can expressed as

$$\left| y'_{\mu j} \right| = f\left(f\left(|2\beta J_\mu| \right) + \sum_{l \in L(\mu)\backslash j} f\left(\left| x'_{\mu l} \right| \right) \right) \;. \tag{23}$$

Using (10) and (11), (23) can be written as

$$|2\beta y_{\mu j}| = f\left(f\left(|2\beta J_\mu| \right) + \sum_{l \in L(\mu)\backslash j} f\left(|2\beta x_{\mu l}| \right) \right) \;. \tag{24}$$

From (18) and (24), one can finally deduce the extrinsic information

$$y_{\mu j} = \prod_{l \in L(\mu)\backslash j} \text{sign}\,(J_\mu x_{\mu l}) \times \frac{1}{2\beta} \times f\left(f\left(|2\beta J_\mu| \right) + \sum_{l \in L(\mu)\backslash j} f\left(|2\beta x_{\mu l}| \right) \right) \;. \tag{25}$$

5.2 BP-Based with TAP Approach

In this section, we detail our second contribution which consists of developing the BP-Based algorithm [3] with the TAP approach. The key idea of this algorithm is based on the fact that $\sum_t f(t) \leq f(\min_t t)$. The equation (24) is then approximated by

$$
|2\beta y_{\mu j}| = f\left(f\left(|2\beta J_\mu|\right) + \sum_{l \in L(\mu)\backslash j} f\left(|2\beta x_{\mu l}|\right)\right) \qquad (26)
$$

$$
\simeq f\left(f\left(\min_{l \in L(\mu)\backslash j}\left(|2\beta J_\mu|, |2\beta x_{\mu l}|\right)\right)\right)
$$

$$
= \min_{l \in L(\mu)\backslash j}\left(|2\beta J_\mu|, |2\beta x_{\mu l}|\right) \ .
$$

Thus, the extrinsic information in the BP-Based algorithm written with the TAP equation can be expressed as

$$
y_{\mu j} = \prod_{l \in L(\mu)\backslash j} \text{sign}\left(J_\mu x_{\mu l}\right) \times \min_{l \in L(\mu)\backslash j}\left(|J_\mu|, |x_{\mu l}|\right) \ . \qquad (27)
$$

The equation (27) reinterpret the BP-Based algorithm when using the TAP approach. This latter allows to replace check node update by a selection of the minimum input value. The other steps of the LLR-BP algorithm remain unchanged.

5.3 λ-min Algorithm with TAP Approach

In this section, we introduce our third contribution which is the development of the λ-min algorithm [4] with the TAP approach. In the extrinsic information (25), the magnitude processing is run using the function f defined by (20). Looking closer, Fig. 4 clearly shows that $\sum_t f(t)$ can be approached by the maximal values of $f(t)$ which is obtained for the minimal values of $|t|$.

Following Guilloud et al. [4] we compute (24) with only the λ bits involving in check μ and having the minimum magnitude.

$$
|2\beta y_{\mu j}| = f\left(f\left(|2\beta J_\mu|\right) + \sum_{l \in L_\lambda(\mu)\backslash j} f\left(|2\beta x_{\mu l}|\right)\right) \ , \qquad (28)
$$

with $L_\lambda(\mu) = \{j_0, \ldots, j_{\lambda-1}\}$ be the subset of $L(\mu)$ which contains the λ bits of check μ which have the smallest magnitude of $x_{\mu l}$.

The extrinsic information in the λ-min algorithm written with the TAP equation is given by (29)

$$
y_{\mu j} = \prod_{l \in L(\mu)\backslash j} \text{sign}\left(J_\mu x_{\mu l}\right) \times \frac{1}{2\beta} \times f\left(f\left(|2\beta J_\mu|\right) + \sum_{l \in L_\lambda(\mu)\backslash j} f\left(|2\beta x_{\mu l}|\right)\right) \ . \quad (29)
$$

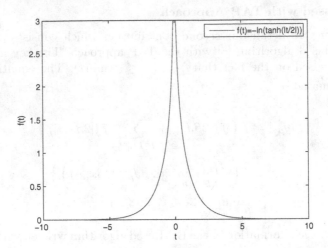

Fig. 4. Shape of the function f

Two cases are possible:

1. If the bit j belongs to the subset $L_\lambda(\mu)$, then $y_{\mu j}$ are processed over $(\lambda - 1)$ values of $L_\lambda(\mu) \backslash j$.
2. Otherwise, $y_{\mu j}$ are processed over the λ values of $L_\lambda(\mu)$. Notice that in this case, the computation have to be performed only once [4].

6   Performance of LDPC Decoding Algorithms

In this section, we evaluate the performance of LDPC codes from statistical physics point of view. Actually, the performance of these codes are presented in terms of the magnetization parameter rather than the usual Bit Error Rate parameter (BER) commonly used in information theory. As mentioned in Sect. 4.2, the magnetization is defined by the overlap between the transmitted codeword and its estimated value which is averaged over the noise ζ and the matrices A of a particular LDPC code.

$$m = \frac{1}{N} \left\langle \sum_{j=1}^{N} S_j \hat{S}_j \right\rangle_{A,\zeta}.$$

In order to evaluate the performance of LDPC decoding algorithms in terms of the magnetization, we proceed in three steps :

1. Generate the check matrices A defined for a particular LDPC(N, C, K) code.
2. Decode the received codeword for each parity check matrix A.
3. Average over the results.

6.1   Proposed Method to Create Matrices

The proposed method aims at swapping the elements of the matrix A in order to create a set of LDPC code. We choose to perform ten regular LDPC codes. The algorithm of this method is the following:

1. Define LDPC(N, C, K) code from the Mackay's online database [15]. The parity check matrix A corresponding to this code will be set as the initial matrix on which the random swap will be performed.
2. Initialize $A1 = A$. Then, generate a random integer representing the number of matrix elements permutations. For each permutation do
   (a) Choose randomly tow rows i and j from the matrix A then compute:

$$\alpha(n) = A(i, n) + A(j, n) \ (\text{mod } 2) \ \text{with } 1 \leq n \leq N \ .$$

   (b) Choose randomly two distinct variables k and l for which $\alpha(k)$ and $\alpha(l)$ are non zero. If $A(i, k) \neq A(i, l)$ and $A(j, k) \neq A(j, l)$ then do

$$A_1(i, k) = \overline{A(i, k)}, \quad A_1(i, l) = \overline{A(i, l)} \ ,$$

$$A_1(j, k) = \overline{A(j, k)}, \quad A_1(j, l) = \overline{A(j, l)} \ .$$

3. Repeat step 2 as much as needed (ten times in our case) to obtain the sought LDPC(N, C, K) codes.

6.2   Simulation Results

We have performed our simulations using regular $(3, 6)$ LDPC code of length $N = 1008$ and with 20 decoding iterations. The results are averaged over ten codes randomly generated. This set of LDPC codes shares the same block length, three ones in each column and six ones in each row. A predefined code have been used to generate 1008 bit codewords from 504 bit message for each run. The received codewords are then decoded with LLR-BP, BP-Based and λ-min decoding algorithms.

In Fig. 5 we compared the magnetization performance for the three decoding algorithms LLR-BP, BP-Based and the λ-min in the case of AWGN channel.

According to Fig. 5, one can conclude that λ-min algorithm is more efficient than the BP-Based algorithm and is closer to the LLR-BP algorithm. At a magnetization value 0.9, the BP-Based algorithm introduces a degradation of 0.5 dB when performing 20 iterations, while the 2-min algorithm leads to a degradation of 0.3 dB only. The 3-min algorithm is more accurate than the previous ones, but is slightly less accurate than the LLR-BP algorithm with a small degradation equal to 0.08 dB.

Figure 5 shows also that all algorithms tend to the value $m - 1$ at high signal to noise ratio. This magnetization value is expected since it characterized the ferromagnetic state in the statistical physics and a very low BER i.e. a reliable communication in the information theory.

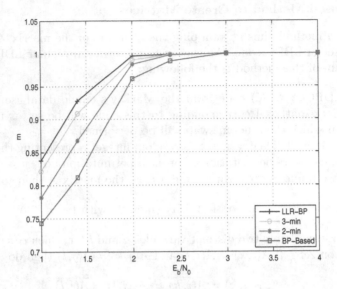

Fig. 5. Magnetization performance function of signal to noise ratio for different LDPC decoding algorithms

6.3   Comparison of Performance

This section is devoted to the comparison of the performance of the LLR-BP algorithm versus the replica method results [14]. Our simulations are evaluated in the case of transmission over an AWGN channel. Nevertheless, almost all results in the literature are given in the case of BSC. Thus, we choose to evaluate the magnetization as a function of signal to noise ratio for both channels in order to make the comparison. Indeed, for the BSC channel the error probability p of the channel is :

$$p = \frac{1}{2}\text{erfc}\sqrt{\frac{RE_b}{N_0}} \; ,$$

where E_b is the energy per information bit, N_0 is the power spectral density and R is the code rate.

In Fig. 6, we depict a comparison between the magnetization performance of the LLR-BP algorithm in the case of AWGN channel and the magnetization performance given by the replica method in the case of BSC channel.

The simulation results show the efficiency of transmission over an AWGN channel comparative to the BSC channel for the LDPC codes having $C = 3$ and $K = 6$. For low signal to noise ratio and at $m = 0.95$, the performance evaluated by the replica method in the case of a BSC channel brings a loss of 0.4 dB compared to the LLR-BP algorithm in the case of an AWGN channel. This result is expected since the Shannon limit on AWGN channel is lower than the Shannon limit on a BSC channel when dealing with the same rate R. On the

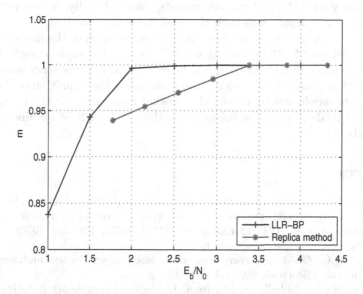

Fig. 6. Magnetization performance of the LLR-BP algorithm in the case of AWGN channel and by replica method in the case of BSC channel with $C = 3$ and $K = 6$

other hand, we notice that for both types of channels, our simulations become similar to those of the replica method when the signal to noise ratio exceeds 3.4 dB.

7   Conclusion

The study of LDPC codes performance by the statistical physics methods have been developed only for the BP algorithm. However, to our knowledge, no studies have been made for the other decoding algorithms. We adapted in this paper a new method based on statistical physics in the development of the LLR-BP algorithm and its simplified versions, the BP-Based algorithm and the λ-min algorithm. We proposed a new method to swap the parity check matrix elements and to create a set of regular LDPC codes having the same number of ones by columns as well as by rows. This intermediate step is necessary in order to evaluate the average value of the overlap between the transmitted codeword ant its estimation. This average parameter, called the magnetization, allows us to evaluate the performance of the decoding algorithms as a function of signal to noise ratio. Simulation results show that the BP-Based algorithm tends to reduce the complexity of updating extrinsic information, with degradation in performance compared to the LLR-BP algorithm. The same conclusion remains valid when using the λ-min algorithm but with more efficient performances compared to the LLR-BP algorithm, especially in the case when λ increases. These results are in adequacy with those of the information theory. Finally, we compared our LLR-BP algorithm simulations with those of the replica method.

A possible continuation to this work is the computation of the magnetization value of the different LDPC decoding algorithms with the replica method. Such a study would be interesting as it allows the evaluation of the system phase transitions for different signal to noise ratio and therefore the definition of the noise threshold that should not be exceeded. Another alternatives is the comparison of the noise threshold given by information theory with those obtained by the replica method.

References

1. European Telecommunications Standards Institute (ETSI). Digital Video Broadcasting (DVB) Second generation framing structure for broadband satellite applications; Part2: DVB-S2 Extensions (DVB-S2X), EN 302 307-2 (V1.1.1), https://www.dvb.org/standards/dvb-s2x
2. Mackay, D.J.C.: Good error-correcting codes based on very sparse matrices. IEEE Trans. Inform. Theory 45, 399–431 (1999)
3. Fossorier, M.P.C., Mihaljevic, M., Imai, I.: Reduced complexity iterative decoding of low density parity check codes based on belief propagation. IEEE Trans. Commun. 47, 673–680 (1999)
4. Guilloud, F., Boutillon, E., Danger, J.L.: λ-min decoding algorithm of regular and irregular LDPC codes. In: Proc. 3rd Int. Symp. on Turbo Codes & Related Topics (ISTC 2003), pp.451–454. Brest (2003)
5. Sourlas, N.: Spin glass models as error correcting codes. Nature 339, 693–695 (1989)
6. Skantzos, N.S., Van Mourik, J., Saad, D.: Magnetization enumerator of real-valued symmetric channels in Gallager error-correcting codes. Phys. Rev. E 67, 037101 (2003)
7. Mezard, M., Montanari, A.: Information, Physics and Computation. Oxford university Press (2009)
8. Huang, H.: Code optimization, frozen glassy phase and improved decoding algorithms for low-density parity-check codes. Commun. Theor. Phys. 63, 115–127 (2015)
9. Neri, I., Skantzos, N.S., Boll, D.: Gallager error-correcting codes for binary asymmetric channels. J. Stat. Mech. 2008, P10018 (2008)
10. Neri, I., Skantzos, N.S.: On the equivalence of Ising models on small-world networks and LDPC codes on channels with memory. J. Phys. A: Math and Theor. 47, 385002 (2014)
11. Murayama, T., Kabashima, Y., Saad, D., Vicente, R.: Statistical physics of regular low-density parity-check error-correcting codes. Phys. Rev. E 62, 1577–1591 (2000)
12. Kabashima, Y., Saad, D.: Belief propagation vs TAP for decoding corrupted messages. Eurphys. Lett. 44, 668–674 (1998)
13. Thouless, D.J., Anderson, P.W., Palmer, R.G.: Solution of solvable model of a spin glass. Phil. Mag. 35, 593–601 (1977)
14. Vicente, R., Saad, D., Kabashima, Y.: Finite-connectivity systems as error-correcting codes. Phys. Rev. E 60, 5352–5366 (1999)
15. Mackay, D.J.C.: Ldpc database, http://www.inference.phy.cam.ac.uk/mackay/codes/data.html

Representation of Dorsal Hand Vein Pattern Using Local Binary Patterns (LBP)

Maleika Heenaye Mamode Khan[✉]

Department of Computer Science and Engineering,
University of Mauritius, Mauritius
m.mamodekhan@uom.ac.mu
http://www.uom.ac.mu

Abstract. In this revolutionized and digital world, the increasing need of security to protect individuals and information has led to a rise in developing biometric systems over traditional security systems such as pincode and password. Finding more reliable, practical and more acceptable biometrics and techniques are attracting the attention of researchers. Recently, hand vein pattern biometrics has gained increasing interest from both research communities and industries. Researchers are exploiting the different biometric phases by applying existing techniques or devising new ones to develop enhanced biometric systems. Up to now, most researchers have thinned the dorsal hand vein pattern and apply corresponding techniques for feature representation and matching. However, not many techniques have been explored with relation to considering the whole hand vein image. In this research work, local binary pattern, which is a powerful technique for representing texture description of an image, have been applied on dorsal hand vein images. This method outperforms existing vein representation techniques by having a recognition rate of 98.4% on a database of more than 1000 images. In addition, this proposed method has no effect on rotated images, which is desirable in any biometric security system.

Keywords: Dorsal hand vein · Biometrics · Local binary pattern

1 Introduction

Biometric access control system measures the physiological or behavioral characteristics of an individual and is replacing traditional methods of security such as pincodes and passwords. Popular biometrics, viz, fingerprints, face recognition and iris have been explored and efficient solutions have been commercialised. However, in this technological era, hackers are devising new ways of manipulating these existing biometrics, urging researchers to come up with new characteristics to be used as biometric. A practical biometric should meet the specified recognition accuracy, speed and resource requirement, be harmless to users, be accepted by intended population and be sufficiently robust to various fraudulent methods and attacks to the system [1]. Factors that may influence the popularity, applicability and performance of biometric verification techniques are uniqueness, repeatability, maximum throughput, whether operable under controlled light or not, invasiveness or non- invasiveness, immunity from forgery, successful identi-fication of dark- skinned subjects, ease of use, user cooperation, cleanliness and so on. Dorsal hand vein pattern is a quite new characteristic that is proving to be a potential candidate for developing biometric security system. Dorsal hand vein pattern is a unique network of blood vessels found below the skin at the back of the hand and is capable of identifying a person. The pattern is invisible to the naked eye and the superficial veins have higher temperature than the surrounding tissues. The vein pattern is best defined when the fist is clenched and can only be captured by using a CCD camera or a thermal camera. The vein pattern is stable, unique and has repeatable features. Up to now, there is no vein pattern database that is available to the research community and thus, researchers have to build their own database [2,3]. A typical hand vein pattern biometric consists of the following five processing stages namely image acquisi-tion, image enhancement, vein segmentation, vein extraction and matching [3,4]. Different researchers explore these different phases. While the concepts of bio-metrics appear to be simple, there are many challenges in the implementation of biometric systems.

2 Research Gap

In biometric security system, images obtained at image capture are subject to preprocessing to enhance their quality. In most dorsal hand vein biometric system deployed so far, vein patterns are aligned, normalized, thresholded, filtered and thinned [5,3,6]. Consequently processing methods like correlations that require squeletonised version has been adopted [7,8]. The idea of using the whole hand with the vein patterns have not explored. Currently, there is no literature which demarcates techniques that can be applied on raw biometric images captured and enhanced biometric images, hence clearly denoting a research gap. There seems to be a lot of scope for further improvement of existing techniques in addition to finding more reliable biometric and techniques for using them for security purposes. In addition, researchers have not paid much attention on methods that can be adopted to overcome the problem of translated images.

3 Proposed Vein Biometric Security System

In this research work, a dorsal hand vein pattern database is built. The Vein images are preprocessed to enhance the quality of the images. Local Binary Pattern (LBP) which is a rotation invariant method has been applied on the vein patterns to extract and represent key features. The extracted features are then concatenated to form a histogram which is considered as the feature vector.

3.1 Image Acquisition and Vein Database

Up to now, there is no dorsal hand vein database available for research. Thus, each researcher has to acquire their own vein images. Since veins are found beneath the skin, they can only be captured using infrared light. In the experimental setup of this research, images were obtained with a digital camera with infrared filters using an appropriate setup. To build the vein database, a Nikon digital camera D3100, a Hoya R-72 infrared filter, LED lights and diffusing papers have been used. The camera and lights were carefully mounted using a closed wooden box with one open side. The infrared light improved the contrast of the hand vein images. The diffusing papers used during the image capture process were to reduce scattering effects and to attenuate the intensity of infrared. Subjects were requested to hold a removable wooded handle specifically mounted to avoid any large image translation. This also helped to obtain a clenched fist which provides a better definition of the vein. Four LED bulbs were mounted in the top surface of the box and are used as sources of near infrared radiation to illuminate the interior of the box. To minimize the reflection of the near infrared lights within the setup, the interior surfaces of the closed box was covered with plain black Bristol paper sheets. A DSLR camera (Nikon D3100) connected with an infrared filter (Hoya R72) are used for image acquisition. The camera was solidly mounted on top of the wooden box with its lens constantly pointing inside the box via another specific opening in the boxs top surface. To minimize any disturbances of the setup during subsequent capture of image instances, the camera was accessed via a remote control (MC-DC2) and photographs acquired were regularly transferred to a computer via a UC-E4 USB cable. During the experimentation, the ideal placement of the hand is deduced after various positioning and orientation. A database of 3000 images has been obtained from 300 subjects. The following diagram illustrates the experiment setup:

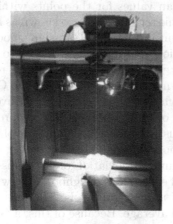

Fig. 1. Image Setup

Subjects involved in image capture were from different ethnic groups, age ranging from 19 years old to 65 years old and have different skin colour. During image capture, various factors namely the positions of the light, the light intensity, background light, the arrangement of the light positions, the angle of orientation of the hand, the distance between the camera and the hand were considered. During the experimentation, the ideal placement of the hand is deduced after various positioning and orientation.

3.2   Vein Preprocessing

In previous research, vein patterns need to be aligned before applying any other processing techniques [7,2]to ensure that the same pattern is obtained each time the processing is carried out. In this work, the vein is not aligned since the proposed local binary pattern technique is a rotation invariant method. The vein images were first normalized so that all the pixel intensities were converted to a domain of $[0, 255]$. The normalized image was then equalized using the adaptive histogram. The function adaptivehisteq() in Matlab was used. This function enhances the contrast of the normalized image. The next step was to obtain the vein from the hand background. Thus, adaptive thresholding was been used. In adaptive thresholding, each pixel was compared to an average of the surrounding pixels. The mean value and the threshold value were set. If the value of the current pixel is lower than the mean, it is set to black else it is set to white. These values were obtained after many tests carried out on the vein images. To eliminate remaining noise and to obtain an enhanced image, different filters were applied on the vein images. The first filter applied, that is, 2-D median filtering performs median filtering of the image in two dimensions. Each output pixel contained the median value in the M-by-N neighborhood around the corresponding pixel in the thresholded vein image. This filtering padded the image with zeros on the edges, so the median values for the points within $[MN]/2$ of the edges might appear distorted. This technique is also known as linear Gaussian filtering. To further enhance the vein images, the Gaussian smoothing was applied. The operator is a 2-D convolution operator that was used to blur images and remove detail and noise. The filter uses a different kernel that represents the shape of a Gaussian hump, also known as bell-shape. The idea of Gaussian smoothing is to use this 2-D distribution which can be achieved by convolution. Since the image was stored as a collection of discrete pixels was required to produce a discrete approximation to the Gaussian function before convolution can be performed. The effect of Gaussian smoothing was to blur an image. The degree of smoothing was determined by the standard deviation of the Gaussian. The Gaussian output a 'weighted average' of each pixel's neighborhood, with the average weighted more towards the value of the central pixels. This was in contrast to the mean filter's uniformly weighted average. Because of this, a Gaussian provides gentler smoothing and preserves edges better than a similarly sized mean filter. Wiener filter was the third filter applied to obtain a better image. The Wiener filter was a stationary linear filter for images degraded by additive noise and blurring. Calculation of the Wiener filter required the assumption that the signal and noise processes are second-order stationary. Wiener filters were usually applied in the frequency domain. Wiener filter was based on a statistical approach. It reduced the amount of noise present in a signal by comparing it with an estimation of the desired noiseless signal.

4 Representing Vein Characteristics Using Local Binary Patterns

4.1 Concept Behind Local Binary Pattern

Local Binary Pattern (LBP) is a powerful technique for representing texture description of an image [9,10]. LBP uses an operator which is defined as a gray scale invariant texture measure. This method is termed as a rotation invariant texture classification technique. It works by dividing the image into small regions and features are extracted. These features are then concatenated to form a histogram which is considered as the feature vector. Images are then recognized by comparing the feature vectors obtained. This first operator of LBP introduced works with eight-neighbors, that is, the 3x3 neighborhood of each pixel, that is, with the eight neighbors with reference to the centre value. Each pixel value is compared with all its neighboring pixel value. If a pixel has a greater value compared to the centre value, then a 1 is assigned to that cell else a 0 is assigned. The LBP code is then obtained by concatenating all the 0s and 1s. For ease of use, the decimal value of the binary codes is then obtained. A histogram over the cell is then built on the frequency of the numbers occurring. A feature vector is then formed by concatenating all the histograms generated from the cells. This approach has many advantages, that is, it is very tolerant to illumination changes, perspective distortions, image blur and image zoom [10]. In another work conducted by Ojala [11], the extended binary operator images was used with textures of different scales to use neighborhoods of different sizes. Instead of choosing a window of 3×3, the local neighborhood was taken as a set of sampling points on a circle. This method allowed the authors to use any radius and number of sampling points in the neighborhood. Bilinear interpolation was applied to ensure that the sampling point does not fall in the centre of a pixel. Thus, the notation LBP (P, R) is used where P represents the sampling points and R as the radius. An example is illustrated below:

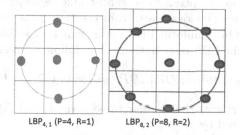

LBP$_{4,1}$ (P=4, R=1)   LBP$_{8,2}$ (P=8, R=2)

Fig. 2. LBP

Ojala [11] explored the local binary pattern and discovered the uniform patterns. A local binary pattern is termed as uniform if it contains at most two bitwise transitions from 0 to 1 or vice- versa when the bit pattern is considered circular. For example, 0000 0000, 0111 0000 and 1100 1111 are uniform [9]. Using this approach for the computation of the LBP histogram, uniform patterns has a separate bin and all non- uniform patterns are represented as a single bin. Recently, in an initial research work, Wang [12] have used partition LBP to represent vein patterns. The image is divided into non- overlapping rectangular or circular regions. The hand dorsal vein image is divided into 64 rectangular region and 16 circular regions. The texture contained in each subimage is then represented using a histogram by grouping the LBP patterns into categories. The feature vector is formed by concatenating the non- overlapping sub- images. The work was further extended to represent the fifth and sixth binary patterns having much higher occurrences than other binary patterns [12]. LBP has not yet been explored on dorsal hand vein patterns. Thus, in this research work, LBP is applied on dorsal hand vein features to determine whether it can be used as a method to develop hand biometrics like face biometrics.

4.2 Vein Representation

The vein representation is constructed by taking a vein image $X_i(x, y)$ to be a 2-dimensional $M \times M$ array of 8-bit intensity values for an individual i where $i = 1, 2, 3, ..., I$. This image is further represented as a vector of dimension M^2 by concatenating the rows. For instance, a vein image for a particular individual of size 256 x 256 becomes a vector of dimension 65,536. Note that this vein image is for a single individual i. Thus, for a set of I individuals, the dimension of the overall vein matrix space X can be represented as

$$X = [X_1, X_2, \ldots, X_I]_{M^2 \times I} \tag{1}$$

4.3 Application of LBP on Hand Vein Features

The dorsal hand vein patterns are considered as an image having circular neighbor sets. Neighborhood with varying number of sample termed as P and radius termed as R can be used. Bilinear interpolation is applied to the samples that do not fall exactly on the pixels. The center pixel is taken as threshold value to obtain the P value. The local pattern in the texture is then obtained by the P-bit binary code. Only uniform LBP codes can be considered to reduce the number of bins for the LBP distributions. The following operator is used:

$$LBP_{P,R}^{ri} = min\left(ROR\left[LBP_{P,R}i | i = 0, 1, 2, \ldots, P-1\right]\right) \tag{2}$$

where $ROR(x, i)$ performs the circular bitwise right shift i times on the P-bit binary number denoted by x.

In this work, $LBP_{8,2}^{riu2}$ has been used. From literature, it was found that $(8, 2)$ neighborhood accounts for more patterns compared to $(16, 2)$ neighborhood.

The texture contained in each sub image is obtained by grouping LBP patterns produced into 10 texture categories and 9 rotations.

Let the number of occurrence of a texture category in the sub image be $X_{i,j}$, where $j = 1, \ldots, 10$, the histogram is generated as follows:

$$H_i = [x_{i,1}, x_{i,2}, \ldots, x_{i,10}] \tag{3}$$

Let the vein pattern be divided into N rectangular regions, then the texture feature is represented by a Vector V of $10 * N$ dimensions. Concatenation of the LBP feature histograms is then as follows:

$$V = [H_1, H_2, \ldots, H_N] \tag{4}$$

Every bin in the histogram represents a pattern. For every region of the image, the non- uniform patterns are represented as a single bin. Every regional histograms consists of $P(P-1)+3$ bins, where P represents the number of sampling points. The total feature vectors contains $k^2 [P(P-1)+3]$ bins, where k^2 is the number of regions in which the image has been broken into.

5   Recognition of Veins

To recognize an image, the feature vector is used. Let the test image be T and the sample image is S. The difference between the feature vectors is computed by using dissimilarity measures for histograms. Histogram intersection, Log- likelihood statistics or Chi square statistic can be used. The equations are as follows: Let T be the test image and S be the template sample, For Histogram intersection:

$$D(T,S) = \sum_{j=1}^{k^2} \sum_{i=1}^{P(P-1)+3} min(T_{i,j}, S_{i,j}) \tag{5}$$

Table 1. Recognition Rate

Number of Images	RR using Histogram	RR using Log-likelihood	RR using Chi square
200	93.2	92.4	93.7
400	94.9	93.6	94.9
600	92.5	92.4	94.9
800	92.3	92.7	93.5
1000	92.9	93.2	94.9

Table 2. Performance of biometric system using dorsal hand vein patterns

	Recognition Rate	False Acceptance Rate	False Rejection Rate
100	94.1	0	5.9
200	93.9	0	6.1
500	95.6	0	4.4

Table 3. Recognition rate on varied images

Angle of deviation (100 test images considered)	Recognition Rate using LBP
2(Right)	96.4
5(Right)	92.5
8(Right)	94.3
8(Right)	92.3
15(Right)	95.6
20(Right)	94.8
25(Right)	98.2
30(Right)	95.5
35(Right)	91.9
40(Right)	97.1
45(Right)	95.7
2(left)	96.6
5(left)	95.3
8(left)	96.2
10(left)	94.7
15(left)	97.1
20(left)	94.2
25(left)	94.8
30(left)	94.4
35(left)	93.2
40(left)	92.9
40(left)	95.2

For Log- likelihood statistics:

$$L(T,S) = -\sum_{j=1}^{k^2} \sum_{i=1}^{P(P-1)+3} T_{i,j} log M_{i,j} \tag{6}$$

For Chi-squared statistics

$$\chi^2(T,S) = \sum_{j=1}^{k^2} \sum_{i=1}^{P(P-1)+3} \frac{(T_{i,j} - S_{i,j})^2}{(T_{i,j} + S_{i,j})} \tag{7}$$

The three distance measures were used for testing. A test set of 100 images are used and a sample set of 1000 images are used. The experiments were conducted 20 times and the average was computed. A sample of the results is displayed below: From the experiment results, chi square statistics works slightly better than histogram intersection and log likelihood statistics by having a better recognition rate. Thus, chi test is adopted for the implementation of the local binary pattern technique.

Table 4. Comparison of Proposed Method with Existing Research

Research	Imaging Technology	Method	Images	Performance in percent
Lin and Fan [2]	Thermal Imaging	Multi-resolution	32	FAR 1.5, FRR 3.5
Wang and Leedham [3]	Thermal Camera	Hausdorff Distance	12	FAR 0, FRR 0
Deepika et al. [14]	Near IR Imaging	Fusion	74	FAR 0, FRR 0.01
Heenaye- Mamode Khan et al. [15]	Near IR Imaging	Dimension Reduction	300	FAR 0, FRR 95.9
Proposed Method (LBP)	Near IR Imaging	LBP	300	FAR 0, FRR 98.4

5.1 Performance of Biometric System Using Dorsal Hand Vein Pattern

To determine the maximum allowable acceptable distance using chi test, different instances of a subject is placed in the test set and sample set. It was found that images captured in a controlled environment do not have many intensity changes. However, in an environment where there is varying lighting and intensity, the pixel values in different instances of an image vary as well. The experiment was conducted using chi test and images captured in a controlled environment. The sample set is varied and the RR, FAR and FRR expressed in percentages are captured.

The RR, FAR and FRR are obtained for both databases. An average RR of 95% is obtained. 0 %FAR is obtained is both cases.

5.2 Experiments on Rotated Images

LBP is termed as a rotation invariant method. To test this concept, oriented dorsal hand vein patterns and palmprints are subject to LBP. 22 varied images were taken for each instance. 100 test images were considered.

On average, the recognition rate is above 90% for images rotated at any angle. Hand images were made to rotate from 20 to 450 to the right and left. The RR proved that LBP was not affected by rotation.

6 Analysis of Results and Conclusion

In this paper, LBP has been applied on dorsal hand vein patterns. LBP is a method that represents texture description of an image and is considered as a rotation invariant texture classification technique. Using LBP, the dorsal hand vein patterns are divided into small regions. Each pixel value is then compared to its neighbors and a binary pattern is generated. The binary pattern is then represented as a decimal value. The number of occurrences of these values is then concatenated to form a histogram which is considered as the feature vector. Images are then recognized by comparing the feature vectors obtained. From the experimental results, it is concluded that this method works well on biometric features and produce a good recognition rate. This method has also been applied on oriented images. Compared to dimensional reduction techniques like principle component analysis (PCA)[13], LBP is a rotation invariant method. The following table provides a summary of comparison of LBP with existing methods.

Though techniques are applied to orient the hand in the same position each time processing takes place, there might be slight difference between instances. This powerful rotation invariant texture classification was applied for the first time on hand images. The system then computed the closest distance and an average recognition rate of 95% was obtained for dorsal hand vein patterns.

References

1. Jain, A., Nandakumar, K., Ross, A.: Score normalization in multimodal biometric systems. The Journal of Pattern Recognition 8, 2270–2285 (2005)
2. Lin, C.L., Fan, L.C.: Biometric Verification Using Thermal Images of Palm- Dorsa Vein Patterns. IEEE Transactions on Circuit and Systems for Video Technology 14(2), 199–213 (2004)
3. Wang, L., Leedham, W., Wang, L.: Near- and- Far- Infrared Imaging for Vein Pattern Biometrics. In: Proceedings of the IEEE International Conference on Video and Signal Based Surveillance (2006)
4. Wang, L., Leedham, W., Cho, D.: Minutiae feature analysis for infrared hand vein pattern biometrics. The Journal of the Pattern Recognition Society 41(3), 920–929 (2008)
5. Wang, L., Leedham, W., Cho, D.: Infrared imaging of hand vein patterns for biometric purposes. IET Computer Vision 1(3-4), 113–122 (2007)
6. Soni, M., Gupta, S., Rao, M.S., Gupta, P.: A New Vein Pattern-based Verification System. (IJCSIS) International Journal of Computer Science and Information Security 8(1) (2010)
7. Badawi, A.: Hand Vein Biometric Verification Prototype: A Testing Performance and Patterns Similarity. In: Proceedings of the 2006 International Conference on Image Processing, Computer Vision, and Pattern Recognition, IPCV 2006, USA, June 26-29 (2006)
8. Shahin, M., Badawi, A., Kamel, M.: Biometric Authentication Using Fast Correlation of Near Infrared Hand Vein Patterns. International Journal of Biomedical Sciences 2(3) (2007)
9. Ahonen, T., Hadid, A., Pietikainen, M.: Description with Local Binary Patterns:Analysis, Application to Face Recognition. IEEE Transactions on Pattern Analysis and Machine Intelligence 28(12) (December 2006)
10. Heikkila, M., Pietikainen, M., Schmid, C.: Description of Interest Regions with Local Binary Patterns. Pattern Recognition 42(3), 425–436 (2009)
11. Ojala, T., Pietinainen, M., Maenpaa, T.: Multiresolution Gray-Scale and Rotation Invariant Texture Classification with Local Binary Patterns. IEEE Transactions on Pattern Analysis and Machine Intelligence 24(7), 971–987 (2002)
12. Wang, W., Li, K., Shark, L., Verley, R.: Hand-Dorsa Vein Recognition Based on Coded and Weighted Partition Local Binary Patterns. In: 2011 International Conference on Hand-Based Biometrics (ICHB), pp. 1– 5 (2011)
13. Turk, M., Pentland, A.: Face Recognition using Eigenfaces. In: The Proceedings of IEEE Computer Society Conference on Computer Vision and pattern Recognition, June 3-6, pp. 586–591 (1991)

14. Deepika, L., Kansaswamy, A., Vimal, C.: Protection of patient identity and privacy using vascular biometrics. International Journal of Security 4(5) (2010)
15. Heenaye-Mamode Khan, M., Subramaniam, R.K., Mamode Khan, N.: Low Dimensional Representation of Dorsal Hand Vein Features Using Principle Component Analysis (PCA). The Proceedings of World Academy of Science and Technology 3 (2009)

Watermarking Based Multi-biometric Fusion Approach

Sanaa Ghouzali[(✉)]

Information Technology Department,
College of Computer and Information Sciences,
King Saud University, Riyadh, Saudi Arabia
sghouzali@ksu.edu.sa

Abstract. In this paper we present a watermarking based multi-biometric fusion method that can embed fingerprint minutia information into host face images in the DCT (Discrete Cosine Transform) domain. This scheme has the advantage that in addition to prevent unauthorized biometric data manipulations, the biometric authentication can be performed efficiently using the fused biometric data without the need to extract the watermark. Orthogonal Locality Preserving Projections (OLPP) method is used in this approach to extract the most pertinent features which are beneficial to identification of the watermarked face images. Preliminarily results using *ORL* and *Yale* face databases, and *FVC2002 DB2* fingerprint database show the effectiveness of the proposed approach in achieving good authentication performance while preventing unauthorized manipulations of biometric data.

Keywords: Biometric authentication · Watermarking · Data fusion

1 Introduction

Biometrics based authentication has attracted researchers' attention in the computer community for its many applications specially related to security and access control. Since biometric modalities are associated permanently with a unique person and cannot be modified, biometrics techniques offer a reliable method for user authentication/identification. However several security breaches have been discovered [1] and more specifically security and integrity of biometric data pose new challenges. For example, if this biometric data is compromised through a biometric attack, either by theft or modification, it is lost forever. In [2], authors provide a critical survey about how watermarking technology can either help to cope with security threats in biometric systems or help to enhance biometric schemes in some other way.

Several studies have been published in the area of using watermarking to enable a multi-biometric approach by embedding a biometric data into another biometric sample of different biometric modalities. Jain et al. [3] proposed a blind watermarking technique which uses a bit stream of eigenface coefficients as a watermark to be embedded into randomly selected fingerprint image pixels

© Springer International Publishing Switzerland 2015
S. El Hajji et al. (Eds.): C2SI 2015, LNCS 9084, pp. 342–351, 2015.
DOI: 10.1007/978-3-319-18681-8_27

using a secret key. In [4], face and text information are embedded in texture regions of fingerprint image using Discrete Wavelet Transform (DWT). Moon et al. [5] presented various watermarking techniques for secure multimodal biometric systems using both fingerprint and face information. In their experimental results, it has been shown that embedding fingerprint features into a face image provides superior performance than embedding facial features into a fingerprint image in terms of user verification accuracy. In [6], a technique based on blockwise image watermarking and cryptography is proposed to embed fingerprint templates into facial images. Their scheme allows to maintain image quality. Park et al. [7] suggest to use robust embedding of iris templates into face image data. In [8] an encrypted palmprint template is embedded into a fingerprint image. The encryption key is derived from palmprint classes. Vatsa et al. [9] employ robust embedding techniques where they embed voice features in color facial images. Kim et al. [10] propose a blind and robust spread spectrum watermarking technique for embedding face template data into fingerprint sample. Authors in [11] propose a block pyramid based adaptive quantization watermarking scheme to embed fingerprint minutiae into face images. Numeric watermark bits with higher priority are embedded into upper pyramid level with a larger embedding strength using first-order statics QIM (quantization index modulation) method. However, in all these different approaches the additional biometric data are not used in multi-biometric fusion scheme but serve as an independent second biometric modality resulting in a two-factor authentication technique.

In this paper we propose a watermarking based multi-biometric fusion approach aiming to protect biometric templates from unauthorized manipulations in addition to provide a good recognition performance. First bits stream of fingerprint minutia is embedded into face image using a block-wise DCT based watermarking approach. The objective of using watermarking in this paper is to develop a multi-biometric fusion technique. Orthogonal Locality preserving projections (OLPP) [12] are then applied on the watermarked face images to extract the most pertinent features which are beneficial to identification. Finally, the cosine distances between feature vectors are calculated to match watermarked face images.

The remainder of this paper is organized as follows. The proposed multi-biometric fusion approach is described in Section 2. Section 3 addresses the experimentations, and Section 4 concludes the paper.

2   Proposed Watermarking Based Multi-biometric Fusion Approach

The proposed multi-biometric authentication system includes two stages, enrollment stage and verification stage. A user presents his/her biometric traits, e.g. face and fingerprint, during the enrollment stage. Then the watermark embedding process is applied and the resulting watermarked image is stored in the database instead of the original templates. Next, we applied OLPP on the watermarked face images to get a low-dimensional subspace.

During the verification stage, the biometric traits are acquired and the same process of watermark embedding is applied. The watermarked face image is then sent to the OLPP projection subspace. Finally, if a matching is found between feature vectors of stored and tested watermarked face images then the user is granted permission to access otherwise he/she will be rejected.

2.1   Watermark Embedding Process

In the literature there exist multiple image watermarking techniques which can be divided into various categories in various ways. The watermark can be applied in spatial domain and frequency domain. In frequency domain, it has been observed that coefficients are slightly modified which makes some unnoticeable changes in the whole image and makes it more robust to attacks compared to spatial based watermarking techniques [13]. In this work we propose to use a watermarking technique based on block-wise Discrete Cosine Transform (DCT) to insert the fingerprint minutia (watermark, W) in the face image (cover, C). To achieve our research objectives stated in this work, the watermarking technique should have the following properties:

1. When acquiring biometric data (face image), these are watermarked, such that sniffed/stollen data cannot be used to fool the system pretending these to be real data. Hence watermarking should be robust against image manipulations (such as compression, noise, etc.).
2. The capacity requirement of the watermarking techniques is very important and should allow carrying the fingerprint template (minutiae points) by the host image (face template).
3. Embedding watermark may result in changing the information of the host image. Therefore, the verification performance based on watermarked images should not be inferior compared to the verification performance of non-watermarked images.

In block-wise DCT based image watermarking, the host image is divided into different blocks. A block size of 8x8, which ideally matches JPEG compression, can be used to provide least distortion of the image against JPEG compression attack [14]. Then DCT of each block is evaluated. Since embedding the watermark may change the inherent characteristics of the cover image, care should be paid not to affect the performance accuracy of the watermarked images authentication. In addition, the DC coefficient should not be modified due to its perceptible effect in the whole image brightness. On the other hand, high frequencies are easily changed under common attacks such as compression. To maintain a good tradeoff between performance accuracy, robustness and imperceptibility, the proposed technique embeds watermark in medium frequency coefficients of the DCT block as shown in Figure 1.

In the proposed approach, first we extract the minutia of the fingerprint image denoted as $W = (W_1, .. W_N)$ where N is the total number of minutia points. W_i is a minutia point represented by the minutia coordinates and orientation.

Fig. 1. 8x8 DCT block where medium-frequency coefficients are shown in gray color

Every fingerprint minutia point W_i is converted to a bit stream and randomly embedded at a different DCT block in the host face image. Selected medium-frequency coefficients are modified according to the following equation [15]:

$$W' = C + (2.b - 1).\alpha.C \tag{1}$$

where W' is the watermarked coefficient, C is the DCT coefficient, α represents the watermarking strength, and b is the watermark bit ($b \in [0,1]$).

The resulting system is vulnerable in principle against all types of attacks in the classical unimodal systems. In particular, an attacker who previously intercepts user biometric data (fingerprint and face image) can embed stollen fingerprint data into stollen face image. In order to prevent this attack, watermark embedding in the proposed approach will be performed using a user key. A random number generator initialized with the user secret key determines the DCT blocks of the image to be watermarked. The block diagram of the proposed approach is given in Figure 2.

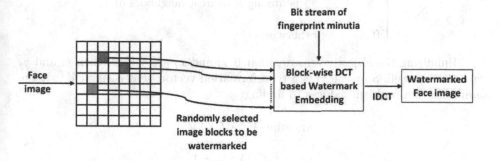

Fig. 2. The proposed DCT Watermarking based multi-biometric approach

2.2   OLPP-Based Feature Extraction and Matching

To extract pertinent information from large volumes of data, it is wise to use methods of dimensionality reduction. These methods can detect and analyze possible structures present but hidden in multidimensional data. Similar to other linear dimensionality reduction methods such as Principal Component Analysis (PCA) and Linear Discriminant Analysis (LDA), Locality Preserving Projection (LPP) has been recently proposed as a way to transform samples into a new small subspace. These methods are theoretically related with different weight matrix. The main objective of LPP is to preserve the local structure of the samples, i.e., samples that were close neighbors in the original space remain so as well in the new space [16]. LPP has been widely used in different applications of pattern recognition [17,18,19].

However, the basis functions obtained by the LPP method are non-orthogonal making it difficult to reconstruct the data. Orthogonal Locality Preserving Projection (OLPP) method called also Orthogonal Laplacianface has been proposed with the same theoretical foundation of LPP method except that it requires the basis functions to be orthogonal. OLPP has been used for face representation and recognition and proven to consistently outperform the Eigenface, Fisherface, and Laplacianface methods [12].

In the proposed approach OLPP is applied on watermarked face images to get a projection matrix. The objective function is as follows:

$$\min \sum_{ij} (y_i - y_j)^2 S_{ij} \tag{2}$$

where y_i is the l-dimensional representation of x_i and the matrix S is a similarity matrix defined as follows when the cosine distance is used to measure the relation between any two points:

$$S_{ij} = \begin{cases} \frac{x_i x_j}{\|x_i\| \|x_j\|} & \begin{array}{l} x_i \text{ is among } k \text{ nearest neighbors of } x_j \text{ or} \\ x_j \text{ is among } k \text{ nearest neighbors of } x_i \end{array} \\ 0 & \text{otherwise} \end{cases} \tag{3}$$

Minimizing the objective ensures that if x_i and x_j are "close" then y_i and y_j are close as well. Suppose V is a transformation vector, that is, $y_i = V^T x_i$, the minimization problem reduces to finding:

$$\arg\min_V V^T X L X^T V$$

$$\text{s.t. } V^T X D X^T V = 1 \tag{4}$$

where $D_{ii} = \sum_j S_{ji}$ is a diagonal matrix which measures the local density around x_i, $L = D - S$ is the Laplacian matrix.

Solution to this problem is to compute the eigenvectors $V = (v_0, v_1, \ldots, v_{l-1})$ associated with the smallest eigenvalues $(\Lambda, \lambda_0 \leq \lambda_1 \ldots \leq \lambda_{l-1})$ of the following generalized eigen problem:

$$X L X^T V = \Lambda X D X^T V \tag{5}$$

Thus the projection of a testing sample x_i is $y_i = V^T x_i$. If V is an orthogonal matrix, then $VV^T = I$ and the metric structure is preserved. Figure 3 shows the first 10 Orthogonal Laplacianfaces.

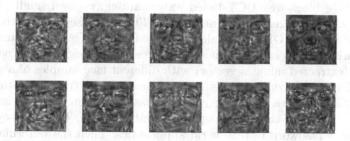

Fig. 3. The first 10 Orthogonal Laplacianfaces obtained from the face images in the *ORL* database

The training and testing watermarked face images are respectively projected onto the subspace V. The matching is proceeded by calculating the cosine distance between feature vectors y_i and y_j as follows:

$$d = \frac{y_i y_j}{||y_i|| ||y_j||} \tag{6}$$

If d is greater than a predetermined threshold value T, the two watermarked face images belong to the same person.

3 Experimentations

To give a first insight of the effectiveness of the proposed approach, we consider in this paper two biometric modalities face and fingerprint. For experimentation, we used *ORL* and *Yale* face databases. All the face images in the *ORL* face database[1] were captured against a dark homogeneous background. These images contain various facial expressions (smiling/not smiling, open/closed eyes) and facial details. The subjects were in an upright, frontal position but there was a tolerance for some tilting and rotation of up to about 20 degrees. Ten different images were obtained for each of the 40 subjects. The *Yale* face database[2] contains 165 gray scale images of 15 individuals. The images demonstrate variations in lighting condition, facial expression (normal, happy, sad, sleepy, surprised, and wink). For both databases, the first five face images of each subject were used as training samples, and the remaining images were used as testing samples.

The fingerprints samples are obtained from *FVC2002 DB2* [20] fingerprint database which contains 800 fingerprint impressions of size 300x480 pixels captured at a resolution of 512 dpi, from 100 distinct fingers (i.e. each person is

---

[1] http://www.cl.cam.ac.uk/research/dtg/attarchive/facedatabase.html
[2] http://vision.ucsd.edu/content/yale-face-database

represented by 8 impressions, 4 images are used for training and the remaining 4 are used for testing). The trial version of the commercial software VeriFinger SDK 6.0[3] has been used to extract the minutiae points. Individual minutia data sets contain between 20 and 40 minutia points.

To apply the block-wise DCT based watermarking, we used small subsets of the *FVC2002 DB2* fingerprint database with 40 and 15 users, respectively. The original face images in ORL and Yale are resized to 64x64 in order to implement OLPP algorithm at a lower computational cost. We assigned each sample of fingerprint (extracted minutia vector) with different face samples of a user. So, we have 20 different combinations for each user in the training databases of *ORL* and *Yale*. First, a user key, generated as a random number, is used in selecting the blocks of the face image to be watermarked. The user key can be derived from a password in real-world applications. Then the watermarking of the selected blocks by the bit stream of the fingerprint minutia is performed as described in the previous section. The resulting watermarked face images are stored in the database for further processing, for example, the extracted watermark (fingerprint minutia) can be used as a second source of authenticity or to prove the ownership of the host face image based on fingerprint verification by comparing the request fingerprint and the extracted watermark.

Next, we applied OLPP on the watermarked face images in the *ORL* and *Yale* training databases, which contain 800 and 300 images respectively, to get the low-dimensional subspaces. Then the watermarked face images in the *ORL* and *Yale* test databases, which contain 800 and 360 images respectively, to be identified are projected into the low-dimensional subspaces to extract the feature vectors. Finally, the matching of projected watermarked face images is performed by a nearest neighbor classifier based on the cosine distance between training and test feature vectors.

The Receiver Operating Characteristic (ROC) curves given in Figure 4 and Figure 5 show the error rate versus the dimensionality reduction (number of eigenvectors) corresponding to face recognition (1) without watermarking and (2) with minutiae- based watermarking for both *ORL* and *Yale* face databases respectively. As can be seen, the similarity of the ROC curves indicates that the proposed watermarking based multi-biometric fusion approach does not introduce any significant degradation in the face recognition accuracy.

To measure the effectiveness of the watermarking approach against compression attack, we have calculated the watermark extraction bit error rate (BER) as:

$$BER = \frac{\text{Number of error bits}}{\text{Total number of embedded bits}} \tag{7}$$

In this experiment, the percentage BER is less than 5% even for compression up to 50%. This result confirms the effectiveness of the use of the medium frequency DCT coefficients in the proposed watermarking approach. Moreover, it

---

[3] http://www.neurotechnology.com/licensing_verifinger_6.html

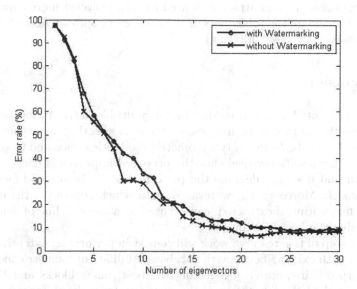

Fig. 4. ROC curves of *ORL* face database.

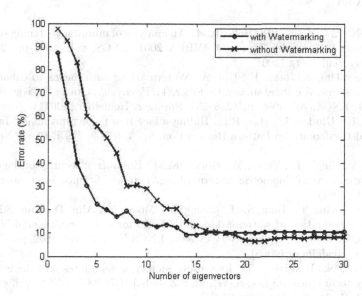

Fig. 5. ROC curves of *Yale* face database.

confirms that the use of block size of 8x8 provides least distortion of the image against JPEG compression attack as stated in [14]. However more experiments are needed to validate the robustness of the proposed approach against different types of attacks (e.g. noise, filtering).

4   Conclusion

The paper presented a watermarking based multi-biometric fusion approach, with an objective to protect biometric data without affecting the authentication performance given the fusion of two biometric modalities (face and fingerprint). The preliminary results revealed that the proposed approach is effective in authentication and does not degrade the performance of the original face recognition approach. Moreover, the watermark data which consists of the minutiae of a user's fingerprint, if extracted, can be used as a "second line of defense" in authenticating the host face image.

The next step of this research work will consist in performing extensive experimentations with existing benchmark databases of different biometric modalities (face, fingerprint, iris, etc ...). Selecting the best image blocks and DCT coefficients to embed the bit stream of the watermark can be enhanced using an optimization technique. In addition, robustness of the watermarking approach needs to be assessed against unintentional/non-malicious image manipulations.

References

1. Ratha, N.K., Connell, J.H., Bolle, R.M.: An analysis of minutiae matching strength. In: Bigun, J., Smeraldi, F. (eds.) AVBPA 2001. LNCS, vol. 2091, pp. 223–228. Springer, Heidelberg (2001)
2. Hämmerle-Uhl, J., Raab, K., Uhl, A.: Watermarking as a means to enhance biometric systems: A critical survey. In: Filler, T., Pevný, T., Craver, S., Ker, A. (eds.) IH 2011. LNCS, vol. 6958, pp. 238–254. Springer, Heidelberg (2011)
3. Jain, A.K., Uludag, U., Hsu, R.L.: Hiding a face in a fingerprint image. In: International Conference on Pattern Recognition, pp. 756–759. IEEE Press, New York (2002)
4. Noore, A., Singh, R., Vatsa, M., Houck, M.M.: Enhancing security of fingerprints through contextual biometric watermarking. Forensic Science International 169, 188–194 (2007)
5. Moon, D., Kim, T., Jung, S.-H., Chung, Y., Moon, K., Ahn, D., Kim, S.K.: Performance evaluation of watermarking techniques for secure multimodal biometric systems. In: Hao, Y., et al. (eds.) CIS 2005. LNCS (LNAI), vol. 3802, pp. 635–642. Springer, Heidelberg (2005)
6. Komninos, N., Dimitriou, T.: Protecting biometric templates with image watermarking techniques. In: Lee, S.-W., Li, S.Z. (eds.) ICB 2007. LNCS, vol. 4642, pp. 114–123. Springer, Heidelberg (2007)
7. Park, K.R., Jeong, D.S., Kang, B.J., Lee, E.C.: A study on iris feature watermarking on face data. In: Beliczynski, B., Dzielinski, A., Iwanowski, M., Ribeiro, B. (eds.) ICANNGA 2007. LNCS, vol. 4432, pp. 415–423. Springer, Heidelberg (2007)

8. Rajibul, M.I., Shohel, M.S., Andrews, S.: Biometric template protection using watermarking with hidden password encryption. In: International Symposium on Information Technology, pp. 296–303. IEEE Press, New York (2008)
9. Vatsa, M., Singh, R., Noore, A.: Feature based RDWT watermarking for multimodal biometric system. Image and Vision Computing 27(3), 293–304 (2009)
10. Kim, W.-G., Lee, H.K.: Multimodal biometric image watermarking using two- stage integrity verification. Signal Processing 89(12), 2385–2399 (2009)
11. Ma, B., Li, C., Wang, Y., Zhang, Z., Wang, Y.: Block Pyramid Based Adaptive Quantization Watermarking for Multimodal Biometric Authentication. In: International Conference on Pattern Recognition, pp. 1277–1280. IEEE Press, New York (2010)
12. Cai, D., He, X., Han, J., Zhang, H.: Orthogonal Laplacianfaces for Face Recognition. IEEE Transactions on Image Processing 15(11), 3608–3614 (2006)
13. Aslantas, V.: A singular value decomposition-based image watermarking using genetic algorithm. International Journal of Electronic Communications 62, 386–394 (2008)
14. Barni, M., Bartolini, F.: Watermarking Systems Engineering. Marcel Dekker Inc., Italy (2004)
15. Cox, J., Kilian, J., Leighton, F.T., Shamoon, T.: Secure spread spectrum watermarking for multimedia. IEEE Transactions on Image Processing 6(12), 1673–1687 (1997)
16. He, X., Niyogi, P.: Locality Preserving Projections. In: Conference on Advances in Neural Information Processing Systems (2003)
17. Cheng, J., Liu, Q., Lu, H., Chen, Y.-W.: Supervised kernel locality preserving projections for face recognition. Neurocomputing 67, 443–449 (2005)
18. Yang, J., Zhang, D., Yang, J.-Y., Niu, B.: Globally maximizing, locally minimizing: unsupervised discriminant projection with applications to face and palm biometrics. IEEE Transactions on Pattern Analysis and Machine Intelligence 29(4), 650–664 (2007)
19. He, X., Yan, S., Hu, Y., Niyogi, P., Zhang, H.-J.: Face recognition using Laplacianfaces. IEEE Transactions on Pattern Analysis and Machine Intelligence 27, 328–340 (2005)
20. Maltoni, D., Maio, D., Jain, A.K., Prabhakar, S.: Handbook of fingerprint recognition. Springer, Heidelberg (2009)

New Attacks on RSA with Moduli $N = p^r q$

Abderrahmane Nitaj[1][(✉)] and Tajjeeddine Rachidi[2]

[1] Laboratoire de Mathématiques Nicolas Oresme,
Université de Caen Basse Normandie, France
abderrahmane.nitaj@unicaen.fr
[2] School of Science and Engineering
Al Akhawayn University in Ifrane, Morocco
T.Rachidi@aui.ma

Abstract. We present three attacks on the Prime Power RSA with modulus $N = p^r q$. In the first attack, we consider a public exponent e satisfying an equation $ex - \phi(N)y = z$ where $\phi(N) = p^{r-1}(p-1)(q-1)$. We show that one can factor N if the parameters $|x|$ and $|z|$ satisfy $|xz| < N^{\frac{r(r-1)}{(r+1)^2}}$ thereby extending the recent results of Sakar [16]. In the second attack, we consider two public exponents e_1 and e_2 and their corresponding private exponents d_1 and d_2. We show that one can factor N when d_1 and d_2 share a suitable amount of their most significant bits, that is $|d_1 - d_2| < N^{\frac{r(r-1)}{(r+1)^2}}$. The third attack enables us to factor two Prime Power RSA moduli $N_1 = p_1^r q_1$ and $N_2 = p_2^r q_2$ when p_1 and p_2 share a suitable amount of their most significant bits, namely, $|p_1 - p_2| < \frac{p_1}{2rq_1q_2}$.

Keywords: RSA · Cryptanalysis · Factorization · Coppersmith's method · Prime Power RSA

1 Introduction

The RSA public-key cryptosystem, invented in 1978 by Rivest, Shamir and Adleman [15], is one of the most popular systems in use today. In the RSA cryptosystem, the public key is (N, e) where the modulus $N = pq$ is a product of two primes of the same bitsize, and the public exponent is a positive integer satisfying $ed \equiv 1 \pmod{\phi(N)}$. In RSA, encryption and decryption require executing heavy exponential multiplications modulo the large integer N. To reduce the decryption time, one may be tempted to use a small private exponent d. However, in 1990 Wiener [18] showed that RSA is insecure if $d < \frac{1}{3}N^{0.25}$, and Boneh and Durfee [2] improved the bound to $d < N^{0.292}$. In 2004, Blömer and May [1] combined both Wiener's method and Boneh and Durfee's method to show that RSA is insecure if the public exponent e satisfies an equation $ex + y = k\phi(N)$ with $x < \frac{1}{3}N^{\frac{1}{4}}$ and $|y| \leq N^{-\frac{3}{4}}ex$.

Partially supported by the French SIMPATIC (SIM and PAiring Theory for Information and Communications security).

© Springer International Publishing Switzerland 2015
S. El Hajji et al. (Eds.): C2SI 2015, LNCS 9084, pp. 352–360, 2015.
DOI: 10.1007/978-3-319-18681-8_28

Concurrent to these efforts, many RSA variants have been proposed in order to ensure computational efficiency while maintaining the acceptable levels of security. One such important variant is the Prime Power RSA. In Prime Power RSA the modulus N is in the form $N = p^r q$ for $r \geq 2$. In [17], Takagi showed how to use the Prime Power RSA to speed up the decryption process when the public and private exponents satisfy an equation $ed \equiv 1 \pmod{(p-1)(q-1)}$. As in the standard RSA cryptosystem, the security of the Prime Power RSA depends on the difficulty of factoring integers of the form $N = p^r q$.

Therefore, a Prime Power RSA modulus must be appropriately chosen, since it has to resist factoring algorithms such as the Number Field Sieve [10] and the Elliptic Curve Method [9]. Table 1, shows the suggested secure Power RSA forms as a function of the size of the modulus back in 2002 (see [4]). Note that, due to the ever increasing development of computing hardware, the form $N = p^2 q$ is no longer recommended for 1024 bit modulus.

Table 1. Optimal number of prime factors of a Prime Power RSA modulus [4]

Modulus size (bits)	1024	1536	2048	3072	4096	8192
Form of the modulus N	pq, p^2q	pq, p^2q	pq, p^2q	pq, p^2q	pq, p^2q, p^3q	pq, p^2q, p^3q, p^4q

In 1999, Boneh, Durfee, and Howgrave-Graham [3] presented a method for factoring $N = p^r q$ when r is large. Furthermore, Takagi [17] proved that one can factor N if $d < N^{\frac{1}{2(r+1)}}$, and May [13] improved the bound to $d < N^{\frac{r}{(r+1)^2}}$ or $d < N^{\frac{(r-1)^2}{(r+1)^2}}$. Very recently, Lu, Zhang and Lin [12] improved the bound to $d < N^{\frac{r(r-1)}{(r+1)^2}}$, and Sarkar [16] improved the bound for $N = p^2 q$ to $d < N^{0.395}$ and gave explicit bounds for $r = 3, 4, 5$.

In this paper, we focus on the Prime Power RSA with a modulus $N = p^r q$, and present three new attacks: In the first attack we consider a public exponent e satisfying an equation $ex - \phi(N)y = z$ where x and y are positive integers. Using a recent result of Lu, Zhang and Lin [12], we show that one can factor N in polynomial time if $|xz| < N^{\frac{r(r-1)}{(r+1)^2}}$. In the standard situation $z = 1$, the condition becomes $d = x < N^{\frac{r(r-1)}{(r+1)^2}}$ which improves the bound of May [13] for $r \geq 3$ and retrieves the bound of Lu, Zhang and Lin [12]. Note that unlike Sarkar [16] who solves $ex - \phi(N)y = 1$, we solve a more general equation $ex - \phi(N)y = z$. This leads to less constraints on the solution space, which in turn leads to an increase in the number of solutions to the equation. Intuitively speaking, our method has higher likelihood of finding solutions; that is, factoring RSA. In section 3, we shall present an example supporting this claim.

In the second attack, we consider an instance of the Prime Power RSA with modulus $N = p^r q$. We show that one can factor N if two private keys d_1 and d_2 share an amount of their most significant bits, that is if $|d_1 - d_2|$ is small enough. More precisely, we show that if $|d_1 - d_2| < N^{\frac{r(r-1)}{(r+1)^2}}$, then N can be factored in polynomial time. The method we present is based on a recent result of [12] with Coppersmith's method for solving an univariate linear equation.

In the third attack, we consider two instances of the Prime Power RSA with two moduli $N_1 = p_1^r q_1$ and $N_2 = p_2^r q_2$ such that the prime factors p_1 and p_2 share an amount of their most significant bits, that is $|p_1 - p_2|$ is small. More precisely, we show that one can factor the RSA moduli N_1 and N_2 in polynomial time if $|p_1 - p_2| < \frac{p_1}{2r q_1 q_2}$. The method we use for this attack is based on the continued fraction algorithm.

The rest of this paper is organized as follows: In Section 2, we briefly review the preliminaries necessary for the attacks, namely Coppersmith's technique for solving linear equations and the continued fractions theorem. In Section 3, we present the first attack on the Prime Power RSA, which is valid with no conditions on the prime factors. In Section 4, we present the second attack in the situation where two decryption exponents share an amount of their most significant bits. In Section 5, we present the third attack on the Prime Power RSA when the prime factors share an amount of their most significant bits. We then conclude the paper in Section 6.

2   Preliminaries

In this section, we present some basics on Coppersmith's method for solving linear modular polynomial equations and an overview of the continued fraction algorithm. Both techniques are used in the crafting of our attacks.

First, observe that if $N = p^r q$ with $q < p$, then $p^{r+1} > p^r q = N$, and $p > N^{\frac{1}{r+1}}$. Hence throughout this paper, we will use the inequality $p > N^\beta$ where $\beta = \frac{1}{r+1}$.

2.1   Linear Modular Polynomial Equations

In 1995, Coppersmith [5] developed powerful lattice-based techniques for solving both modular polynomial diophantine equations with one variable and two variables. These techniques have been generalized to more variables, and have served for cryptanalysis of many instances of RSA. More on this can be found in [14,8]. In [7], Herrmann and May presented a method for finding the small roots of a modular polynomial equation $f(x_1, \ldots, x_n) \equiv 0 \pmod{p}$ where $f(x_1, \ldots, x_n) \in \mathbb{Z}[x_1, \ldots, x_n]$ and p is an unknown divisor of a known integer N. Their method is based on the seminal work of Coppersmith [5]. Very recently, Lu, Zhang and Lin [12] presented a generalization for finding the small roots of a modular polynomial equation $f(x_1, \ldots, x_n) \equiv 0 \pmod{p^v}$, where p^v is a divisor of some composite integer N. For the bivariate case, they proved the following result, which we shall use in the crafting of our attacks.

Theorem 1 (Lu, Zhang and Lin). *Let N be a composite integer with a divisor p^u such that $p \geq N^\beta$ for some $0 < \beta \leq 1$. Let $f(x, y) \in \mathbb{Z}[x, y]$ be a homogenous linear polynomial. Then one can find all the solutions (x, y) of* the equation $f(x, y) = 0 \mod p^v$ with $\gcd(x, y) = 1$, $|x| < N^{\gamma_1}$, $|y| < N^{\gamma_2}$, in polynomial time if

$$\gamma_1 + \gamma_2 < uv\beta^2.$$

2.2 The Continued Fractions Algorithm

We present here the well known result of Legendre on convergents of a continued fraction expansion of a real number. The details can be found in [6]. Let ξ be a positive real number. Define $\xi_0 = \xi$ and for $i = 0, 1, \ldots, n$, $a_i = \lfloor \xi_i \rfloor$, $\xi_{i+1} = 1/(\xi_i - a_i)$ unless ξ_i is an integer. This expands ξ as a continued fraction in the following form:

$$\xi = a_0 + \cfrac{1}{a_1 + \cfrac{1}{\ddots + \cfrac{1}{a_n + \cfrac{1}{\ddots}}}}, \quad a_0 \in \mathbb{N}, \text{ and } a_i \in \mathbb{N}^* \text{ for } i \geq 1,$$

which is often rewritten as $\xi = [a_0, a_1, \ldots, a_n, \ldots]$. For $i \geq 0$, the rational numbers $[a_0, a_1, \ldots, a_i]$ are the convergents of ξ. If $\xi = \frac{a}{b}$ is a rational number, then $\xi = [a_0, a_1, \ldots, a_n]$ for some positive integer n, and the continued fraction expansion of ξ is finite with the total number of convergents being polynomial in $\log(b)$. The following result enables one to determine if a rational number $\frac{a}{b}$ is a convergent of the continued fraction expansion of a real number ξ (see Theorem 184 of [6]).

Theorem 2 (Legendre). *Let ξ be a positive real number. Suppose $\gcd(a, b) = 1$ and*

$$\left| \xi - \frac{a}{b} \right| < \frac{1}{2b^2}.$$

Then $\frac{a}{b}$ is one of the convergents of the continued fraction expansion of ξ.

Note that the continued fractions expansion process is polynomial in time.

3 The First Attack on Prime Power RSA with Modulus $N = p^r q$

In this section, we present an attack on the Prime Power RSA when the public key (N, e) satisfies an equation $ex - \phi(N)y = z$ with small parameters x and $|z|$.

Theorem 3. *Let $N = p^r q$ be a Prime Power RSA modulus and e a public exponent satisfying the equation $ex - \phi(N)y - z$ with $1 < e < \phi(N)$ and $\gcd(e, \phi(N)) = 1$. Then one can factor N in polynomial time if*

$$|xz| < N^{\frac{r(r-1)}{(r+1)^2}}.$$

Proof. Suppose that $e < N$ satisfies an equation $ex - \phi(N)y = z$ with $|x| < N^\delta$ and $|z| < N^\gamma$. Then, since $\phi(N) = p^{r-1}(p-1)(q-1)$, we get $ex - z \equiv 0$ (mod p^{r-1}). Applying Theorem 1 with $u = r$, $v = r - 1$ and $\beta = \frac{1}{r+1}$, we can solve the equation in polynomial time if

$$\delta + \gamma < uv\beta^2 = \frac{r(r-1)}{(r+1)^2},$$

that is $|xz| < N^{\frac{r(r-1)}{(r+1)^2}}$. Since $\frac{e}{\phi(N)} < 1$, then, using x and z in the equation $ex - \phi(N)y = z$, we get for sufficiently large N comparatively to r,

$$y = \frac{ex - z}{\phi(N)} < \frac{e|x|}{\phi(N)} + \frac{|z|}{\phi(N)} < |x| + |z| \leq 1 + |xz| < 1 + N^{\frac{r(r-1)}{(r+1)^2}} < N.$$

Hence

$$\gcd(ex - z, N) = \gcd(p^{r-1}(p-1)(q-1)y, p^r q) = g,$$

with $g = p^{r-1}$, $g = p^r$ or $g = p^{r-1}q$. If $g = p^{r-1}$, then $p = g^{\frac{1}{r-1}}$, if $g = p^r$, then $p = g^{\frac{1}{r}}$ and if $g = p^{r-1}q$, then $p = \frac{N}{g}$. This leads to the factorization of N.   □

Example 1. For $r = 2$ and $N = p^r q$, let us take for N and e the 55 digit numbers

$$N = 8138044578297117319482018441148072252199996769522371021,$$

$$e = 1199995230601021126201343651611107957480251354355883029.$$

In order to solve the diophantine equation $ex - \phi(N)y = z$, we transformed it into the equation $ex - z \equiv 0$ (mod p^{r-1}) using Theorem 3. To be able to apply Coppersmith's technique via Theorem 1, we chose the parameters $m = 7$, $t = 6$ so that the dimension of constructed the lattice is 36, and $X = \left[N^{\frac{r(r-1)}{(r+1)^2}} \right] = 1592999974064$. We built the lattice using the polynomial $f(x_1, x_2) = x_1 + ex_2$, then applied the LLL algorithm [11], and used Gröbner basis method to find the smallest solution $x_1 = -11537$ and $x_2 = 7053$ to $f(x_1, x_2) \equiv 0$ (mod p^{r-1}) in 174 seconds using an off-the-shelf computer. From this solution, we deduced $p = \gcd(x_1 + ex_2, N) = 2294269585934949239$, and finally recovered $q = \frac{N}{p^2} = 1546077175000723901$. We then computed $\phi(N)$ and $d \equiv e^{-1}$ (mod $\phi(N)$) as follows:

$$\phi(N) = 8138044578297117310671227668089561946257896925261579800,$$

$$d = 2015994747748388772982436393811213317361971865510756269.$$

Observe that $d \approx N^{0.98}$ which is out of range of Sarkar's bound [16] which can only retrieve private keys $d < N^{0.395}$ for $r = 2$.

4   The Second Attack on Prime Power RSA Using Two Decryption Exponents

In this section, we present an attack on the Prime Power RSA when two private exponents d_1 and d_2 share an amount of their most significant bits, that is $|d_1 - d_2|$ is small.

Theorem 4. *Let $N = p^r q$ be an RSA modulus and d_1 and d_2 be two private exponents. Then, one can factor N in polynomial time, if*

$$|d_1 - d_2| < N^{\frac{r(r-1)}{(r+1)^2}}.$$

Proof. Suppose that $e_1 d_1 - k_1 \phi(N) = 1$ and $e_2 d_2 - k_2 \phi(N) = 1$ with $e_1 > e_2$. Hence $e_1 d_1 \equiv 1 \pmod{\phi(N)}$ and $e_2 d_2 \equiv 1 \pmod{\phi(N)}$. Multiplying the first equation by e_2 and the second by e_1 and subtracting, we get

$$e_1 e_2 (d_1 - d_2) \equiv e_2 - e_1 \pmod{\phi(N)}.$$

Since $\phi(N) = p^{r-1}(p-1)(q-1)$, we get $e_1 e_2 (d_1 - d_2) \equiv e_2 - e_1 \pmod{p^{r-1}}$. Now, consider the modular linear equation

$$e_1 e_2 x - (e_2 - e_1) \equiv 0 \pmod{p^{r-1}},$$

$d_1 - d_2$ is a root of such equation. Suppose further that $|d_1 - d_2| < N^\delta$, then applying Theorem 1 with $u = r$, $v = r - 1$ and $\beta = \frac{1}{r+1}$ will lead to the solution $x = d_1 - d_2$ obtained in polynomial time if

$$\delta < uv\beta^2 = \frac{r(r-1)}{(r+1)^2}.$$

That is if $|d_1 - d_2| < N^{\frac{r(r-1)}{(r+1)^2}}$. Computing

$$\gcd(e_1 e_2 x - (e_2 - e_1), N) = \gcd\left(p^{r-1}(p-1)(q-1)y, p^r q\right) = g,$$

will lead to determining p, hence factoring N as follows: $p = g^{\frac{1}{r-1}}$ when $g = p^{r-1}$, or $p = g^{\frac{1}{r}}$ when $g = p^r$, or $p = \frac{N}{g}$ if $g = p^{r-1} q$. □

Example 2. Let us present an example corresponding to Theorem 4. Consider $N = p^2 q$ with

$$N = 60932538514861208788594719583997377258859465265536 26219,$$

$$e_1 = 27496003818474873897159647672356188025296758556063 77411,$$

$$e_2 = 35750812449524140093163965015123722265458925588982 76551.$$

The polynomial equation is $f(x) = e_1 e_2 x - (e_2 - e_1) \equiv 0 \pmod{p^{r-1}}$, which can be transformed into $g(x) = x - a \equiv 0 \pmod{p^{r-1}}$ where $a \equiv (e_2 - e1)(e_1 e_2)^{-1} \pmod{N}$. Using $m = 8$ and $t = 6$, we built a lattice with dimension $\omega = 9$. Applying the LLL algorithm [11] and solving the first reduced polynomials, we get the solution $x_0 = 1826732340$. Hence $\gcd(f(x_0), N) = p = 1789386140116417697$ and finally $q = \frac{N}{p^2} = 1903010275819064491$. The whole process took less than 4 seconds using an off-the-shelf computer. Then, using $\phi(N) = p(p-1)(q-1)$, we retrieved the private exponents $d_1 \equiv e_1^{-1} \pmod{\phi(N)}$ and $d_2 \equiv e_2^{-1} \pmod{\phi(N)}$. Note that again $d_1 \approx d_2 \approx N^{0.99}$ which Sarkar's method with the bound $d < N^{0.395}$ could not possibly retrieve.

5   The Third Attack on Prime Power RSA with Two RSA Moduli

In this section, we consider two Prime Power RSA moduli $N_1 = p_1^r q_1$ and $N_2 = p_2^r q_2$, where p_1 and p_2 share an amount of their most significant bits.

Theorem 5. *Let $N_1 = p_1^r q_1$ and $N_2 = p_2^r q_2$ be two RSA moduli with $p_1 > p_2$. If*

$$|p_1 - p_2| < \frac{p_1}{2rq_1q_2},$$

then, one can factor N in polynomial time.

Proof. Suppose that $N_1 = p_1^r q_1$ and $N_2 = p_2^r q_2$ with $p_1 > p_2$. Then $q_2 N_1 - q_1 N_2 = q_1 q_2 (p_1^r - p_2^r)$. Hence

$$\left| \frac{N_2}{N_1} - \frac{q_2}{q_1} \right| = \frac{q_1 q_2 |p_1^r - p_2^r|}{q_1^2 p_1^r}.$$

In order to apply Theorem 2, we need that $\frac{q_1 q_2 |p_1^r - p_2^r|}{q_1^2 p_1^r} < \frac{1}{2q_1^2}$, or equivalently

$$|p_1^r - p_2^r| < \frac{p_1^r}{2q_1 q_2}. \tag{1}$$

Observe that

$$|p_1^r - p_2^r| = |p_1 - p_2| \sum_{i=0}^{r-1} p_1^{r-1-i} p_2^i < r|p_1 - p_2| p_1^{r-1}.$$

Then (1) is fulfilled if $r|p_1 - p_2| p_1^{r-1} < \frac{p_1^r}{2q_1 q_2}$, that is if

$$|p_1 - p_2| < \frac{p_1}{2rq_1 q_2}.$$

Under this condition, we get $\frac{q_2}{q_1}$ among the convergents of the continued fraction expansion of $\frac{N_2}{N_1}$. Using q_1 and q_2, we get $p_1 = \left(\frac{N_1}{q_1} \right)^{\frac{1}{r}}$ and $p_2 = \left(\frac{N_2}{q_2} \right)^{\frac{1}{r}}$.   □

Example 3. We present here an example corresponding to Theorem 5. Consider $N_1 = p_1^2 q_1$ and $N_2 = p_2^2 q_2$ with

$N_1 = 170987233913769420505896917437304719816691353833034482461,$

$N_2 = 120532911819726882881630714003135237766675602824250965921.$

We applied the continued fraction algorithm to compute the first 40 convergents of $\frac{N_2}{N_1}$. Every convergent is a candidate for the ratio $\frac{q_2}{q_1}$ of the prime factors. One of the convergents is $\frac{36443689}{51698789}$ leading to $q_2 = 36443689$ and $q_1 = 51698789$. This gives the prime factors p_1 and p_2

$$p_1 = \sqrt{\frac{N_1}{q_1}} = 1818618724382942951460443,$$

$$p_2 = \sqrt{\frac{N_2}{q_2}} = 1818618724382943035672683.$$

6   Conclusion

In this paper, we have considered the Prime Power RSA with modulus $N = p^r q$ and public exponent e. We presented three new attacks to factor the modulus in polynomial time. The first attack can be applied if small parameters x, y and z satisfying the equation $ex - \phi(N)y = z$ can be found . The second attack can be applied when two private exponents d_1 and d_2 share an amount of their most significant bits. The third attack can be applied when two Prime Power RSA moduli $N_1 = p_1^r q_1$ and $N_2 = p_2^r q_2$ are such that p_1 and p_2 share an amount of their most significant bits.

References

1. Blömer, J., May, A.: A Generalized Wiener Attack on RSA. In: Bao, F., Deng, R., Zhou, J. (eds.) PKC 2004. LNCS, vol. 2947, pp. 1–13. Springer, Heidelberg (2004)
2. Boneh, D., Durfee, G.: Cryptanalysis of RSA with private key d less than $N^{0.292}$. In: Stern, J. (ed.) EUROCRYPT 1999. LNCS, vol. 1592, pp. 1–11. Springer, Heidelberg (1999)
3. Boneh, D., Durfee, G., Howgrave-Graham, N.: Factoring tex2html_wrap_inline127 for Large r. In: Wiener, M. (ed.) CRYPTO 1999. LNCS, vol. 1666, pp. 326–337. Springer, Heidelberg (1999)
4. Compaq Computer Corperation. Cryptography using Compaq multiprime technology in a parallel processing environment (2002), ftp://ftp.compaq.com/pub/solutions/CompaqMultiPrimeWP.pdf
5. Coppersmith, D.: Small solutions to polynomial equations, and low exponent RSA vulnerabilities. Journal of Cryptology 10(4), 233–260 (1997)
6. Hardy, G.H., Wright, E.M.: An Introduction to the Theory of Numbers. Oxford University Press, London (1975)
7. Herrmann, M., May, A.: Solving linear equations modulo divisors: On factoring given any bits. In: Pieprzyk, J. (ed.) ASIACRYPT 2008. LNCS, vol. 5350, pp. 406–424. Springer, Heidelberg (2008)
8. Hinek, M.J.: Cryptanalysis of RSA and its variants. Chapman & Hall/CRC Cryptography and Network Security. CRC Press, Boca Raton (2010)
9. Lenstra, H.: Factoring integers with elliptic curves. Annals of Mathematics 126, 649–673 (1987)
10. Lenstra, A.K., Lenstra Jr., H.W.: The Development of the Number Field Sieve. Lecture Notes in Mathematics, vol. 1554. Springer, Heidelberg (1993)
11. Lenstra, A.K., Lenstra, H.W., Lovász, L.: Factoring polynomials with rational coefficients. Mathematische Annalen 261, 513–534 (1982)
12. Lu, Y., Zhang, R., Lin, D.: New Results on Solving Linear Equations Modulo Unknown Divisors and its Applications, Cryptology ePrint Archive, Report 2014/343 (2014), https://eprint.iacr.org/2014/343
13. May, A.: Secret Exponent Attacks on RSA-type Schemes with Moduli $N = p^r q$. In: Bao, F., Deng, R., Zhou, J. (eds.) PKC 2004. LNCS, vol. 2947, pp. 218–230. Springer, Heidelberg (2004)
14. May, A.: Using LLL-reduction for solving RSA and factorization problems: a survey. In: LLL+25 Conference in Honour of the 25th Birthday of the LLL Algorithm. Springer, Heidelberg (2007)

15. Rivest, R., Shamir, A., Adleman, L.: A Method for Obtaining digital signatures and public-key cryptosystems. Communications of the ACM 21(2), 120–126 (1978)
16. Sarkar, S.: Small secret exponent attack on RSA variant with modulus $N = p^r q$. Designs, Codes and Cryptography 73(2), 383–392 (2015)
17. Takagi, T.: Fast RSA-type cryptosystem modulo $p^k q$. In: Krawczyk, H. (ed.) CRYPTO 1998. LNCS, vol. 1462, pp. 318–326. Springer, Heidelberg (1998)
18. Wiener, M.: Cryptanalysis of short RSA secret exponents. IEEE Transactions on Information Theory 36, 553–558 (1990)

Factoring RSA Moduli
with Weak Prime Factors

Abderrahmane Nitaj[1]([✉]) and Tajjeeddine Rachidi[2]

[1] Laboratoire de Mathématiques Nicolas Oresme,
Université de Caen Basse Normandie, France
abderrahmane.nitaj@unicaen.fr
[2] School of Science and Engineering,
Al Akhawayn University in Ifrane, Morocco
T.Rachidi@aui.ma

Abstract. In this paper, we study the problem of factoring an RSA modulus $N = pq$ in polynomial time, when p is a weak prime, that is, p can be expressed as $ap = u_0 + M_1 u_1 + \ldots + M_k u_k$ for some k integers $M_1, \ldots, M_k$ and $k+2$ suitably small parameters $a, u_0, \ldots u_k$. We further compute a lower bound for the set of weak moduli, that is, moduli made of at least one weak prime, in the interval $[2^{2n}, 2^{2(n+1)}]$ and show that this number is much larger than the set of RSA prime factors satisfying Coppersmith's conditions, effectively extending the likelihood for factoring RSA moduli. We also prolong our findings to moduli composed of two weak primes.

Keywords: RSA · Cryptanalysis · Factorization · LLL algorithm · Weak primes

1 Introduction

The RSA cryptosystem, invented in 1978 by Rivest, Shamir and Adleman [17] is undoubtedly one of the most popular public key cryptosystems. In the standard RSA [17], the modulus $N = pq$ is the product of two large primes of the same bit-size. The public exponent e is an integer such that $1 \leq e < \phi(N)$ and $\gcd(e, \phi(N)) = 1$ where $\phi(N) = (p-1)(q-1)$ is the Euler totient function. The corresponding private exponent is the integer d such that $ed \equiv 1 \pmod{\phi(N)}$. In RSA, the encryption, decryption, signature generation, and signature verification require substantial CPU cycles because the time to perform these operations is proportional to the number of bits in public or secret exponents [17]. To reduce CPU time necessary for encryption and signature verification, one may be tempted to use a small public exponent e. This situation has been proven to be insecure against some small public exponent attacks (see [8] and [9]). To reduce the decryption and signature generation time, one may also be tempted to use a small private

Partially supported by the French SIMPATIC (SIM and PAiring Theory for Information and Communications security).

© Springer International Publishing Switzerland 2015
S. El Hajji et al. (Eds.): C2SI 2015, LNCS 9084, pp. 361–374, 2015.
DOI: 10.1007/978-3-319-18681-8_29

exponent d. Unfortunately, RSA is also vulnerable to various powerful short secret exponent attacks such as, the attack of Wiener [20], and the attack of Boneh and Durfee [4] (see also [3]). An alternate way for increasing the performance of encryption, decryption, signature generation, and signature verification, without reverting to small exponents, is to use the multi-prime variant of RSA. The multi-prime RSA is a generalization of the standard RSA cryptosystem in which the modulus is in the form $N = p_1 p_2 \cdots p_k$ where $k \geq 3$ and the p_i's are distinct prime numbers. Combined with the Chinese Remainder Theorem, a multi-prime RSA is much more efficient than the standard RSA (see [5]).

In Section 4.1.2 of the X9.31-1998 standard for public key cryptography [1], some recommendations are presented regarding the generation of the prime factors of an RSA modulus. For example, it is recommended that the modulus should have $1024+256x$ bits for $x \geq 0$. This requirement deters some factorization attacks, such as the Number Field Sieve (NFS) [12] and the Elliptic Curve Method (ECM) [11]. Another recommendation is that the prime difference $|p - q|$ should be large, and $\frac{p}{q}$ should not be near the ratio of two small integers. These requirements guard against Fermat factoring algorithm [19], as well as Coppersmith's factoring attack on RSA [6] when one knows half of the bits of p. For example, if $N = pq$ and p, q are of the same bit-size with $|p - q| < N^{1/4}$, then $\left| p - \left[\sqrt{N} \right] \right| < N^{1/4}$ (see [16]) where $\left[\sqrt{N} \right]$ is the nearest integer to $\sqrt{N}$, which means that half of the bits of p are those of $[\sqrt{N}]$ which leads to the factorization of N (see [6] and [19]). Observe that the factorization attack of Coppersmith applies provided that one knows half of the bits of p, that is p is in one of the forms

$$p = \begin{cases} M_1 + u_0 & \text{with known } M_1 \text{ and unknown } u_0 \leq N^{\frac{1}{4}}, \\ M_1 u_1 + M_0 & \text{with known } (M_1, M_0) \text{ and unknown } u_1 \leq N^{\frac{1}{4}}. \end{cases}$$

Such primes are called Coppersmith's weak primes. In the case of $p = M_1 u_1 + M_0$ with known M_1 and M_0, the Euclidean division of q by M_1 is in the form $q = M_1 v_1 + v_0$. Hence $N = pq = (M_1 u_1 + M_0)(M_1 v_1 + v_0)$ which gives $M_0 v_0 \equiv N$ (mod M_1). Hence, since $gcd(M_0, M_1) = 1$, then $v_0 \equiv N M_0^{-1}$ (mod M_1). This means that when p is in the form $p = M_1 u_1 + M_0$ with known M and u_0, then q is necessarily in the form $q = M_1 v_1 + v_0$ with known v_0. Coppersmith's attack is therefore applicable only when small enough parameters M_0 and v_0 can be found such that $p = M_1 u_1 + M_0$ and $q = M_1 v_1 + v_0$. This reduces the applicability of the attack to the set of moduli such that p and q are of the form defined above.

In this paper, we consider the generalization of Coppersmith's attack by considering a more satisfiable decomposition of any of the multipliers of p or q, i.e., ap or aq not just p or q, effectively leading to an increased set of moduli that can be factored. We describe two new attacks on RSA with a modulus $N = pq$. The first attack applies in the situation that, for given positive integers $M_1, \ldots, M_k$, one of the prime factors, p say, satisfies a linear equation $ap = u_0 + M_1 u_1 + \ldots + M_k u_k$ with suitably small integers a and $u_0, \ldots, u_k$. We call such prime factors *weak primes* for the integers $M_1, \ldots, M_k$. The second attack applies when both factors p and q are weak for the integers $M_1, \ldots, M_k$.

We note that, for $k = 1$, the weak primes are such that $ap = u_0 + M_1 u_1$. This includes the class of Coppersmith's weak primes. For both attacks, we give an estimation of the RSA moduli $N = pq$ with a prime factor $p \in [2^n, 2^{n+1}]$ which is weak for the integers $M, M^2, \ldots, M^k$ where $M = \lceil 2^{\frac{n}{2k}} \rceil$. We show that the number of moduli with a weak prime factor is much larger than the number of moduli with a Coppersmith's weak prime factor.

The rest of the paper is organized as follows. In Section 2, we give some basic concepts on integer factorization and lattice reduction as well as an overview of Coppersmith's method. In Section 3, we present an attack on an RSA modulus $N = pq$ with one weak prime factor. In Section 4, we present the second attack an RSA modulus $N = pq$ with two weak prime factors. We conclude the paper in Section 5.

2   Preliminaries

In this section we give the definitions and results that we need to perform our attacks. These preliminaries include basic concepts on integer factorization and lattice reduction techniques.

2.1   Integer Factorization: The State of the Art

Currently, the most powerful algorithm for factorizing large integers is the Number Field Sieve (NFS) [12]. The heuristic expected time $T_{NFS}(N)$ of the NFS depends on the bitsize of the integer N to be factored:

$$T_{NFS}(N) = \exp\left((1.92 + o(1))(\log N)^{1/3}(\log\log N)^{2/3}\right).$$

If the integer N has small factors, the Elliptic Curve Method (ECM) [11] for factoring is substantially faster than the NFS. It can compute a non-trivial factor p of a composite integer N in an expected runtime T_{ECM}:

$$T_{ECM}(p) = \exp\left(\left(\sqrt{2} + o(1)\right)(\log p)^{1/2}(\log\log p)^{1/2}\right),$$

which is sub-exponential in the bitsize of the factor p. The largest factor found so far with the ECM is a 83 decimal digits (275 bits) prime factor of the special number $7^{337} + 1$ (see [18]).

2.2   Lattice Reduction

Let m and n be positive integers with $m \leq n$. Let $u_1, \ldots, u_m \in \mathbb{R}^n$ be m linearly independent vectors. The lattice $\mathcal{L}$ spanned by $u_1, \ldots, u_m$ is the set

$$\mathcal{L} = \left\{ \sum_{i=1}^{m} a_i u_i \mid a_i \in \mathbb{Z} \right\}.$$

The set $\{u_1, \ldots, u_m\}$ is called a lattice basis for $\mathcal{L}$. The dimension (or rank) of the lattice $\mathcal{L}$ is $\dim(\mathcal{L}) = m$, and $\mathcal{L}$ is called full rank if $m = n$. It is often useful to represent the lattice $\mathcal{L}$ by the $m \times n$ matrix M whose rows are the coefficients of the vectors $u_1, \ldots, u_m$. The determinant (or volume) of $\mathcal{L}$ is defined as $\det(\mathcal{L}) = \sqrt{M \cdot M^t}$. When $\mathcal{L}$ is full rank, the determinant reduces to $\det(\mathcal{L}) = |\det(M)|$. The Euclidean norm of a vector $v = \sum_{i=1}^{m} a_i u_i \in \mathcal{L}$ is defined as $\|v\| = \sqrt{\sum_{i=1}^{m} a_i^2}$. As a lattice has infinitely many bases, some bases are better than others, and a very important task is to find a basis with small vectors $\{b_1, \ldots, b_m\}$ called the reduced basis. This task is very hard in general, however, the LLL algorithm proposed by Lenstra, Lenstra, and Lovász [13] finds a basis of a lattice with relatively small vectors in polynimial time. The following theorem determines the sizes of the reduced basis vectors obtained with LLL (see [15] for more details).

Theorem 1. *Let $\mathcal{L}$ be a lattice spanned by a basis $\{u_1, \ldots, u_m\}$. The LLL algorithm applied to $\mathcal{L}$ outputs a reduced basis $\{b_1, \ldots, b_m\}$ with*

$$\|b_1\| \le \|b_2\| \le \ldots \le \|b_i\| \le 2^{\frac{m(m-1)}{4(m-i+1)}} \det(L)^{\frac{1}{m+i-1}}, \text{ for } i = 1, 2, \ldots, m.$$

The existence of a short nonzero vector in a lattice is guaranteed by a result of Minkowski stating that every m-dimensional lattice $\mathcal{L}$ contains a non-zero vector v with $\|v\| \le \sqrt{m} \det(L)^{\frac{1}{m}}$. On the other hand, the Gaussian Heuristic asserts that the norm γ_1 of the shortest vector of a random lattice satisfies

$$\gamma_1 \approx \sqrt{\frac{\dim(\mathcal{L})}{2\pi e}} \det(\mathcal{L})^{\frac{1}{\dim(\mathcal{L})}}.$$

Hereafter, we will use this result as an estimation for the expected minimum norm of a non-zero vector in a lattice.

2.3   Coppersmith's Method

In 1996, Coppersmith [6] presented two techniques based on LLL to find small integer roots of univariate modular polynomials or of bivariate integer polynomials. Coppersmith showed how to apply his technique to factorize an RSA modulus $N = pq$ with $q < p < 2q$ when half of the least or the most significant bits of p is known.

Theorem 2. *Let $N = pq$ be an RSA modulus with $q < p < 2q$. Let M_0 and M_1 be two positif integers. If $p = M_1 + u_0$ with $u_0 < N^{\frac{1}{4}}$ or if $p = M_1 u_1 + M_0$ with $u_1 < N^{\frac{1}{4}}$, then N can be factored in time polynomial in $\log N$.*

Coppersmith's technique extends to polynomials in more variables, but the method becomes heuristic. The problem of finding small roots of linear modular polynomials $f(x_1, \ldots, x_n) = a_1 x_1 + a_2 x_2 + + a_n x_n + a_{n+1} \pmod{p}$ for some unknown p that divides the known modulus N has been studied using Coppersmith's technique by Herrmann and May [10]. The following result, due to Lu, Zhang and Lin [14] gives a sufficient condition under which modular roots can be found efficiently.

Theorem 3 (Lu, Zhang, Lin). *Let N be a composite integer with a divisor p^u such that $p \geq N^\beta$. Let $f(x_1, \ldots, x_n) \in \mathbb{Z}[x_1, \ldots, x_n]$ be a homogenous linear polynomial. Then one can find all the solutions $(y_1, \ldots, y_n)$ of the equation $f(x_1, \ldots, x_n) = 0 \mod p^v$, $v \leq u$ with $\gcd(y_1, \ldots, y_n) = 1$ and $|y_1| < N^{\delta_1}, \ldots, |y_n| < N^{\delta_n}$ if*

$$\sum_{i=1}^{n} \delta_i \leq \frac{u}{v}\left(1 - \left(1 - \frac{u}{v}\beta\right)^{\frac{n}{n-1}} - n\left(1 - \sqrt[n-1]{1 - \frac{u}{v}\beta}\right)\left(1 - \frac{u}{v}\beta\right)\right).$$

The time complexity of the algorithm for finding such sulution $(y_1, \ldots, y_n)$ is polynomial in $\log N$.

3   The Attack with One Weak Prime Factor

3.1   The Attack

In this section, we present an attack to factor an RSA modulus $N = pq$ when p satisfies a linear equation in the form $ap = u_0 + M_1 u_1 + \ldots M_k u_k$ for a suitably small positive integer a and suitably small integers $u_0, u_1, \ldots, u_k$ where $M_1, \ldots, M_k$ are given positive integers. Such prime factor p is called a weak prime for the integers $M_1, \ldots, M_k$.

Theorem 4. *Let $N = pq$ be an RSA modulus such that $p > N^\beta$ and $M_1, \ldots, M_k$ be k positive integers with $M_1 < M_2 < \ldots < M_k$. Suppose that there exists a positive integer a, and $k+1$ integers u_i, $i = 0, \ldots, k$ such that $ap = u_0 + M_1 u_1 + \ldots + M_k u_k$ with $\max(u_i) < N^\delta$ and*

$$\delta < \frac{1}{k+1}\left(1 - (1 - \beta)^{\frac{k+1}{k}} - (k+1)\left(1 - \sqrt[k]{1 - \beta}\right)(1 - \beta)\right).$$

Then one can factor N in polynomial time.

Proof. Let $M_1, \ldots, M_k$ be k positive integers such that $M_1 < M_2 < \ldots < M_k$. Suppose that $ap = u_0 + M_1 u_1 + \ldots + M_k u_k$, that is $(u_0, \ldots, u_k)$ is a solution of the modular polynomial equation

$$x_0 + M_1 x_1 + \ldots + M_k x_k = 0 \pmod{p}. \tag{1}$$

Suppose that $|u_i| < N^\delta$ for $i = 0, \ldots, k$. Using $n = k + 1$, $u = 1$ and $v = 1$ in Theorem 3, means that the equation (1) can be solved in polynomial time, i.e., finding $(u_0, \ldots, u_k)$ if

$$(k+1)\delta < \left(1 - (1 - \beta)^{\frac{k+1}{k}} - (k+1)\left(1 - \sqrt[k]{1 - \beta}\right)(1 - \beta)\right),$$

which gives the bound

$$\delta < \frac{1}{k+1}\left(1 - (1 - \beta)^{\frac{k+1}{k}} - (k+1)\left(1 - \sqrt[k]{1 - \beta}\right)(1 - \beta)\right).$$

This terminates the proof.   □

Remark 1. For a balanced RSA modulus, the prime factors p and q are of the same bit size. Then $p > N^\beta$ with $\beta = \frac{1}{2}$. Hence, the condition on δ becomes

$$\delta < \frac{1}{k+1}\left(1 - \left(\frac{1}{2}\right)^{\frac{k+1}{k}}\right) - \frac{1}{2}\left(1 - \left(\frac{1}{2}\right)^{\frac{1}{k}}\right). \tag{2}$$

In Table 1, we give the bound for δ for given β and k.

Table 1. Upper bounds for δ by Theorem 4

	$k=1$	$k=2$	$k=3$	$k=4$	$k=5$	$k=6$	$k=7$	$k=8$	$k=9$	$k=10$
$\beta = 0.5$	0.125	0.069	0.047	0.036	0.029	0.024	0.021	0.018	0.016	0.015
$\beta = 0.6$	0.180	0.101	0.071	0.054	0.044	0.037	0.032	0.028	0.025	0.022
$\beta = 0.7$	0.245	0.142	0.100	0.077	0.063	0.053	0.046	0.046	0.036	0.032

Remark 2. We note that Coppersmith's weak primes correspond to moduli $N = pq$ with $q < p < 2q$ where one of the prime factors is of the form $p = M_1 + u_0$ or $p = M_1 u_1 + M_0$ with $u_0, u_1 < N^{0.25}$ as mentioned in Theorem 2. This a special case of the equation of Theorem 4. Indeed, we can solve the equations $p = M_1 + u_0$ and $p = M_1 u_1 + M_0$ when $|u_0|, |u_1| < N^{\frac{1}{4}}$. Alternatively, Coppersmith's weak primes correspond to the cell $(k, 2\beta) = (1, 0.25)$ in Table 1.

3.2 Numerical Examples

Example 1. Let

$N =10009752886312109988022778227550577837081215192005129864784685$
$\qquad 1857440468018795774211860316385574268129624076883575119637091 41,$

be a 412-bit RSA modulus with $N = pq$ where $q < p < 2q$. Then p and q are balanced and $p \approx N^{\frac{1}{2}} \approx 2^{206}$. Hence for $\beta = 0.5$, we have $p > N^\beta$. Suppose that p satisfies an equation of the form $ap = u_0 + Mu_1 + M^2 u_2$. Typically, $M^2 \approx N^{\frac{1}{2}}$, that is $M \approx N^{\frac{1}{4}}$. So let $M = 2^{100}$. For $\beta = 0.5$ and $k = 2$, Table (1) gives the bound $\delta < 0.069$. Assume therefore that the parameters u_i satisfy $|u_i| < N^{0.069} \approx 2^{28}$ for $i = 0, 1, 2$. By applying Theorem 4 we should find u_0, u_1 and u_2 as long as $u_0, u_1, u_2 < 2^{28}$. We apply the method of Lu et al. [14] with $m = 4$ and $t = 1$. This gives a 35-dimensional lattice. Applying the LLL algorithm [13], we find a reduced basis with multivariate polynomials $f_i(x_1, x_2, x_3) \in \mathbb{Z}[x_1, x_2, x_3]$, $i = 1, \ldots, 3$. Applying the Gröbner basis technique for solving a system of polynomial equations, we get $u_0 = 9005$, $u_1 = 7123$, $u_2 = 3915$. Using these values, we can compute $ap = u_0 + Mu_1 + M^2 u_2$ from which we deduce $p = \gcd(u_0 + Mu_1 + M^2 u_2, N)$, that is

$p = 1233561263387048417401329723828368836098009882095391170026821 43.$

Finally, we can compute $q = \frac{N}{p}$, that is

$q = 8114516225021407246598039619256282180269797066143262376503 8987.$

Note here that there is no linear decomposition of p in the form $p = M_1 + u_0$ nor $p = M_1u_1 + M_0$ with $u_0, u_1 < N^{0.25}$ that makes p vulnerable to the attack of Coppersmith. This shows that the modulus N is vulnerable to our attack, while it is not vulnerable to Coppersmith's attack. Finally, the overall recorded execution time for our attack using an off-the-shelf computer was 17 seconds.

Example 2. In [2], Bernstein et al. discovered many prime factors with special forms. Many of these primes were found by computing the greatest common divisor of a collection of RSA moduli. Others were found by applying Coppersmith's technique. We show below that our attack can find some primes among the list of Bernstein et al. One of these primes is

$p =$ 0xc0000000000000000000000000000000000000000000000000000000000000

0000000000000000000000000000000000000000000000000000000000002$f9$,

$=$ 1005585594745694782468051874865438459560952436544429503329267 1082

7913230225551602326014057236251775707675238936398645381403154 1210

8959927459825236754563833.

Using $M = 2^{510}$, we get $p = 3M + 761 = Mu_1 + u_0$ where $u_1 = 3$ and $u_0 = 761$. We have $u_1, u_0 < N^{\delta}$ with $\delta \approx 0.007$ which is less than the bound 0.125 in Table 1 for a 1024 bit-size RSA modulus N with $\beta = 0.5$, and $k = 1$. This implies that the conditions for Theorem 4 are satisfied and our method finds p when used in any RSA modulus.

Example 3. Now, consider this other example from the list of Bernstein et al. [2]

$p =$ 0xc000b8000000000000000000000000000000000000000000000000000000000

0000068000000000000000000000000000000000000000000000000000000251

$=$ 1005600299430066190917858574741029677291519034741120712409376 115

2520749216065545598886037221777994938111659319232428746318812 487

6095138372637727117017009393

Then p has the form $p = 3145774M^7 + 27262976M^3 + 593 = M^7u_7 + M^3u_3 + u_0$ where $M = 2^{70}$. The coefficients u_7, u_3 and u_0 satisfy $u_7, u_3, u_0 < N^{\delta}$ with $\delta \approx 0.016$ while the bound of Theorem 4 is 0.021 (see Table 1 for $k = 7$ and $\beta = 0.5$). Again, this shows that our method will find the factorization of any RSA modulus that is a multiple of p.

3.3   The Number of Single Weak Primes in an Interval

In this section, we consider two positive integers n and M and present a study of the weak primes with M, that is the primes $p \in [2^n, 2^{n+1}]$ such that there exists a positive integer a that gives the decomposition

$$ap = \sum_{i=0}^{k} M^i u_i$$

where $|u_i| < N^\delta$ and δ satisfies Theorem 4. We show that the number of the RSA moduli N in the interval $[2^{2n}, 2^{2(n+1)}]$ with a weak prime factor $p \in [2^n, 2^{n+1}]$ is polynomial in 2^n. That is, this number is lower bounded by 2^η where $\eta > \frac{1}{2}$. We call such a class weak RSA Moduli in the interval $[2^{2n}, 2^{2(n+1)}]$.

Theorem 5. *Let n be a positive integer. For $k \geq 1$, define $M = \lceil 2^{\frac{n}{k}} \rceil$. Let $\mathcal{N}$ be the set of the weak RSA moduli $N \in [2^{2n}, 2^{2(n+1)}]$ such that $N = pq$, p and q are of the same bitsize, $p > q$, and $p = \left\lfloor \frac{\sum_{i=0}^{k} M^i u_i}{a} \right\rfloor + b \in [2^n, 2^{n+1}]$ for some small integers b, $a < N^\delta$ and $|u_i| < N^\delta$ for $i = 0, \ldots, k$ with*

$$\delta = \frac{1}{k+1}\left(1 - \left(\frac{1}{2}\right)^{\frac{k+1}{k}}\right) - \frac{1}{2}\left(1 - \left(\frac{1}{2}\right)^{\frac{1}{k}}\right).$$

Then the cardinality of $\mathcal{N}$ satisfies $\#\mathcal{N} \geq 2^\eta$ where

$$\eta = (1 + 2(k+1)\delta)n + \log_2\left(\frac{(n-1)}{n(n+1)\log(2)}\right).$$

Proof. Let N be an RSA moduli. Suppose that $N \in [2^{2n}, 2^{2(n+1)}]$ with $N = pq$ where p and q are of the same bitsize. Since $p \approx N^{\frac{1}{2}}$, then $p \in [2^n, 2^{n+1}]$. Suppose further that for some positive integer a, we have $ap = \sum_{i=0}^{k} M^i u_i$. Then

$$M^k = \frac{ap - \sum_{i=0}^{k-1} M^i u_i}{u_k} \approx \frac{a}{u_k}p,$$

which implies $M \approx p^{\frac{1}{k}} \approx N^{\frac{1}{2k}}$. Now, define

$$M = \left\lceil N^{\frac{1}{2k}} \right\rceil = \left\lceil 2^{\frac{n}{k}} \right\rceil,$$

where $\lceil x \rceil$ is the integer greater or equal to x. This yields $2^n \leq M^k \leq 2^{n+1}$. Consider the set

$$\mathcal{P} = \left\{ p = \left\lfloor \frac{\sum_{i=0}^{k} M^i u_i}{a} \right\rfloor + b, \ p \text{ is prime}, \ | \ p \in [2^n, 2^{n+1}], \ a < N^\delta, \ |u_i| < N^\delta \right\},$$

where δ satisfies (2). Here b is as small as possible so that $\left\lfloor \frac{\sum_{i=0}^{k} M^i u_i}{a} \right\rfloor + b$ is prime. Also, since M^k is the leading term, then observe that

$$\frac{\sum_{i=0}^{k} M^i u_i}{a} - M^k = \frac{u_k - a}{a}M^k + \frac{\sum_{i=1}^{k} M^i u_i}{a}.$$

To ensure $p \in [2^n, 2^{n+1}]$, we consider only the situation where $u_k \geq a$. Hence, using the bounds $a < N^\delta$ and $|u_i| < N^\delta$ for $i = 0, \ldots, k-1$, we get a lower bound for the number of possibilities for a and for u_i, which themselves set a lower bound for the cardinality of $\mathcal{P}$ as follows:

$$\#\mathcal{P} \geq \lfloor N^\delta \rfloor \lfloor N^\delta \rfloor^k \approx N^{(k+1)\delta} \approx 2^{2(k+1)n\delta}. \tag{3}$$

On the other hand, the prime number theorem asserts that the number $\pi(x)$ of the primes less than x is

$$\pi(x) \approx \frac{x}{\log(x)}.$$

Hence, the number of primes in the interval $[2^n, 2^{n+1}]$ is approximately

$$\pi(2^{n+1}) - \pi(2^n) \approx \frac{2^{n+1}}{\log(2^{n+1})} - \frac{2^n}{\log(2^n)} = \frac{(n-1)2^n}{n(n+1)\log(2)}. \tag{4}$$

It follows that the number of RSA moduli $N = pq \in [2^{2n}, 2^{2(n+1)}]$ with a weak factor $p \in \mathcal{P}$ and $q \in [2^n, 2^{n+1}]$ is at least $\#(\mathcal{N}) \geq \#\mathcal{P} \times (\pi(2^{n+1}) - \pi(2^n))$. Using 3 and 4, we get

$$\#(\mathcal{N}) \geq 2^{2(k+1)n\delta} \times \frac{(n-1)2^n}{n(n+1)\log(2)}$$

$$= \frac{(n-1)}{n(n+1)\log(2)} \times 2^{(1+2(k+1)\delta)n}$$

$$= 2^\eta,$$

where

$$\eta = (1 + 2(k+1)\delta)n + \log_2\left(\frac{(n-1)}{n(n+1)\log(2)}\right).$$

This terminates the proof.   □

Table 2 presents a list of values of the bound η in terms of k and n. In Table 2,

Table 2. Lower bounds for η under Theorem 5

	$k=1$	$k=2$	$k=3$	$k=4$	$k=5$	$k=6$	$k=7$
$n = \frac{1}{2}\log_2(N) = 512$	759	715	698	689	684	680	677
$n = \frac{1}{2}\log_2(N) = 1024$	1526	1438	1404	1386	1375	1368	1362
$n = \frac{1}{2}\log_2(N) = 2048$	3061	2885	2818	2782	2759	2744	2733

we see that in the situation $(\beta, k) = (0.5, 1)$, the number $\#(\mathcal{N})$ of 1024-bits RSA moduli $N = pq \in [2^{1024}, 2^{1026}]$ with a weak factor p is at least $\#(\mathcal{N}) \geq 2^{759}$ This is much larger than the number of RSA moduli with a weak Coppersmith prime factor in the same interval, which is actually $N^{0.25} \approx 2^{256}$. This remark is also valid for 2048-bits and 4096-bits RSA moduli.

4   The Attack with Two Weak Prime factors

4.1   The Attack

In this section, we present an attack on RSA with a modulus $N = pq$ when both the prime factors p and q are weak primes.

Theorem 6. *Let $N = pq$ be an RSA modulus and M be a positive integer. Let $k \geq 1$. Suppose that there exist integers a, b, u_i and v_i, $i = 1, \ldots, k$ such that $ap = \sum_{i=0}^{k} M^i u_i$ and $bq = \sum_{i=0}^{k} M^i v_i$ with $|u_i|, |v_i| < N^\delta$ and*

$$\delta < \frac{1}{2k+1} + \frac{\log\left(2k^3\right)}{2(2k+1)\log(N)} + \frac{\log(2k+1) - \log(2\pi e)}{4\log(N)} - \frac{\log\left(4k^3\right)}{4\log(N)}.$$

Then one can factor N in polynomial time.

Proof. Suppose that $ap = \sum_{i=0}^{k} M^i u_i$ and $bq = \sum_{i=0}^{k} M^i v_i$. Then multiplying ap and bq, we get

$$abN = \sum_{i=0}^{2k} M^i w_i, \quad \text{with} \quad w_i = \sum_{j=0}^{i} u_j v_{i-j}.$$

This can be transformed into the equation

$$M^{2k} x_{2k} + M^{2k-1} x_{2k-1} + \ldots + M x_1 - yN = -x_0, \tag{5}$$

with the solution $(x_{2k}, x_{2k-1}, \ldots, x_1, y, x_0) = (w_{2k}, w_{2k}, \ldots, w_1, ab, u_0 v_0)$. For $i = 0, \ldots, k$, suppose that $|u_i|, |v_i| < N^\delta$. Since for $i = 0, \ldots, 2k$, the maximal number of terms in w_i is k, we get

$$|x_i| = |w_i| \leq k \max_j(|u_j|) \cdot \max_j(|v_j|) < kN^{2\delta}. \tag{6}$$

Let C be a constant to be fixed later. Consider the lattice $\mathcal{L}$ generated by the row vectors of the matrix

$$M(\mathcal{L}) = \begin{bmatrix} 1 & 0 & 0 & \ldots & 0 & CM^{2k} \\ 0 & 1 & \ldots & 0 & 0 & CM^{2k-1} \\ \vdots & \vdots & \ddots & \vdots & \vdots & \vdots \\ 0 & 0 & 0 & \ldots & 1 & CM \\ 0 & 0 & 0 & \ldots & 0 & -CN \end{bmatrix}. \tag{7}$$

The dimension of the lattice $\mathcal{L}$ is $\dim(\mathcal{L}) = 2k+1$ and its determinant is $\det(\mathcal{L}) = CN$. According to the Gaussian Heuristic, the length of the shortest non-zero vector of the lattice $\mathcal{L}$ is approximately $\sigma(\mathcal{L})$ with

$$\sigma(\mathcal{L}) \approx \sqrt{\frac{\dim(\mathcal{L})}{2\pi e}} \det(\mathcal{L})^{\frac{1}{\dim(\mathcal{L})}} = \sqrt{\frac{2k+1}{2\pi e}} (CN)^{\frac{1}{2k+1}}.$$

Consider the vector $v = (x_{2k}, x_{2k-1}, \ldots, x_1, -Cx_0)$. Then, using (5), we get

$$(x_{2k}, x_{2k-1}, \ldots, x_1, -Cx_0) = (x_{2k}, x_{k-1}, \ldots, x_1, y) \cdot M(\mathcal{L}).$$

This means that $v \in \mathcal{L}$. Consequently, if C satisfies $\|v\| \leq \sigma(\mathcal{L})$, then, by the Gaussian Heuristic, v is the shortest vector of L. Using the bound (6), the length of the vector v satisfies

$$\|v\|^2 = C^2 x_0^2 + \sum_{i=1}^{2k} x_i^2 \leq \left(C^2 + \sum_{i=1}^{2k} k^2 \right) N^{4\delta} = \left(C^2 + 2k^3 \right) N^{4\delta}.$$

Let C be a positive integer satisfying $C \leq \sqrt{2k^3}$. Then the norm of the vector v satisfies $\|v\|^2 < 4k^3 N^{4\delta}$. Hence, using the Gaussian approximation $\sigma(\mathcal{L})$, the inequality $\|v\| \leq \sigma(\mathcal{L})$ is satisfied if

$$2k^{\frac{3}{2}} N^{2\delta} \leq \sqrt{\frac{2k+1}{2\pi e}} \left(2^{\frac{1}{2}} k^{\frac{3}{2}} N \right)^{\frac{1}{2k+1}}.$$

Solving for δ, we get

$$\delta < \frac{1}{2k+1} + \frac{\log\left(2k^3\right)}{2(2k+1)\log(N)} + \frac{\log(2k+1) - \log(2\pi e)}{4\log(N)} - \frac{\log\left(4k^3\right)}{4\log(N)}.$$

If δ satisfies the former bound, then the LLL algorithm, applied to the lattice $\mathcal{L}$ will output the vector $v = (x_{2k}, x_{2k-1}, \ldots, x_1, -Cx_0)$ from which, we deduce

$$w_{2k} = |x_{2k}|, \ w_{2k-1} = |x_{2k-1}|, \ldots, w_1 = |x_1|, \ w_0 = \frac{|-Cx_0|}{C}.$$

Using the coefficients w_i, $i = 1, \ldots, 2k$, we construct the polynomial $P(X) = w_{2k} X^{2k} + w_{2k-1} X^{2k-1} + \ldots + w_1 X + w_0$. Factoring $P(X)$, we get

$$P(X) = \left(\sum_{i=0}^{k} M^i u_i \right) \left(\sum_{i=0}^{k} M^i v_i \right),$$

from which we deduce all the values u_i and v_i for $i = 1, \ldots, k$. Using each u_i and v_i for $i = 1, \ldots, k$, we get $ap = \sum_{i=0}^{k} M^i u_i$ and finally obtain $p = \gcd\left(\sum_{i=0}^{k} M^i u_i, N \right)$ which in turn gives $q = \frac{N}{q}$. This terminates the proof. $\square$

In Table 3, we give the bound for δ for a given k and a given size of the RSA modulus.

4.2 Examples

Example 4. Consider the 234 bits RSA modulus

$N = 18128727522177729435347634587168292968987318316812435932174117774340029.$

Table 3. Upper bounds for δ with Theorem 6

	$k = 1$	$k = 2$	$k = 3$	$k = 4$	$k = 5$
$\log_2(N) = 1024$	0.332	0.199	0.141	0.109	0.089
$\log_2(N) = 2048$	0.333	0.199	0.142	0.110	0.090

Let $M = 2^{50}$. Suppose further that the prime factors p and q are such that $ap = M^2 u_2 + M u_1 + u_0$ and $bq = M^2 v_2 + M v_1 + v_0$, that is $k = 2$ with the notation of Theorem 6. We built the matrix (7) with $C = \sqrt{2k^3} = 4$ and applied the LLL algorithm [13]. We got a new basis, where the last row is:

$$(w_4, w_3, w_2, w_1, -C w_0) = (30223231819936, 68646317659290, 109044283791446,$$
$$80821741694637, -162291153390444).$$

From this, we form the polynomial $P(X) = w_4 X^4 + w_3 X^3 + w_2 X^2 + w_1 X^1 + w_0$. which factors as:

$$P(X) = \left(4678994 X^2 + 5832048 X + 4871673\right)\left(6459344 X^2 + 6620037 X + 8328307\right).$$

From this, we deduce

$$u_2 = 4678994, \quad u_1 = 5832048, \quad u_0 = 4871673,$$
$$v_2 = 6459344, \quad v_1 = 6620037, \quad v_0 = 8328307.$$

Using these values, we compute

$$ap = M^2 u_2 + M u_1 + u_0 = 5931329552564290566528965219451557369,$$
$$bq = M^2 v_2 + M v_1 + v_0 = 8188191298680619668680362464158618739.$$

and obtain

$$p = \gcd(ap, N) = 1261985011183891609899779833925863 27,$$
$$q = \gcd(bq, N) = 1436524789242213976961467098975196 27.$$

This leads to the factorization of $N = pq$. We note that the first attack described in Section 3 does not succeed to factor N. Indeed, we have $\frac{\log(\max_i(|v_i|))}{\log N} \approx 0.098$ which is larger than the value $\delta = 0.069$ for $k = 2$ and $\beta = 0.5$ in Table 1. Finally, the overall recorded execution time for our attack using an off-the-shelf computer was 12 seconds.

4.3 The Number of Double Weak Primes in an Interval

In this section, we consider two positive integers n and M and present a study of the double weak primes with M, that is the primes $p, q \in \left[2^n, 2^{n+1}\right]$ such that there exists positive integer a and b that give the decompositions:

$$ap = \sum_{i=0}^{k} M^i u_i, \quad bq = \sum_{i=0}^{k} M^i v_i$$

where $|u_i| < N^\delta$, $|v_i| < N^\delta$ and δ satisfies Theorem 6. We show that the number of the RSA moduli N in the interval $[2^{2n}, 2^{2(n+1)}]$ with a weak prime factors $p, q \in [2^n, 2^{n+1}]$ is lower bounded by 2^{η_2} where $\eta_2 > \frac{1}{2}$.

Theorem 7. *Let n be a positive integer. For $k \geq 1$, define $M = \lceil 2^{\frac{n}{k}} \rceil$. Let $\mathcal{N}$ be the set of the weak RSA moduli $N \in [2^{2n}, 2^{2(n+1)}]$ such that $N = pq$ with*
$$p = \left\lfloor \frac{\sum_{i=0}^{k} M^i u_i}{a} \right\rfloor + u, \quad q = \left\lfloor \frac{\sum_{i=0}^{k} M^i v_i}{b} \right\rfloor + v, \; p, q \in [2^n, 2^{n+1}] \; \text{for some small}$$
integers $u, v, a < N^\delta$, $b < N^\delta$, $|u_i| < N^\delta$ and $|v_i| < N^\delta$ for $i = 0, \ldots, k$ with

$$\delta = \frac{1}{k+1}\left(1 - \left(\frac{1}{2}\right)^{\frac{k+1}{k}}\right) - \frac{1}{2}\left(1 - \left(\frac{1}{2}\right)^{\frac{1}{k}}\right).$$

Then the cardinality of $\mathcal{N}$ is at least $\#\mathcal{N} \geq 2^{\eta_2}$ where $\eta_2 = 4(k+1)n\delta$.

Proof. As in the proof of Theorem 5, the number of prime numbers $p \in [2^n, 2^{n+1}]$ such that $p = \frac{\sum_{i=0}^{k} M^i u_i}{a} + u$ with $|u_i| < 2^{2n\delta}$ is

$$\#\mathcal{P} \geq 2^{2(k+1)n\delta}.$$

Then, the number $\mathcal{N}_2$ of RSA modulus $N \in [2^{2n}, 2^{2(n+1)}]$ with $N = pq$, where both p and q are weak primes is at least

$$\#\mathcal{N}_2 \geq 2^{4(k+1)n\delta} = 2^{\eta_2},$$

where $\eta_2 = 4(k+1)n\delta$. This terminates the proof.   □

In Table 3, we present a list of values of the bound η_2 in terms of k and n.

Table 4. Lower bounds for η_2 under Theorem 7

	$k = 1$	$k = 2$	$k = 3$	$k = 4$	$k = 5$	$k = 6$	$k = 7$
$n = 512$	512	424	390	372	361	353	348
$n = 1024$	1024	848	780	744	722	707	696
$n = 2048$	2048	1696	1560	1489	1444	1414	1392

5   Conclusions

In this paper we presented and illustrated two attacks based on factoring RSA moduli with weak primes. We further computed lower bounds for the sets of weak moduli -that is, moduli made of at least one or two weak prime respectively- in the interval $[2^{2n}, 2^{2(n+1)}]$ and showed that these sets are much larger than the set of RSA prime factors satisfying Coppersmith's conditions, which effectively extending the likelihood for factoring RSA moduli.

References

1. ANSI Standard X9.31-1998, Digital Signatures Using Reversible Public Key Cryptography for the Financial Services Industry (rDSA)
2. Bernstein, D.J., Chang, Y.-A., Cheng, C.-M., Chou, L.-P., Heninger, N., Lange, T., van Someren, N.: Factoring RSA keys from certified smart cards: Coppersmith in the wild. In: Sako, K., Sarkar, P. (eds.) ASIACRYPT 2013, Part II. LNCS, vol. 8270, pp. 341–360. Springer, Heidelberg (2013)
3. Boneh, D.: Twenty years of attacks on the RSA cryptosystem. Notices Amer. Math. Soc. 46(2), 203–213 (1999)
4. Boneh, D., Durfee, G.: Cryptanalysis of RSA with private key d less than $N^{0.292}$. In: Stern, J. (ed.) EUROCRYPT 1999. LNCS, vol. 1592, pp. 1–11. Springer, Heidelberg (1999)
5. Compaq Computer Corperation. Cryptography using Compaq multiprime technology in a parallel processing environment (2002), ftp://ftp.compaq.com/pub/solutions/CompaqMultiPrimeWP.pdf
6. Coppersmith, D.: Small solutions to polynomial equations, and low exponent RSA vulnerabilities. Journal of Cryptology 10(4), 233–260 (1997)
7. Hardy, G.H., Wright, E.M.: An Introduction to the Theory of Numbers. Oxford University Press, London (1975)
8. Håstad, J.: On Using RSA with Low Exponent in a Public Key Network. In: Williams, H.C. (ed.) CRYPTO 1985. LNCS, vol. 218, pp. 403–408. Springer, Heidelberg (1986)
9. Hastad, J.: Solving simultaneous modular equations of low degree. SIAM J. of Computing 17, 336–341 (1988)
10. Herrmann, M., May, A.: Solving linear equations modulo divisors: On factoring given any bits. In: Pieprzyk, J. (ed.) ASIACRYPT 2008. LNCS, vol. 5350, pp. 406–424. Springer, Heidelberg (2008)
11. Lenstra, H.: Factoring integers with elliptic curves. Annals of Mathematics 126, 649–673 (1987)
12. Lenstra, A.K., Lenstra Jr., H.W.: The Development of the Number Field Sieve. Lecture Notes in Mathematics, vol. 1554. Springer, Berlin (1993)
13. Lenstra, A.K., Lenstra, H.W., Lovász, L.: Factoring polynomials with rational coefficients. Mathematische Annalen 261, 513–534 (1982)
14. Lu, Y., Zhang, R., Lin, D.: New Results on Solving Linear Equations Modulo Unknown Divisors and its Applications, Cryptology ePrint Archive, Report 2014/343 (2014), https://eprint.iacr.org/2014/343
15. May, A.: New RSA Vulnerabilities Using Lattice Reduction Methods. PhD thesis, University of Paderborn (2003)
16. Nitaj, A.: Another generalization of wiener's attack on RSA. In: Vaudenay, S. (ed.) AFRICACRYPT 2008. LNCS, vol. 5023, pp. 174–190. Springer, Heidelberg (2008)
17. Rivest, R., Shamir, A., Adleman, L.: A Method for Obtaining digital signatures and public-key cryptosystems. Communications of the ACM 21(2), 120–126 (1978)
18. Zimmermann, P.: 50 largest factors found by ECM, http://www.loria.fr/~zimmerma/records/top50.html
19. de Weger, B.: Cryptanalysis of RSA with small prime difference. Applicable Algebra in Engineering, Communication and Computing 13(1), 17–28 (2002)
20. Wiener, M.: Cryptanalysis of short RSA secret exponents. IEEE Transactions on Information Theory 36, 553–558 (1990)

Author Index

Printed in the United States
By Bookmasters